BIG IDEAS MATH®

BLUE

AF732643

Resources by Chapter

- Family and Community Involvement
- Start Thinking! and Warm Up
- Extra Practice (A and B)
- Enrichment and Extension
- Puzzle Time
- School-to-Work
- Graphic Organizers / Study Help
- Financial Literacy
- Technology Connection
- Life Connections
- Stories in History
- Cumulative Practice
- Unit Project with Rubric

BIG IDEAS LEARNING®

Erie, Pennsylvania

ISBN 13: 978-1-60840-041-6
ISBN 10: 1-60840-041-7

3456789-VLP-13 12 11 10

Contents

About the Resources by Chapter

Family and Community Involvement (English, Spanish, and Haitian Creole)

The Family and Community Involvement letters provide a way to quickly communicate to family members how they can help their student with the material of the chapter. They make the mathematics less intimidating and provide suggestions for helping students see mathematical concepts in common activities.

Start Thinking! and Warm Up

Each Start Thinking! and Warm Up includes two options for getting the class started. The Start Thinking! questions provide students with an opportunity to discuss thought-provoking questions and analyze real-world situations. The Warm Up questions review prerequisite skills needed for the lesson.

Extra Practice

The Extra Practice exercises provide additional practice on the key concepts taught in the lesson. There are two levels of practice provided for each lesson: A (basic) and B (average).

Enrichment and Extension

Each Enrichment and Extension extends the lesson and provides a challenging application of the key concepts.

Puzzle Time

Each Puzzle Time provides additional practice in a fun format in which students use their mathematical knowledge to solve a riddle. This format allows students to self-check their work.

School-to-Work

The School-to-Work projects show students how the concepts of the chapter are applied in various careers.

Graphic Organizers / Study Help

The ready-to-use Graphic Organizers assist students with organizing key ideas and concepts of the chapter

About the Resources by Chapter (continued)

Financial Literacy

The Financial Literacy projects help students develop an awareness of good financial practices. Students will consider risks and trade offs as they practice decision making.

Technology Connection

Each Technology Connection provides opportunities for students to explore mathematical concepts using tools such as scientific and graphing calculators, spreadsheets, geometry software, and the Internet.

Life Connections

The Life Connections connect the mathematics of the chapter to real-life situations.

Stories in History

The Stories in History allow students to learn about the development of mathematical ideas in an entertaining and engaging way.

Cumulative Practice

The Cumulative Practice exercises help students maintain skills through ongoing review.

Unit Project with Rubric

The Unit Projects summarize the key concepts of the unit. They require students to investigate a concept, gather and analyze data, and summarize the results. Scoring rubrics are provided.

Chapter 1

Name____ Date__________

Chapter 1 Solving Equations

Dear Family,

The Internet has made more information available to people than ever before. Because much of this information comes from other countries, you are likely to see more metric units of measure than ever before. The dimensions for many products are often given using centimeters (cm) or meters (m). You may find a recipe online that uses milliliters (ml) and grams (g) instead of cups and ounces. How will you convert these measurements to more familiar units?

You may know some ways to approximate some familiar customary units. An inch is about the width of your thumb. A cup of flour fits in two cupped hands. "Room temperature" is about 72° Fahrenheit ($^\circ$F). To develop a way to approximate metric measurements, you will need to convert an unfamiliar unit to a familiar one.

Work with your student to find some analogies for some common metric units. For example, one centimeter (cm) is about four tenths of an inch. A common approximation for a centimeter is the width of the nail on your index finger.

Figure out these analogies with your student:

- Find out what "room temperature" is on the Celsius scale. Normal body temperature is 98.6°F—what is normal body temperature on the Celsius scale?
- A mile is about 12 city blocks. Find out how many city blocks are in a kilometer (km).
- A yard is roughly the distance from your nose to your outstretched fingertips. About how long is a meter?
- About how many grams of flour fit in two cupped hands?

There may be other units you have encountered. By finding common analogies, these new units of measure may become more familiar.

A pinch of understanding is worth a pound of information!

Nombre ______________________________ Fecha ________

Capítulo 1 Resolviendo Ecuaciones

Estimada Familia:

La Internet ha hecho posible que hoy más que nunca, las personas tengamos más información disponible. Dado que la mayoría de esta información proviene de otros países, estará viendo más unidades métricas que nunca antes. Las dimensiones de muchos productos a menudo se expresan en centímetros (cm) o metros (m). En Internet podrá encontrar una receta que use mililitros (ml) y gramos (g) en vez de tazas y onzas. ¿Cómo convertiría estas medidas a unidades más conocidas?

Pueda que conozca algunas formas de aproximar algunas unidades conocidas. Una pulgada es aproximadamente el ancho de su pulgar. Una taza de harina cabe en dos manos cóncavas. La "temperatura ambiente" es alrededor de 72° Fahrenheit (°F). Para desarrollar una forma de aproximar las medidas métricas, necesitará convertir una unidad no conocida en una conocida.

Trabaje con su estudiante para hallar algunas analogías para ciertas unidades métricas comunes. Por ejemplo, un centímetro (cm) equivale aproximadamente a cuatro décimas de pulgada. Una aproximación común para un centímetro es el ancho de la uña de su dedo índice.

Averigüe estas analogías con su estudiante:

- Averigüen cuánto es la "temperatura ambiente" en la escala de Celsius. La temperatura normal del cuerpo es 98.6°F—¿cuál es la temperatura normal del cuerpo en la escala de Celsius?
- Una milla es alrededor de 12 cuadras de la ciudad. Averigüen cuántas cuadras de la ciudad equivalen a un kilómetro (km).
- Una yarda es más o menos la distancia desde su nariz hasta las puntas estiradas de sus dedos. Aproximadamente, ¿cuánto será un metro?
- Aproximadamente, ¿cuántos gramos de harina caben en dos manos cóncavas?

Pueda ser que hayan encontrado otras unidades. Al encontrar analogías comunes, estas nuevas unidades de medida se hacen más conocidas.

¡Un poquito de comprensión equivale a una libra de información!

Non ______________________________ Dat __________

Chapit 1

Ann Rezoud Ekwasyon

Chè Fanmi:

Entènèt fè plis enfòmasyon disponib pou moun pase jan sa te janm ye anvan. Kòm anpil nan enfòmasyon sa yo sòti nan lòt peyi, ou siman ap wè plis inite mezi metrik pase sa ou te janm wè anvan. Yo souvan bay dimansyon anpil pwodwi an santimèt (cm) oswa mèt (m). Ou gen dwa jwenn yon resèt sou Entènèt ki sèvi ak milimèt (ml) ak gram (g) olyede tas ak ons. Kijan ou pwal konvèti mezi sa yo an inite ou pi abitye avèk yo?

Ou gen dwa konnen kèk jan pou fè estimasyon kèk inite abityèl. Yon pous se apeprè lajè gwo dwèt ou. Yon tas farin kenbe nan de men ou mete an koup. "Tanperati milye a" se apaprè 72° Fahrenheit (°F). Pou devlope yon jan pou fè estimasyon yon mezi metrik, w'ap bezwen konvèti yon inite ou pa abitye avèk li an youn ou abitye avèk li.

Travay avèk elèv ou a pou jwenn kèk analoji pou kèk inite metrik komen. Paregzanp, yon santimèt (cm) se apeprè kat dizyèm yon pous. Yon estimasyon komen pou yon santimèt se lajè zong ki sou endèks ou.

Jwenn analoji sila yo avèk elèv ou a:

- Jwenn sa "tanperati milye a" ye nan echèl Celsius. Tanperati kò nòmal la se 98.6°F—kisa tanperati kò nòmal la ye nan echèl Celsius la?
- Yon mil se apeprè 12 gwoup kay. Jwenn konbyen gwoup kay ki genyen nan yon kilomèt (km).
- Yon yad se apeprè distans ant nen ou ak pwent dwèt ou lè ou lonje yo. Ki longè yon mèt ye apeprè?
- Apeprè konbyen gram farin kenbe nan de men ou mete an koup?

Gen dwa gen lòt inite ou rankontre. Lè ou jwenn analoji komen, ou gen dwa vin pi abitye ak nouvo inite mezi sa yo.

Yon pense konpreyansyon vo yon liv enfòmasyon!

Activity 1.1 Start Thinking!
For use before Activity 1.1

What are some rules in your classroom or school? How are they similar to rules in mathematics?

Activity 1.1 Warm Up
For use before Activity 1.1

Use a protractor to find the measure of the angle.

1. 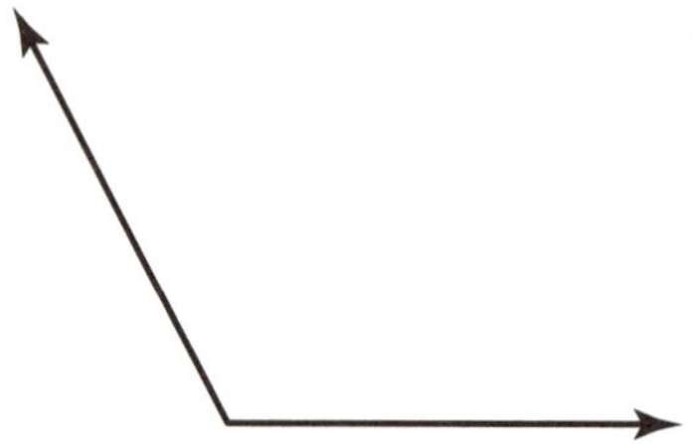

2. 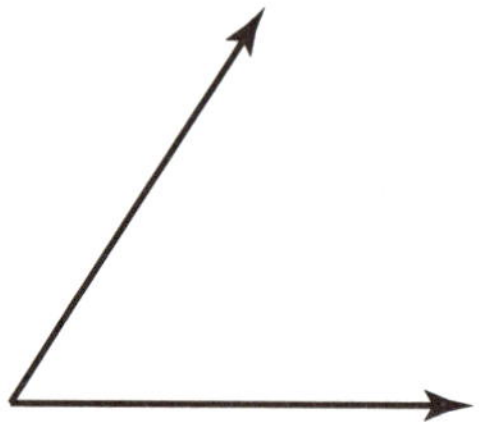

3. 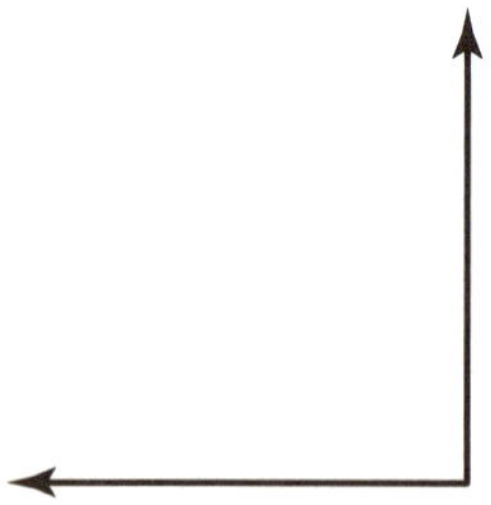

4. 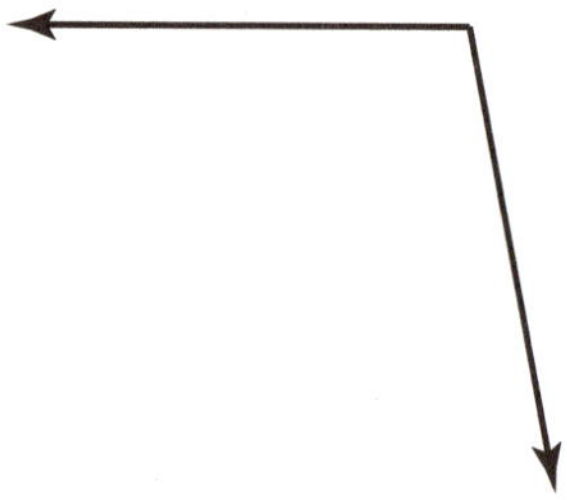

5. 

6. 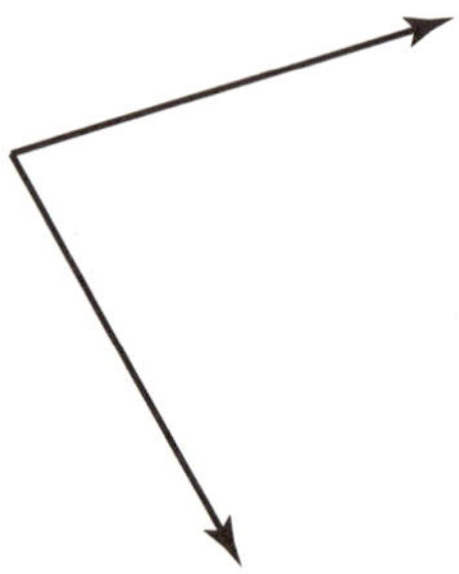

Lesson 1.1 Start Thinking!

For use before Lesson 1.1

The Addition Property of Equality states that adding the same number to each side of an equation produces an equivalent equation. What do you think the Subtraction, Multiplication, and Division Properties of Equality state?

Describe a real-life situation that you can relate to one of the properties of equality.

Lesson 1.1 Warm Up

For use before Lesson 1.1

Find the value of *x*. Use a protractor to check the reasonableness of your answer.

1.

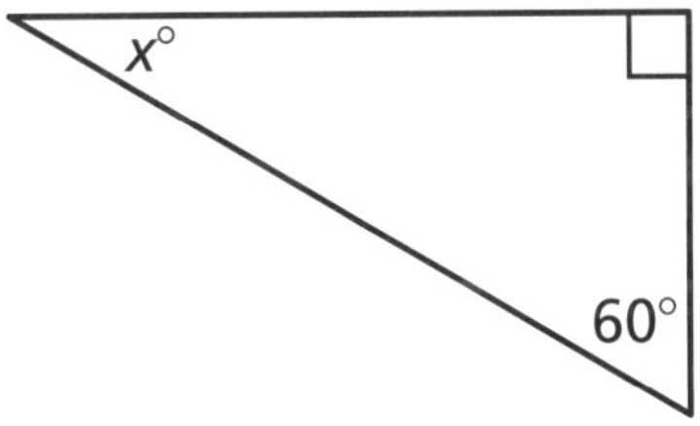

2.

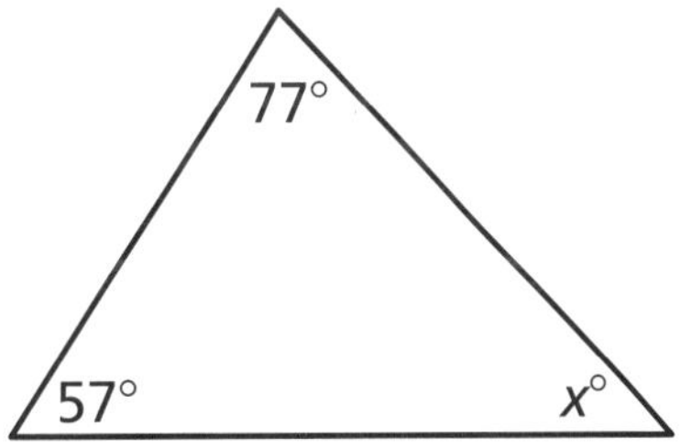

3.

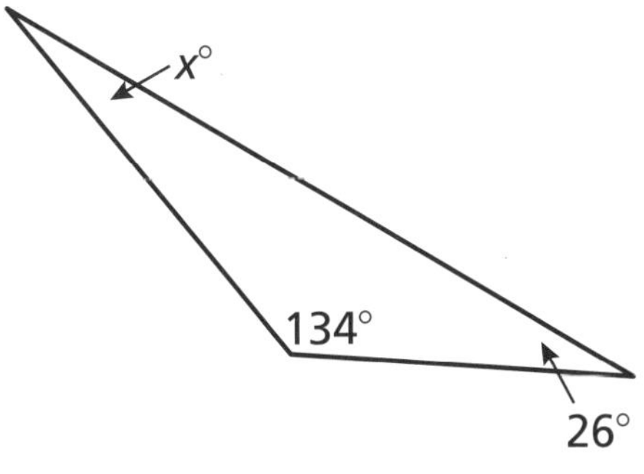

4.

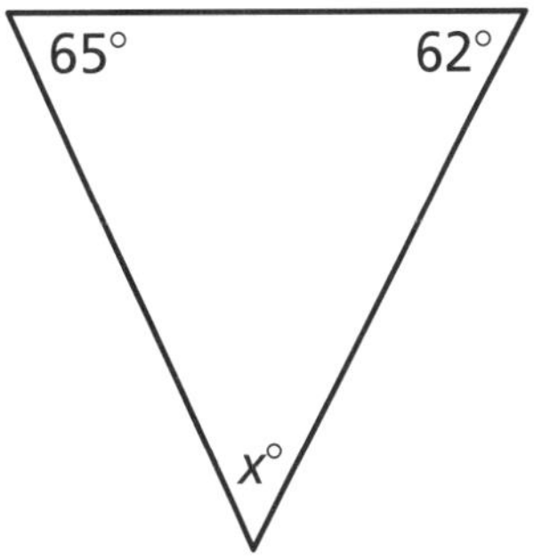

Name ______________________________ Date __________

1.1 Practice A

Solve the equation. Check your solution.

1. $x + 4 = 11$
2. $n - 14 = 20$
3. $-6 + k = -9$
4. $2\pi + d = 5\pi$
5. $y - 1.4 = -2.7$
6. $\frac{2}{5} = w - \frac{3}{2}$
7. Your school's football team scored 49 points. Your team's score was 19 points more than the opponent's score s. Write and solve an equation to find the opponent's score.

Solve the equation. Check your solution.

8. $5y = 40$
9. $\frac{d}{9} = -2$
10. $1.2 = -3b$
11. $\frac{x}{4.1} = -2$
12. $\frac{2}{5}p = \frac{3}{5}$
13. $-4.5 = -1.2k$
14. You earn \$7.50 per hour to help your uncle in his shop. You earn \$33.75. Write and solve an equation to find how many hours you worked.

Solve the equation. Check your solution.

15. $s - |-4| = 7.3$
16. $p + 1.6 \div (-0.4) = -12$
17. Is the solution of $\frac{1}{3}x = 21$ *greater than* or *less than* 21?
18. The volume V of the cylinder is 65π cubic centimeters. The height of the cylinder is 5 centimeters. Use the formula $V = Bh$ to find the area of the base of the cylinder.
19. The total area of this shape is 44 square inches. The area of the triangle is 20 square inches. Write and solve an equation to find the area of the rectangle.

Name______________________________ Date__________

1.1 Practice B

Solve the equation. Check your solution.

1. $-13 + x = -25$

2. $h + 4\pi = 30\pi$

3. $4.5 = m + 2.75$

4. $a - \frac{3}{4} = \frac{2}{3}$

5. $\frac{1}{6} = \frac{5}{12} + p$

6. $c - 2.3 = -5.1$

7. You shopped online and found your MP3 player for \$9.75 less than the store price p. The online price was \$64. Write and solve an equation to find the store price.

Solve the equation. Check your solution.

8. $-1.6x = 8$

9. $\frac{h}{2\pi} = 4.3$

10. $\frac{4}{3} = \frac{2}{15}j$

11. $-23.6 = 5.9t$

12. $6\pi = -2\pi q$

13. $\frac{3}{7}w = -4$

14. The area of a rectangle 55.8 square inches. The width of the rectangle is 4.5 inches. Write and solve an equation to find the length of the rectangle.

Solve the equation. Check your solution.

15. $5.6 \div 0.4 - r = -8$

16. $n - 5 \bullet \frac{2}{3} = \frac{3}{4}$

17. Write an addition equation and a multiplication equation that each has a solution of -5.

18. A fruit basket contains oranges and grapefruits. One-third of the oranges and one-fourth of the grapefruits were spoiled. You threw away 4 oranges and 7 grapefruits. How many pieces of fruit were in the basket?

19. You and two friends pay \$40 for tickets. The cost was divided three ways in the ratio 1 : 3 : 6. How much did each person pay?

Name ______________________________ Date __________

1.1 Enrichment and Extension

Properties of Equality

There are four properties of equality. These properties are usually used to solve equations. They can also be used to write an equivalent form of a given equation.

Determine which property of equality can be used to transform the first equation into the second equation. Then use the color key to shade the square the appropriate color.

Color Key:

Addition Property of Equality - Yellow

Subtraction Property of Equality - Green

Multiplication Property of Equality - Red

Division Property of Equality - Blue

$2x = 4$ $4x = 8$	$10x = -100$ $x = -10$	$x + 1 = 0$ $x + 6 = 5$	$7 - 2x = -2$ $14 - 2x = 5$	$15x = 45$ $x = 3$	$5x = 10$ $6x = 12$
$9x = 81$ $x = 9$	$x + 5 = 1$ $x + 9 = 5$	$x + 4 = 5$ $x = 1$	$x + 7 = 10$ $x + 2 = 5$	$x = 8$ $x + 2 = 10$	$2.2x = 22$ $x = 10$
$x = 0$ $x + 2 = 2$	$x + 3 = 3$ $x = 0$	$3x = 11$ $-3x = -11$	$12.5x = 7$ $25x = 14$	$x + 2 = 19$ $x = 17$	$3 - x = -4$ $5 - x = -2$
$x + 3 = 6$ $x + 6 = 9$	$x = 2$ $x - 1 = 1$	$4x = 5$ $20x = 25$	$2x = 10$ $5x = 25$	$x - 2 = 14$ $x - 4 = 12$	$x + 1 = 2$ $x + 6 = 7$
$-12x = 60$ $2x = -10$	$2x - 1 = 5$ $2x + 2 = 8$	$x + 13 = 15$ $x + 4 = 6$	$x = 10$ $x - 8 = 2$	$3x + 1 = -4$ $3x + 6 = 1$	$1.5x = 20$ $0.15x = 2$
$-4x = 1$ $16x = -4$	$100x = 50$ $20x = 10$	$x + 1 = 3$ $x + 8 = 10$	$x + 1.5 = 1.7$ $x + 2.8 = 3$	$-14x = 28$ $x = -2$	$x = 3$ $9x = 27$

1. Is it possible to complete the board so that there are no blue squares? Explain your reasoning.

2. Is it possible to complete the board so that there are no green squares? Explain your reasoning.

Name____ Date__________

1.1 Puzzle Time

Did You Hear About...

A	B	C	D	E	F
G	H	I	J	K	L
M	N	O	P		

Complete each exercise. Find the answer in the answer column. Write the word under the answer in the box containing the exercise letter.

5 WHO
7.2 HOW
−9 MAN
−13 SOUP
$-\frac{4}{3}$ IT
−14 HE
$\frac{3}{4}$ ARE
$\frac{7}{8}$ IN
42 SOMETHING
4.5 WASH

Solve the equation.

A. $x + 9 = -12$

B. $-8 + k = -17$

C. $-0.3 + r = 4.7$

D. $f + 3\pi = 12\pi$

E. $d - 6\pi = -\pi$

F. $\frac{1}{8} = s - \frac{3}{4}$

G. $m - \frac{1}{4} = -\frac{1}{3}$

H. $-0.8b = 10.4$

I. $46.2 = 4.2p$

J. $-7\pi h = 98\pi$

K. $12.5 = \frac{t}{\pi}$

L. $3.9r = 17.55$

M. $\frac{3}{8}q = -\frac{1}{2}$

N. $-\frac{5}{6} = \frac{2}{3}w$

O. Your dog weighs 28.5 pounds more than your cat. Your dog weighs 37.8 pounds. What is your cat's weight?

P. Nathaniel's sister has $\frac{5}{6}$ as many songs on her MP3 player as he has on his MP3 player. His sister has 35 songs. How many songs does Nathaniel have?

9.3 WITH
5π SOAP
12.5π COULD
8π WAS
$-\frac{1}{12}$ HIS
9π PUT
$-\frac{5}{4}$ DOWN
11 SO
−21 THE
$-\frac{3}{5}$ THAT

Activity 1.2 Start Thinking!
For use before Activity 1.2

Why do you think it is important to check your answer to a math problem?

Activity 1.2 Warm Up
For use before Activity 1.2

Simplify the expression.

1. $2n + 5 + 3n$

2. $x - 7 - 4x$

3. $4f + f + 6f$

4. $(9 - m) + 4m + 7$

5. $17 + 2t - 9 + 2t$

6. $(y + 7) + (2y - 5)$

Lesson 1.2

Start Thinking!

For use before Lesson 1.2

A multi-step equation requires two or more operations to solve the equation. Explain why the following situation can be modeled by a multi-step equation.

> A plumber charges \$80 per hour for labor plus \$60 for parts.

Come up with your own scenario that can be modeled by a multi-step equation.

Lesson 1.2

Warm Up

For use before Lesson 1.2

Find the value of *x*. Use a protractor to check the reasonableness of your answer.

1.
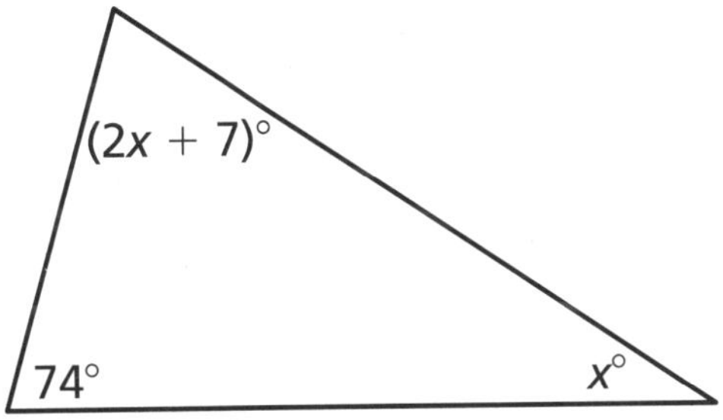

Sum of angle measures: 180°

2.
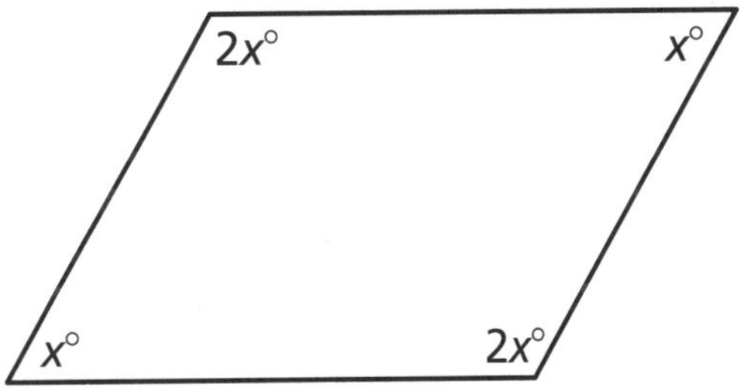

Sum of angle measures: 360°

3.
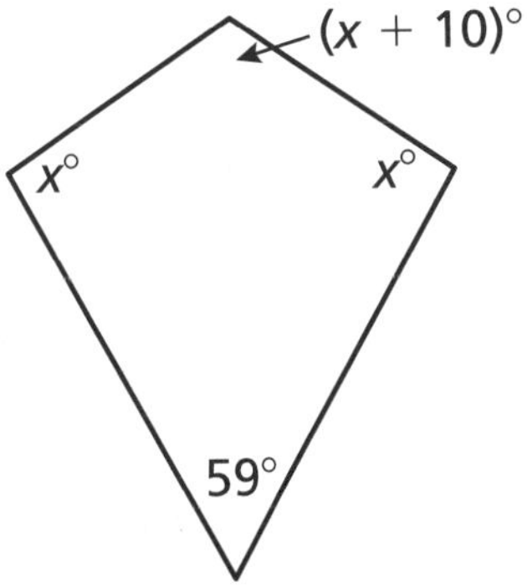

Sum of angle measures: 360°

4.
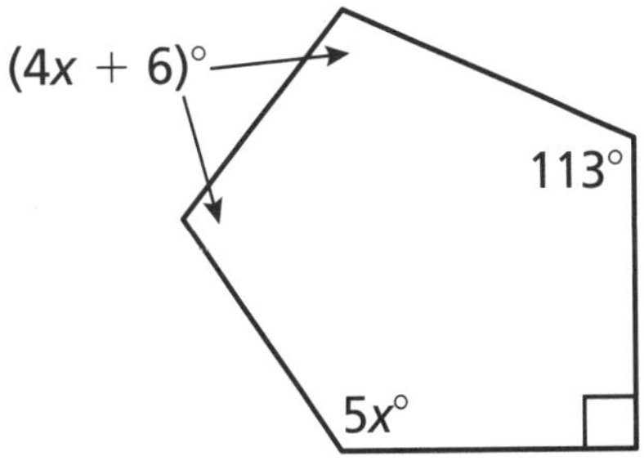

Sum of angle measures: 540°

Name ______________________________ Date __________

1.2 Practice A

Solve the equation. Check your solution.

1. $8y - 7 = 9$

2. $14 - 3m = -1$

3. $30 + 2k + 5k = 100$

4. $z + (z - 6) - 2 = -10$

5. $3.2x - 1.7x + 5.5 = 10$

6. $\frac{3}{4}x - \frac{1}{4}x + 14 = 3$

7. The cost (in dollars) of making n feet of cabinet is represented by $C = 18n + 45$. How many feet of cabinet are made when the cost is \$441?

8. The sum of the measures of the interior angles of the parallelogram is 360°. Write and solve an equation to find the value of the variable.

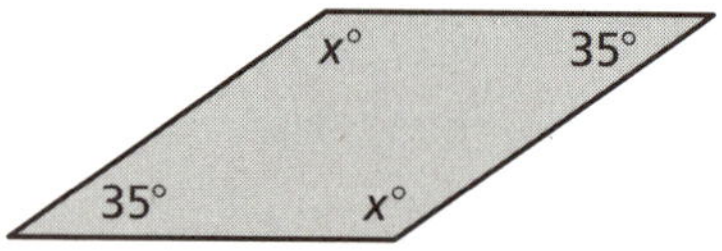

9. At the movies, you order 3 boxes of popcorn and a bottle of water. The cost of the bottle of water is \$1.75. Your total cost is \$9.25. Write and solve an equation to find the cost of one box of popcorn.

10. You and your friend each purchase an equal number n of magazines. Your magazines cost \$1.50 each and your friend's magazines cost \$2 each. The total cost for you and your friend is \$10.50. Write and solve an equation to find the number of magazines you purchased.

11. A rope 25 feet long is cut into 3 pieces. The first piece is $2x$ feet long, the second piece is $5x$ feet long, and the third piece is 4 feet long.

a. Write and solve an equation to find x.

b. Find the lengths of the first and second pieces.

12. The average of your 3 quiz grades is 17 points. Two of your quiz grades are 14 points and 19 points. Write and solve an equation to find the third quiz grade.

13. You had \$26 in your pocket. You purchased x pens at \$3.50 each. You now have \$8.50 in your pocket. Write and solve an equation to find the number of pens purchased.

Name____ Date__________

1.2 Practice B

Solve the equation. Check your solution.

1. $4k - 1.5k = 50$

2. $\frac{5}{7}p - \frac{2}{7}p + 12 = 6$

3. $\frac{2}{3}y + y - 4 = 31$

4. $2.1x + 1.3x - 4.6 = 2.2$

5. $3(5 - 2h) + 9 = \quad 30$

6. $14(x - 3) - 22x = -18$

7. The sum of the measures of the interior angles of the triangle is 180°. Write and solve an equation to find the value of the variable.

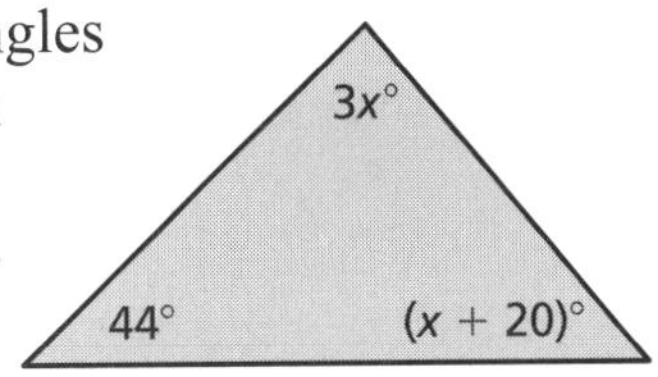

8. A rectangular field has an area of 2100 square feet. The length of the field is 50 feet.

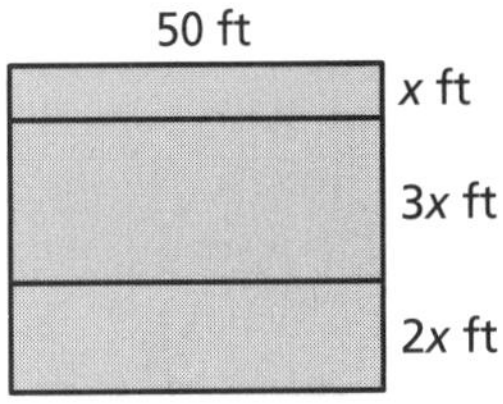

a. How wide is the field?

b. The field is divided in to 3 rectangles, as shown. Write and solve an equation to find x.

c. Determine the dimensions of each rectangle.

9. You are researching the price of MP3 players. You found an average price of \$58.80. One MP3 player cost \$56 and another cost \$62. Find the price of the third MP3 player.

10. The perimeter of a triangle is 42 inches. One side measures 18 inches. The shortest side measures x inches. The longest side measures 1 inch less than four times the length of the shortest side. Write and solve an equation to find the length of the longest side.

11. You order 4 fish sandwiches and a hamburger. The cost of the hamburger is \$2.50. Your total bill before tax is \$14.30. Write and solve an equation to find the cost of a fish sandwich.

Name ______________________________ Date __________

1.2 Enrichment and Extension

Solving Equations

Work with a partner to design a game board. Cut out each of the cards, fold them in half, and place them in a bag.

Game Rules:

Each player puts their game piece on the start square. The youngest student goes first, selects a card from the bag, and solves the equation on a separate piece of paper. If the answer is positive, move forward that many spaces. If the answer is negative, move backward that many spaces. If there are no spaces to move back, the player loses a turn. If a player suspects his competitor's answer is incorrect, he or she can challenge by solving the equation to check the solution. If a player challenges and wins, he moves forward 3 spaces. If a player challenges and loses, he forfeits a turn. Place the card back in the bag and it is the next person's turn. The first player to the end square wins.

$2x - 12 = 8$	$\frac{1}{2}x = 3$	$2x + 6 = 0$
$-14x + 7 = -21$	$0.75x + 0.5 = 2$	$4x + 2 = 22$
$15x + 2 = 62$	$-30x = 90$	$3x = 21$
$x + 10 = 7$	$2x - 4 = 10$	$\pi x = 6\pi$
$4x - 10 = 6$	$\frac{3}{4}x - 0.25 = 0.5$	$13x - 2x - 3 = 41$
$25x - 20 = 5$	$-\frac{2}{5}x = -2$	$4x - 11x + 32 = 4$

Name____ Date__________

1.2 Puzzle Time

Where Was The Declaration Of Independence Signed?

Circle the letter of each correct answer in the boxes below. The circled letters will spell out the answer to the riddle.

Solve the equation.

1. $8 - 3x = 17$

2. $5a - 6 - 2a = 12$

3. $4.3t - 2.1t - 2.3 = 7.6$

4. $8.1 + 3.8h - 5.6h = -7.2$

5. $\frac{2}{5}c + 4 - \frac{1}{5}c = -9$

6. $2(4s - 16) - 5s = -5$

7. $3g - 6(g - 8) = 42$

8. $1.3(8 - b) + 3.7b = -5.2$

9. For the past three months, Grace used her cell phone for 43 minutes, 62 minutes, and 57 minutes. How many minutes would she have to use her cell phone this month for the average usage over the four months to be 55 minutes?

10. A triangle has one angle measuring $3x$ degrees. A second angle measures $2x + 20$ degrees and the third angle measures $4x - 20$ degrees. What is the value of x?

11. You and a friend buy two fruit smoothies and leave a tip. You split the total and your half comes to \$3.60. What percent tip (in decimal form) did you and your friend leave if the fruit smoothies cost \$3 each?

A	**M**	**T**	**E**	**S**	**R**	**T**	**I**	**H**	**P**	**E**	**D**	**G**	**B**	**Y**
2	60	6	8.2	0.10	5	9	0.18	8.5	55	–3	10.5	–2	0.15	–62
F	**B**	**I**	**R**	**O**	**H**	**T**	**C**	**U**	**T**	**N**	**O**	**L**	**M**	**S**
9.5	0.20	3	10	–65	15	58	90	8	20	–55	4.5	68	–6.5	12.5

Activity 1.3 Start Thinking!
For use before Activity 1.3

Study the equation below.

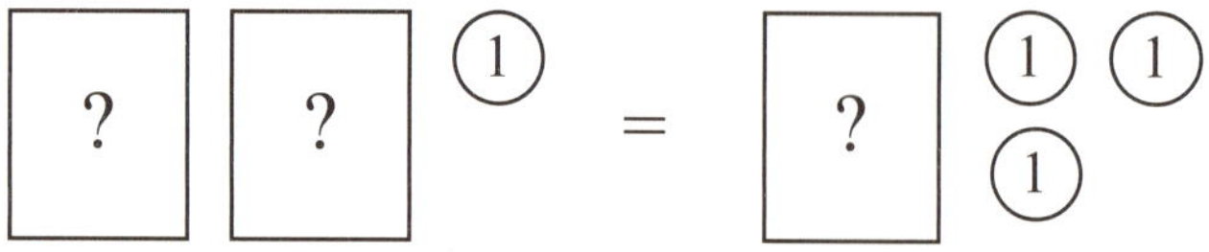

The same number of 1-chips are in each box with a question mark. How many chips are in each box?

Activity 1.3 Warm Up
For use before Activity 1.3

Find the width of the rectangle.

1. Area $= 32\ \text{m}^2$

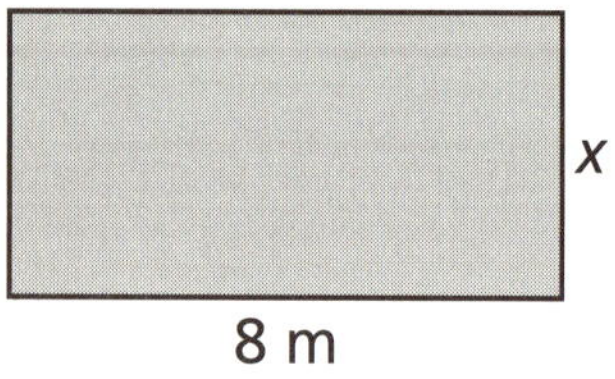

2. Area $= 150\ \text{cm}^2$

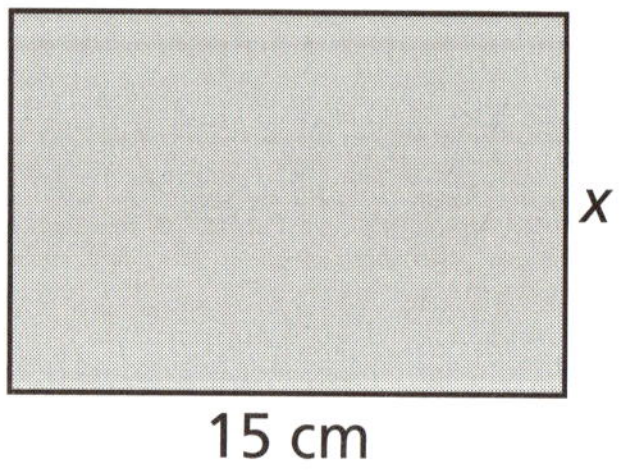

3. Perimeter $= 35$ in.

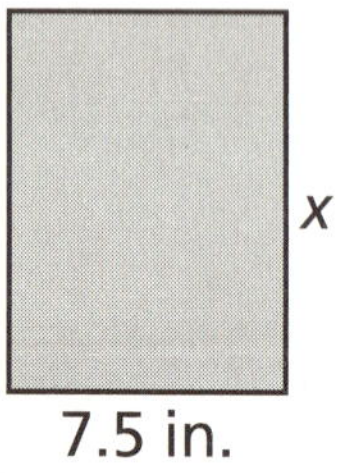

4. Perimeter $= 40$ ft

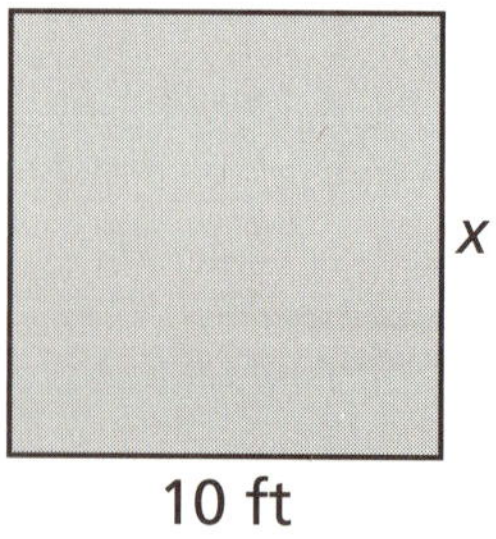

Lesson 1.3

Start Thinking!

For use before Lesson 1.3

Try solving the equation $2x + 20 = 12x$ by first subtracting 20 from each side. What property allows you to do so? Does it help you get closer to finding a solution?

What is a better first step? Explain.

Lesson 1.3

Warm Up

For use before Lesson 1.3

The value of the figure's surface area is equal to the value of the figure's volume. Find the value of *x*.

1.

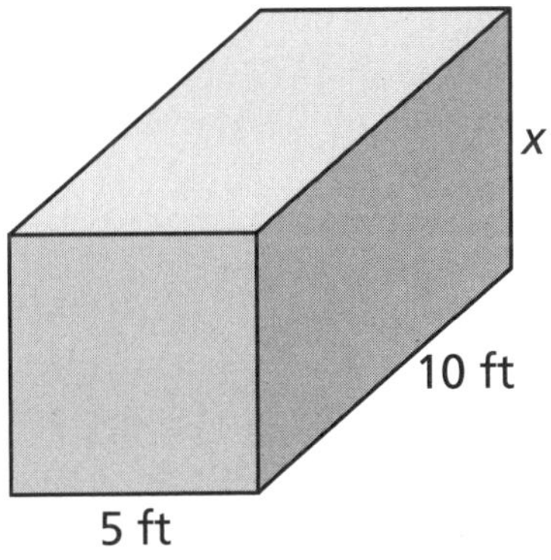

2.

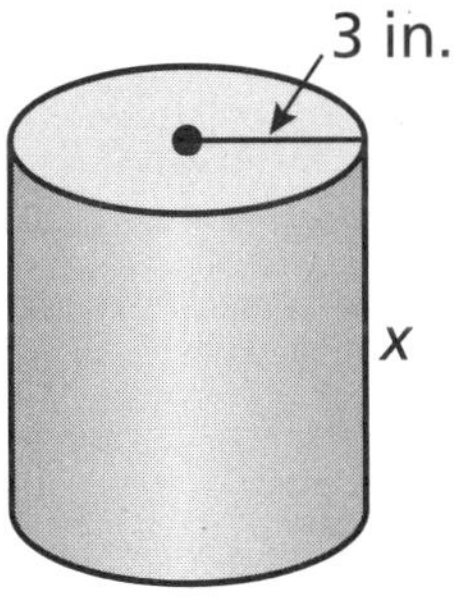

Name ______________________________ Date ________

1.3 Practice A

The value of the perimeter of the figure is equal to the value of the area. Find the value of *x*.

1. 

2. 

Solve the equation. Check your solution.

3. $y - 12 = 4y$

4. $6n - 12 = n + 3$

5. $\frac{1}{5}q = 9 - \frac{2}{5}q$

6. $4.3d + 7.5 = 5.8d$

7. $6(h + 4) = -2h$

8. $3(b - 4) = 5b - 2$

9. Your long distance telephone provider offers two plans. Plan A has a monthly fee of \$15 and \$0.25 per minute. Plan B has a monthly fee of \$20 and \$0.05 per minute. Write and solve an equation to find the number of minutes that you must talk to have the same cost for each of the plans.

10. One-third of a number x is equal to 22 less than the number. Write and solve an equation to find the number.

11. Find the perimeter of the regular polygon.

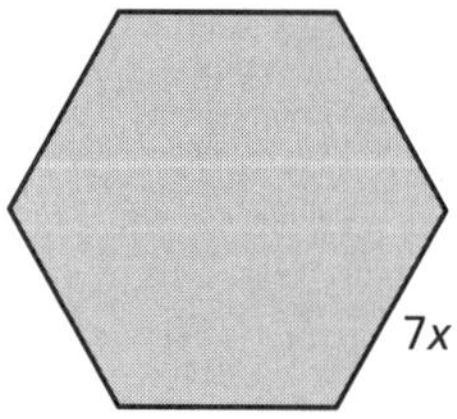

12. You purchase a desk for 60% of the original price p. This price is \$32 less than the original price. Write and solve an equation to find the original price of the desk.

Name______________________________ Date__________

1.3 Practice B

The value of the figure's surface area is equal to the value of the figure's volume. Find the value of *x*.

1.

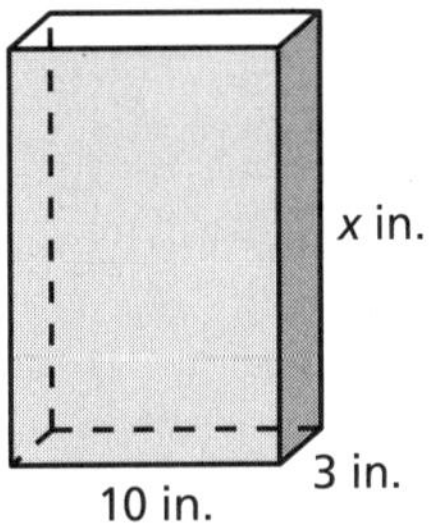

2.

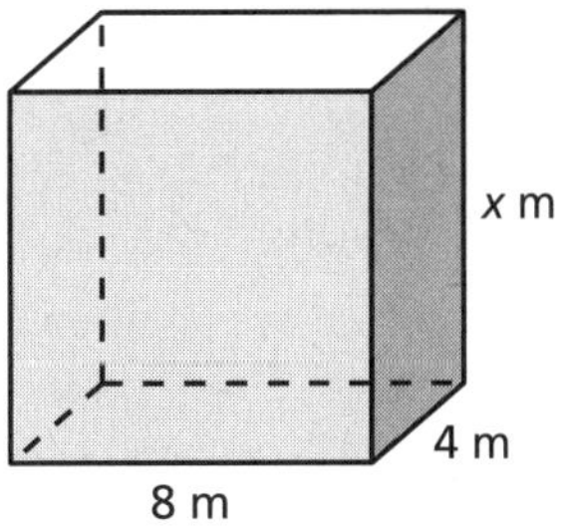

Solve the equation. Check your solution.

3. $5y - 14 = 2y - 2$

4. $\frac{4}{7}m = 18 - \frac{2}{7}m$

5. $16(p - 2) = 7p + 4$

6. $4(2s - 3) = 3(s + 1)$

7. $0.3(t - 2) = 0.4t$

8. $\frac{2}{9}n + \frac{1}{2} = \frac{2}{3}(n + 3)$

9. Describe and correct the error in solving the equation.

$$\begin{aligned} 0.4x &= 0.2(x - 8) \\ 0.4x &= 0.2x - 8 \\ 0.2x &= -8 \\ x &= -40 \end{aligned}$$

10. A parcel of land is in the shape of a rectangle with length ℓ (in feet). Sixty-five percent of the length is waterfront and 70 feet of the length is not waterfront. Write and solve an equation to find the length of the parcel of land.

11. The original price p for a necklace is the same at both jewelry stores. At Store A, the sale price is 60% of the original price. Last month, at Store B the sale price was $40 less than the original price. This month, Store B is selling the necklace for 80% of last month's reduced price. The current sale prices are the same for both stores. Write and solve an equation to find the original price of the necklace.

Name ______________________________ Date __________

1.3 Enrichment and Extension

Where can you buy a ruler that is three feet long?

Solve the equations. Order the solutions from least to greatest. Once ordered, the variables will spell the answer to the riddle.

1. $5d - 4 = 4 - d$

2. $-10e + 15 = 95 - 30e$

3. $15t + 17 = 13t + 14$

4. $-12 - a = 4a - 7$

5. $4a - 16 = a - 15$

6. $4y + 12 = 6y + 12$

7. $0.25r - 0.25 + 0.25r = 0.5 - 0.25r$

8. $-4a + 7 = a + 32$

9. $13s - 31 = 2s - 9$

10. $a + 1.25 = 2a - 1$

11. $3\ell + 4 + \ell = 13 + \ell$

Solution														
Variable														

Name______________________________ Date__________

1.3 Puzzle Time

What Happens When A Frog Double-Parks On A Lily Pad?

Write the letter of each answer in the box containing the exercise number.

Solve the equation.

1. $x + 36 = 4x$

2. $5a + 12 = 3a - 8$

3. $\frac{3}{2}p - 14 = p + 13$

4. $7 - 4.9t = 15 + 7.6t$

5. $3(12f - 3) = 4f - 1$

6. $\frac{1}{3}(b + 6) = \frac{1}{4}b + 8$

7. $\frac{3}{5}(2m - 10) = \frac{2}{3}m + 10$

8. $8.2(s + 4) = 6.7s + 5.2$

9. On Monday, you run on a treadmill for $\frac{1}{2}$ hour at x miles per hour. On Tuesday, you walk the same distance on the treadmill, at 2 miles per hour slower, and it takes you $\frac{3}{4}$ hour. How many miles did you run on the treadmill on Monday?

10. Jess spent $7x$ minutes on the computer. Her sister spent $5x + 10$ minutes on the computer, which was the same amount of time Jess spent. How many minutes was Jess on the computer?

11. A rectangle is 6 units wide and $x - 8$ units long. It has the same area as a triangle with a height of 7 units and a base of $x - 3$ units. What is the area of the rectangle?

Answers

Y. 72

A. -18.4

T. 42

O. -10

A. 35

S. 54

A. 12

D. -0.64

W. 3

I. 0.25

T. 30

5	7		3		11	2	10	4		8	9	1	6
		,											

Activity 1.4 Start Thinking!

For use before Activity 1.4

List all of the formulas for area and perimeter (including circumference) that you know. Draw and label a sketch for each shape.

Activity 1.4 Warm Up

For use before Activity 1.4

Find the volume of the solid.

1. 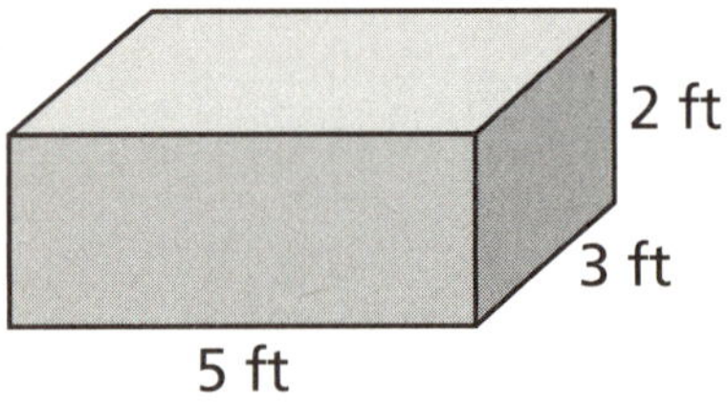

2. 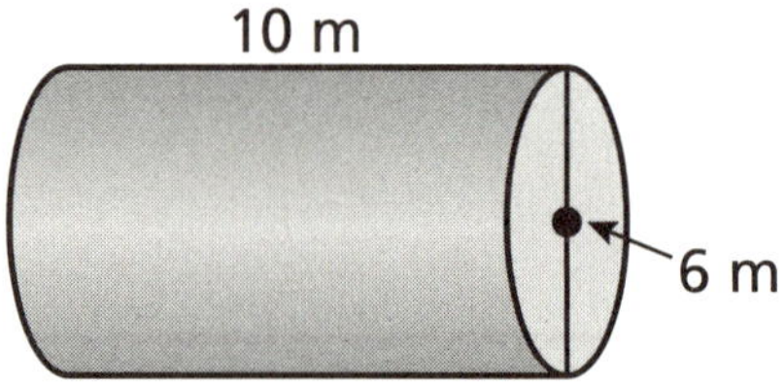

3. 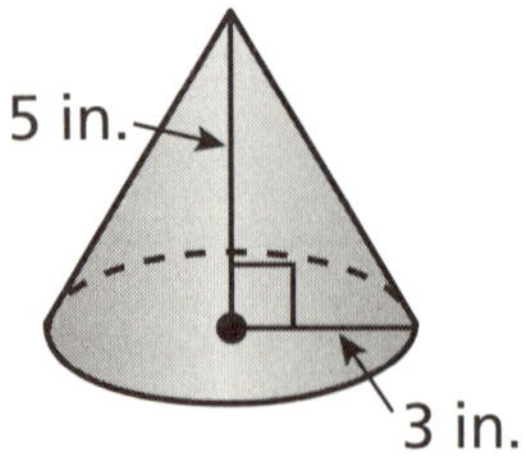

4. 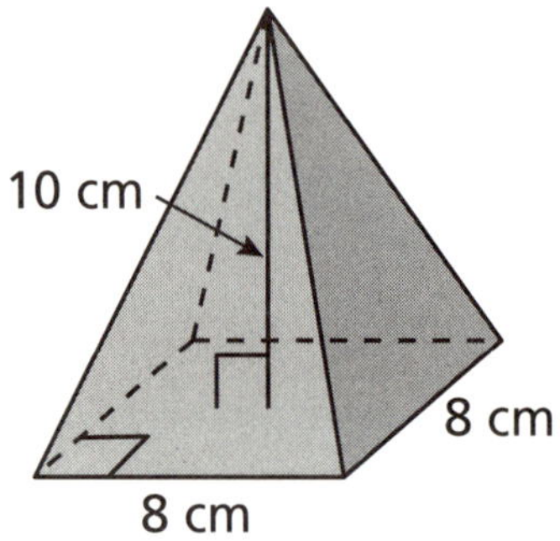

Lesson 1.4 Start Thinking!

For use before Lesson 1.4

How does solving the equation $5x + 4y = 14$ for x compare to solving the equation $5x + 20 = 14$ for x? Describe the steps involved in each solution.

Lesson 1.4 Warm Up

For use before Lesson 1.4

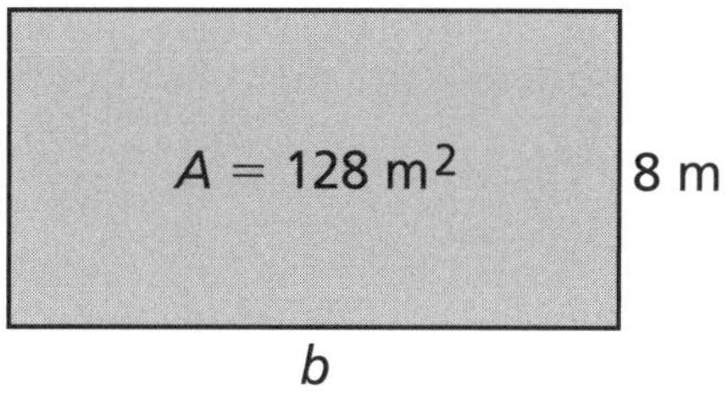

1. a. Write a formula for the area A of a rectangle.

   b. Solve the formula for b.

   c. Use the new formula to find the base of the rectangle.

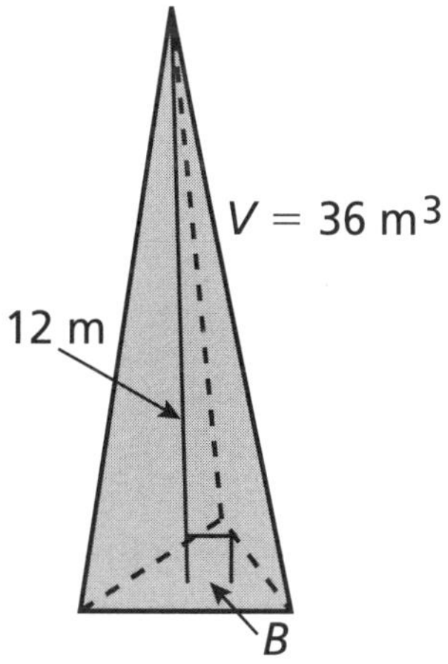

2. a. Write a formula for the volume V of a pyramid.

   b. Solve the formula for B.

   c. Use the new formula to find the area of the base of the pyramid.

Name ________________________________ Date __________

1.4 Practice A

Solve the equation for *y*.

1. $\frac{2}{5}x + y = 7$

2. $24 = 4x + 6y$

3. $5x - \frac{1}{2}y = 3$

4. $6\pi = x + 2y$

5. The formula $d = 2r$ can be used to find the distance d traveled after 2 hours when driving at rate r.

   a. Solve the formula for r.

   b. Find the rate r when the distance traveled is 130 miles.

Solve the equation for the bold variable.

6. $P = \boldsymbol{R} + C$

7. $p = \frac{\boldsymbol{X}}{N}$

8. $V = \frac{1}{3}\pi r^2 \boldsymbol{h}$

9. $A = \frac{1}{2}\boldsymbol{b}h$

10. The formula for the circumference of a circle with diameter d is $C = \pi d$.

    a. Solve the formula for d.

    b. The circumference of a circle is 8 inches. What is the diameter of the circle?

    c. The circumference of a circle is 6π inches. What is the radius of the circle?

11. The formula for the area of a rhombus with diagonals of lengths c and d is $A = \frac{1}{2}cd$.

    a. Solve the formula for c.

    b. The area of a rhombus is 35 square feet. The length of one of the diagonals is 10 feet. What is the length of the other diagonal?

    c. In a square (which is a rhombus), the lengths of the diagonals are the same. If a square has an area of 32 square feet, then what is the length of each diagonal?

Name____ Date__________

1.4 Practice B

Solve the equation for *y*.

1. $3x - \frac{1}{4}y = -2$

2. $5x + 8y = 6\pi$

3. $4y - 3.2x = 6$

4. $4.5x - 1.5y = 5.4$

5. The formula for the volume of a rectangular prism is $V = \ell wh$.

a. Solve the formula for w.

b. Use the new formula to find the value of w when $V = 210$ cubic feet, $\ell = 10$ feet, and $h = 3$ feet.

Solve the equation for the bold variable.

6. $T = hP + 2\boldsymbol{B}$

7. $C = 1000 + 80\boldsymbol{x}$

8. $S = \pi r^2 + 2\pi r\boldsymbol{h}$

9. $A = \frac{1}{2}\boldsymbol{P}a$

10. The formula $F = \frac{9}{5}C + 32$ converts temperatures from Celsius C to Fahrenheit F.

a. Solve the formula for C.

b. The boiling point of water is $212°F$. What is the temperature in Celsius?

c. If a house thermostat is set at $80°F$, what is the setting in Celsius? Round your answer to the nearest tenth.

11. The formula for the area of a sector of a circle is $A = \frac{m}{360}\pi r^2$, given the measure m of the angle and the radius r of the circle.

a. Solve the formula for m.

b. Find the measure of the angle when the area of the sector is 5 square centimeters and the radius is 2 centimeters. Round your answer to the nearest tenth.

c. If the area of the sector in part (b) is greater than 5 square centimeters, is the measure of the angle *greater than* or *less than* the answer to part (b)? Explain.

Name ______________________________ Date ________

1.4 Enrichment and Extension

Rewriting Formulas

Neon is a gas that is used in signs. Neon is usually transported in large cylindrical tanks from laboratories to manufacturers.

The volume of Neon is measured in cubic centimeters. A laboratory wishes to transport 22,222.22 cubic centimeters of Neon using a cylindrical tank. The tank is 15 centimeters in diameter.

In feet, how tall will the tank need to be in order to accommodate the volume of the gas? Round your answer to the nearest hundredth.

There are two methods that can be used to solve this problem. Complete each of the methods and answer the questions.

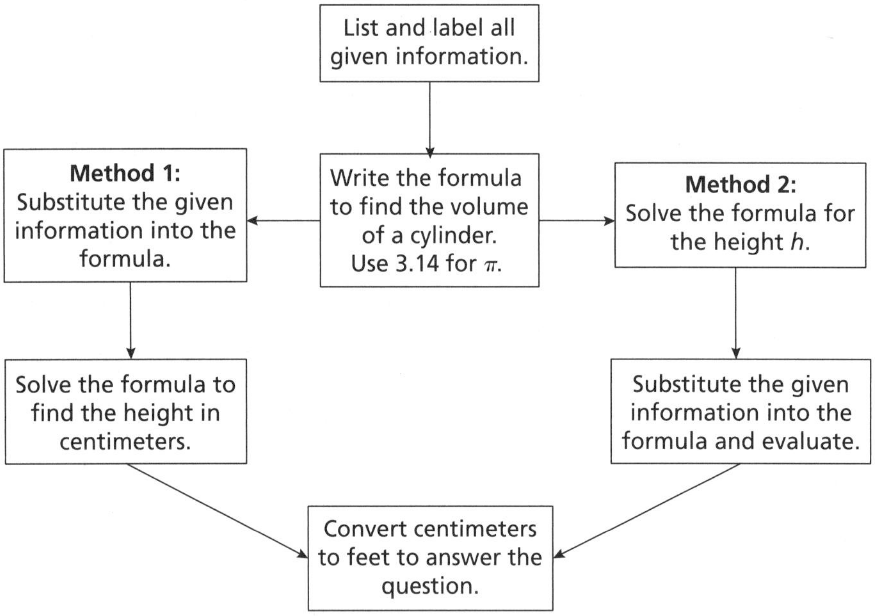

1. What is your solution using Method 1? What is your solution using Method 2?
2. Which method do you prefer? Why? Explain your reasoning.

Name____ Date__________

1.4 Puzzle Time

What Can't Walk, But Can Run?

Write the letter of each answer in the box containing the exercise number.

Solve the equation for *y*.

1. $16 - 12x = 4y$

A. $y = -3x + 4$ **B.** $y = 3x + 4$ **C.** $y = 3x - 4$

2. $2x + \frac{6}{5} = 2y$

D. $y = 2x + \frac{6}{5}$ **E.** $y = x + \frac{3}{5}$ **F.** $y = 2x + \frac{3}{5}$

3. $7 = 14x - 42y$

G. $y = -\frac{2}{3}x - \frac{1}{6}$ **H.** $y = \frac{1}{3}x + \frac{1}{6}$ **I.** $y = \frac{1}{3}x - \frac{1}{6}$

4. $8.1x - 4.5y = 5.4$

R. $y = 1.8x - 1.2$ **S.** $y = -1.8x + 1.2$ **T.** $y = 1.8x + 1.2$

5. The formula for Body Mass Index is $BMI = \frac{w}{h^2} \times 703$, where w is a person's weight (in pounds) and h is a person's height (in inches). Find the weight of a 13-year-old boy that is 60 inches tall and has a BMI of 20.5.

Q. 98 pounds **R.** 105 pounds **S.** 112 pounds

6. The formula $d = rt$ relates distance d to rate r and time t. Find how long it takes an airplane to fly 375 miles at 500 miles per hour.

T. 35 minutes **U.** 40 minutes **V.** 45 minutes

1		5	3	6	2	4

Activity 1.5 Start Thinking!

For use before Activity 1.5

What are some careers in which a person may need to convert units of measure?

Activity 1.5 Warm Up

For use before Activity 1.5

Copy and complete the statement.

1. 1 ft = __?__ in.

2. 1 yd = __?__ ft

3. 1 min = __?__ sec

4. 1 day = __?__ h

5. 1 mi = __?__ ft

6. 1 gal = __?__ qt

Lesson 1.5

Start Thinking!

For use before Lesson 1.5

If you know that a door is 30 inches wide and 7 feet tall, how can you find the area of the door?

Is there more than one correct answer?

Lesson 1.5

Warm Up

For use before Lesson 1.5

Find the perimeter or circumference in feet and in yards.

1.

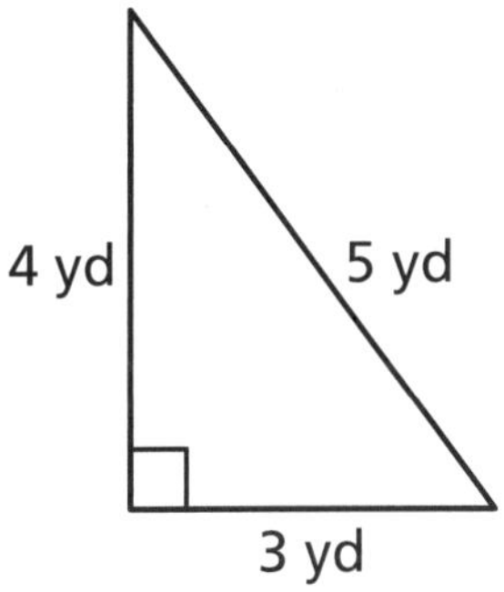

2.

3.

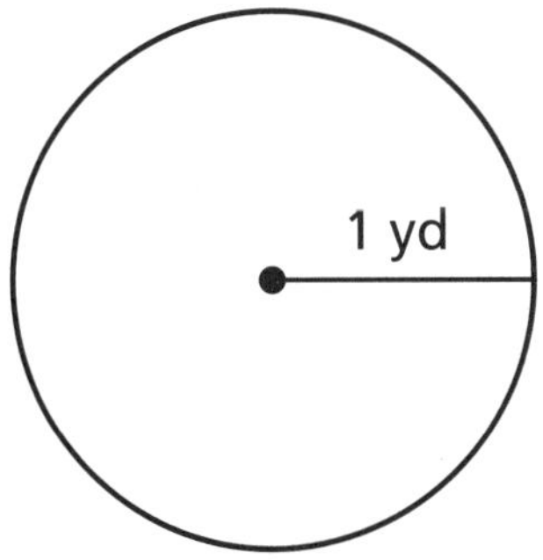

Name ________________________________ Date __________

1.5 Practice A

Find the perimeter in inches and in centimeters. Round your answer to the nearest tenth.

1.

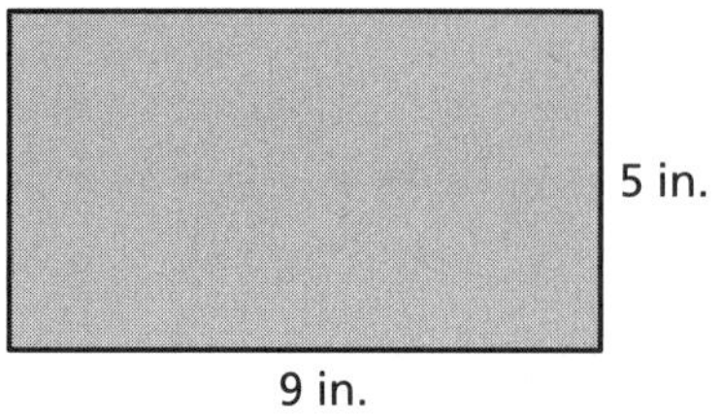

2.

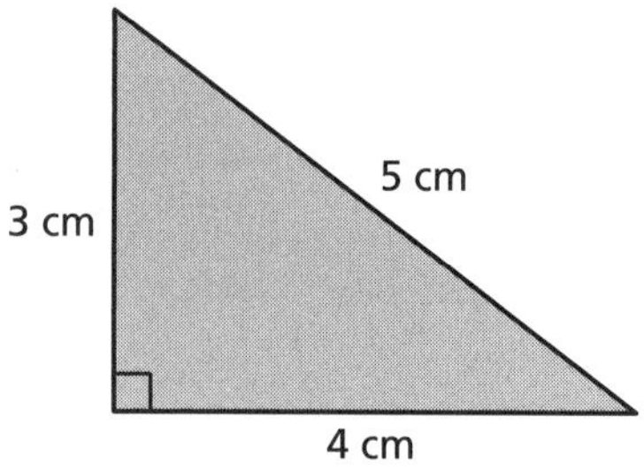

Copy and complete the statement. Round your answer to the nearest tenth.

3. 20 lb ≈ __?__ kg

4. 9 m ≈ __?__ ft

5. 10,000 ft ≈ __?__ mi

6. 2.5 qt ≈ __?__ L

7. A rectangular stone measures 400 centimeters by 90 centimeters.

a. What are the measurements of the stone in meters?

b. What are the measurements of the stone in yards?

Copy and complete the statement. Round your answer to the nearest tenth.

8. $\frac{30 \text{ mi}}{\text{h}} \approx \frac{\square \text{ km}}{\text{h}}$

9. $\frac{14 \text{ cm}}{\text{min}} \approx \frac{\square \text{ in.}}{\text{min}}$

10. $\frac{32 \text{ ft}}{\text{min}} \approx \frac{\square \text{ ft}}{\text{sec}}$

11. $\frac{360 \text{ mm}}{\text{h}} \approx \frac{\square \text{ mm}}{\text{sec}}$

12. $9 \text{ ft}^2 \approx$ __?__ yd^2

13. $12 \text{ m}^2 \approx$ __?__ cm^2

14. $7 \text{ km}^3 \approx$ __?__ m^3

15. $640 \text{ in.}^3 \approx$ __?__ ft^3

16. One gallon of paint will cover approximately 350 square feet of space.

a. How many gallons of paint will cover 130 square yards of space?

b. How many square feet of space will one quart of paint cover?

c. The paint costs $23.99 per gallon and $8.99 per quart. What is the most economical purchase to cover 130 square yards?

Name ____________________ Date __________

1.5 Practice B

Find the perimeter in inches and in centimeters. Round your answer to the nearest tenth.

1.

2.

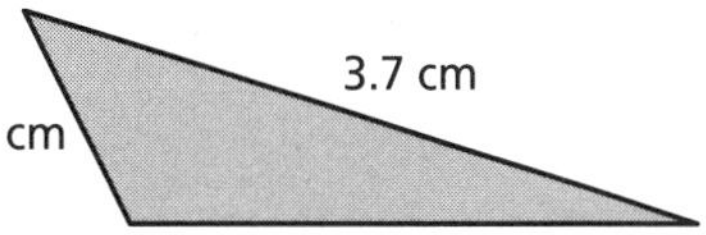

Copy and complete the statement.

3. 70 kg ≈ __?__ lb

4. 21 mi ≈ __?__ km

5. 10 L ≈ __?__ qt

6. 0.2 ft ≈ __?__ m

7. A rectangular table measures 53 inches by 36 inches. Round your answer to the nearest tenth.

a. What are the measurements of the table in feet?

b. What are the measurements of the table in meters?

Copy and complete the statement. Round your answer to the nearest tenth.

8. $\frac{15 \text{ cm}}{\text{sec}} \approx \frac{\square \text{ in.}}{\text{min}}$

9. $\frac{15 \text{ qt}}{\text{sec}} \approx \frac{\square \text{ L}}{\text{sec}}$

10. $\frac{8 \text{ yd}}{\text{min}} \approx \frac{\square \text{ m}}{\text{min}}$

11. $\frac{270 \text{ kg}}{\text{h}} \approx \frac{\square \text{ kg}}{\text{min}}$

12. $0.00025 \text{ km}^2 \approx$ __?__ m^2

13. $40 \text{ ft}^2 \approx$ __?__ in.^2

14. $3 \text{ yd}^3 \approx$ __?__ ft^3

15. $75{,}000{,}000 \text{ cm}^3 \approx$ __?__ m^3

16. A bedroom measures 12 feet by 14 feet.

a. How many square feet of carpet are required for the bedroom?

b. The installed price for the carpet is $21.99 per square yard. How much will it cost to install the carpet?

c. It is recommended that you purchase an additional 10% of carpet to cover any waste. How much does this add to the cost of the carpet?

Name ______________________________ Date __________

1.5 Enrichment and Extension

Converting Units of Measure

Architects are often asked to build scale models of their building designs before the actual construction of the building is approved.

You are asked to make a scale model of the Eiffel Tower. The actual measurements of the Eiffel Tower are given on the blueprint.

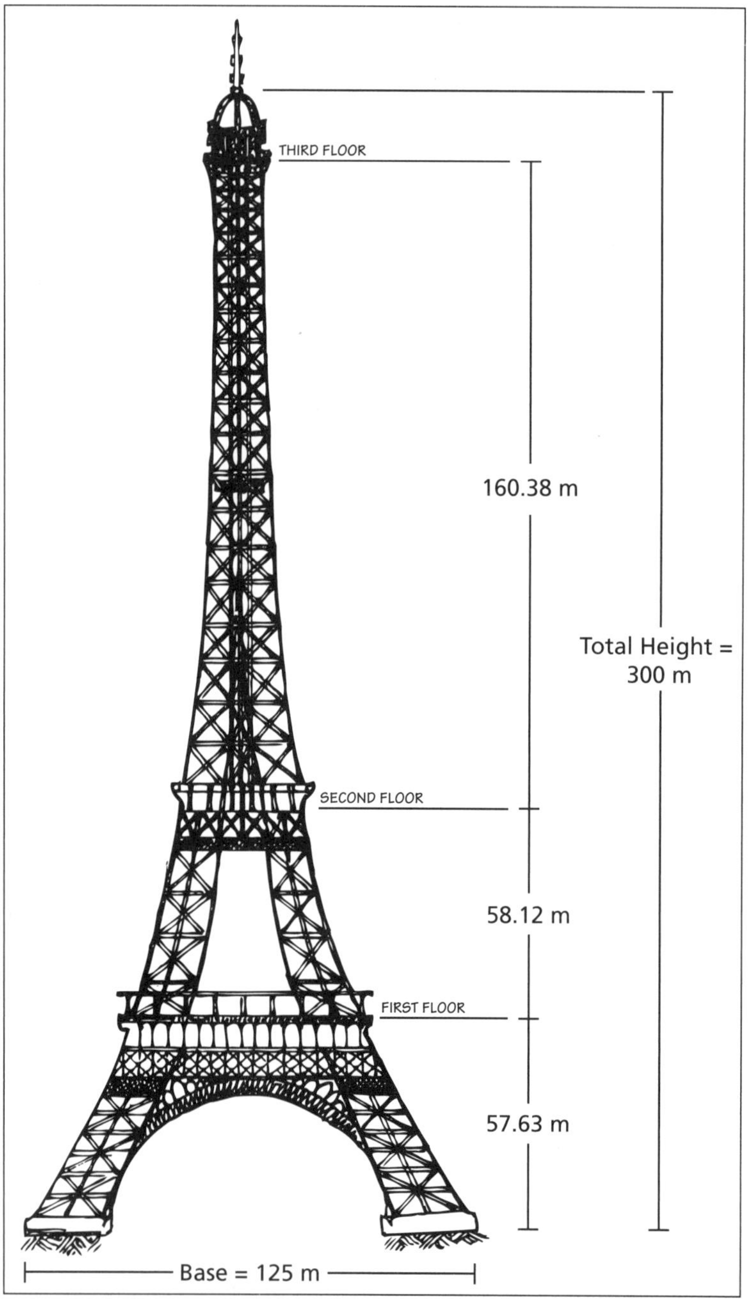

1. The Eiffel Tower is located in Paris, France. France uses the International System of Units. The United States uses the Standard System. Because your model will be built in the United States, convert each measurement on the blueprint from meters to feet. Round all answers to the nearest hundredth.

2. Building a full size Eiffel Tower is not practical. Instead, you want to build a scaled down model of the tower. Your model should be 12 inches tall. What is the scale factor you will use to build your model?

3. Use the measurements on the blueprint and the scale factor to determine the dimensions of your model in inches. Round your answers to the nearest hundredth.

4. Why is the construction of scale models useful? Besides architects, can you think of any other careers that might utilize scale models?

Name____ Date__________

1.5 Puzzle Time

Did You Hear About...

A	B	C	D	E	F
G	H	I	J	K	L
M	N	O	P		

Complete each exercise. Find the answer in the answer column. Write the word under the answer in the box containing the exercise letter.

456	COULD
1.5	IN
1500	SLEPT
125.7	IS
0.068	UP
21,120	CAR
100.4	MORNING
2045.5	SO
29.7	WHO
1728	WAKE

Find the missing value. Round to the nearest tenth if necessary.

A. $7 \text{ mi} \approx \square \text{ km}$

B. $48 \text{ lb} \approx \square \text{ kg}$

C. $28 \text{ L} \approx \square \text{ qt}$

D. $15 \text{ m} = \square \text{ cm}$

E. $\dfrac{32 \text{ kg}}{\text{h}} \approx \dfrac{\square \text{ lb}}{\text{h}}$

F. $\dfrac{13 \text{ gal}}{\text{min}} \approx \dfrac{\square \text{ L}}{\text{min}}$

G. $\dfrac{4 \text{ mi}}{\text{h}} = \dfrac{\square \text{ ft}}{\text{h}}$

H. $\dfrac{75 \text{ lb}}{\text{min}} \approx \dfrac{\square \text{ kg}}{\text{h}}$

I. $\dfrac{3 \text{ km}}{\text{sec}} \approx \dfrac{\square \text{ mi}}{\text{min}}$

J. $\dfrac{8 \text{ qt}}{\text{sec}} \approx \dfrac{\square \text{ L}}{\text{min}}$

K. $12 \text{ ft}^2 = \square \text{ in.}^2$

L. $680 \text{ cm}^2 = \square \text{ m}^2$

M. $351 \text{ ft}^3 = \square \text{ yd}^3$

N. $25 \text{ cm}^3 \approx \square \text{ in.}^3$

O. Noah can run 2 miles in 15 minutes. About how many kilometers can he run in that same time?

P. The goal area on a soccer field is 120 square yards. How many square meters is the goal area?

49.0	HIS
58.6	ENGINE
13	OILY
112.5	HE
21.8	MECHANIC
3.2	THE
70.4	UNDER
15.9	THAT
190	ARE
11.2	THE

Name ______________________________ Date __________

Chapter 1

School-to-Work

For use after Section 1.5

Landscape Designer

You are working as a landscaper. A potential customer has asked for an estimate of the cost of spreading bark mulch in a large bed. The bed is shaded in the diagram below.

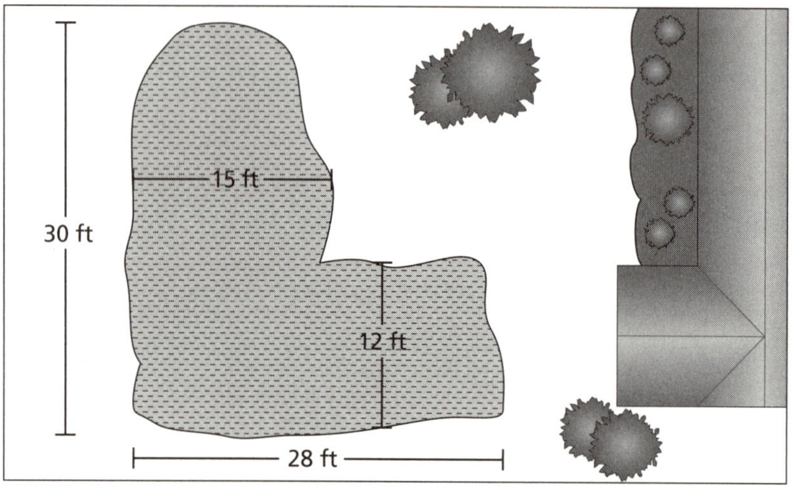

1. To estimate the cost of filling the bed with bark mulch, you first need to estimate the area of the bed. How can you estimate the area using rectangles? What is the approximate area?

2. You normally spread bark mulch in a layer that is 3 inches thick. Use this information to calculate the volume (in cubic feet) of bark mulch needed for the bed.

3. Bark mulch is sold in cubic yards, so you need to convert cubic feet to cubic yards. How many cubic feet are in 1 cubic yard? How many cubic yards of mulch do you need for the bed?

4. The prices for three types of bark mulch are given in the table. How much will it cost to fill the bed with each type of mulch?

5. You charge $16 per cubic yard to spread bark mulch. How does this affect the total costs for the customer?

Bark mulch	Cost per yd^3
pine	$15
cedar	$17
hemlock	$20

6. The customer decides that he wants hemlock bark and he wants to spend no more than $135. How thick could you spread the mulch and still stay within the customer's budget? Explain your reasoning.

Name____ Date__________

Chapter 1 Study Help

You can use a **Y Chart** to compare two topics. List differences in the branches and similarities in the base of the Y.

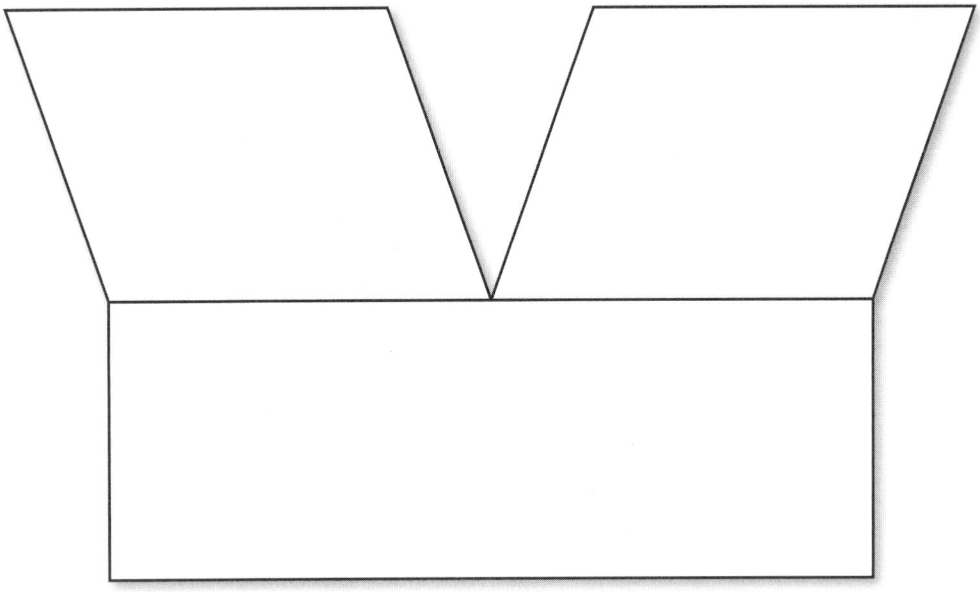

On Your Own

Make a Y chart to help you study and compare these topics.

1. solving equations with the variable on one side and solving equations with variables on both sides

2. solving multi-step equations and solving equations with variables on both sides

After you complete this chapter, make Y charts for the following topics.

3. solving multi-step equations and rewriting literal equations

4. converting meters to feet and converting feet to meters

5. converting one unit to another and converting one rate to another

Name ______________________________ Date __________

Chapter 1

Financial Literacy

For use after Section 1.3

Planning a Fundraiser

Your social studies class is raising money for a trip to Washington, D.C. For one fundraiser, you are planning on having your school's name and image printed on T-shirts. The T-shirts will then be sold at school events, such as sports competitions, dances, and concerts.

1. How many T-shirts do you think it will be possible to sell? To help you answer this question, think about how many students are enrolled at your school. What other people associated with your school might buy T-shirts? Explain the reasoning behind your estimate.

2. Research the cost of having a T-shirt design made and the cost for printing one shirt using a silk screen. The price per shirt decreases as the number of shirts ordered increases, so you will need to use your estimate for the number of shirts you will sell.

3. At what price should you sell each T-shirt? How much greater than the cost of producing each shirt do you want the price to be? How much would you be willing to pay for a T-shirt?

4. Write an expression for the costs associated with producing the T-shirts. Let n represent the number of T-shirts.

5. Write an expression for the money made by selling the T-shirts. Let n represent the number of T-shirts.

6. The *break-even point* for the T-shirts is the point at which the costs associated with producing the T-shirts is equal to the money made by selling the T-shirts. Write and solve an equation relating these two quantities. How many T-shirts do you need to produce and sell to break even? Why is this number important in this situation?

7. Suppose you raised the price of each T-shirt. What affect does this have on the break-even point? How might raising the price per T-shirt help your fundraising effort? How might it hurt your fundraising effort? Explain.

8. How much money do you think is reasonable for your class to raise by selling T-shirts? Based on your answer, how many T-shirts do you need to sell, and at what price, to meet your fundraising goal? Explain.

Name________________________________ Date__________

Chapter 1 Cumulative Practice

Solve the equation. Check your solution.

1. $x - 7 = 15$

2. $1.2b - 4.2 = -10.2$

3. $\frac{2}{5}h - 2\frac{1}{5} = 3$

4. $3n = 27$

5. $6k - 13 = 5k$

6. $\frac{1}{2}(a - 16) = -9$

7. $8g - 14g + 22 = 37$

8. $\frac{2}{3}(v - 3) = \frac{1}{3}(v + 6)$

9. $9 - p = 11$

10. $\frac{y}{8} = 5$

11. $5 - 4m = 17$

12. $2w + 7w - 28 = 26$

13. A movie theater sells tickets for \$4.50 and boxes of popcorn for \$2.50. You have \$12. How many boxes of popcorn can you buy if you buy one ticket at the theater?

Copy and complete the statement.

14. 16 m $\approx$ $\square$ ft

15. 108 in. = $\square$ yd

16. $\frac{14 \text{ cm}}{\text{min}} \approx \frac{\square \text{ in.}}{\text{min}}$

17. 0.05 mi^2 = $\square$ yd^2

18. $\frac{150 \text{ qt}}{\text{h}} \approx \frac{\square \text{ L}}{\text{h}}$

19. 1500 km = $\square$ m

20. 60 lb $\approx$ $\square$ kg

21. $\frac{40 \text{ cm}}{\text{min}} = \frac{\square \text{ mm}}{\text{h}}$

22. 6000 mm^3 = $\square$ cm^3

23. A birdbath contains 6 gallons of water. How many quarts of water does the birdbath contain?

Solve the equation for the indicated variable.

24. $7xy = 21$; y

25. $6k + 8m = 20$; m

26. $\frac{bh}{2} = A$; h

27. $4p - 3q = 2d$; p

28. $\frac{1}{5}(z - v) = w$; z

29. $\frac{ad}{bc} = 1$; d

30. The volume of the rectangular prism is 10.8 cubic meters. Solve the volume equation for ℓ. Use the new formula to find the length of the prism.

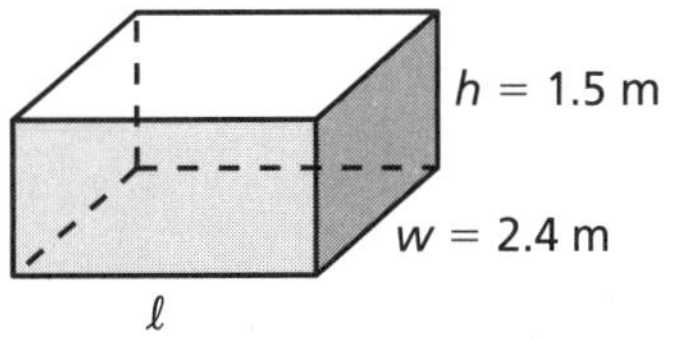

Chapter 2

Name ______________________________ Date __________

Chapter 2 Graphing Linear Equations and Linear Systems

Dear Family,

Some people have a side business to supplement their income: delivering papers, selling crafts, or running a website, to name a few. The goal is to make a *profit*—to have more income than expenses.

The *break-even point* is where the income equals the expenses. Making a graph is a good way to keep track of income and expenses and will show at a glance when the business will break even. On a graph, the break-even point is where the income line crosses the expense line.

My Woodworking Business

Selling Stools at Craft Fair	
Craft Fair Booth:	$160.00
Business Cards:	$85.00
New Tools:	$125.00
Initial Investment:	**$370.00**
Electricity:	$0.50
Rough Lumber:	$3.00
Fasteners & Glue:	$0.75
Stain & Sandpaper:	$1.25
My Time:	$5.00
Unit Cost:	**$10.50**
Unit Price:	**$14.95**

Units	Expense	Income
0	$370.00	$0.00
10	$475.00	$149.50
20	$580.00	$299.00
30	$685.00	$448.50
40	$790.00	$598.00
50	$895.00	$747.50
60	$1,000.00	$897.00
70	$1,105.00	$1,046.50
80	$1,210.00	$1,196.00
90	$1,315.00	$1,345.50
100	$1,420.00	$1,495.00
110	$1,525.00	$1,644.50
120	$1,630.00	$1,794.00
130	$1,735.00	$1,943.50

Selling Stools at Craft Fair

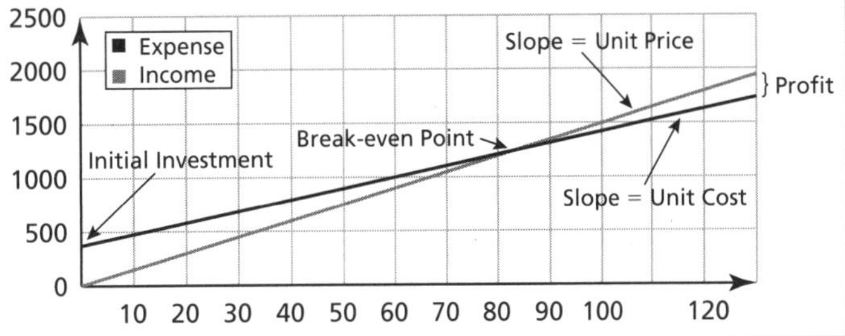

Have your student help you make a plan for a small business. A few basic steps will get your business plan started.

- How much money will you need for supplies to get started? This represents your *initial investment*. Plot this point on your graph.
- How much does it cost you to produce each item? This is the *unit cost*. Use this to plot more points on the graph to make an expense line.
- What price will you charge for each item? Starting at the origin of the graph, use this unit price to make an income line.
- Do the two lines cross? This is your break-even point—the number of items you must sell to pay for your expenses.

If the two lines do not cross, you will have to make some changes. Can you increase your unit price? You may not be able to charge more than your competitors. In that case, you will have to find a way to cut expenses.

What effect do changes to your initial investment have on the break-even point? What effect do changes to the unit cost have on the break-even point? Ask your student which one has a greater impact over time.

May your collaboration be a profitable one!

Nombre ______________________________ Fecha ________

Capítulo 2 Graficando Ecuaciones y Sistemas Lineales

Estimada Familia:

Algunas personas tienen un negocio para complementar sus ingresos: repartir el diario, vender artesanías, administrar un sitio web, por nombrar algunos ejemplos. El objetivo es generar utilidades—tener más ingresos que gastos.

El punto de equilibrio se da cuando los ingresos son iguales a los egresos. Hacer un gráfico es una buena manera de monitorear los ingresos y gastos y muestra a simple vista cuando un negocio alcanza este punto. En un gráfico el punto de equilibrio se da cuando la línea de ingresos cruza la línea de gastos.

Mi negocio de carpintería

Vender bancos en feria artesanal	
Stand en la feria:	$160.00
Tarjetas de presentación:	$85.00
Herramientas nuevas:	$125.00
Inversión inicial:	**$370.00**
Electricidad:	$0.50
Madera:	$3.00
Grapas y adhesivos:	$0.75
Barniz y lijas:	$1.25
Tiempo:	$5.00
Costo unitario:	**$10.50**
Precio unitario:	**$14.95**

Unidades	Gastos	Ingresos
0	$370.00	$0.00
10	$475.00	$149.50
20	$580.00	$299.00
30	$685.00	$448.50
40	$790.00	$598.00
50	$895.00	$747.50
60	$1,000.00	$897.00
70	$1,105.00	$1,046.50
80	$1,210.00	$1,196.00
90	$1,315.00	$1,345.50
100	$1,420.00	$1,495.00
110	$1,525.00	$1,644.50
120	$1,630.00	$1,794.00
130	$1,735.00	$1,943.50

Vender bancos en feria artesanal

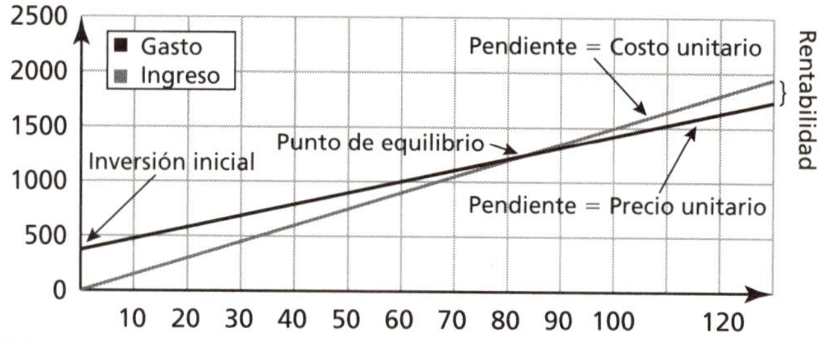

Haga que su estudiante lo ayude a preparar un plan para un pequeño negocio. Con unos cuantos pasos básicos empezarán su plan de negocios.

- ¿Cuánto dinero necesitarán para comenzar? Esto representa su inversión inicial. Ubiquen este punto en su gráfico.
- ¿Cuánto les costará producir cada artículo? Este es el costo unitario. Con este dato, coloquen más puntos con el fin de trazar una línea de gastos.
- ¿Qué precio cobrarán por cada artículo? Comenzando en el origen del gráfico, utilicen este precio unitario para dibujar la línea de ingresos.
- ¿Se cruzan estas dos líneas? Este es su punto de equilibrio—el número de artículos que deben vender para cubrir sus gastos.

Si las dos líneas no se cruzan deberán hacer ciertos cambios. ¿Pueden aumentar su precio unitario? Tal vez no puedan cobrar más que la competencia. En ese caso, tendrán que hallar un modo de recortar los gastos.

¿Cómo afectan al punto de equilibro las modificaciones en su inversión inicial? ¿Cómo afectan al punto de equilibro las modificaciones del costo unitario? Pregunte a su estudiante cuál tiene mayor impacto a largo plazo.

Non ______________________________ Dat __________

Chapit 2 Ann Fè Graf Ekwasyon Lineyè ak Sistèm Lineyè

Chè Fanmi:

Gen moun ki gen yon biznis sou kote pou ajoute yon sipleman sou revni yo: livre jounal, vann pwodwi atizanal, oswa fè sit wèb fonksyone, pou nonmen detwa. Bi a se fè yon *pwofi*—pou gen plis revni pase depans.

Pwen mò a se kote revni a egal depans yo. Fè yon graf se yon bon mannyè pou swiv revni ak depans yo epi sa ap montre nan yon koudèy lè biznis la ap rantre frè li. Sou yon graf, pwen mò a se kote liy revni an kwaze liy depans la.

Biznis Menwizri Mwen an

Vant Taboure nan Fwa Travo Manyèl la	
Estann Fwa Travo	
Manyèl la:	$160.00
Kat Vizit:	$85.00
Nouvo Zouti:	$125.00
Envestisman Inisyal:	**$370.00**
Kouran:	$0.50
Bwa Brit Siye:	$3.00
Fèmti ak Lakòl:	$0.75
Koulè ak Papye Sable:	$1.25
Tan Mwen:	$5.00
Kou Inite:	**$10.50**
Pri Inite:	**$14.95**

Inite	Depans	Revni
0	$370.00	$0.00
10	$475.00	$149.50
20	$580.00	$299.00
30	$685.00	$448.50
40	$790.00	$598.00
50	$895.00	$747.50
60	$1,000.00	$897.00
70	$1,105.00	$1,046.50
80	$1,210.00	$1,196.00
90	$1,315.00	$1,345.50
100	$1,420.00	$1,495.00
110	$1,525.00	$1,644.50
120	$1,630.00	$1,794.00
130	$1,735.00	$1,943.50

Vant Taboure nan Fwa Travo Manyèl la

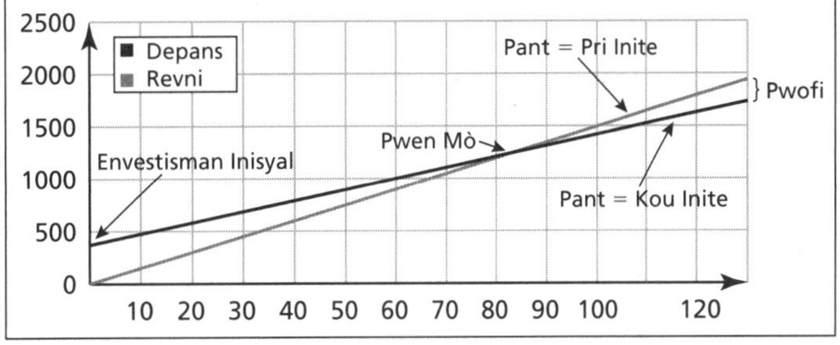

Fè elèv ou a ede ou fè yon pwojè pou yon ti biznis. Detwa pwen fondamantal ap ede ou kòmanse pwojè biznis ou an.

- Ki kantite lajan w'ap bezwen pou materyèl pou kòmanse? Sa reprezante *envestisman inisyal* ou a. Mete pwen sa a sou graf ou a.
- Konbyen sa koute ou pou pwodwi chak atik? Sa se *kou inite* a. Sèvi ak sa pou mete plis pwen sou graf la pou fè yon liy depans.
- Ki pri w'ap mande pou chak atik? Kòmanse nan orijin graf la, sèvi ak pri inite sa a pou fè yon liy revni.
- Èske de liy yo kwaze? Se pwen mò ou a—kantite atik ou dwe vann pou peye pou depans ou yo.

Si de liy yo pa kwaze, w'ap oblije fè kèk chanjman. Èske ou kapab ogmante pri inite ou a? Ou gen dwa pa kapab mande plis pase konpetitè ou yo. Nan ka sa a, w'ap oblije jwenn yon jan pou diminye depans yo.

Ki efè chanjman nan envestisman inisyal ou a gen sou pwen mò a? Ki efè chanjman nan kou inite a gen sou pwen mò a? Mande elèv ou a kilès ki gen pi gwo enpak avèk letan.

Mwen swete pou kolaborasyon nou an pwofitab!

Activity 2.1 Start Thinking!
For use before Activity 2.1

Which of the following graphs do you think show *linear* equations? Explain.

A.

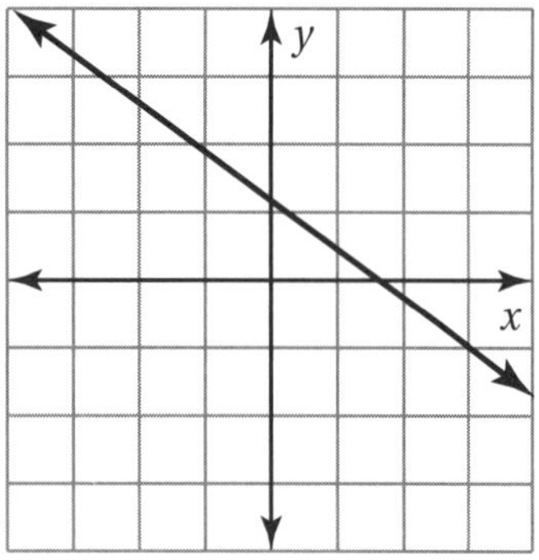

B.

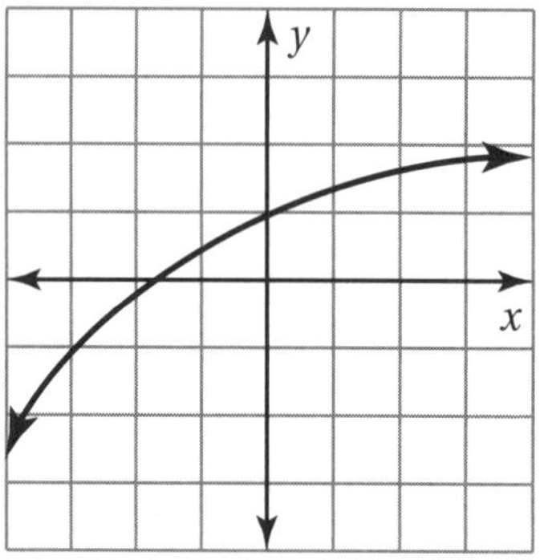

C.

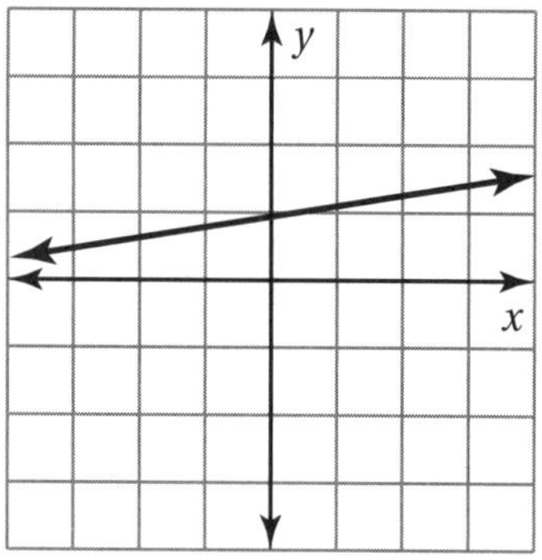

Activity 2.1 Warm Up
For use before Activity 2.1

Copy and complete the table using the given equation.

1. $y = 2x - 3$

x	−1	0	1	2
y				

2. $y = -x + 1$

x	−1	0	1	2
y				

3. $y = x + 5$

x	−1	0	1	2
y				

4. $y = \frac{1}{2}x + 2$

x	−1	0	1	2
y				

Lesson 2.1 Start Thinking!

For use before Lesson 2.1

Think about how much energy you have on an average day.

Graph your energy level (on a scale of 0 to 10) throughout an average day.

Are any sections of your graph linear?

Lesson 2.1 Warm Up

For use before Lesson 2.1

Copy and complete the table. Plot the two solution points and draw a line *exactly* through the two points. Find a different solution point on the line.

1.

x		
$y = 4x - 2$		

2.

x		
$y = -x + 5$		

Name ______________________________ Date __________

2.1 Practice A

Copy and complete the table. Plot the two solution points and draw a line *exactly* through the two points. Find a different solution point on the line.

1.

x		
$y = 4x + 3$		

2.

x		
$y = \frac{3}{2}x - 1$		

Graph the linear equation.

3. $y = -2x$

4. $y = \frac{2}{5}x$

5. $y = -4$

6. $y = x + 2$

7. $y = -5x + 3$

8. $y = \frac{x}{2} + 1$

9. The equation $y = \frac{2}{3}x$ represents the cost y (in dollars) for x pounds of bananas.

   a. Graph the equation.

   b. Use the graph to estimate the cost of 8 pounds of bananas.

   c. Use the equation to find the exact cost of 8 pounds of bananas.

Solve for *y*. Then graph the equation.

10. $y - 2x = 5$

11. $6x + 5y = 15$

12. You have \$110 in your lunch account and plan to spend \$2.75 each school day.

    a. Write and graph a linear equation that represents the balance in your lunch account.

    b. How many school days will it take to spend all of the money in your lunch account?

Name____ Date__________

2.1 Practice B

Determine which points in the table are solution points. Plot the solution points and draw a line *exactly* through the two points.

1.

x	-1	0	1	2
$y = -x + 3$	4	2	4	1

2.

x	-3	0	3	6
$y = \frac{1}{3}x + 5$	6	5	4	7

Graph the linear equation.

3. $y = 3.5$

4. $y = \frac{4}{5}x$

5. $y = \frac{2}{3}x - 2$

6. $y = -\frac{1}{10}x + 4$

7. $y = \frac{10}{3}x$

8. $y = -\frac{x}{2} + \frac{3}{2}$

9. The equation $y = 1.5x + 35$ represents the cost y (in dollars) of the family meal when the food costs $35 and x beverages are purchased.

a. Graph the equation.

b. Use the graph to estimate the cost of the family meal when 5 beverages are purchased.

c. Use the equation to find the exact cost of the family meal when 5 beverages are purchased.

Solve for *y*. Then graph the equation.

10. $2y + 3x = -6$

11. $x + 0.25y = 1.5$

12. There are 10 coconuts at the base of your tree. The coconuts are falling off the tree at a rate of 6 coconuts per week. Assume that you do not pick up any coconuts.

a. Write and graph a linear equation that represents the number of coconuts at the base of your tree after x weeks.

b. The tree will have no coconuts on it when there are 52 coconuts at the base of the tree. After how many weeks will this occur?

13. The sum s of the first n integers is $s = \frac{1}{2}n(n + 1)$. Plot four points (n, s) that satisfy the equation. Is the graph of the equation a line?

Name ______________________________ Date __________

2.1 Enrichment and Extension

Graphing Equations

You have studied the properties of linear equations. Now you will learn about two additional types of equations, the quadratic equation and the absolute value equation.

Copy and complete the table. Then plot the points and draw a graph of the equation.

1. $y = -2x + 3$

x	*y*
−2	7
−1	
0	
1	
2	

2. $y = x^2 - 1$

x	*y*
−2	3
−1	
0	
1	
2	

3. $y = -|x| + 2$

x	*y*
−2	0
−1	
0	
1	
2	

4. Linear equations have the form $y = mx + b$. Which of the given equations is linear?

5. An equation of the form $y = ax^2 + b$ is called a quadratic equation. Which of the given equations is quadratic? Describe the graph of the quadratic equation.

6. An equation of the form $y = a|x| + b$ is called an absolute value equation. Which of the given equations is an absolute value equation? Describe the graph of the absolute value equation.

7. Compare and contrast the three different types of graphs.

8. How would each graph change if a different *b* value was selected?

Name ______________________________ Date __________

2.1 Puzzle Time

What Arctic Bird Can Be Found In A Bakery?

Write the letter of each answer in the box containing the exercise number.

Find the values of *y* that correspond to the given values of *x* for the linear function.

1. $y = 4x + 3$ for $x = -1, 0, 1$
2. $y = -\frac{3}{2}x + 5$ for $x = 0, 2, 4$
3. $y = -9$ for $x = 0, 1, 2$
4. $y = -7x + 8$ for $x = -1, 0, 1$
5. $y = \frac{5}{3}x - 6$ for $x = -3, 0, 3$
6. $y = 1.4x - 9$ for $x = 0, 1, 2$

Solve for *y*. Then find the values of *y* that correspond to the given values of *x* for the linear function.

7. $y + 8x = -2$ for $x = 0, 1, 2$
8. $12x + 3y = 15$ for $x = -1, 0, 1$
9. $\frac{1}{4}y - 3x = 9$ for $x = -2, 0, 2$
10. $0.4y + 2x = 1.2$ for $x = -3, 0, 3$
11. The equation $22 = 2y + x$ represents the perimeter of a flower garden with length y (in feet) and width x (in feet). Solve for y. Then find the length of the flower bed when the width is 2 feet, 3 feet, and 4 feet.
12. The equation $0.60 = 0.05x + 0.10y$ represents the number of nickels x and dimes y needed to add up to 60 cents. Solve for y. Then find the number of dimes that are needed to make 60 cents when the number of nickels is 0, 2, and 4.

Answers

E. $y = -9, -7.6, -6.2$

F. $y = 12, 36, 60$

P. $y = 18, 3, -12$

I. $y = -2, -10, -18$

A. $y = 15, 8, 1$

F. $y = 6, 5, 4$

M. $y = 10, 9.5, 9$

U. $y = -11, -6, -1$

C. $y = 9, 5, 1$

N. $y = 5, 2, -1$

R. $y = -9, -9, -9$

A. $y = -1, 3, 7$

1		8	3	6	4	11		10	5	9	12	7	2

Activity 2.2

Start Thinking!

For use before Activity 2.2

Use the definition of ski slope to help you think of what the slope of a line is in mathematics.

Activity 2.2

Warm Up

For use before Activity 2.2

Write the fraction in simplest form.

1. $\frac{6}{2}$

2. $\frac{8}{28}$

3. $\frac{10}{25}$

4. $\frac{10}{8}$

5. $\frac{6}{9}$

6. $\frac{16}{12}$

Lesson 2.2 Start Thinking!
For use before Lesson 2.2

1. Each student must choose an ordered pair.
2. Choose a partner and work together to find the slope of the line joining your two points. Use a graph to help you.
3. Repeat the process several times with different partners.

Were any of the slopes positive? negative? zero?

Lesson 2.2 Warm Up
For use before Lesson 2.2

Draw a line through each point using the given slope. What do you notice about the two lines?

1. Slope = 2

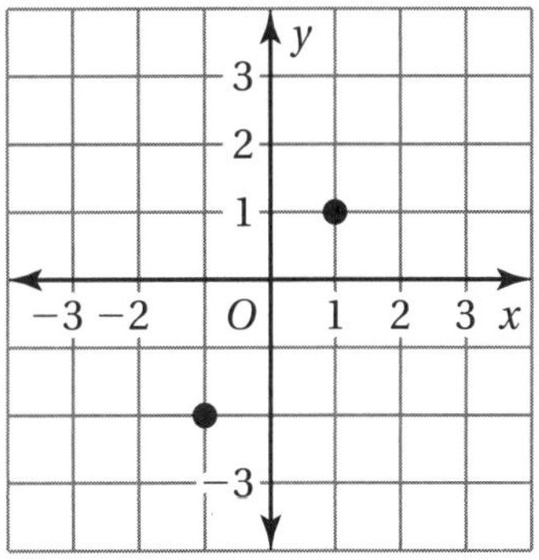

2. Slope = −2

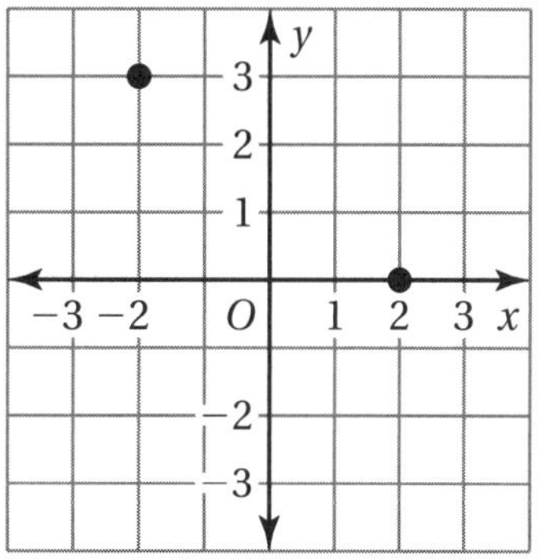

Name ______________________________ Date __________

2.2 Practice A

1. Refer to the graph.

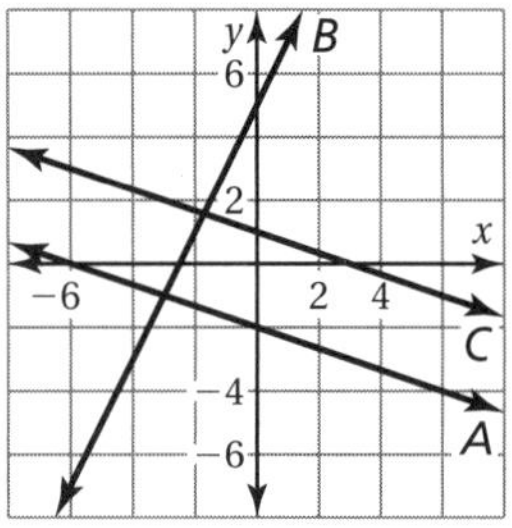

a. Which lines have negative slopes?

b. Which line has the steepest slope?

c. Are any two of the lines parallel? Explain.

Draw a line through each point using the given slope. What do you notice about the two lines?

2. Slope $= -2$

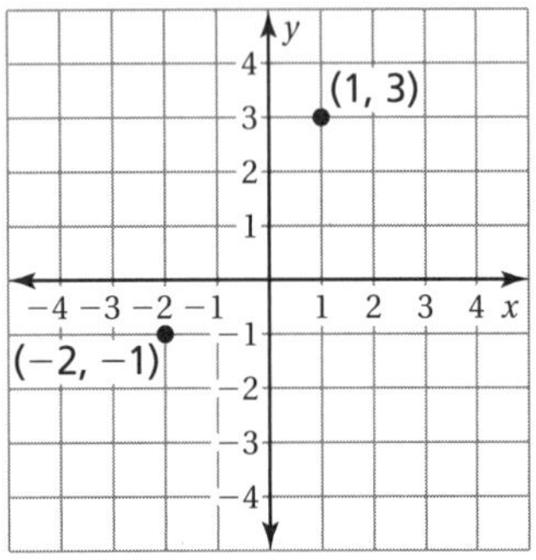

3. Slope $= \dfrac{1}{2}$

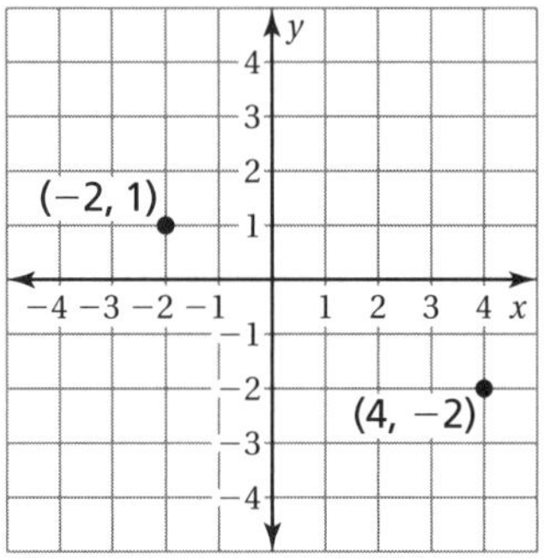

Find the slope of the line.

4.

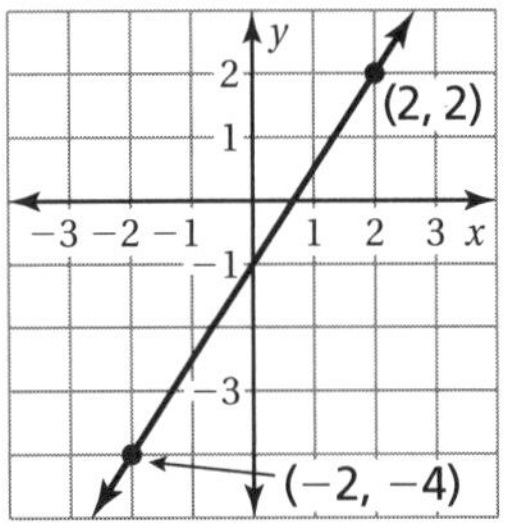

5.

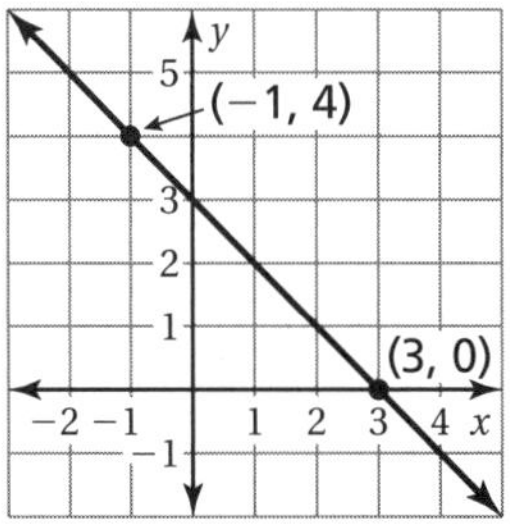

6.

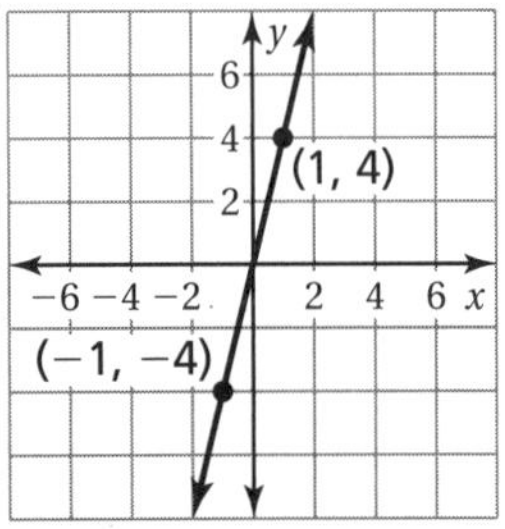

7.

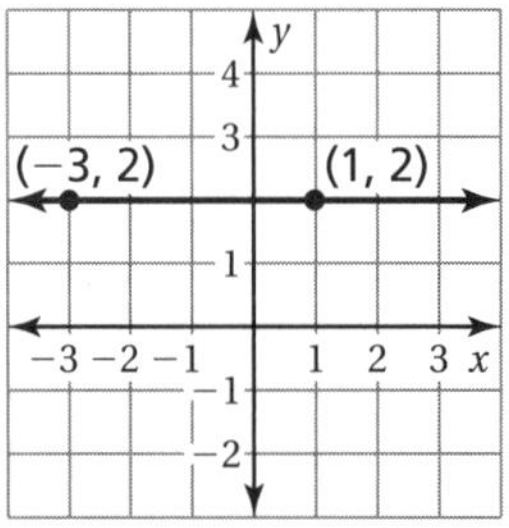

8. An awning covers a window that is 4 feet high. When the awning is opened, it extends 2 feet from the base of the window. Find the slope of the awning.

Name______________________________ Date__________

2.2 Practice B

Find the slope of the line.

1.

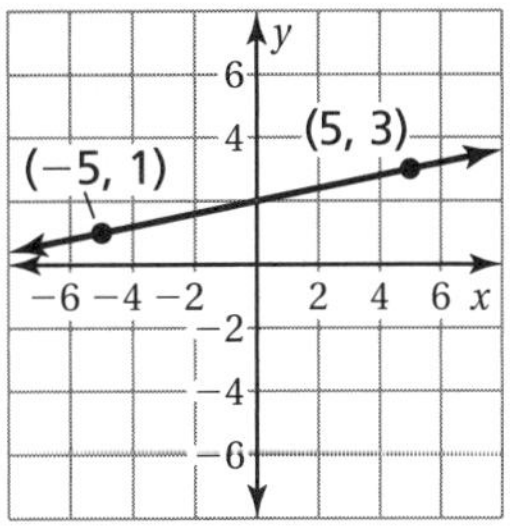

2.

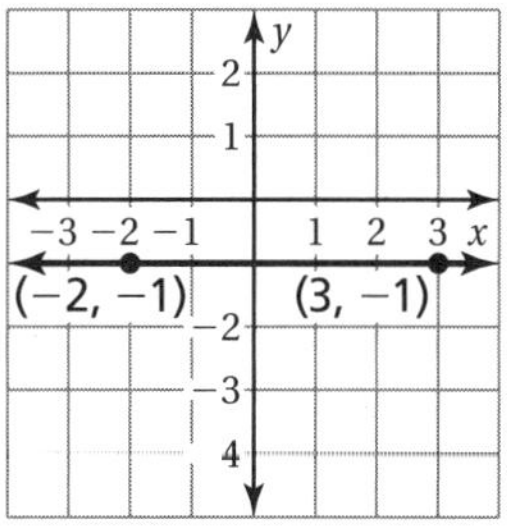

3.

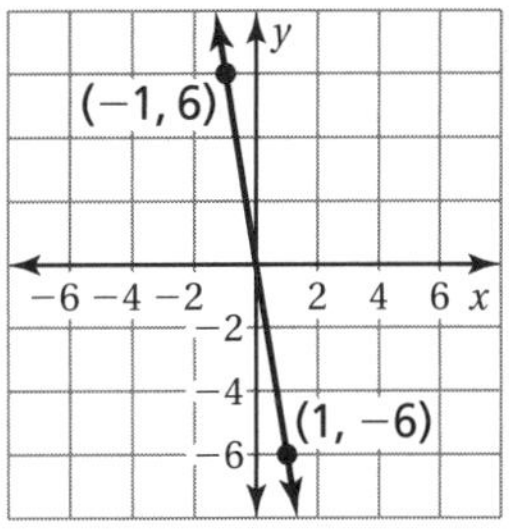

4.

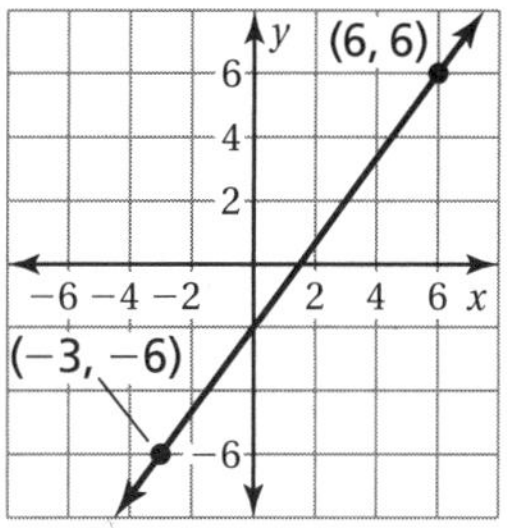

The points in the table lie on a line. Find the slope of the line. Then draw its graph.

5.

x	0	2	4	6
y	−4	−1	2	5

6.

x	−4	−1	0	3
y	7	4	3	0

7. A ramp used to remove furniture from a moving truck has a slope of $\frac{2}{5}$. The height of the ramp is 4 feet. How far does the base of the ramp extend from the end of the truck?

8. The graph shows the cost of a long distance phone call.

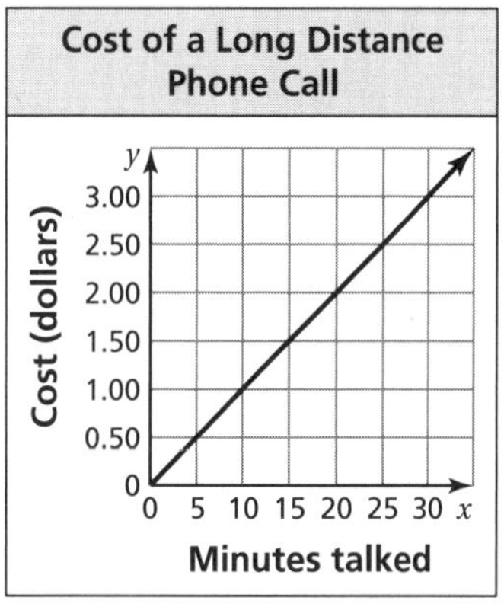

a. Find the slope of the line.

b. Explain the meaning of the slope as a rate of change.

c. How much money is added to the phone bill if you talk for 5 extra minutes?

d. How many minutes did you talk if the phone call costs $3?

Name ______________________________ Date __________

2.2 Enrichment and Extension

Perpendicular Lines

Lines that are parallel do not intersect and have the same slope. Lines that are perpendicular form right angles when they intersect and have slopes that are negative reciprocals of each other. For example, if one line has a slope of $\frac{2}{3}$, any line with a slope of $-\frac{3}{2}$ would be perpendicular to it.

Find the slopes of the lines. Determine if the lines are parallel, perpendicular, or neither.

1.

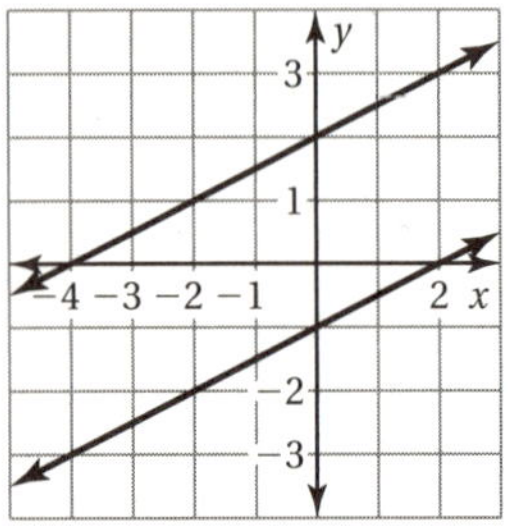

2.

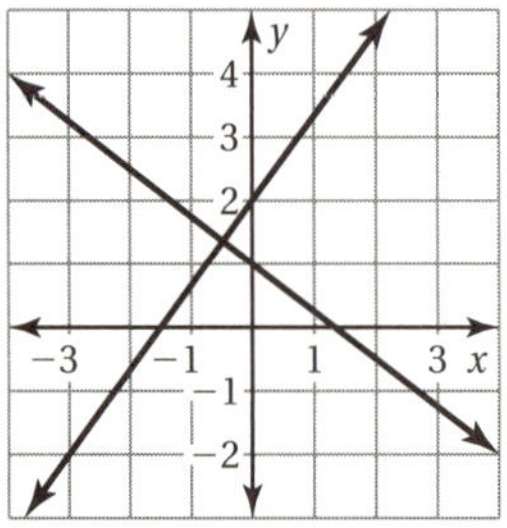

3.

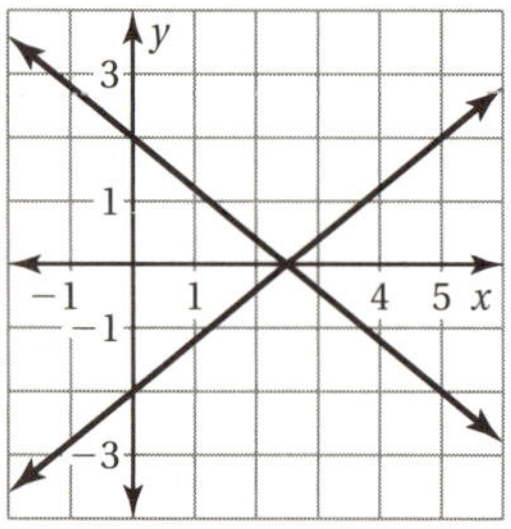

4.

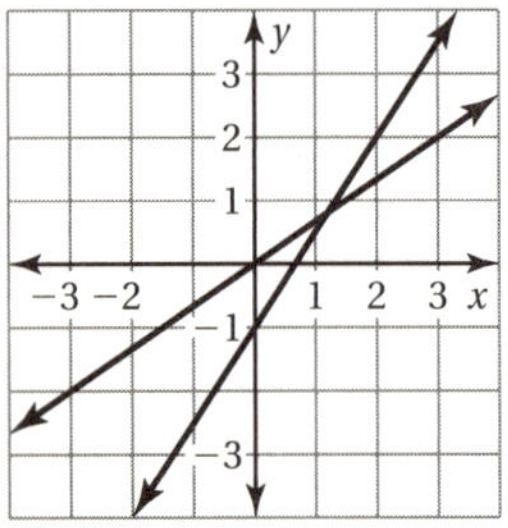

5.

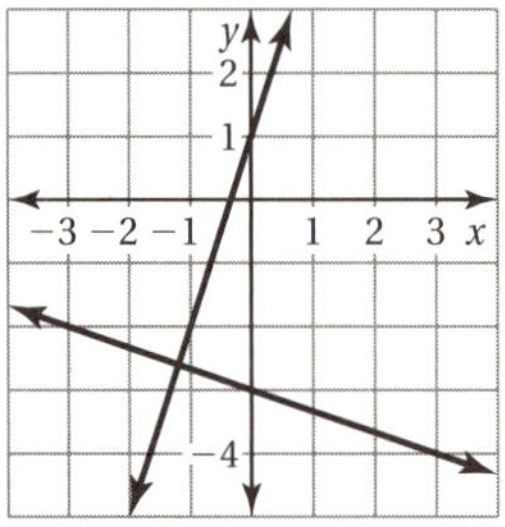

6.

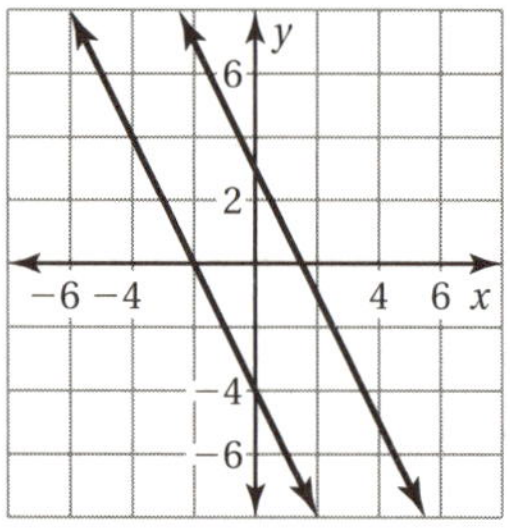

7. The equation $y = 4$ represents a horizontal line that passes through $(0, 4)$. The equation $x = 2$ represents a vertical line that passes through $(2, 0)$.

a. Draw both lines on the same graph. Use the graph to determine if the two lines are perpendicular. Explain your reasoning.

b. What is the slope of the line $y = 4$?

c. A vertical line has an undefined slope. Why do you think that is? Use two different methods of finding the slope to explain your reasoning.

Name____ Date__________

2.2 Puzzle Time

What Did One Poppy Seed Say To The Other?

Circle the letter of each correct answer in the boxes below. The circled letters will spell out the answer to the riddle.

Find the slope of the line.

1.

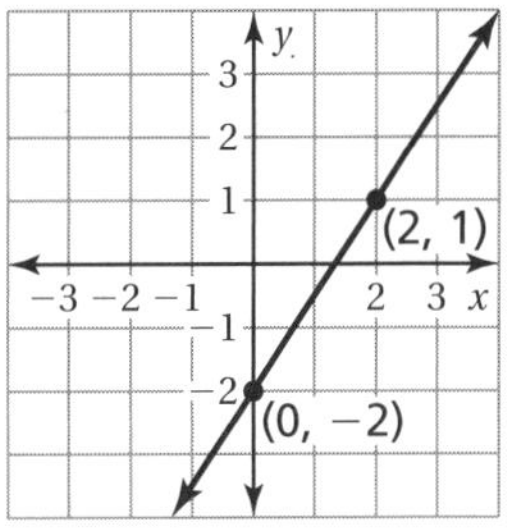

2.

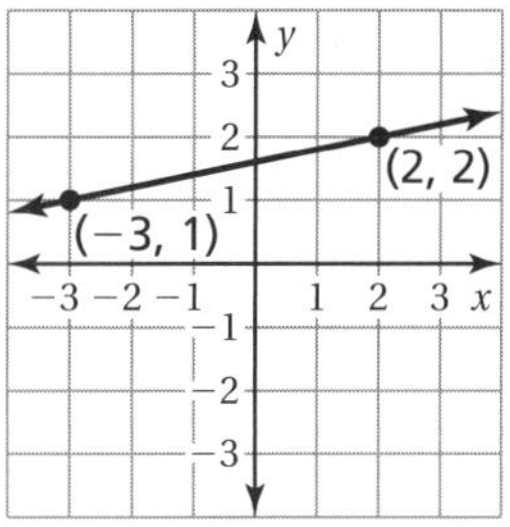

3.

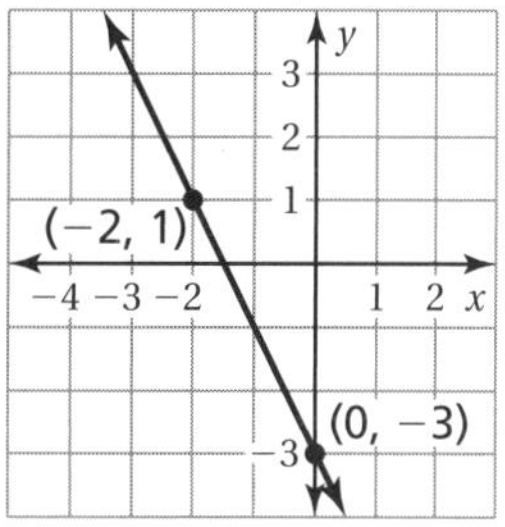

4.

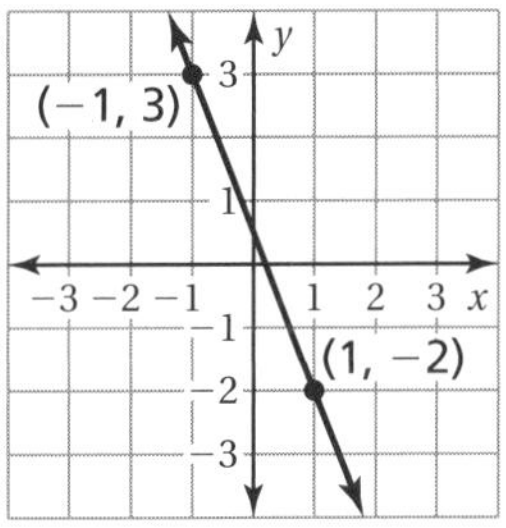

5.

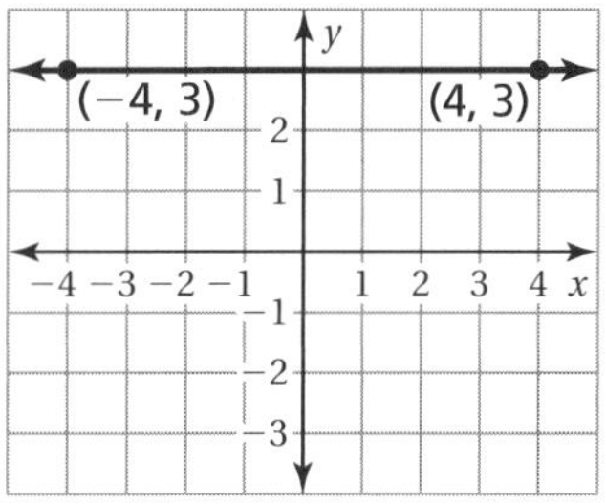

6.

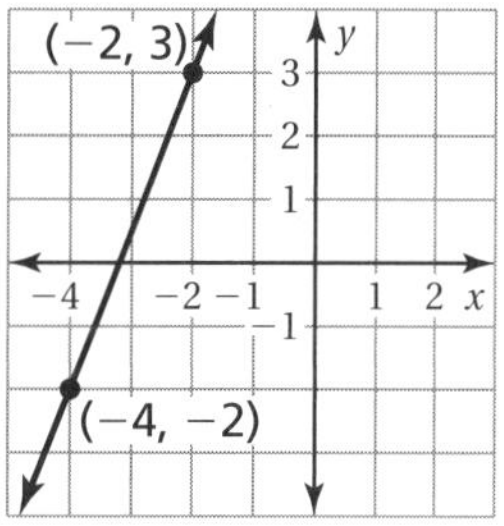

7.

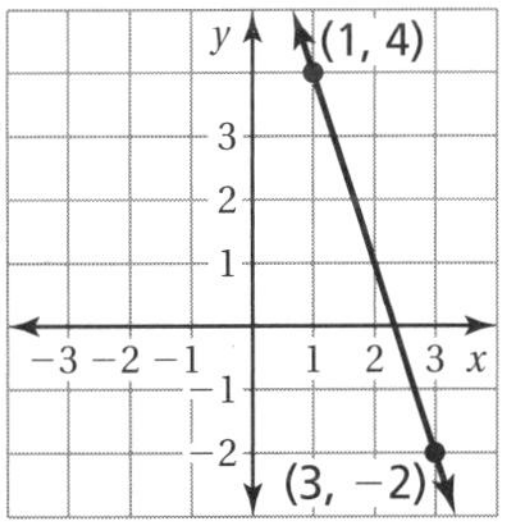

8.

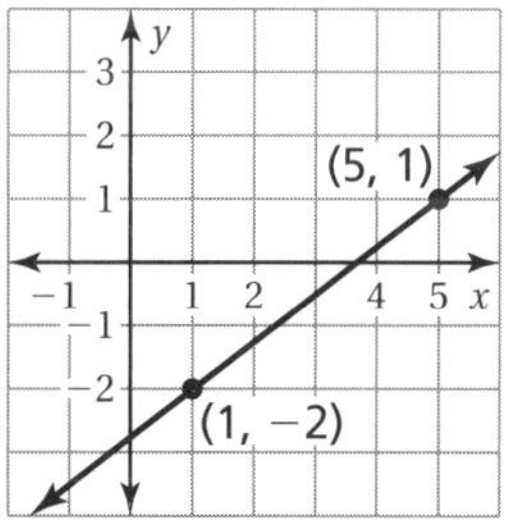

9. The slope of a ramp from the sidewalk to the street has to be $\frac{1}{20}$. The height of a ramp is 2.5 inches. What is the horizontal length (in inches) of the ramp?

I	T	M	S	A	O	N	H	A	P	L	R	O	M	E	L	S	L
-2	$\frac{2}{5}$	0	$\frac{1}{6}$	$\frac{4}{3}$	$\frac{1}{5}$	-3	$\frac{1}{4}$	$\frac{3}{2}$	$\frac{1}{50}$	1	50	$\frac{5}{2}$	$-\frac{1}{3}$	2	$-\frac{5}{2}$	5	$\frac{3}{4}$

Activity 2.3 Start Thinking!

For use before Activity 2.3

The graph shows the population of the United States from 2000 to 2007.

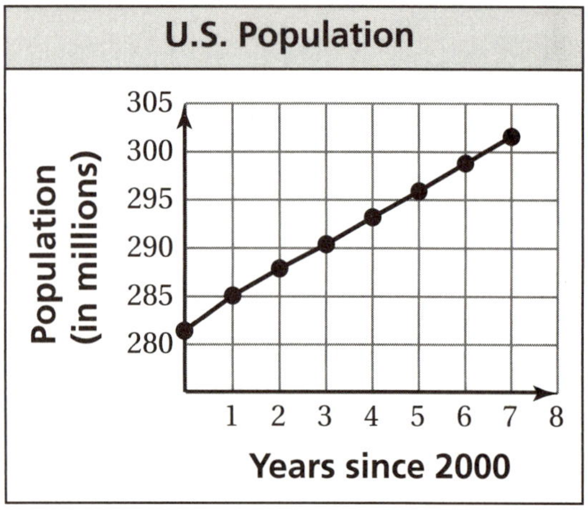

Explain what some points on the graph mean.

Is the graph linear?

What is the approximate slope? What does it represent?

What is the approximate point where the line crosses the y-axis? What does it represent?

Activity 2.3 Warm Up

For use before Activity 2.3

Graph the linear equation using the input-output table.

1. $y = x - 1$

x	0	1	2	3
y				

2. $y = \frac{1}{2}x + 2$

x	–2	0	2	4
y				

Lesson 2.3 Start Thinking!

For use before Lesson 2.3

Describe a situation involving online shopping that can be modeled with a linear equation.

What is the slope?

What is the y-intercept?

Lesson 2.3 Warm Up

For use before Lesson 2.3

Match the equation with its graph. Identify the slope and y-intercept.

1. $y = 2x - 1$

2. $y = -x + 2$

3. $y = -2x - 1$

A.

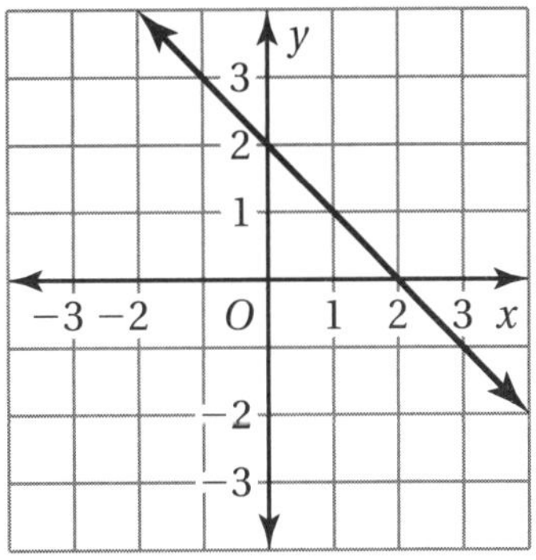

B.

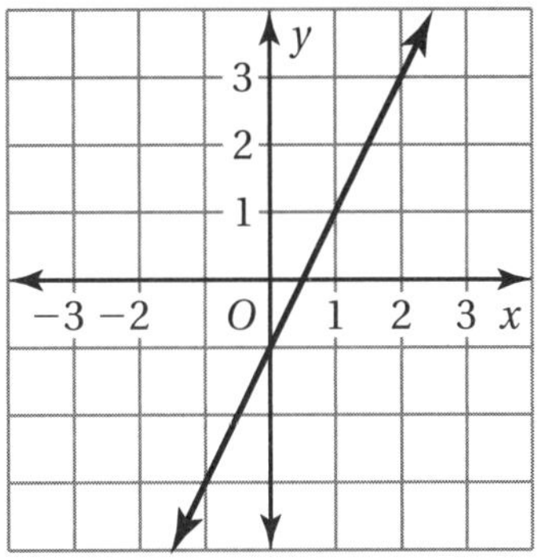

C.

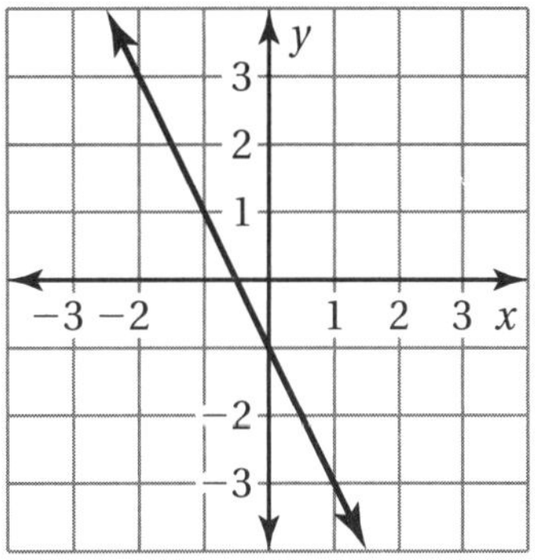

Name ________________________________ Date __________

2.3 Practice A

Match the equation with its graph. Identify the slope and *y*-intercept.

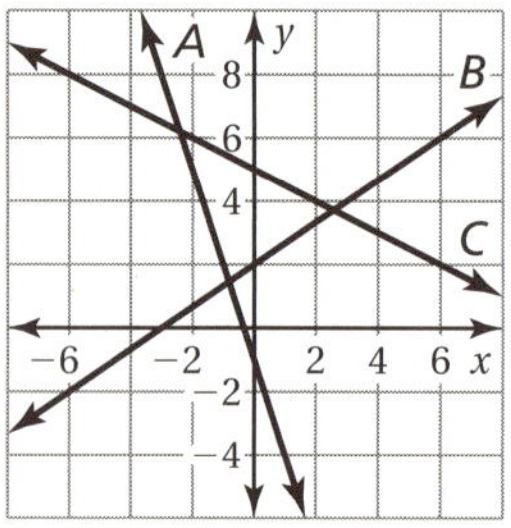

1. $y = -\frac{1}{2}x + 5$

2. $y = -3x - 1$

3. $y = \frac{2}{3}x + 2$

Find the slope and *y*-intercept of the graph of the linear equation.

4. $y = x + 4$

5. $y = -8x + 3$

6. $y = -\frac{5}{7}x - 2$

7. $y = 1.75x - 1$

8. $y - 2 = 6x$

9. $y + 7 = \frac{1}{9}x$

10. The depreciated value y (in dollars) of a business car after x years is $y = -4200x + 21{,}000$.

    a. Graph the equation.

    b. Interpret the slope.

    c. Interpret the y-intercept.

    d. Interpret the x-intercept.

Graph the linear equation. Identify the *x*-intercept.

11. $y = 3x - 6$

12. $y = -\frac{1}{4}x + 12$

13. $y = 3.2x + 9.6$

14. $y - 2 = 5x$

15. The amount of fertilizer y (in cups) that is needed for x square feet of grass is $y = \frac{1}{4}x$.

    a. Graph the equation.

    b. Interpret the slope.

Name____ Date__________

2.3 Practice B

Find the slope and *y*-intercept of the graph of the linear equation.

1. $y = -\frac{3}{8}x + 10$

2. $y = 4.5x + 7$

3. $y = -\frac{4}{5}x - \frac{1}{5}$

4. $y + 2.5 = 5.5x$

5. $y - \frac{2}{7} = 4x$

6. $y + 5 = \frac{2}{3}x$

Graph the linear equation. Identify the *x*-intercept.

7. $y = \frac{5}{3}x - 2$

8. $y = -1.2x + 9$

9. $y - 6.6 = 1.1x$

10. $y + 3 = -\frac{6}{7}x$

11. There is a $10 monthly membership fee to download music. There is a $0.50 fee for each song downloaded.

a. Write a linear equation that models the cost of downloading x songs per month.

b. Graph the equation.

c. What is the cost of downloading 15 songs?

12. An entrepreneur is opening a business to market pies and pie fillings based on her family's recipes. The price of every item in the store is $6.

a. Write a linear equation that models the amount of revenue y (in dollars) taken in for selling x items.

b. Graph the equation.

c. The monthly cost of rent and utilities for the store space is $1100. What is the minimum number of items that must be sold each month in order to make a profit?

d. Assuming 4 weeks in a month, what is the average number of items that need to be sold each week in order to turn a profit?

Name ______________________________ Date __________

2.3 Enrichment and Extension

Matching Slopes and Intercepts

Each of the equations below is missing a *b*-value. Use each value in the B-hive exactly once to complete the seven linear equations below. The graph of each linear equation should pass through two of the given points on the graph. No two equations pass through the same given point.

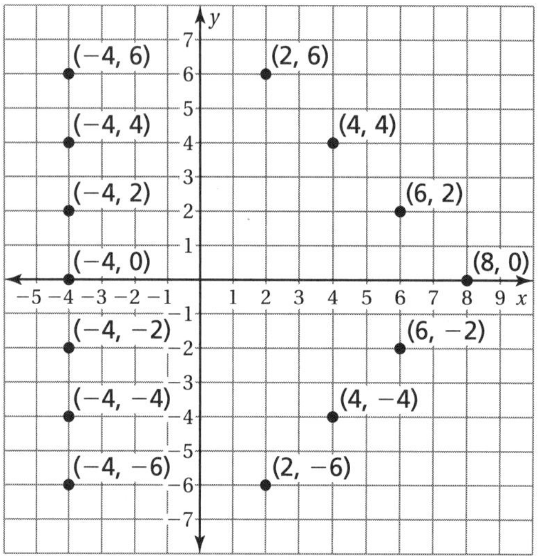

1. $y = -\frac{5}{4}x + \square$

2. $y = x + \square$

3. $y = \frac{3}{5}x + \square$

4. $y = \frac{1}{2}x + \square$

5. $y = \frac{1}{4}x + \square$

6. $y = \square$

7. $y = -\frac{5}{3}x + \square$

B-Hive

Name____ Date __________

2.3 Puzzle Time

Did You Hear About...

A	B	C	D	E	F
G	H	I	J	K	L
M	N	O	P		

Complete each exercise. Find the answer in the answer column. Write the word under the answer in the box containing the exercise letter.

$\frac{3}{2}$ MANY
$-\frac{1}{2}$ WHO
-12 THAT
-7 SO
0.25 SEASON
-5 GAVE
-2 FOR
16 TICKETS
$\frac{2}{3}$ THE
-8 DRIVING

Find the slope of the graph of the linear equation.

A. $3y = 2x + 3$ **B.** $y = -x - 2$

C. $4y = -2x + 12$ **D.** $5y - 10 = x$

Find the *y*-intercept of the graph of the linear equation.

E. $y = 4x - 4$ **F.** $2y = x - 4$

G. $y - 12 = -9x$ **H.** $7 + y = 4.3x$

Find the *x*-intercept of the graph of the linear equation.

I. $y = 6x - 9$ **J.** $3y = 2x + 36$

K. $2y = -5x + 6$ **L.** $3y - 9 = 4x$

M. $y = 1.6x + 8$ **N.** $y + 15 = 12.5x$

O. Shannon's hair is 12 inches long and grows 0.25 inch per month. In an equation that represents the length y of her hair after x months, what number represents the slope?

P. You have a \$20 gift card to a coffee shop. Each time you go there, you get chai tea for \$1.25. The equation $y = -1.25x + 20$ represents how much you have left on the gift card after x visits. How many chai teas can you purchase before the balance on your card runs out?

-4 STOPPED
5 HAD
$-\frac{9}{4}$ POLICE
-1 LADY
22 CAR
$\frac{6}{5}$ THE
-18 TIMES
$\frac{1}{5}$ GOT
1.2 HER
12 SPEEDING

Activity 2.4 Start Thinking!
For use before Activity 2.4

Consider the equation $y = -\frac{2}{5}x + 2$.

Which of the following is an equivalent equation?

$5x + 2y = 10$ or $2x + 5y = 10$

How do you know?

Activity 2.4 Warm Up
For use before Activity 2.4

Solve the equation for *y*.

1. $x + y = 4$

2. $2x + y = 10$

3. $3x + 4y = 12$

4. $-5x + 10y = 8$

5. $-4x + 2y = 10$

6. $-x + 2y = 4$

Lesson 2.4

Start Thinking!

For use before Lesson 2.4

You have \$40 to spend on turkey and cheese for a party. At the deli, turkey is \$10 per pound and cheese is \$6 per pound.

Is it easier to write an equation to represent the situation in *slope-intercept form* or *standard form*? Why?

Lesson 2.4

Warm Up

For use before Lesson 2.4

Define two variables for the verbal model. Write an equation in slope-intercept form that relates the variables. Graph the equation.

1. $\dfrac{\text{25 miles}}{\text{hour}} \bullet \text{Non-highway hours} + \dfrac{\text{60 miles}}{\text{hour}} \bullet \text{Highway hours} = \text{240 miles}$

2. $\dfrac{\$5.00}{\text{hat}} \bullet \text{Number of hats} + \dfrac{\$10.00}{\text{T-shirt}} \bullet \text{Number of T-shirts} = \$30$

Name ______________________________ Date __________

2.4 Practice A

Write the linear equation in slope-intercept form.

1. $4x + y = 10$

2. $3x - y = 7$

Graph the linear equation.

3. $2x - 3y = 6$

4. $5x - 3y = 15$

Use the graph to find the *x*- and *y*-intercepts.

5.

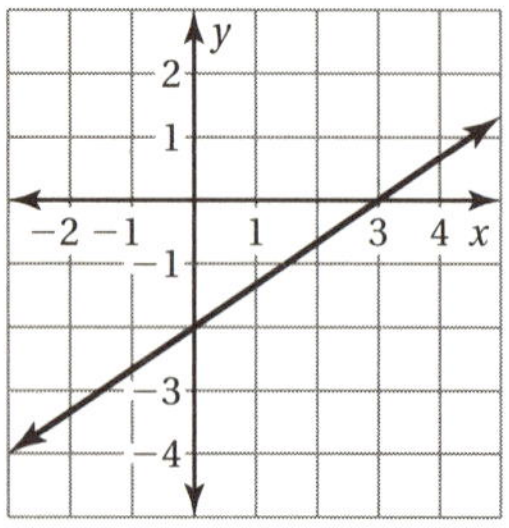

6.

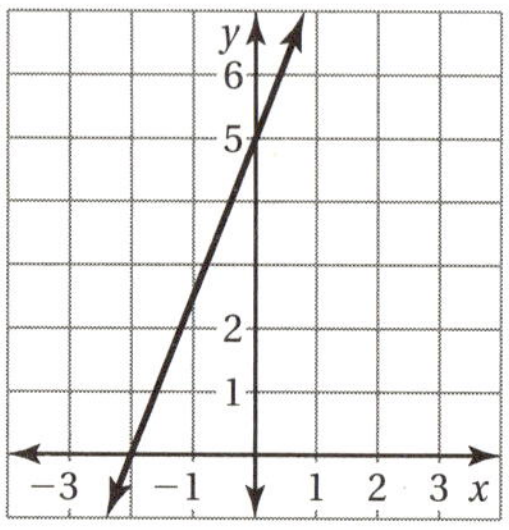

Graph the linear equation using intercepts.

7. $4x + y = 8$

8. $3x - 2y = 12$

9. The total amount of fiber (in grams) in a package containing x apples and y oranges is given by the equation $5x + 10y = 110$.

   a. Find and interpret the y-intercept.

   b. Find and interpret the x-intercept.

   c. How many grams of fiber does an orange contain?

   d. How many grams of fiber does an apple contain?

   e. Is it possible for the package to contain 15 apples? Explain.

10. You have two jobs. You earn \$8 for each hour x that you work as a restaurant host and \$6 for each hour y that you work as a hair washer. Your earnings for the pay period is \$144.

   a. Write an equation in standard form that models your earnings.

   b. Find the x- and y-intercepts.

   c. Graph the equation.

   d. You worked 10 hours as a hair washer. How many hours did you work as a host?

Name________________________________ Date__________

2.4 Practice B

Write the linear equation in slope-intercept form.

1. $\frac{2}{3}x + y = 4$

2. $4x - 2y = 10$

Graph the linear equation.

3. $4.5x - 0.5y = 3$

4. $\frac{2}{3}x + \frac{1}{3}y = 2$

Use the graph to find the *x*- and *y*-intercepts.

5.

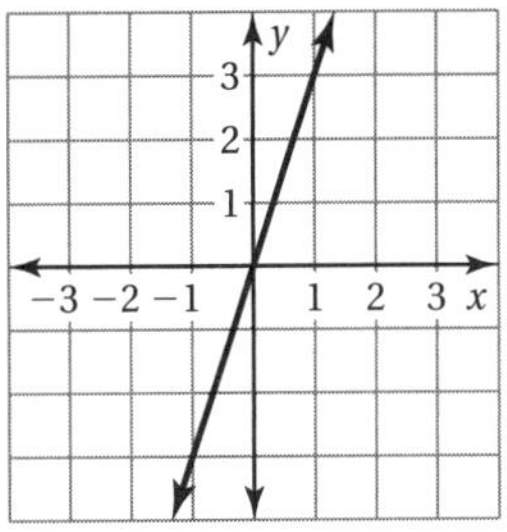

6.

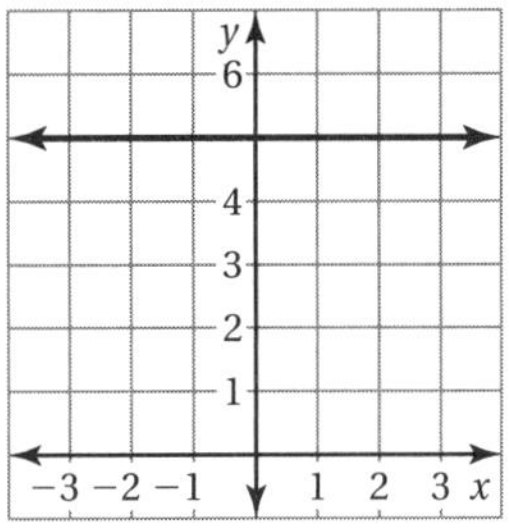

Graph the linear equation using intercepts.

7. $\frac{1}{5}x + \frac{1}{10}y = \frac{2}{5}$

8. $2.5x - 1.25y = 5$

9. Your family is on a ski vacation. Lift tickets for the family cost \$80 per day. Snowboard rentals cost \$40 per day. You purchase lift tickets for x days and snowboard rentals for y days and spend \$480.

a. Write an equation in standard form that represents the situation.

b. Find the x- and y-intercepts.

c. Graph the equation.

d. You rent snowboards for 2 days. How many days did you purchase lift tickets?

10. An electrician charges \$80 plus \$32 per hour.

a. Write an equation that represents the total fee y (in dollars) charged by the electrician for a job lasting x hours.

b. Find the x- and y-intercepts.

c. Graph the equation.

d. Is the value of the x-intercept applicable to the electrician? Explain.

Name ______________________________ Date __________

2.4 Enrichment and Extension

Interpreting Intercepts

The graph shows the horizontal position y (in feet) of a trapeze artist after x seconds during an act.

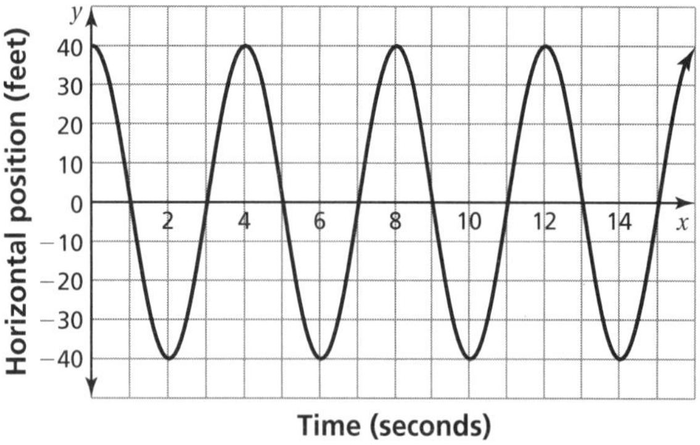

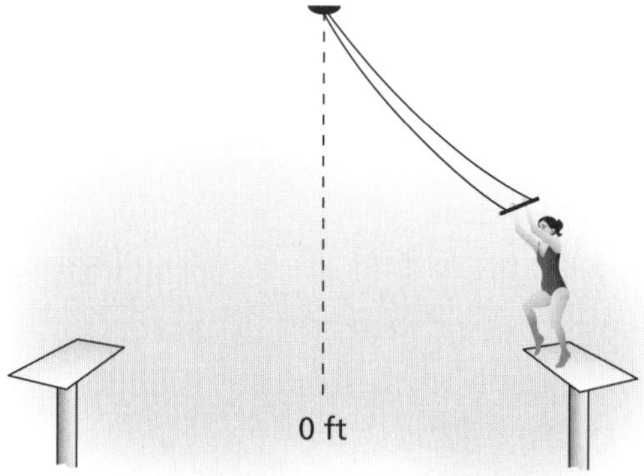

1. Does the graph represent a linear equation? Explain.
2. What is the y-intercept of the graph?
3. Interpret the y-intercept.
4. How many x-intercepts does the graph have?
5. Why does the graph have more than one x-intercept?
6. How many seconds does the trapeze act last?
7. How many feet apart are the platforms?

Name____ Date__________

2.4 Puzzle Time

How Do Kangaroos Travel Across The Ocean?

Write the letter of each answer in the box containing the exercise number.

Find the *x*- and *y*-intercepts for the linear equation.

1. $-3x + 5y = 15$
2. $2x - y = 4$
3. $7x + 9y = 63$
4. $3x + 2y = 6$
5. $4x - 9y = 36$
6. $x + \frac{1}{3}y = -3$
7. $\frac{1}{2}x + \frac{1}{3}y = 15$
8. $\frac{2}{5}x - \frac{3}{4}y = 12$
9. $3.5x + 4.2y = 35.7$
10. $7.6x + 15.2y = 38$
11. The booster club sells popcorn at basketball games for \$0.75 per bag. Their cost for supplies is \$12. The equation $-0.75x + y = -12$ represents the booster club's income y after selling x bags of popcorn. Find the x- and y-intercepts of the linear equation.
12. You upload digital photos to an online photo processing website. You can print 4-inch by 6-inch photos for \$0.30 each and 5-inch by 7-inch photos for \$0.75 each. The linear equation $0.30x + 0.75y = 15$ represents the ways you can print x 4-inch by 6-inch photos and y 5-inch by 7-inch photos for \$15. Find the x- and y-intercepts of the linear equation.

Answers

H. x-intercept: 2; y-intercept: 3

S. x-intercept: 5; y-intercept: 2.5

H. x-intercept: –3; y-intercept: –9

J. x-intercept: 50; y-intercept: 20

M. x-intercept: 2; y-intercept: –4

Y. x-intercept: 30; y-intercept: –16

U. x-intercept: 9; y-intercept: –4

P. x-intercept: 16; y-intercept: –12

T. x-intercept: 30; y-intercept: 45

E. x-intercept: –5; y-intercept: 3

P. x-intercept: 10.2; y-intercept: 8.5

I. x-intercept: 9; y-intercept: 7

7	4	1	8		12	5	2	9		10	6	3	11

Activity 2.5 Start Thinking!

For use before Activity 2.5

Let x be your age.

Let y be the age of a relative or friend.

What is the sum of your ages? Write an equation that models the sum.

What is the difference of your ages? Write an equation that models the difference.

Graph both equations on the same coordinate plane. What does the intersection represent?

Activity 2.5 Warm Up

For use before Activity 2.5

Solve for *y*. Then graph the equation.

1. $y - 2x = 6$

2. $2x + y = 10$

3. $\frac{1}{2}y - x = 4$

4. $x + 3y = 9$

5. $x - 2y = 2$

6. $5x + 3y = 30$

Lesson 2.5 Start Thinking!
For use before Lesson 2.5

Write a word problem that can be solved using a system of linear equations.

Exchange problems with a classmate and solve your classmate's problem.

What method did you use to solve the problem?

Lesson 2.5 Warm Up
For use before Lesson 2.5

Use the table to find the break-even point. Check your solution.

1. $C = 10x + 25$
$R = 15x$

x	0	1	2	3	4	5	6
C							
R							

2. $C = 20x + 90$
$R = 35x$

x	0	1	2	3	4	5	6
C							
R							

Name ______________________________ Date ________

2.5 Practice A

1. Use the table to find the break-even point. Check your solution.

$C = 24x + 48$

$R = 36x$

x	0	1	2	3	4	5	6
C							
R							

Solve the system of linear equations using a table.

2. $y = x + 1$
$y = 2x - 3$

3. $y = x - 3$
$y = -2x + 6$

Solve the system of linear equations using a graph.

4. $y = x - 3$
$y = -\frac{1}{2}x$

5. $y = -x - 2$
$y = \frac{5}{2}x + 5$

Solve the system of linear equations algebraically.

6. $x + y = 19$
$y = x - 5$

7. $y - x = 14$
$y = 3x + 6$

8. You and your friend caught 36 sunfish in Lake Okeechobee. The number y of sunfish caught by your friend is 8 more than the number x of sunfish caught by you.

a. Write a system of linear equations.

b. Solve the system of linear equations.

9. You and your friend are traveling through an obstacle course. You left before your friend. The distance you have traveled y (in yards) is represented by $y = 30x + 75$, where x is in minutes. Your friend's distance is represented by $y = 45x$.

a. How long will it take your friend to catch up with you?

b. How far will you each have traveled when your friend catches up with you?

Name_____________________________________ Date__________

2.5 Practice B

1. Use the table to find the break-even point. Check your solution.

   $C = 65x + 150$

   $R = 90x$

x	0	1	2	3	4	5	6
C							
R							

Solve the system of linear equations using a table.

2. $y = 3x - 1$

   $y = \frac{3}{2}x + 2$

3. $y = -\frac{3}{2}x$

   $y = -x - 2$

Solve the system of linear equations using a graph.

4. $y = -x - 1$

   $y = -\frac{1}{3}x + 1$

5. $y = \frac{1}{2}x - 6$

   $y = -\frac{3}{2}x + 2$

Solve the system of linear equations algebraically.

6. $x + y = 8$

   $y = 3.5x - 1$

7. $x - y = 2$

   $y = \frac{5}{2}x - \frac{1}{2}$

8. The sum of two numbers is 45. The larger number y is 6 less than twice the smaller number x.

   a. Write a system of linear equations.

   b. What is the smaller number?

9. You sell your bicycle to your friend for \$65. Your friend gives you x five-dollar bills and y ten-dollar bills. There are a total of 9 bills.

   a. Write a system of linear equations.

   b. How many five-dollar and ten-dollar bills did your friend give you?

   c. Could your friend pay you \$65 using 6 five-dollar bills and the rest ten-dollar bills? Explain.

Name ______________________________ Date __________

2.5 Enrichment and Extension

Solving Systems of Linear Equations

An alternative way to solve a system of linear equations is through the method of elimination. Elimination uses subtraction to eliminate one of the variables.

Example: Solve the system.

$$3x + 4y = 10 \quad \text{Equation 1}$$

$$3x + 2y = 8 \quad \text{Equation 2}$$

Step 1: Equation 1 and Equation 2 both have a $3x$ term. Because both equations have a $3x$ term, subtracting Equation 2 from Equation 1 will cause the $3x$ term to be eliminated.

$3x$	+	$4y$	=	10
$-\ 3x$	+	$-\ 2y$	=	$-\ 8$
0	+	$2y$	=	2

Step 2: Solve the resulting equation.

$$2y = 2$$

$$y = 1$$

Step 3: Substitute the solution into one of the original equations and solve for the other variable.

$$3x + 2y = 8$$

$$3x + 2(1) = 8$$

$$3x = 6$$

$$x = 2$$

► The solution is $(2, 1)$.

Solve the system of linear equations using elimination.

1. $2x + 5y = 15$
$2x + 2y = 6$

2. $x + 3y = 17$
$x - y = 1$

3. $3x - 4y = 7$
$3x + 4y = -1$

4. $2x + 4y = 20$
$x + 4y = 16$

5. $x + y = 13$
$-3x + y = -7$

6. $6x - 2y = 10$
$4x - 2y = 8$

7. When you subtract Equation 2 from Equation 1, what property of equality are you using? Explain.

Name______________________________ Date__________

2.5 Puzzle Time

What Did The Lumberjack Do After He Bought A Computer?

Write the letter of each answer in the box containing the exercise number.

Solve the system of linear equations.

1. $y = 4x + 15$
$y = -2x + 9$

2. $y = 7x - 12$
$y = 3x + 8$

3. $y = -6x - 5$
$y = 14x + 3$

4. $y = \frac{1}{2}x + 4$
$y = -4x + 1$

5. $y = \frac{3}{4}x - 2$
$y = 6x - 1$

6. $x + y = 24$
$y = 3x + 8$

7. $6x - y = 16$
$y = 8x + 14$

8. $x - 2y = 12$
$\frac{1}{2}x + 3y = 18$

9. Tyler's height x and Paul's height y combined are 120 inches. Tyler is 4 inches taller than Paul. What are their heights, expressed as a solution to a system of linear equations?

10. Your baseball practice lasts 45 minutes. You spend x minutes batting and y minutes fielding. The number of minutes spent batting is 7 more than the number of minutes spent fielding. How many minutes are spent on each, expressed as a solution to a system of linear equations?

Answers

O. $\left(-\frac{2}{5}, -\frac{13}{5}\right)$

H. $(5, 23)$

D. $(4, 20)$

L. $(18, 3)$

O. $(26, 19)$

E. $(-1, 11)$

G. $(-15, -106)$

N. $\left(-\frac{2}{3}, \frac{11}{3}\right)$

G. $(62, 58)$

E. $\left(-\frac{4}{21}, -\frac{15}{7}\right)$

2	5		8	3	7	9	1	6		10	4

Activity 2.6 Start Thinking!

For use before Activity 2.6

Write a linear equation in standard form.

Multiply both sides of the equation by the same nonzero number to create a second equation.

Graph the system of two equations on the same coordinate plane.

What do you notice about the two lines?

What do you think is the solution of the system?

Activity 2.6 Warm Up

For use before Activity 2.6

Solve the system of linear equations algebraically.

1. $x + y = 19$

$y = x + 5$

2. $x - y = 10$

$x + y = 6$

3. $y = 2x + 1$

$y = 3x$

4. $x + 2y = 17$

$y = 2x + 6$

Lesson 2.6

Start Thinking!

For use before Lesson 2.6

A car is traveling at an average rate of 60 miles per hour.

Another car starts from the same location one hour later and travels on the same route at the same average rate.

Will the second car ever catch up?

How does this situation relate to solving a system of equations?

Lesson 2.6

Warm Up

For use before Lesson 2.6

Let *x* and *y* be two numbers. Find the solution of the puzzle.

1. The value of x is 3 more than twice the value of y.

The difference of $2x$ and $4y$ is 6.

2. The difference of $3x$ and $2y$ is 14.

The product of x and $\frac{3}{2}$ is 8 more than y.

Name ____________________ Date ________

2.6 Practice A

Solve the system of linear equations.

1. $6x - 2y = -2$
$y = 3x - 4$

2. $x - 2y = -8$
$y = \frac{1}{2}x + 4$

3. $y = 3x$
$y = -2x$

4. $x - 4y = 3$
$-3x + 12y = -9$

5. You and your friend are saving money to go on an end-of-the-year school trip. You have $25 in your savings account and add $15 each week. Your friend has $10 in a savings account and adds $15 each week.

a. Write a system of linear equations that represents the amount of money y (in dollars) in each savings account after x weeks.

b. Will there be a day when you and your friend have the same amount of money in your savings accounts? Explain.

c. If your friend were able to add $20 each week, would there be a day when you and your friend have the same amount of money in your savings accounts? Explain.

Solve the system of linear equations.

6. $3.6x - 1.2y = 2.4$
$3y - 4.2x = 4.8x - 6$

7. $1.5x + y = 2.6$
$x - 3y = 5.4$

8. Is it possible for a system of linear equations to have exactly two solutions? Explain.

9. One equation in a system of linear equations is $y = 2x + 1$.

a. Write a second equation so that $(1, 3)$ is the only solution of the system.

b. Write a second equation so that the system has infinitely many solutions.

c. Write a second equation so that the system has no solution.

Name________________________________ Date__________

2.6 Practice B

Solve the system of linear equations.

1. $3x + 2y = 10$

$y = -\frac{3}{2}x + 5$

2. $7x + 3y = -6$

$y = \frac{1}{3}x - 2$

3. $y = \frac{3}{8}x - \frac{1}{4}$

$8y - 3x = -2$

4. $y = \frac{1}{4}x + 3$

$3x - 12y = 7$

5. Use the system of linear equations to answer the questions.

$$10x - 12y = 36$$
$$y = \frac{5}{6}x - 3$$

a. Is $(6, 2)$ a solution to the system of linear equations?

b. Is $(0, 1)$ a solution to the system of linear equations?

c. Is $(-12, -13)$ a solution to the system of linear equations?

d. Does this system of linear equations have *no solution*, *one solution*, or *infinitely many solutions*? Explain.

Solve the system of linear equations.

6. $4.8y + 20 = 13.2x + 3.2$

$y = 2.75x - 3.5$

7. $y = \frac{4}{\pi}x - 3$

$\pi y - 4x = -3\pi$

8. One equation in a system of linear equations is $y = -\frac{3}{4}x + 2$.

a. Write a second equation so that $(8, -4)$ is the only solution of the system.

b. Write a second equation so that the system has no solution.

c. Write a second equation so that the system has infinitely many solutions.

d. Is there more than one possible answer to part (c)? Explain your reasoning.

Name ______________________________ Date __________

2.6 Enrichment and Extension

Solutions of Systems of Linear Equations

Solve each system of linear equations in the grid below using any of the methods you have learned. Shade each square in the grid according to the nature of the solution(s) of each system.

 One solution in Quadrant I

 One solution in Quadrant II

 One solution in Quadrant III

 One solution in Quadrant IV

 No solution

 Infinitely many solutions

$4x - 6y = -6$ $-2x + y = 5$	$x + y = 4$ $3x - y = -12$	$y = 2x + 14$ $y = -x - 10$
$x - y = 5$ $2y = 2x - 10$	$x + 2y = 14$ $y = -\frac{1}{2}x + 7$	$4x - 8y = 20$ $2y = x - 11$
$3x - 1.5y = 7.5$ $y = 5x - 8$	$4x + 3y = 25$ $y = x - 1$	$y = -x$ $6x + 5y = 2$

Name____ Date__________

2.6 Puzzle Time

What Is Luke Skywalker's Favorite Type of Car?

Write the letter of each answer in the box containing the exercise number.

Solve the system of linear equations.

1. $9x - 3y = 12$ and $y = 3x - 4$

A. infinitely many solutions **B.** no solution **C.** $(-8, -28)$

2. $-5y + 4x = 15$ and $-4x + 5y = 15$

N. infinitely many solutions **O.** no solution **P.** $(30, 19)$

3. $6x + y = 9$ and $18x + 2y = 12$

M. infinitely many solutions **N.** no solution **O.** $(-1, 15)$

4. $y + 2.5x = 4.8$ and $x + 0.4y = 3$

C. infinitely many solutions **D.** no solution **E.** $(-0.9, 9.75)$

5. $-5x + y = -6.5$ and $x - 0.2y = 1.3$

T. infinitely many solutions **U.** no solution **V.** $(1.3\ 0)$

6. You and your friend are riding go-karts and your friend starts around the track first. Your friend's distance y around the track at time x is $y = 15x + 0.25$. Your distance y at time x is $y = 15x$. What solution represents when you will catch up to your friend?

X. infinitely many solutions **Y.** no solution **Z.** $(3.5, 4)$

5	3		6	2	4	1
		–				

Activity 2.7 Start Thinking!

For use before Activity 2.7

How does the graph below relate to the equation $-2x + 4 = x - 5$?

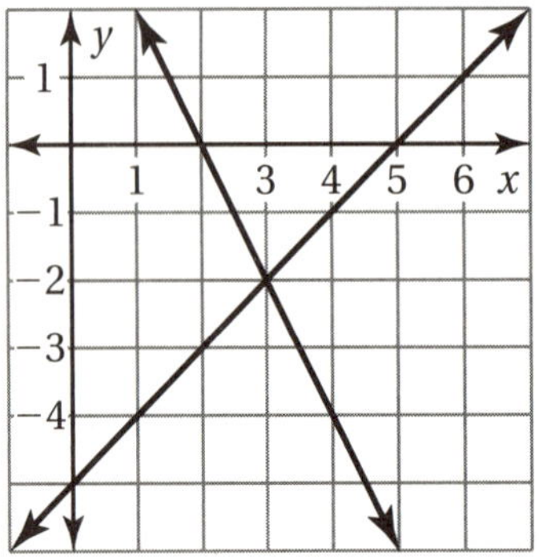

Activity 2.7 Warm Up

For use before Activity 2.7

Solve the equation. Check your solution.

1. $m - 8 = 2m$

2. $3x - 5 = 2x + 1$

3. $12 - 4p = p - 3$

4. $2(x - 8) = -x + 5$

5. $-2r - 9 = -3(r + 2)$

6. $4(x + 8) = -2(x - 1)$

Lesson 2.7 Start Thinking!

For use before Lesson 2.7

You have used two ways to solve an equation with variables on both sides: using algebra and using a graph.

Which method do you prefer? Why?

Lesson 2.7 Warm Up

For use before Lesson 2.7

Solve the equation algebraically and graphically.

1. $\frac{1}{2}x + 4 = -x + 1$

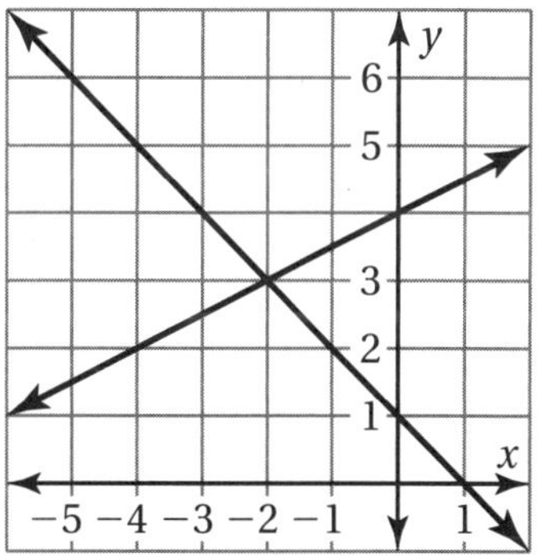

2. $-x + 2 = \frac{1}{4}x - 3$

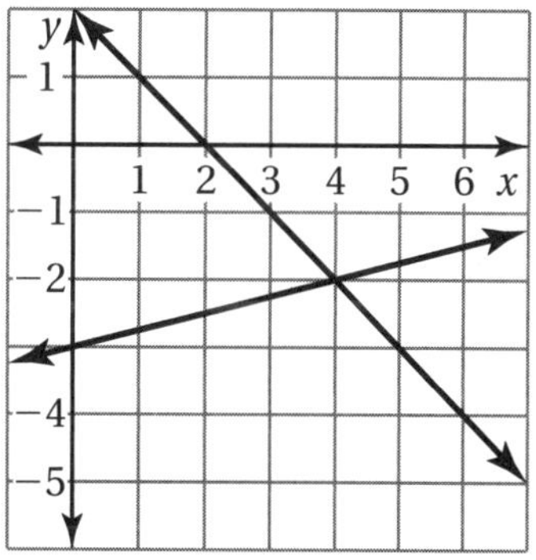

Name ____ Date __________

2.7 Practice A

Solve the equation algebraically and graphically.

1. $\frac{1}{2}x - 2 = -\frac{1}{4}x + 1$

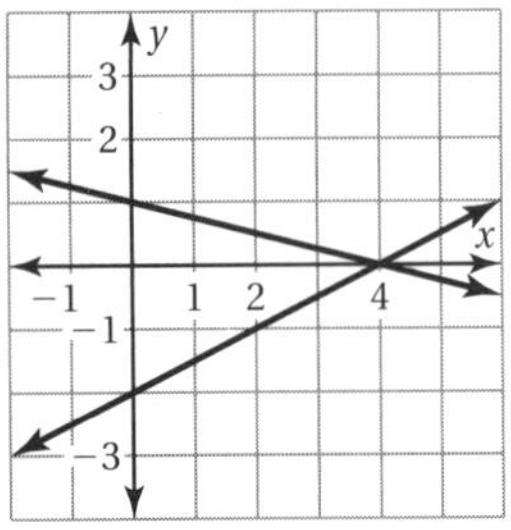

2. $\frac{2}{3}x + 4 = \frac{1}{6}x + \frac{5}{2}$

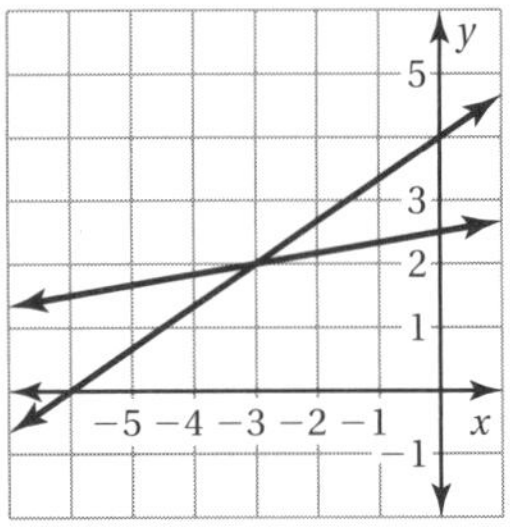

Use a graph to solve the equation. Check your solution.

3. $x + 4 = -2x + 7$

4. $\frac{1}{2}x + \frac{9}{2} = -5x - 1$

5. You have $4x + 10$ CDs and your friend has $5x + 22$ CDs. Is it possible that you both have the same number of CDs? Explain your reasoning.

Use a graph to solve the equation. Check your solution.

6. $0.75x = 2x - 10$

7. $1.2x + 4.8 = -0.8x - 0.2$

8. One tank contains 50 gallons of water and another tank contains 75 gallons of water. Water is draining out of both tanks at a rate of 6 gallons per minute. Do the tanks ever have the same amount of water before either is empty? Explain.

9. You and your friend sell lemonade. You get $\frac{1}{3}$ of the total money collected. Your friend gets \$20 less than the total money collected.

a. Write an equation to represent that you both earn the same amount of money.

b. What is the total amount of money collected?

c. How much does each of you earn?

d. Describe a situation to explain the unclaimed money.

Name____ Date__________

2.7 Practice B

Solve the equation algebraically and graphically.

1. $\frac{2}{3}x = \frac{4}{3}x + 2$

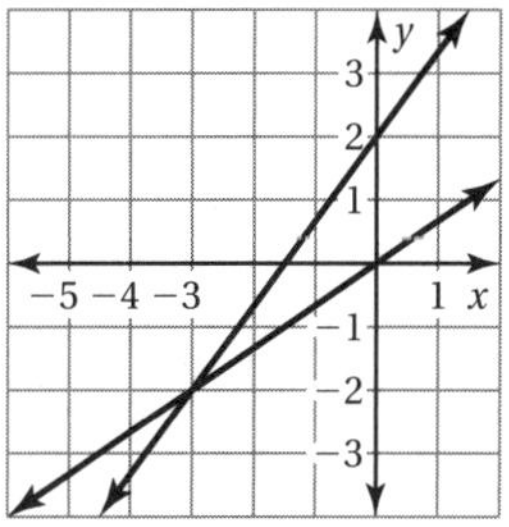

2. $-\frac{1}{2}x + 1 = \frac{1}{3}x - 4$

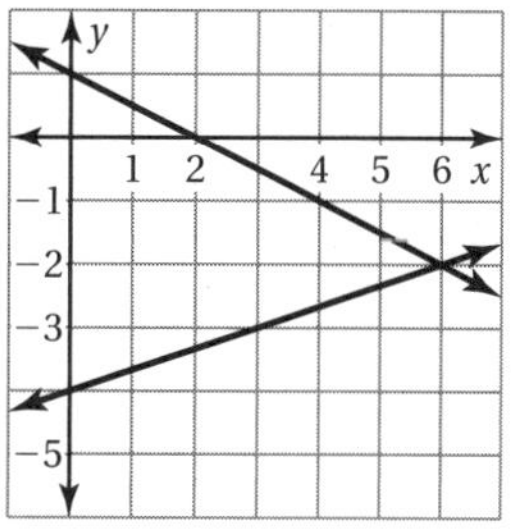

Use a graph to solve the equation. Check your solution.

3. $\frac{1}{4}x + 2 = \frac{3}{4}x + \frac{1}{2}$

4. $-\frac{5}{8}x + \frac{5}{4} = \frac{3}{4}x + 4$

5. One gym charges a \$22 membership fee and \$4.50 per visit. Another gym charges a \$14 membership fee and \$3.50 per visit. Is it possible for you to pay the same amount at the two gyms? Explain your reasoning.

Use a graph to solve the equation. Check your solution.

6. $-1.5x + 7.5 = 2.5x - 0.5$

7. $1.4x + 6.4 = -0.6x - 3.6$

8. Tank A contains 150 gallons of water and water is draining out at a rate of 6 gallons per minute. Tank B contains 200 gallons of water and water is draining at a rate of 10 gallons per minute.

a. After how many minutes do the tanks have the same amount of water?

b. How much water is in the tanks at that time?

c. Which tank will empty first? Explain.

9. Consider the equation $\frac{3}{4}x - \frac{3}{2} = -1.25x + 4.5$.

a. Use a graph to solve the equation.

b. Convert the fractions to decimals and solve the equation algebraically. Do you get the same answer?

Name ______________________________ Date __________

2.7 Enrichment and Extension

Solving Systems of Linear Equations by Graphing

Not all systems of equations are linear. In the given systems, one equation is linear and one equation is nonlinear. The graph of the nonlinear equation is given. To solve the system, graph the linear equation and find the point(s) of intersection. Check your solution(s).

1. $y = x^2$
$y = 2x - 1$

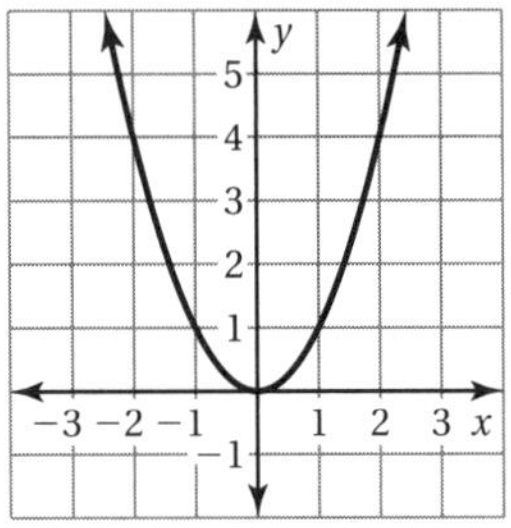

a. How many solutions does the system have?

b. What is the solution(s) to the system of equations?

2. $y = x^3 - x$
$y = 3x$

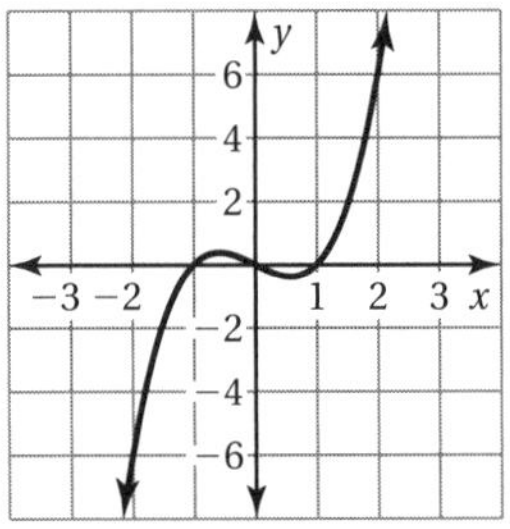

a. How many solutions does the system have?

b. What is the solution(s) to the system of equations?

3. $y = -2|x + 3|$
$y = x + 5$

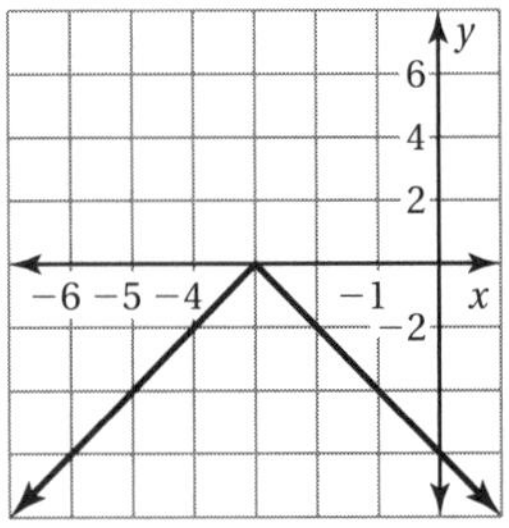

a. How many solutions does the system have?

b. Are the lines parallel? Explain

Name ______________________________ Date __________

2.7 Puzzle Time

Did You Hear About...

A	B	C	D	E	F
G	H	I	J	K	L
M					

Complete each exercise. Find the answer in the answer column. Write the word under the answer in the box containing the exercise letter.

$\frac{5}{2}$ WAS
-9 WHO
-6 BECAUSE
8 KNEE
10 A
25 HAD
9 WAND
15 TRICK
6 MAGICIAN

Use a graph to solve the equation.

A. $x + 7 = 3x - 3$

B. $4x - 6 = x + 12$

C. $5x - 3 = 4x - 12$

D. $x - 4 = \frac{1}{2}x + 6$

E. $\frac{2}{3}x + 6 = -\frac{4}{3}x + 9$

F. $\frac{3}{5}x - 2 = -\frac{2}{5}x + 8$

G. $\frac{1}{4}x + 5 = \frac{5}{4}x - 7$

H. $\frac{3}{2}x + 7 = -x - 8$

I. $\frac{1}{6}x - 9 = \frac{2}{3}x + 6$

J. $\frac{2}{5}x + \frac{1}{2} = \frac{1}{2}x - 2$

K. $1.5x - 8 = 3x + 4$

L. $2.6x + 3 = 3.4x - 9$

M. Andrew is on page 26 of a book and his friend Michael is on page 42 of the same book. Andrew reads 7 pages per day and Michael reads 5 pages per day. After how many days will they be on the same page?

-8 A
-10 RAN
20 WALKED
$-\frac{3}{2}$ HAD
5 THE
-5 HAT
12 CANE
$\frac{3}{2}$ WITH
-30 HE

Name ____ Date __________

Chapter 2 School-to-Work
For use after Section 2.3

Personal Trainer

You are a personal trainer at a heath club. The manager of the club has asked you to create a graphic to educate clients about target heart rate. Your target heart rate is an optimal range of beats per minute that your heart should perform while exercising. You know the following information about target heart rate:

- Your maximum heart rate can be found by subtracting your age from 220.
- Your target heart rate is between 50% and 85% of your maximum heart rate.

1. Write an equation, in slope-intercept form, that relates your maximum heart rate to your age. What does each variable represent?
2. Write equations, in slope-intercept form, for the lower and upper bounds of your target heart rate.
3. Graph your three equations in a large coordinate plane. Let age be on the horizontal axis and let heart rate be on the vertical axis. Label the axes and label the meaning of each graphed line.
4. Shade and label the section for target heart rate using a color that will draw the attention of clients. Explain how a client can use the graph to quickly find his or her target heart rate. Add this information to your graph.
5. What is your maximum heart rate? Does your maximum heart rate increase or decrease as you age? Explain.
6. If you are exercising and you find that your heart rate is 170 beats per minute, are you within your target heart rate? Would the same be true for a 42-year-old? Explain.

Name____ Date__________

Chapter 2 Study Help

You can use a **process diagram** to show the steps involved in a procedure.

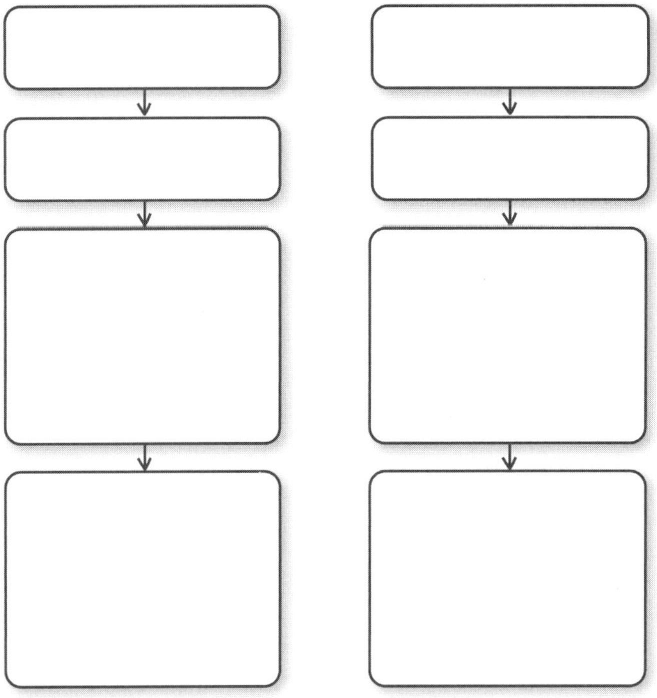

On Your Own

Make a process diagram with an example to help you study these topics.

1. finding the slope of a line
2. graphing a linear equation using
   a. slope and y-intercept
   b. x- and y-intercepts

After you complete this chapter, make process diagrams for the following topics.

3. solving a linear system
   a. using a table
   b. using a graph
   c. algebraically
4. finding the number of solutions of a linear system
5. solving an equation by graphing

Name ______________________________ Date __________

Chapter 2

Financial Literacy

For use after Section 2.6

Choosing a Cell Phone Plan

You are deciding on a new cell phone plan with the help of your parents. You have researched different options and are going to choose one of the following plans.

Plan	Monthly fee	Allowed minutes	Rate per minute after allowed minutes
A	$40	450	$0.45
B	$60	900	$0.40

1. A graph is a good way to better understand the cell phone plans. Sketch a set of axes on a coordinate plane. Let the horizontal axis represent minutes m used per month and let the vertical axis represent monthly cost C. Decide on the scales to use for each axis.

2. Sketch the graph for plan A for $0 \leq m \leq 450$ and the graph for Plan B for $0 \leq m \leq 900$. Which is the more economical plan for using up to 450 minutes per month? Explain.

3. Write an equation, in slope-intercept form, for plan A for $m > 450$. Graph the equation. What is the significance of the intersection of the lines representing plans A and B? Find the point of intersection algebraically. Show your work.

4. How can you use the point of intersection to help you choose between plans A and B?

5. How many minutes per month do you tend to use your cell phone? Do you think this number will stay relatively the same for each month of the next year? If not, how many minutes do you predict you will use per month?

6. Which plan would be the best for you? Explain your reasoning.

7. Research cell phone plans online. Find a plan that differs from plans A and B. Add this plan as plan C on your graph and describe how it relates to plans A and B. For what number of minutes per month is plan C a good choice? Explain.

Name________________________________ Date__________

Chapter 2 Technology Connection

For use after Section 2.5

Solving a Linear System

You can use the *intersect* feature of a graphing calculator to solve a system of two linear equations.

EXAMPLE Solve the system: $2x + 5y = 10$

$x - 3y = 16$

SOLUTION

Step 1 Rewrite each equation in slope-intercept form.

$$y = -\frac{2}{5}x + 2$$

$$y = \frac{x}{3} - \frac{16}{3}$$

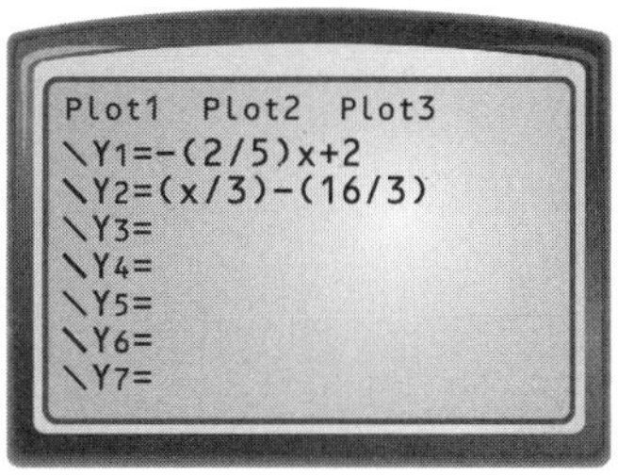

Step 2 Press [Y=] and enter the equations in Y1 and Y2.

Step 3 Press [GRAPH] to graph the system. Adjust the viewing window until the point of intersection is visible.

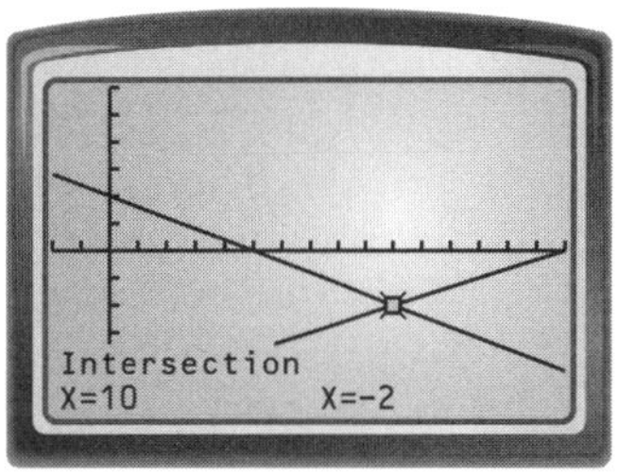

Step 4 Press [2nd] [CALC] 5 to use the *intersect* feature.

The solution is $(10, -2)$.

Solve the system using a graphing calculator.

1. $y = 5x - 8$

   $y = \frac{x}{3} + 6$

2. $y = -3$

   $2x + 4y = -7$

3. $-2x + 6y = 2$

   $9x - 7y = -9$

4. $y - \frac{2}{3}x = 4$

   $5x + \frac{y}{2} = -62$

Name ____ Date __________

Chapter 2

Life Connections

For use after Section 2.2

Using Slope to Describe Stairs

Math can help us to see everyday objects in a new way. Have you ever climbed a staircase and found it harder to climb than other staircases? Do you find some stairs more awkward to climb than others? Describing stairs using slope may help you understand why.

1. Find several staircases. Look in a variety of places, such as at your school, your home, and library. Compare the ease with which you climb each staircase. Record your observations.

2. For each staircase, measure the *tread* and the *riser* as shown below.

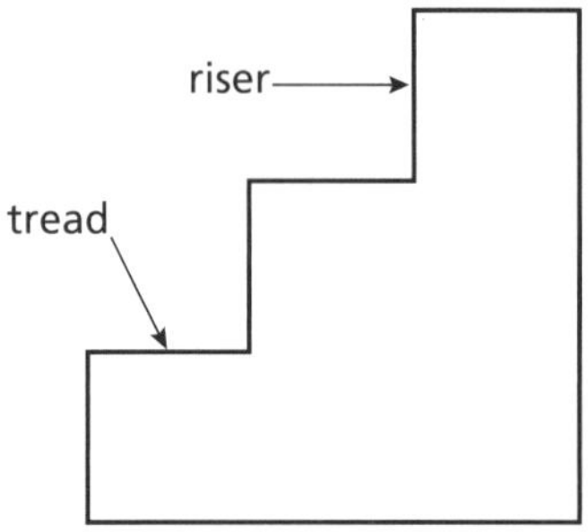

3. Imagine a straight line connecting the tip of each tread. Find the slope of such a line for each staircase. What do you think the slopes tell you about the staircases?

4. Order the stairs from least to greatest slope. Is the staircase with the least slope also the staircase that you found the easiest to climb?

5. Suppose that you are building a staircase with a total height of 4 feet. Describe two different ways of building the staircase using different numbers of steps. How many steps are in each staircase? What is the slope of each staircase?

6. Cities and towns have building codes that give rules about the maximum riser height and minimum tread length. Why do you think these codes were established? Try to find out the codes regarding the building of staircases in your city or town.

7. If you were writing the building codes for staircases, what rules would you set? Explain your reasoning.

Name________________________________ Date__________

Chapter 2 Cumulative Practice

Tell whether x and y show *direct variation*, *inverse variation*, or *neither*. Explain your reasoning.

1. $y = \frac{7}{x}$

2. $y + x = 3$

3.

x	−1	0	1	2
y	4	0	−4	−8

4.

x	2	4	6	8
y	3	5	7	9

5. The number of pictures you can hang on your wall varies inversely with the width of the pictures. You can hang 15 pictures that are 10 inches wide. How many pictures can you hang if you use pictures that are 6 inches wide?

Find the slope and *y*-intercept of the line.

6.

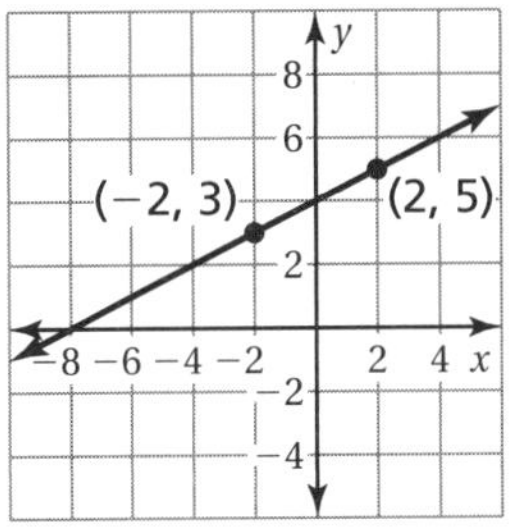

7.

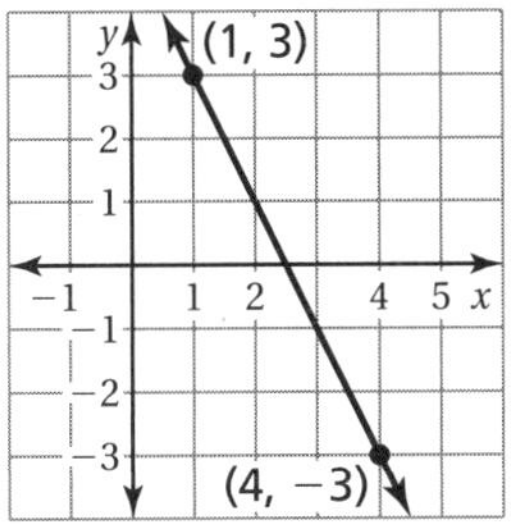

8.

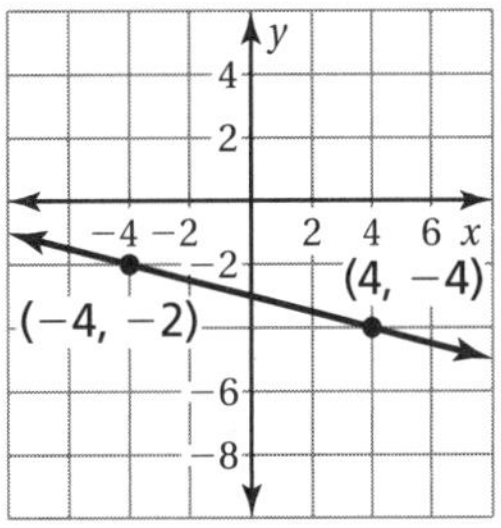

9. The total cost y (in dollars) of getting a car fixed at a repair shop is represented by the equation $y = 15.5x + 20$, where x is the number of hours it takes to make the repairs.

a. What is the slope of the graph of the equation?

b. Interpret the slope.

Solve the system of linear equations.

10. $y = 0.5x - 5$
$y = 1.5x + 1$

11. $\frac{1}{2}y + x = -10$
$y = -2x - 20$

12. $y - x = 14$
$x - y = 5$

13. The cost C (in dollars) of running a gift shop for x days is represented by the equation $C = 300x + 1500$. The revenue R (in dollars) is represented by the equation $R = 600x$. How many days will it take for the gift shop to break even?

Chapter 3

Name______________________________ Date__________

Chapter 3 Writing Linear Equations and Linear Systems

Dear Family,

In the nineteenth century, railroads were the fastest form of transportation—the jet airplanes of their day. Yet, because there were no railroads connecting the East Coast to the West Coast, it could take half a year to travel across the country. Congress decided a transcontinental railroad had to be built.

The line connecting the eastern railroads to the West Coast was built from both ends, finally meeting at Promontory Point in northern Utah. By 1872, it was possible to travel from New York to San Francisco in just one week—about 30 times faster than before.

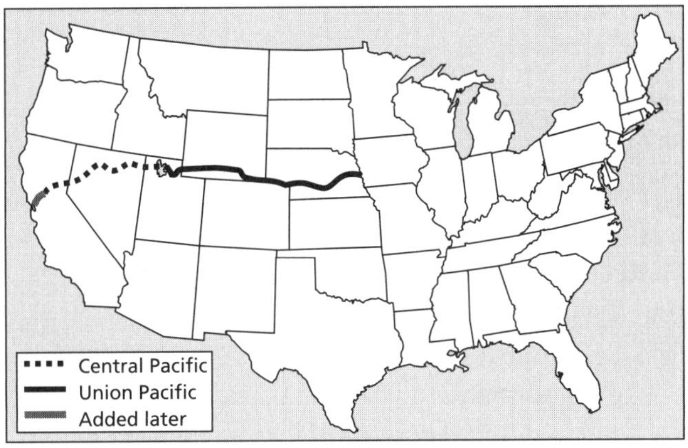

The Central Pacific Railroad started building the new track in Sacramento on January 8, 1863 and averaged 0.30 mile of track per day. The Union Pacific Railroad started 40 miles outside Council Bluffs, Iowa on July 4, 1865 and averaged 0.74 mile of track per day.

Ask your student to use this information and a map of the United States to find the following information.

- When did the Central Pacific track meet the Union Pacific track? Why did it take several years to complete the first transcontinental railroad?
- How much track did each railroad company lay?
- Why do you think the Union Pacific Railroad was able to work so much more quickly than the Central Pacific Railroad?

What were some of the effects the railroad had on the United States? Find another transcontinental building project that Congress ordered during this period in history. How did the builders complete their project?

Have fun using math to *travel* through history!

Nombre ______________________________ Fecha ________

Capítulo 3 Escribiendo Ecuaciones y Sistemas Lineales

Estimada Familia:

En el siglo XIX, el ferrocarril era el medio de transporte más rápido—los aviones jet de hoy en día. Aun así, debido a que no existían vías férreas que conectaran la Costa Este con la Costa Oeste, atravesar el país podía tomar 6 meses. El Congreso decidió que tenía que construirse una línea de ferrocarril transcontinental.

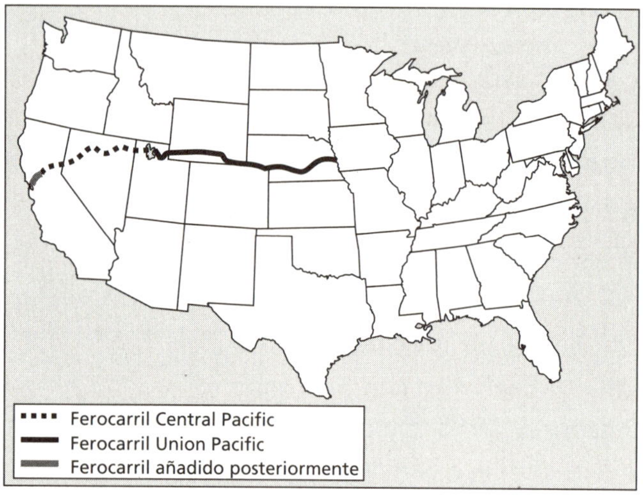

La línea que conectaría los ferrocarriles orientales con la Costa Oeste fue construida desde ambos lados y finalmente se encontraron en el Punto Promontory al norte de Utah. Para 1872, era posible viajar de Nueva York a San Francisco en tan sólo una semana—cerca de 30 veces más rápido que antes.

El Ferrocarril Central Pacific empezó a construir las nuevas vías en Sacramento, el 8 de enero de 1863 y se construía un promedio de 0.30 millas de tramo por día.

El Ferrocarril Union Pacific empezó 40 milla fuera de Council Bluffs, Iowa el 4 de julio de 1865 y se construía un promedio de 0.74 milla de tramo por día.

Haga que su estudiante utilice estos datos y un mapa de los Estados Unidos para encontrar la siguiente información:

- ¿Cuándo se encontraron los tramos del Central Pacific con los del Union Pacific? ¿Por qué tomó tantos años completar el primer ferrocarril transcontinental?
- ¿Cuánta longitud de tramo construyó cada compañía?
- ¿Por qué crees que el Ferrocarril de Union Pacific pudo funcionar mucho más rápido que el Ferrocarril Central Pacific?

¿Cuáles son algunos de los efectos que el ferrocarril tuvo en los Estados Unidos? Encuentren otro proyecto de construcción transcontinental que el Congreso haya ejecutado durante este período histórico. ¿Cómo es que los constructores completaron el proyecto?

¡Diviértanse usando las matemáticas para *viajar* a través de la historia!

Non ______________________________ Dat ____________

Chapít 3 Ann Ekri Ekwasyon Lineyè ak Sistèm Lineyè

Chè Fanmi:

Pandan diznevyèm syèk la, chemennfè se te fòm transpò ki te pi rapid la—se te avyon djèt epòk sa a. Poutan, poutèt pat gen chemennfè ki te relye Kòt Lès ak Kòt Lwès la, sa t'ap pran mwatye yon ane pou travèse peyi a. Kongrè deside yo te dwe konstwi yon chemennfè ki travèse kontinan an.

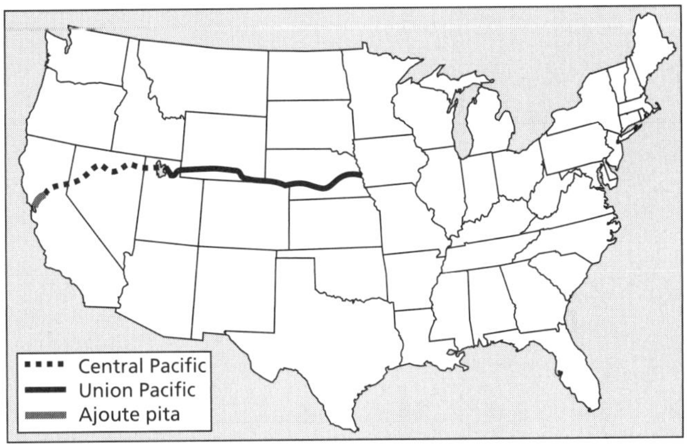

Liy ki relye chemennfè lès yo ak Kòt Lwès la te konstwi apatide toude bò yo, pou finalman rankontre nan Promontory Point nan nò Utah. An 1872, sa te posib pou vwayaje sòti New York pou al San Francisco nan sèlman yon semèn—apeprè 30 fwa pi vit pase anvan.

Central Pacific Railroad te kòmanse konstwi nouvo ray nan Sacramento 8 Janvye 1863 epi te poze an mwayèn 0.30 mil ray pa jou. Union Pacific Railroad te kòmanse 40 mil andeyò Council Bluffs, Iowa 4 Jiyè 1865 epi te poze an mwayèn 0.74 mil ray pa jou.

Mande elèv ou a sèvi avèk enfòmasyon sa yo ak yon kat Lèzetazini pou jwenn enfòmasyon sila yo.

- Kilè ray Central Pacific la rankontre ray Union Pacificla? Poukisa sa pran plizyè ane pou fini premye chemennfè ki travèse kontinan an?
- Ki kantite ray chak konpayi chemennfè poze?
- Dapre ou, poukisa Union Pacific Railroad te kapab travay pi vit konsa pase Central Pacific Railroad?

Bay kèk efè chemennfè a te gen sou Lèzetazini? Jwenn yon lòt pwojè konstriksyon ki travèse kontinan an ke Kongrè te òdone pandan peryòd sa a nan listwa. Kijan konstriktè yo te fini pwojè yo a?

Amize nou sèvi ak matematik pou *vwayaje* atravè listwa!

Activity 3.1

Start Thinking!

For use before Activity 3.1

Use the graph to answer the questions.

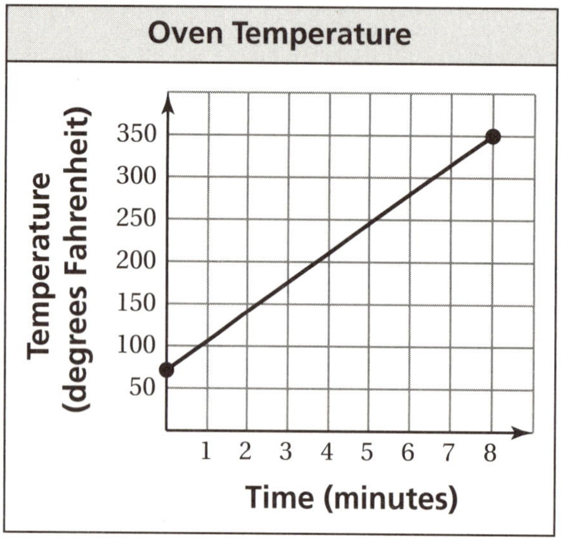

What does the graph show?

Estimate the y-intercept. What does it represent?

Estimate the slope. What does it represent?

Activity 3.1

Warm Up

For use before Activity 3.1

Find the slope of the line.

1.

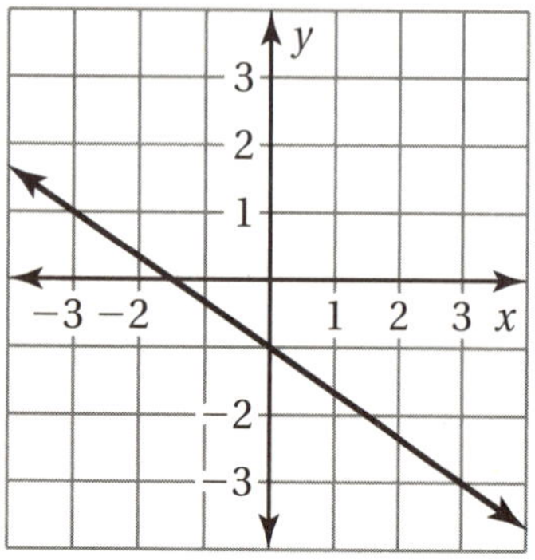

2.

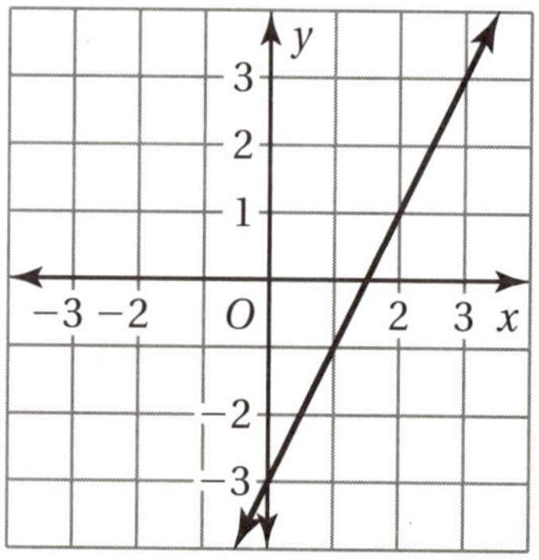

Lesson 3.1 Start Thinking!

For use before Lesson 3.1

A gym membership has a $20 enrollment fee and costs $40 per month.

Write an equation in slope-intercept form that represents the cost y after x months of joining the gym.

What does the slope represent?

What does the y-intercept represent?

Lesson 3.1 Warm Up

For use before Lesson 3.1

Write an equation for each side of the figure.

1.

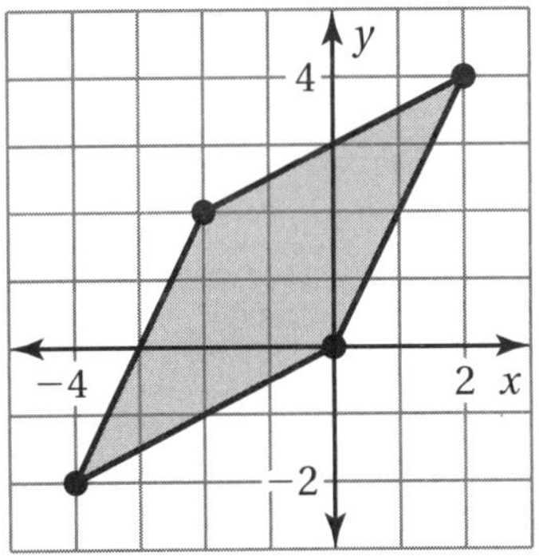

2.

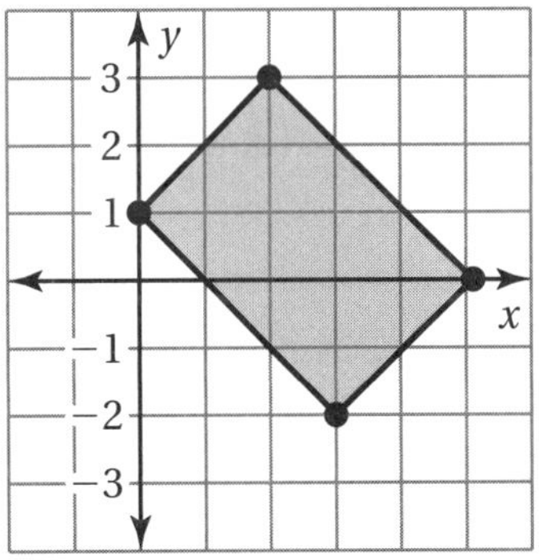

Name ______________________________ Date __________

3.1 Practice A

1. Write an equation for each side of the figure.

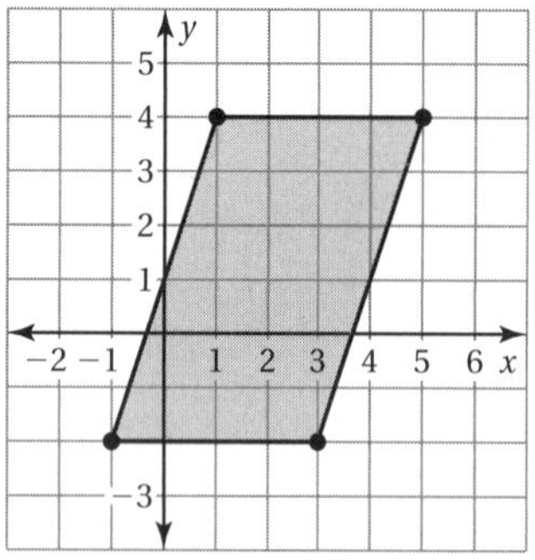

Write an equation of the line in slope-intercept form.

2. 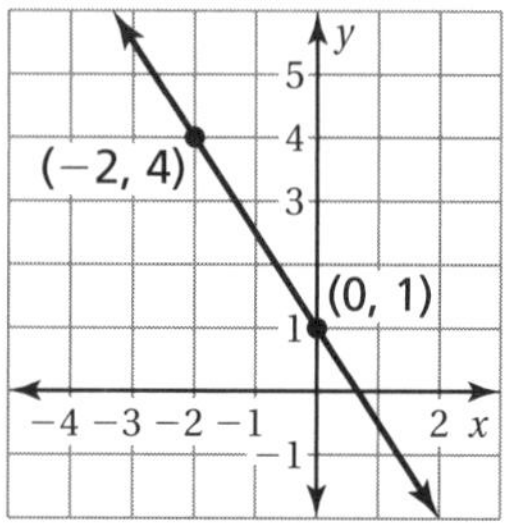

3. 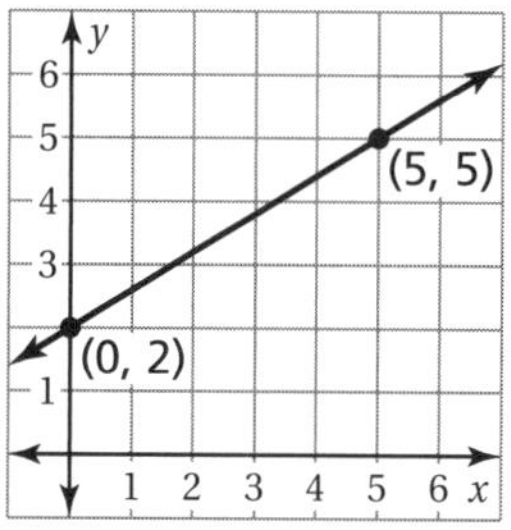

4. A plant is 3 inches tall when you purchase it and grows 2 inches per month. Write an equation that represents the length y (in inches) of a plant that you purchased x months ago.

Write an equation of the line that passes through the points.

5. $(0, 0), (4, -2)$

6. $(-2, 6), (0, 3)$

7. A bucket is empty. You are filling the bucket with water at a rate of 3 inches per second.

   a. Plot the points $(0, 0)$ and $(5, 15)$.

   b. What do the points in part (a) represent?

   c. Draw a line through the points.

   d. What does the line represent?

   e. Write an equation of the line.

Name____ Date__________

3.1 Practice B

1. Write an equation for each side of the figure.

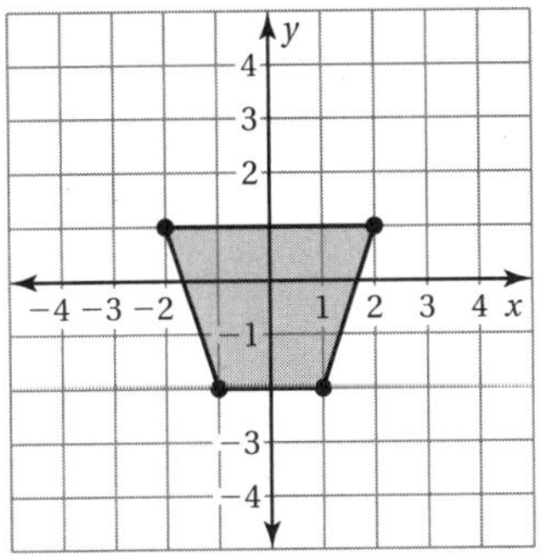

Write an equation of the line in slope-intercept form.

2.

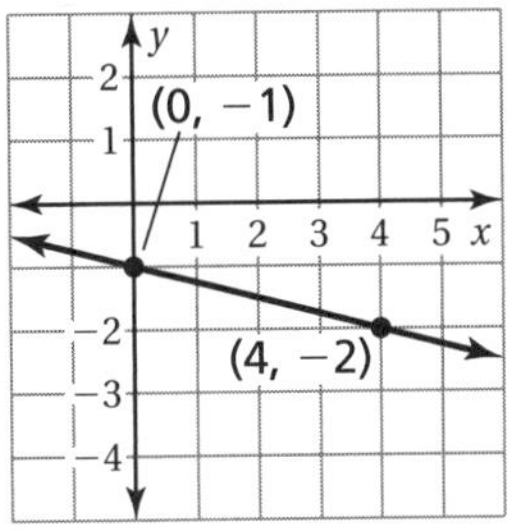

3.

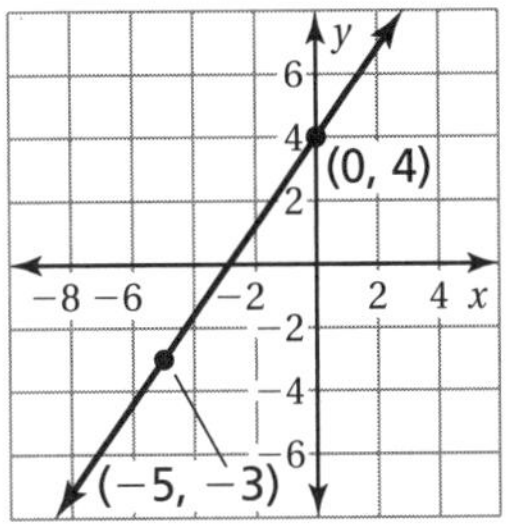

4. Your hair is 6 inches long and grows at a rate of 144 millimeters per year.

   a. Convert 144 millimeters per year to inches per year. Round your answer to the nearest tenth.

   b. Write an equation that represents the length y (in inches) of your hair after x years.

   c. How long is your hair after 4 years?

Write an equation of the line that passes through the points.

5. $(-4, -1), (0, 5)$

6. $(0, -3), (1, -5)$

7. Yesterday, you typed 8 pages in 48 minutes. Today, you typed 20 pages in 2 hours.

   a. Plot the two points (x, y), where x is the time (in minutes) and y is the number of pages.

   b. What is the rate of typing?

   c. Write an equation that represents the number of pages in terms of the number of minutes.

Name ______________________________ Date __________

3.1 Enrichment and Extension

Matching Equations and Graphs

Copy the equations and graphs onto index cards. Mix the cards up and lay them face down. With a friend, take turns turning over pairs of cards. If you find a matching graph and equation, remove the pair and take another turn. If the pair doesn't match, turn both cards face down again. Continue until all pairs are removed from the board. The player with the most pairs wins.

$y = \frac{1}{3}x + 2$	$y = 2x - 2$	$y = -x + 3$
$y = 4x$	$y = -\frac{2}{5}x + 1$	$y = \frac{3}{4}x - 3$
$y = -\frac{1}{2}x - 1$	$y = -3x$	$y = \frac{3}{5}x - 4$

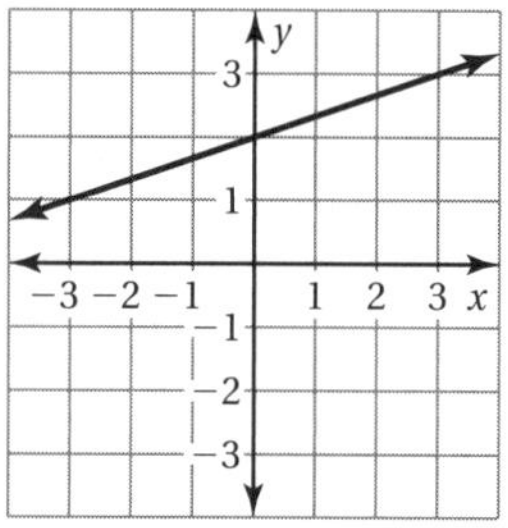

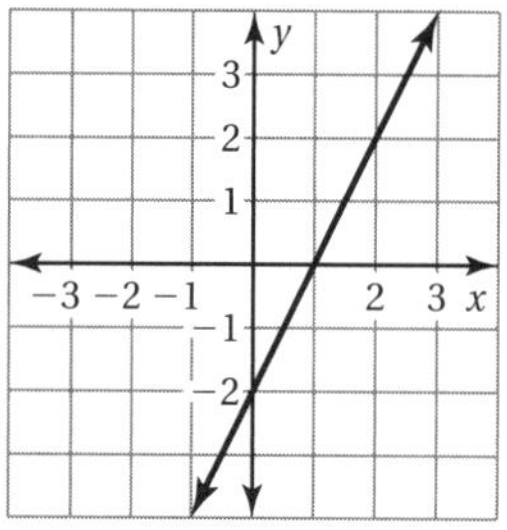

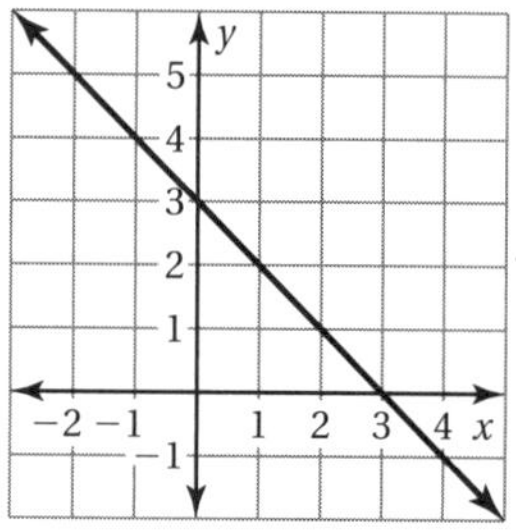

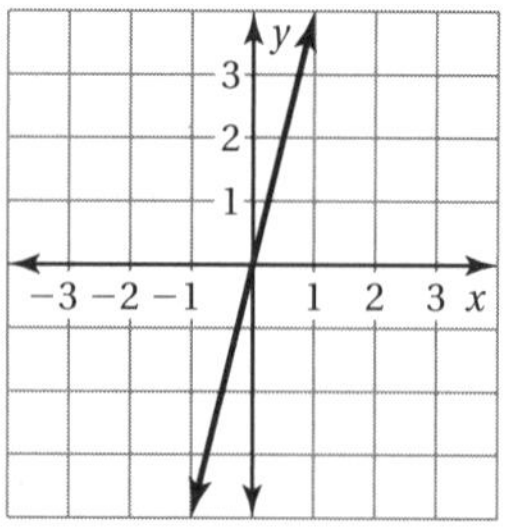

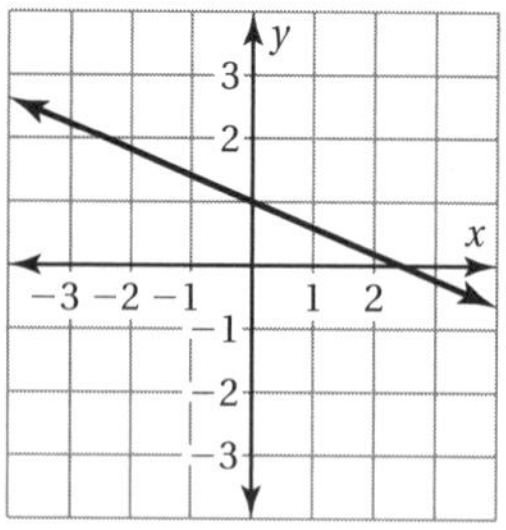

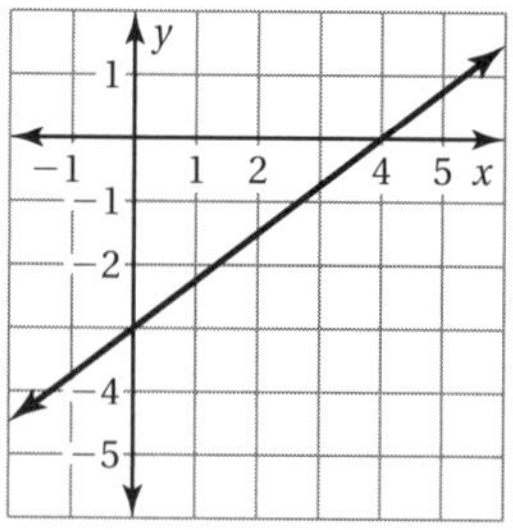

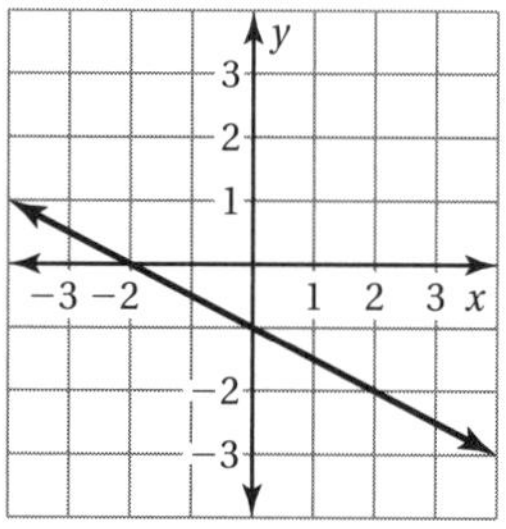

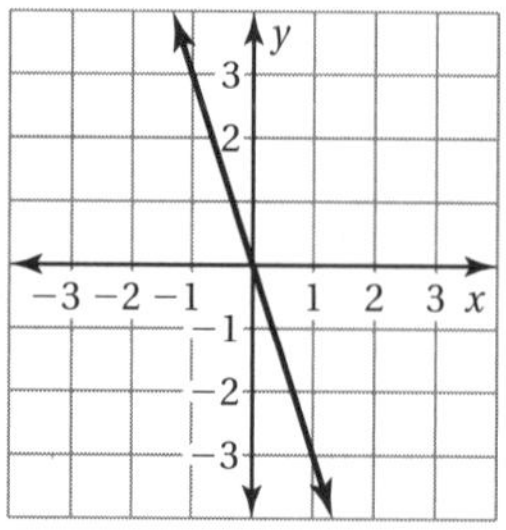

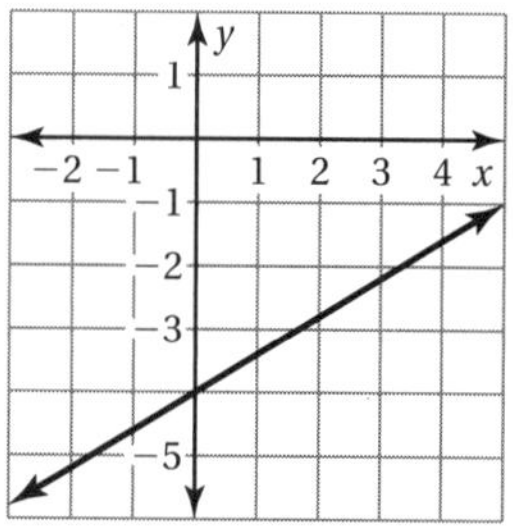

Name ______________________ Date __________

3.1 Puzzle Time

What Should You Know If You Want To Become A Lion Tamer?

Write the letter of each answer in the box containing the exercise number.

Write an equation of the line that passes through the points.

1. $(0, 3), (1, 4)$
2. $(0, 0), (5, -2)$
3. $(-2, 0), (0, 4)$
4. $(-3, 2), (0, -3)$
5. $(-7, 4), (0, 4)$
6. $(0, -8), (4, 8)$
7. $(0, -2), (-5, -2)$
8. $(-12, -9), (0, -3)$
9. $(0, 10), (5, 0)$
10. $(-14, 12), (0, 6)$
11. $(0, -6), (6, -24)$
12. $(0, -15), (5, 0)$
13. You are planning to make a scrapbook. The album costs $20 and each of the scrapbook papers cost an additional $1. Write an equation that represents the cost of the completed scrapbook where x represents the number of scrapbook papers you purchase.
14. A hot tub that holds 300 gallons of water drains at a rate of 8 gallons per minute. Write an equation that represents how many gallons of water are left in the tub after it has drained for x minutes.
15. An elevator in a tall building is at a point 180 above the ground. The elevator descends at a rate of 12 feet per second. Write an equation that represents how far above the ground the elevator is after descending for x seconds.

Answers

T. $y = -2x + 10$

N. $y = \frac{1}{2}x - 3$

H. $y = x + 3$

R. $y = -12x + 180$

E. $y = 2x + 4$

A. $y = -\frac{2}{5}x$

H. $y = -\frac{3}{7}x + 6$

N. $y = x + 20$

E. $y = 4$

T. $y = 4x - 8$

I. $y = 3x - 15$

L. $y = -3x - 6$

M. $y = -8x + 300$

O. $y = -2$

O. $y = -\frac{5}{3}x - 3$

14	7	15	3		6	10	2	13		9	1	5		11	12	4	8

Activity 3.2 Start Thinking!
For use before Activity 3.2

Choose an ordered pair that is not on the y-axis. Choose a slope.

Can you write an equation of a line with the slope you chose that goes through the point you chose? If so, explain the method you used. If not, explain why not.

Activity 3.2 Warm Up
For use before Activity 3.2

Graph the linear equation.

1. $y = 2x + 3$

2. $y = -x + 2$

3. $y = \frac{2}{3}x - 1$

4. $y = -\frac{1}{2}x + 4$

5. $y = 5x - 10$

6. $y = -\frac{5}{7}x + 14$

Lesson 3.2

Start Thinking!

For use before Lesson 3.2

How is writing the equation of a line given the slope and a point on the line similar to writing the equation of a line given the slope and y-intercept? How is it different?

Lesson 3.2

Warm Up

For use before Lesson 3.2

Write an equation of the line with the given slope that passes through the given point.

1. $m = 2$

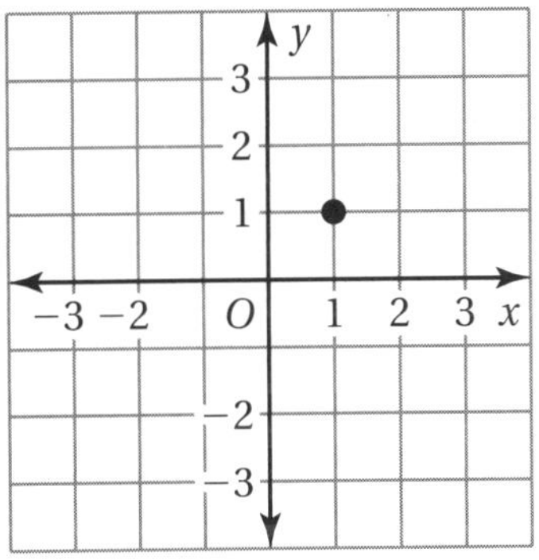

2. $m = -1$

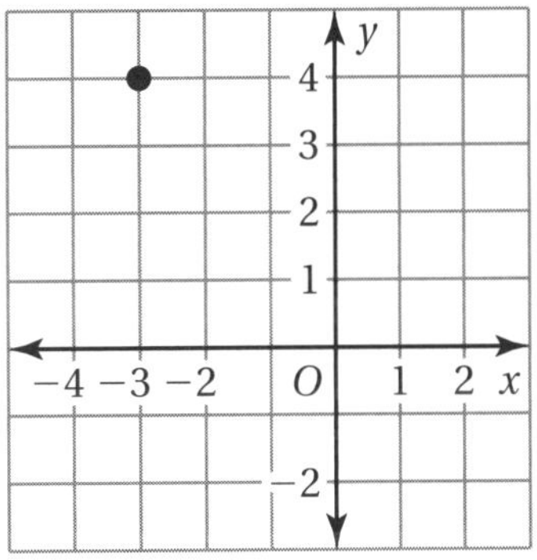

3. $m = -\frac{1}{2}$

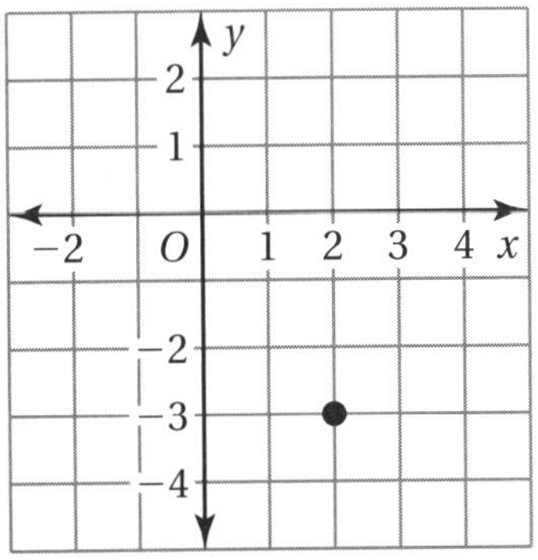

4. $m = \frac{2}{3}$

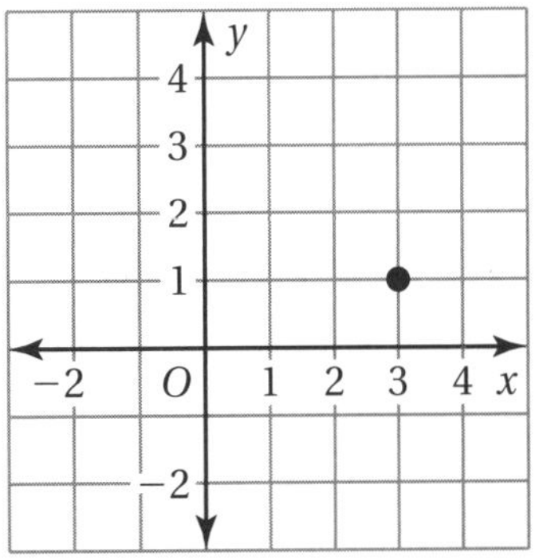

Name ________________________________ Date __________

3.2 Practice A

Write an equation of the line with the given slope that passes through the given point.

1. $m = 3$

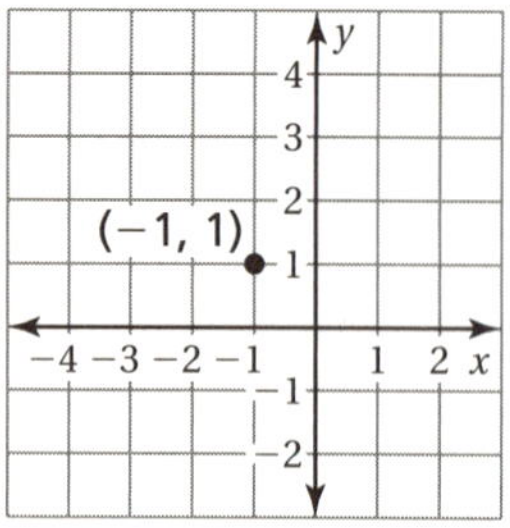

2. $m = -\frac{2}{3}$

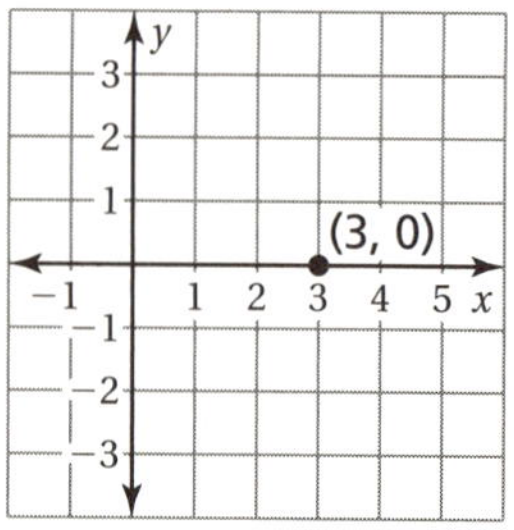

3. $m = \frac{1}{4}$; $(4, -2)$

4. $m = -\frac{4}{3}$; $(-3, 5)$

5. $m = -1$; $(2, 2)$

6. $m = 4$; $(-1, -5)$

7. After a laptop is purchased, its value decreases by \$150 each year. After 2 years, the laptop is worth \$600.

a. Write an equation that represents the value V (in dollars) of the laptop x years after it is purchased.

b. What was the original value of the laptop?

c. What is the value of the laptop 5 years after it is purchased?

8. For the past 4 weeks, you have saved \$5 per week. You now have \$65 in your savings account.

a. Write an equation that represents your balance A after t weeks.

b. What was the balance in the account before you started saving?

c. How many more weeks will it take to get a balance of \$120 in your account if you continue to save at the same rate?

d. For the past 3 weeks, your friend has saved \$8 per week. Your friend now has \$40 in her savings account. Who will have a balance of \$120 first, *you* or *your friend*?

Name____ Date__________

3.2 Practice B

Write an equation of the line with the given slope that passes through the given point.

1. $m = \frac{5}{4}$

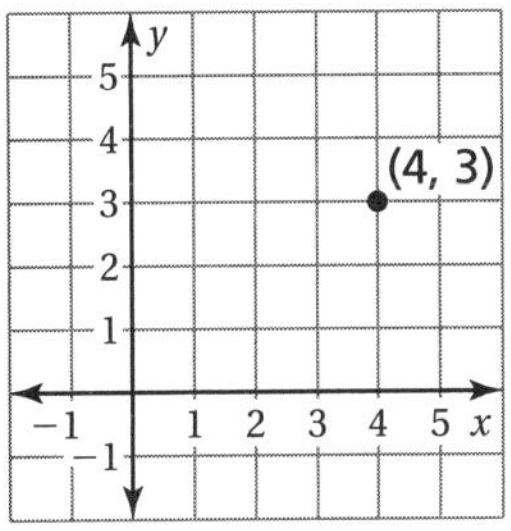

2. $m = -4$

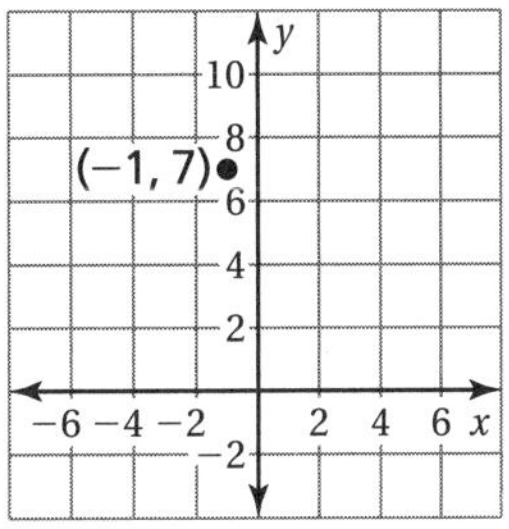

3. $m = \frac{1}{3}$; $(-6, 3)$

4. $m = -\frac{3}{4}$; $(8, -7)$

5. $m = 2$; $(-1, -5)$

6. $m = -3$; $(-2, 8)$

7. You are pulling a kite back to the ground at a rate of 2 feet per second. After 4 seconds, the kite is 16 feet above the ground.

a. Write an equation that represents the height y (in feet) above the ground after x seconds.

b. At what height was the kite when you started pulling it in?

c. When does the kite touch the ground?

8. You rent a moving truck. The total cost includes a flat fee plus a mileage fee of $2 per mile.

a. After 60 miles, the total cost is $160. Write an equation that represents the total cost y after x miles.

b. Interpret the y-intercept.

c. You have $300 to spend on the moving truck rental. The number of miles to your destination is the same as the number of miles returning. You have already driven 60 miles. How many more miles can you drive before you must turn around to stay within your budget?

Name ______________________________ Date __________

3.2 Enrichment and Extension

Ski Slopes

The grade of a ski trail describes the steepness of the ski slope. A skier can calculate the grade by dividing the vertical decrease (in meters) of the slope by the horizontal distance (in meters) it covers.

Frosty's Ski Center	
Trail	**Grade of Trail**
Bunny Slope	4%
Medium Trail	15%
Expert Route	30%

1. How does the grade of a trail relate to the slope of a line?

2. What is the slope of each of the ski trails?

3. There is a First Aid Station located at the point $(750, 28)$ on the Bunny Slope. Write an equation that describes the Bunny Slope.

4. What is the y-intercept of your equation? What does the y-intercept mean in terms of the Bunny Slope?

5. What is the x-intercept of your equation? What does the x-intercept mean in terms of the Bunny Slope?

6. The halfway point on the Medium Trail is located at $(5100, 765)$. Write an equation that describes the Medium Trail.

7. The Expert Route passes by the ski lodge located at the point $(15{,}000, 1500)$. Write an equation that describes the Expert Route.

8. Beginner slopes for new skiers have a maximum grade of 5%. A ski trail has a slope of $-\frac{2}{25}$. Is the trail a beginner slope? Explain.

Name____ Date__________

3.2 Puzzle Time

What's A Pumpkin's Favorite Sport?

Write the letter of each answer in the box containing the exercise number.

Write an equation of the line with the given slope that passes through the given point.

1. $m = 2; (1, 5)$

A. $y = 2x + 3$ **B.** $y = 2x + 5$ **C.** $y = 2x - 3$

2. $m = 6; (-3, -13)$

R. $y = 6x - 5$ **S.** $y = 6x + 5$ **T.** $y = 6x + 3$

3. $m = -4; (-12, -15)$

T. $y = -4x - 60$ **U.** $y = -4x - 63$ **V.** $y = -4x - 75$

4. A sports park charges a flat rate for birthday parties, plus \$5 per guest. A party with 17 guests costs \$155. Which equation represents the cost of a birthday party for x guests?

F. $y = 5x + 85$ **G.** $y = 5x + 155$ **H.** $y = 5x + 70$

5. You are running laps at a track at a pace of 0.8 lap per minute. After 10 minutes, you have 16 laps left to run. Which equation represents the number of laps left to run after running for x minutes?

S. $y = -0.8x + 24$ **T.** $y = 0.8x + 8$ **U.** $y = -0.8x + 16$

6. A tennis center charges a membership fee plus \$20 per lesson. After 8 lessons, the total cost would be \$280. Which equation represents the cost of a tennis membership and x lessons at the center?

O. $y = 20x + 280$ **P.** $y = 20x + 160$ **Q.** $y = 20x + 120$

5	6	3	1	2	4

Activity 3.3 Start Thinking!

For use before Activity 3.3

Estimate the height of a helium balloon 1 second after a child has let go of the string.

The height of the balloon 30 seconds after the child has let go is about as high as a 10 story building. How high is that?

Graph the height y of the balloon x seconds after it was let go. Use estimates from your graph to write an equation.

Do you think that everyone in your class will end up with the same equation? Explain why or why not.

Activity 3.3 Warm Up

For use before Activity 3.3

Write an equation of the line given slope *m* and *y*-intercept *b*.

1. $m = \frac{1}{2}, b = 3$

2. $m = -2, b = 0$

3. $m = 1, b = -3$

4. $m = \frac{1}{3}, b = -1$

5. $m = -\frac{3}{5}, b = \frac{1}{2}$

6. $m = -\frac{7}{5}, b = -\frac{1}{5}$

Lesson 3.3

Start Thinking!

For use before Lesson 3.3

The table shows the U.S. Postal Service cost for sending media mail (small and large packages of books, film, manuscripts, sound recordings, video tapes, and computer readable media such as CDs, DVDs, and diskettes).

How can you write an equation to represent the cost y of sending x pounds of media mail?

Weight Not Over (pounds)	Cost
1	$2.23
2	$2.58
3	$2.93
4	$3.28
5	$3.63
6	$3.98

Lesson 3.3

Warm Up

For use before Lesson 3.3

Find the slope and *y*-intercept of the line that passes through the points. Then write an equation of the line.

1.

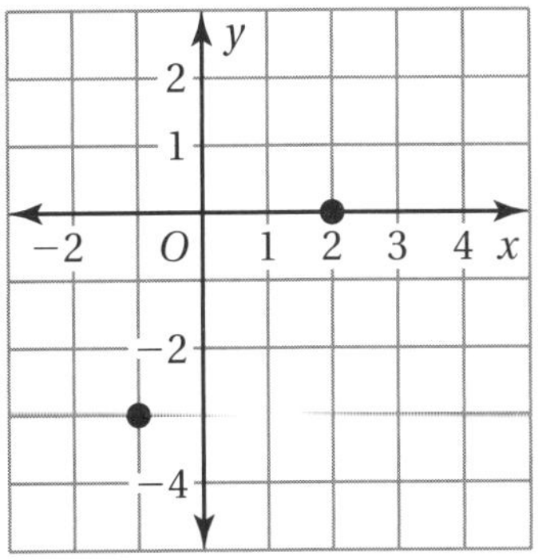

2.

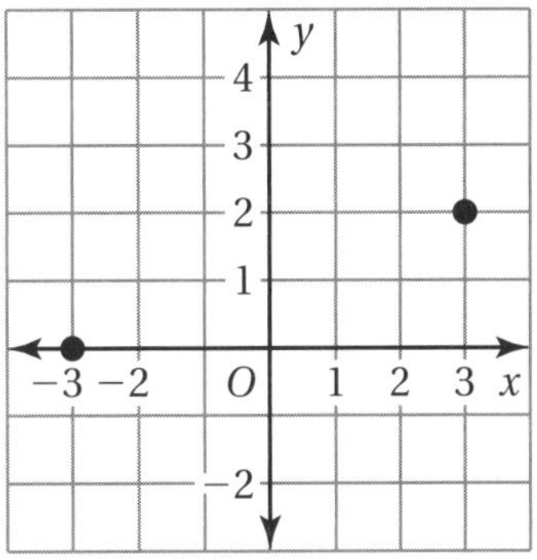

Name ________________________________ Date __________

3.3 Practice A

Find the slope and *y*-intercept of the line that passes through the points. Then write an equation of the line.

1.

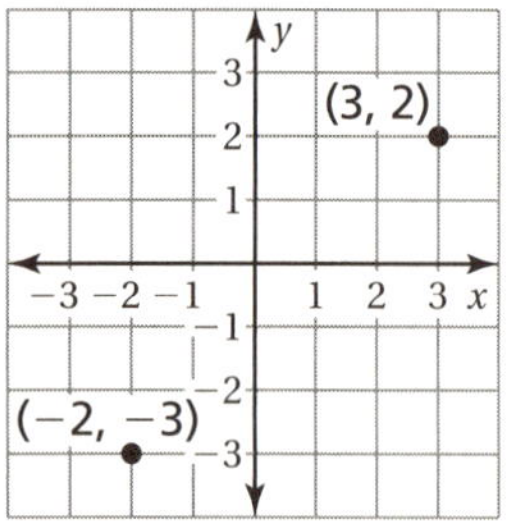

2.

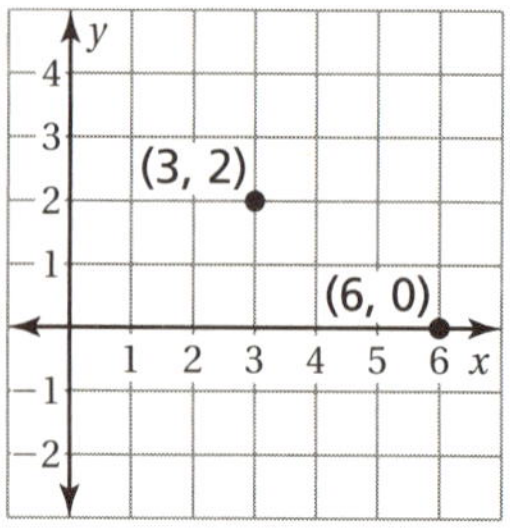

Write an equation of the line that passes through the points.

3. $(-3, -4), (6, -1)$

4. $(-4, 12), (2, -3)$

5. $(-1, -2), (1, -6)$

6. $(-2, -9), (1, 6)$

7. It costs \$45 to rent a surfboard for 3 days. It costs \$70 to rent a surfboard for 8 days.

a. Write an equation that represents the cost y (in dollars) of renting a surfboard for x days.

b. What is the flat fee for renting the surfboard?

c. What is the rate per day to rent the surfboard?

d. Suppose there is a 40% increase in the flat fee for renting the surfboard. Write an equation that represents the new cost y (in dollars) of renting a surfboard for x days.

8. The value of a diamond ring increases at a constant rate. After 2 years, the value of the ring is \$1500. After 3 more years, the value of the ring is \$2025.

a. Write an equation that represents the value y (in dollars) of the ring after x years.

b. Interpret the y-intercept of the graph of the equation.

c. What is the constant rate of increase in the value of the ring?

d. Would you expect this constant rate to continue forever? Explain.

Name____ Date__________

3.3 Practice B

Find the slope and *y*-intercept of the line that passes through the points. Then write an equation of the line.

1.

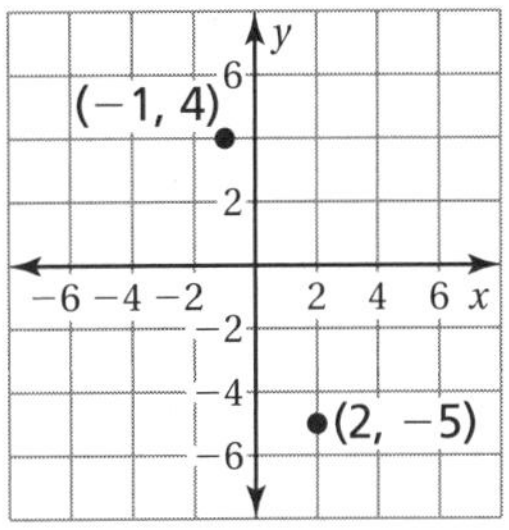

2.

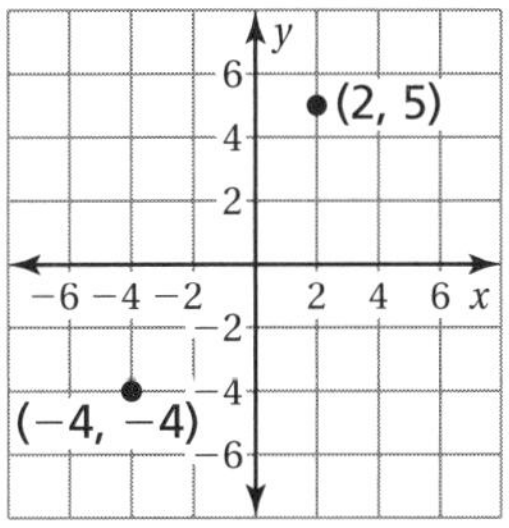

Write an equation of the line that passes through the points.

3. $(2, 3), (3, 7)$

4. $(-5, -8), (10, 4)$

5. $(-6, 4), (6, 0)$

6. $(2, 4), (4, 9)$

7. Water is draining out of the bathtub at a constant rate. After 2 minutes, the water in the bathtub is 8 inches deep. Eight minutes later, the water is 4 inches deep.

a. Write an equation that represents the depth y (in inches) of the water in the bathtub after x minutes.

b. How much water was in the bathtub when you started draining the tub?

c. When is the bathtub empty?

d. Does this seem like a reasonable amount of time to empty a bathtub? Explain your answer.

8. It costs \$5 to ship 3 pounds of chocolate. It costs \$11 to ship 12 pounds of chocolate. It costs \$14 to ship 21 pounds of chocolate.

a. Plot the points for each cost y (in dollars) and pounds x.

b. Does the cost to ship the chocolate increase at a constant rate? Explain.

c. Write an equation that represents the cost y (in dollars) for shipping x pounds of chocolate when you ship up to 12 pounds. Write a second equation to represent the shipping cost when you ship 12 pounds or more.

d. What is the constant rate in dollars per pound when you ship more than 12 pounds of chocolate? Round to the nearest cent.

Name ______________________________ Date __________

3.3 Enrichment and Extension

Writing Equations Using Two Points

Write an equation of the line that passes through the points. Check your solution by evaluating the given points in the equation.

1.

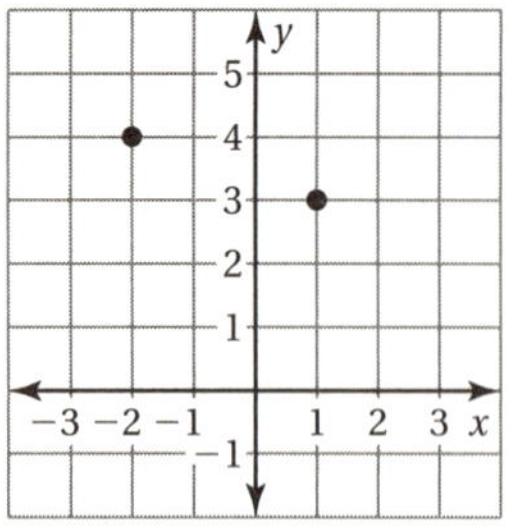

2.

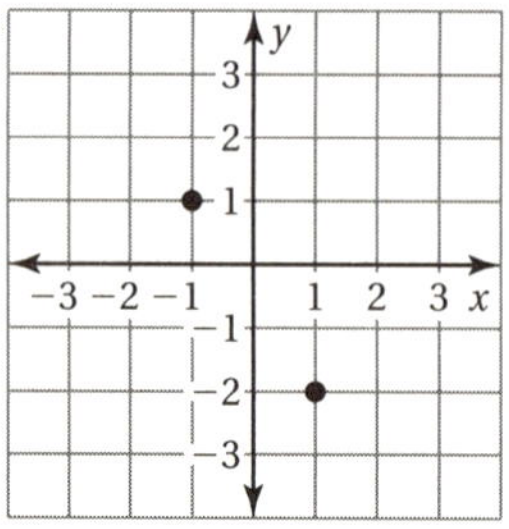

3.

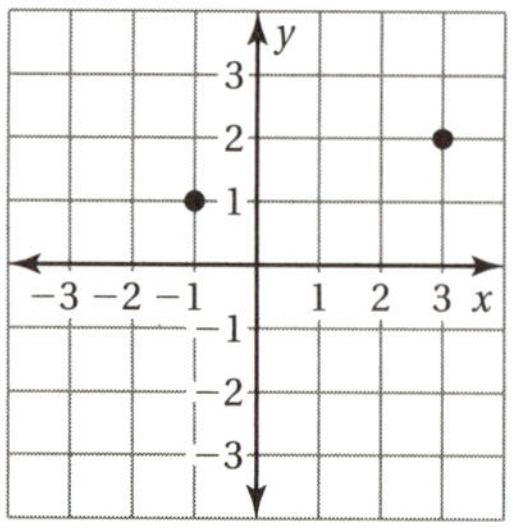

4.

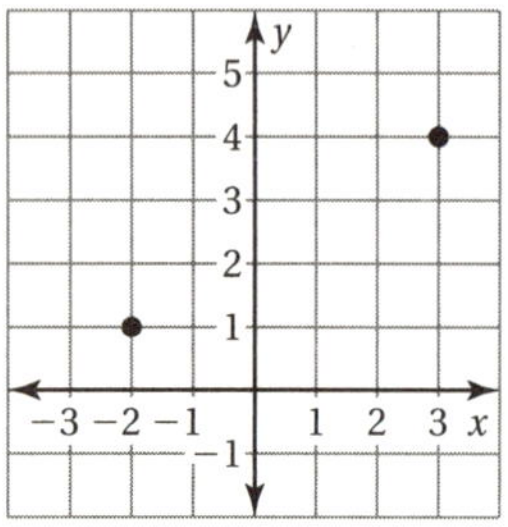

5.

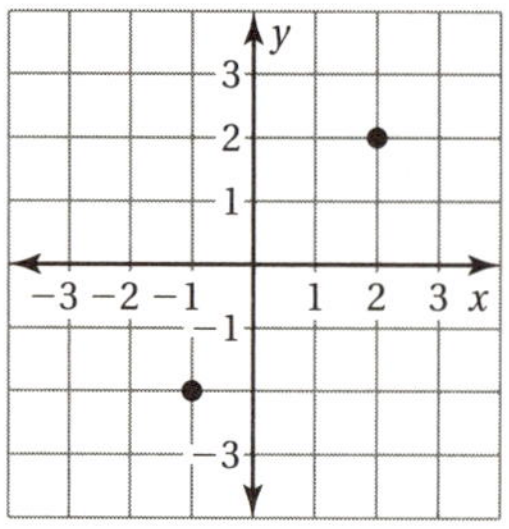

6.

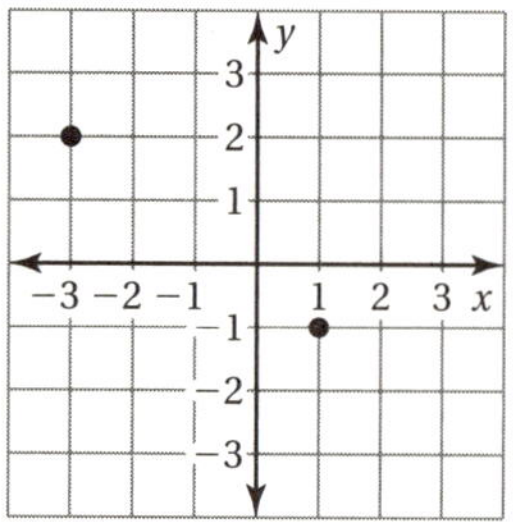

Name____ Date __________

3.3 Puzzle Time

What Do You Call A Ghost Cheerleader?

Write the letter of each answer in the box containing the exercise number.

Write an equation of the line that passes through the points.

1. $(3, 2), (6, 3)$
2. $(2, 3), (4, 1)$
3. $(-6, 1), (-4, 2)$
4. $(-4, -2), (-1, 4)$
5. $(-2, 2), (-1, -1)$
6. $(-8, 7), (-4, 6)$
7. $(-5, -5), (5, -7)$
8. $(-3, -4), (3, 0)$
9. $(-2, -7), (2, -1)$
10. $(-6, -4), (6, 4)$
11. Brian goes to an arcade and purchases a card with game credits. After playing 5 arcade games, he has 33 credits. He plays 4 more games and has 21 credits. What equation represents the number of credits y on the card after x games?
12. You go to a school dance. There is an entrance fee and they are selling slices of pizza. After having one slice of pizza, you have spent \$6. After having 2 more slices of pizza, you have spent \$10. What equation represents the total cost y after buying x slices of pizza?
13. Jenna is making headbands out of ribbon. She makes two headbands and has 6 feet of ribbon remaining. She makes one more headband and has 4 feet of ribbon left. What equation represents the amount of ribbon y Jenna has left after making x headbands?

Answers

R. $y = -x + 5$

E. $y = 2x + 4$

P. $y = -2x + 10$

I. $y = -3x + 48$

A. $y = \frac{1}{2}x + 4$

E. $y = \frac{1}{3}x + 1$

T. $y = -\frac{1}{4}x + 5$

M. $y = \frac{2}{3}x$

T. $y = \frac{2}{3}x - 2$

S. $y = \frac{3}{2}x - 4$

I. $y = -3x - 4$

T. $y = 2x + 6$

H. $y = -\frac{1}{5}x - 6$

4	7	1		6	12	3	10		9	13	5	2	11	8

Activity 3.4 Start Thinking!
For use before Activity 3.4

Follow the steps to write a linear equation for the amount of money y you have in your savings after x weeks.

1. What do you think is a reasonable weekly allowance? Use this in your model.
2. What other amount do you need to know before you can write an equation? Make up an amount and use it in your model.
3. Write a linear equation using the numbers you chose in steps 1 and 2.

What does the slope represent? What does the y-intercept represent? Graph the linear equation.

Activity 3.4 Warm Up
For use before Activity 3.4

Find the slope and y-intercept of the line.

1.

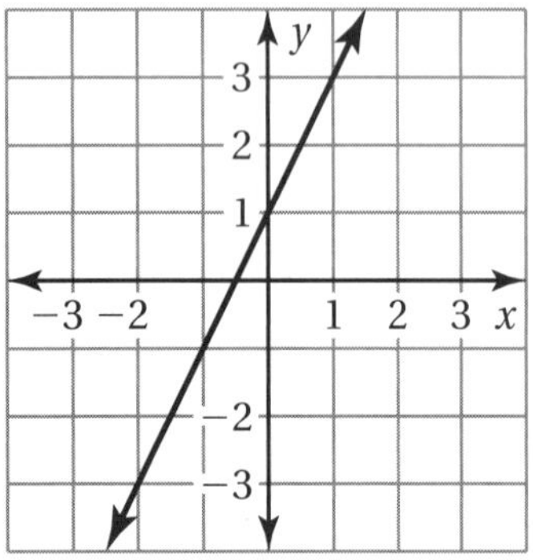

2.

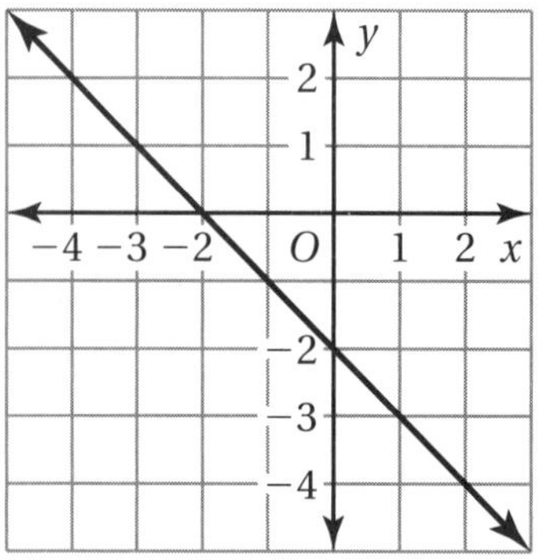

Lesson 3.4 Start Thinking!

For use before Lesson 3.4

How can you use linear equations to solve problems about taking a road trip?

Lesson 3.4 Warm Up

For use before Lesson 3.4

Describe a real-life problem that has the given rate and intercepts. Draw a line that represents the problem.

1. Rate: -1 dollar per song

y-intercept: 25 dollars

x-intercept: 25 songs

2. Rate: -10 feet per second

y-intercept: 60 feet

x-intercept: 6 seconds

Name ______________________________ Date __________

3.4 Practice A

Describe a real-life problem that has the given rate and intercepts. Draw a line that represents the problem.

1. Rate: $-\$7$ per week

y-intercept: \$154

x-intercept: 22 weeks

2. Rate: -3 trees per day

y-intercept: 27 trees

x-intercept: 9 days

3. You are playing a board game that uses an hourglass timer filled with sand. The percent y (in decimal form) of sand remaining in the timer after x minutes is $y = -0.25x + 1$.

a. Graph the equation.

b. Interpret the x- and y-intercepts.

c. After how many minutes is the timer 25% full?

d. Your friend finds another timer to use and this timer is 0% full in 2 minutes. Find an equation for this new timer.

4. You are cooling water for a science experiment. The graph relates the temperature y (in degrees Fahrenheit) and the cooling time x (in minutes).

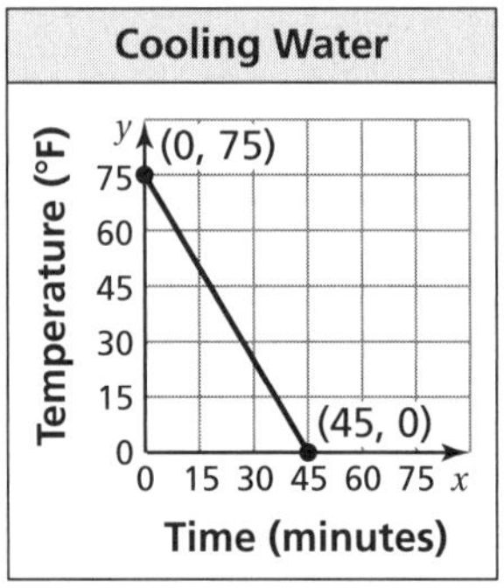

a. Interpret the x- and y-intercepts.

b. What is the slope? What does the slope represent in this situation?

c. Write an equation of the line.

d. When is the temperature of the water 55°F?

e. Water freezes at 32°F. After how many minutes is the water frozen?

Name____ Date__________

3.4 Practice B

Describe a real-life problem that has the given rate and intercepts. Draw a line that represents the problem.

1. Rate: \$15 per week

y-intercept: \$30

x-intercept: −2 weeks

2. Rate: −1.5 gallons per week

y-intercept: 15 gallons

x-intercept: 10 days

3. Your choir is presenting a musical at the city theater to raise money for a trip to New York. The Parents Association is donating \$400. In addition, the choir will receive 75% of each \$20 ticket sold. The maximum capacity of the theater is 500 people.

a. Write an equation that represents the choir's revenue R when x tickets are sold.

b. What is the maximum amount of money that the choir can raise?

c. The choir director has calculated that if they can fill the theater to 60% capacity, then they will raise enough money for the trip. How much money is required for the trip?

4. Water is slowly leaking from a large bottle. The percent y (in decimal form) of water remaining in the bottle after x hours is $y = -0.02x + 1$.

a. Graph the equation.

b. Interpret the x- and y-intercepts.

c. Interpret the slope.

d. When will the bottle be one-quarter full?

5. You are riding your bicycle from school to home. The graph relates your distance from home y (in city blocks) and travel time x (in minutes).

a. How far away from home are you when you leave school?

b. How long will it take you to get home?

c. Write an equation of the line.

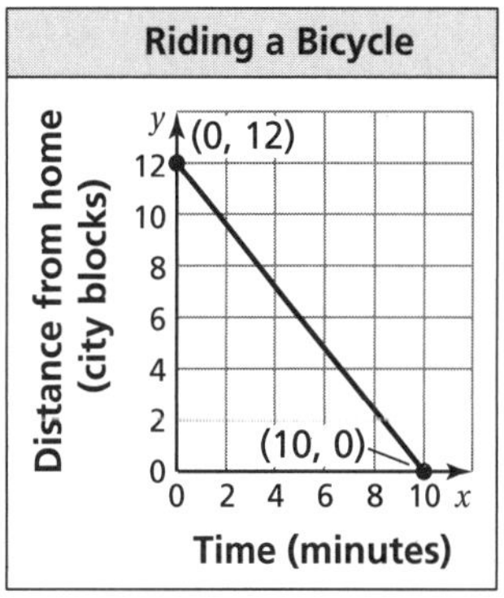

Name ______________________________ Date __________

3.4 Enrichment and Extension

Maze

Each square contains the information necessary to write the equation of a line. Find a path from start to finish in the maze. You may not travel in any of the following squares.

- Lines that have a slope of 2.
- Lines that pass through the point (1, 4).
- Lines that have an x-intercept greater than 5.
- Lines that have a y-intercept less than –4.

Start	(2, 1) (3, 2)	(−2, 3) $m = \frac{1}{2}$	x-intercept: 3 y-intercept: 2	(2, 5) (1, 0)
$m = 3$ y-intercept: 1	(−3, −10) (2, −5)	x-intercept: –2 y-intercept: 4	(3, 3) $m = \frac{2}{3}$	x-intercept: 6 y-intercept: 3
(−6, 0) $m = -1$	$m = -2$ y-intercept: 3	(2, 2) $m = \frac{3}{2}$	$m = 4$ y-intercept: 1	(2, 5) $m = 1$
x-intercept: –1 y-intercept: 2	(−4, −1) $m = \frac{1}{4}$	x-intercept: 1 y-intercept: –5	(−1, −1) (1, 3)	$m = \frac{3}{5}$ y-intercept: –6
$m = -4$ y-intercept: 8	(3, 5) (1, −1)	x-intercept: –2 y-intercept: 1	(−5, −6) (−1, −2)	**Finish**

Name____ Date__________

3.4 Puzzle Time

What Is A Computer's Favorite Kind of Music?

Write the letter of each answer in the box containing the exercise number.

1. The student government is selling T-shirts for a fundraiser. They spent \$280 on the shirts and will sell them for \$8 each. The profit y from selling x shirts is $y = 8x - 280$. How many shirts will the student government have to sell before breaking even?

R. 15 **S.** 35 **T.** 45

2. The number of feet y a jet ski will still have left to glide to a stop after x seconds is $y = -25y + 300$. How many seconds does it take for the jet ski to come to a complete stop?

I. 12 **J.** 15 **K.** 25

3. You are saving money to go on your class trip. You already have saved \$35 and you plan to add \$5 to that each week. Which equation represents the amount y you will have after x weeks?

M. $y = 5x - 35$ **N.** $y = 35x + 5$ **O.** $y = 5x + 35$

4. A swimmer is competing in the 50-meter butterfly. The number of meters y that she has left to swim after x seconds is $y = -\frac{5}{3}x + 50$. After how many seconds will the swimmer reach the halfway point in the race?

J. 30 **K.** 15 **L.** 25

5. A craft fair promises to donate \$100 plus 3% of every \$12 ticket sold to a charity. Which equation represents the amount y of money the craft fair will give to the charity if x tickets are sold?

B. $y = 0.03x + 100$ **C.** $y = 0.03x + 112$ **D.** $y = 0.36x + 100$

5	2	1	4		3
				-	

Activity 3.5 Start Thinking!
For use before Activity 3.5

A runner is jogging around a track. A second runner starts 5 minutes after the first, jogging at a faster pace.

How can the situation be modeled using a system of linear equations?

Make a general sketch of the situation on a coordinate grid. What does the intersection of the graphs of the equations represent?

Activity 3.5 Warm Up
For use before Activity 3.5

Solve the linear system algebraically.

1. $x + y = 10$
$y = 2x + 1$

2. $x - y = 2$
$x + y = 18$

3. $y = -2x + 9$
$y = x - 3$

4. $y = x + 5$
$y = 6x$

Lesson 3.5 Start Thinking!

For use before Lesson 3.5

1. Pick any two integers. Use them to copy and complete the statements.
   - The sum of the two numbers is __?__.
   - The difference of the first number and twice the second number is __?__.
2. Exchange your two statements with a classmate. Write two linear equations from your classmate's statements.
3. Solve the system of equations using any method to find your classmate's numbers.

Did you guess the two numbers correctly?

Lesson 3.5 Warm Up

For use before Lesson 3.5

1. The perimeter of a rectangle is 60 feet. The length is twice the width.

Write a system of linear equations to represent the situation. Find the dimensions of the rectangle using a graph. Check your answer using algebra.

Name ______________________________ Date __________

3.5 Practice A

1. The sum of two numbers is 29. The second number y is 5 more than twice the first number x. Write and solve a system of linear equations to find the two numbers.

Words:	First number	plus	second number	is	29.

Words:	Second number	is	two times	first number	plus	5.

2. The length ℓ of a rectangle is 4 times the width w. The perimeter is 30 feet. Write and solve a system of linear equations to find the dimensions of the rectangle.

3. The varsity football team consists of x seniors and y juniors. The number of seniors is 7 less than the number of juniors. There are 43 members on the team. Write and solve a system of linear equations to find the numbers of seniors and juniors on the team.

4. Your purchase of CDs and DVDs has a total of 7 items. CDs cost $9 each and DVDs cost $15 each. The total purchase costs $87. Write and solve a system of linear equations to find the number x of CDs and the number y of DVDs that you purchased.

5. Two pears and three apples cost $3. One pear and 2 apples cost $1.75. Write and solve a system of linear equations to find the cost x (in dollars) of a pear and the cost y (in dollars) of an apple.

6. You have dimes and quarters in your pocket. There are 12 coins that total $2.25. Write and solve a system of linear equations to find the number x of dimes and the number y of quarters.

Name____ Date__________

3.5 Practice B

1. You have quarters and dimes in your pocket. There are 21 coins that total \$2.85. Write and solve a system of linear equations to find the number x of quarters and the number y of dimes.

   Words: Number of quarters plus number of dimes is 21.

   Words: \$0.25 times number of quarters plus \$0.10 times number of dimes is \$2.85.

2. The difference of two numbers is 45. The larger number x is equal to 30 less than twice the smaller number y. Write and solve a system of linear equations to find the two numbers.

3. The seventh grade consists of 105 students. The number x of girls is equal to $\frac{3}{4}$ the number y of boys. Write and solve a system of linear equations to find how many girls x and boys y are in seventh grade?

4. The table shows the distance (in miles) from a warehouse of two trucks after x hours. They are both headed to the same destination.

x	Truck A	Truck B
2	110	20
3	165	90
4	220	160
5	275	230

   a. Write a system of linear equations to represent this situation.

   b. After how many hours will the trucks meet?

   c. Which truck is traveling at a faster rate?

   d. How did the trucks leave the warehouse? Did Truck A leave first, Truck B leave first, or did both trucks leave at the same time? Explain.

5. A system of two linear equations has no solution. Describe the graph of the system.

Name ______________________________ Date __________

3.5 Enrichment and Extension

Writing Systems of Linear Equations

Write a system of linear equations to represent the situation. Then solve the system and answer the question. You may use any of the methods you have learned to solve the system.

1. There are 41 total cats and chickens on a farm. There is a total of 116 legs. How many of each type of animal are on the farm?

2. There are 28 total spiders and mice in a field. There is a total of 156 legs. How many of each type of animal are in the field?

3. There are 19 total frogs and salamanders in a pond. There is a total of 76 legs. How many of each type of animal are in the pond?

4. There are 24 total sparrows and horses in a barn. There is a total of 66 legs. How many of each type of animal are in the barn?

5. There are 35 total ants and rabbits in a driveway. There is a total of 210 legs. How many of each type of animal are in the driveway?

6. There are 14 tables in a room. There is a total of 50 legs. Is this possible? Explain your reasoning.

Name_______________________________________ Date__________

3.5 Puzzle Time

How Did Benjamin Franklin Feel When He Discovered Electricity?

Write the letter of each answer in the box containing the exercise number.

Answers

K. $3.75

D. $5.25

H. $2.95

E. 17

O. 26

C. 11

S. 22

1. The eighth grade class orders roast beef subs and turkey subs for a class picnic. A total of 53 subs are ordered. The number of roast beef subs ordered is 2 more than twice the number of turkey subs that are ordered. How many turkey subs are ordered?

2. The price of 2 adult tickets and 3 child tickets to a movie is $24. The price for 1 adult ticket and 2 child tickets is $14.25. What is the price of one adult ticket?

3. Jake and his brother have 45 baseball cards altogether. Jake's brother has 6 more than half the number of cards that Jake has. How many baseball cards does Jake have?

4. Lisa's and her dad's ages combined are 51. Her dad's age is 4 less than 4 times her age. How old is Lisa?

5. You buy 2 body lotions and 2 bath gels for $12.90. Your friend buys 3 body lotions and 1 bath gel for $12.35. What is the cost of one body lotion?

6. You watched the weather for the last 30 days to see how many days were sunny and how many days were rainy. You observed that the number of sunny days was 2 less than 3 times the number of rainy days. How many days were sunny?

7. Steven buys 2 magazines and a CD for $18.50. Rachel buys 1 magazine and 2 CDs for $25.75. What is the cost of one magazine?

6	5	3	4	7	1	2

Name ______________________________ Date __________

Chapter 3 School-to-Work

For use after Section 3.1

Animal Hospital Employee

You work at an animal hospital. The veterinarian has asked you to monitor the intravenous therapy (IV) of a dehydrated 50-pound black lab. The following chart is on the wall of the examination room.

Weight of dog (lb)	IV drip rate (mL/min)
25	16.5
50	34.5
75	51
100	67.5

1. The veterinarian is giving the black lab 600 milliliters of fluid. Write an equation in slope-intercept form that relates the amount of fluid left in the IV bag and the number of minutes since the IV was started.

2. Graph your equation in a coordinate plane. What does each intercept represent?

3. How long will it take for the IV to finish dripping? If the veterinarian started the IV at 2:15 P.M., at what time will the IV be finished?

4. A 75-pound golden retriever receives 700 milliliters of fluid through an IV at the same time as the black lab. Write an equation, in slope-intercept form, to model the IV drip for the golden retriever. Graph the equation on the same coordinate plane as for the black lab's IV.

5. Which dog's IV finishes first? By how many minutes? How can you tell by the graph? Justify your answer algebraically.

6. What do you think the intersection of the two lines tells you about the amount of fluid left in the IV bags? Explain.

Name______________________________ Date__________

Chapter 3 Study Help

You can use a **notetaking organizer** to write notes, vocabulary, and questions about a topic.

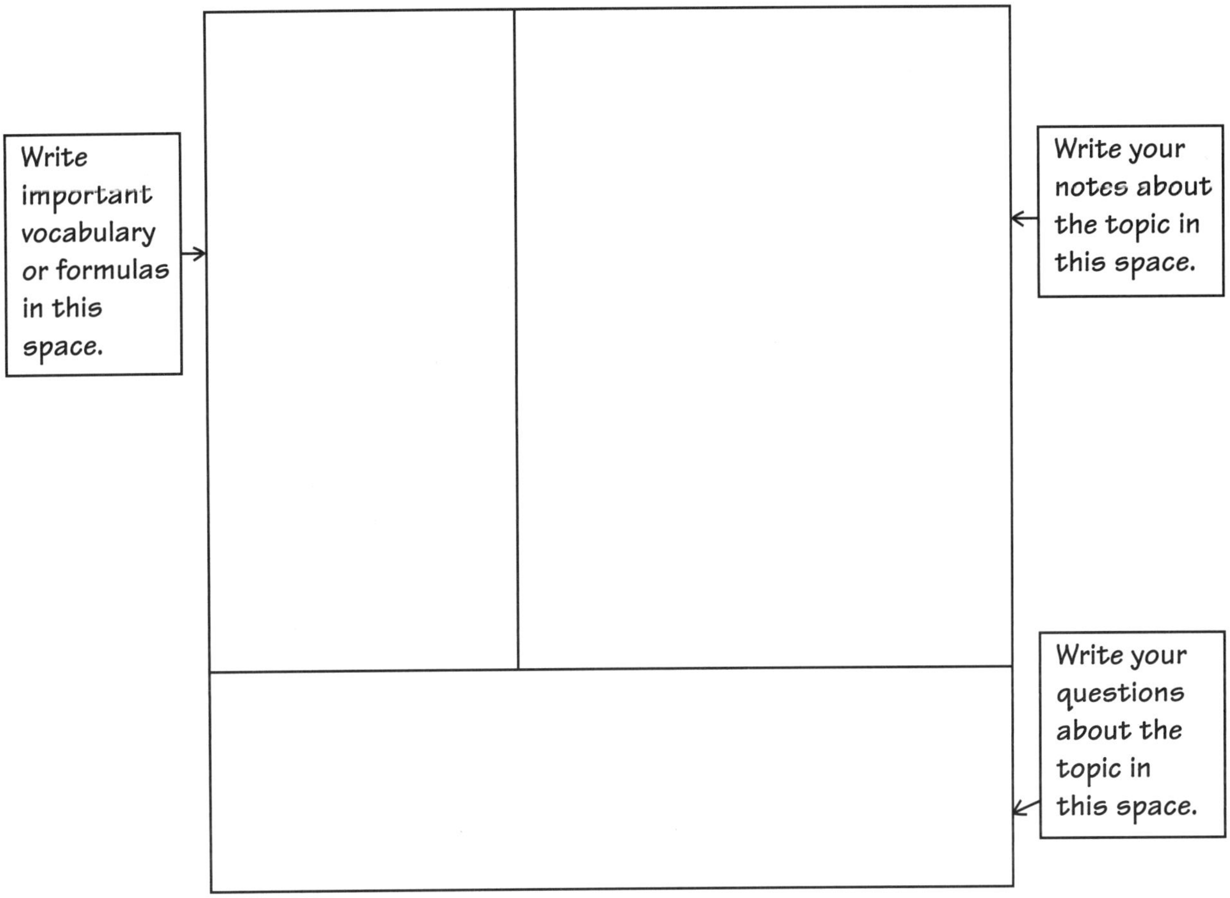

On Your Own

Make a notetaking organizer to help you study these topics.

1. writing equations in slope-intercept form
2. writing equations using two points

After you complete this chapter, make notetaking organizers for the following topics.

3. writing systems of linear equations
4. Think of a real-life application that can be modeled using a linear equation. Then make a notetaking organizer to help you study this application.

Name ______________________________ Date __________

Chapter 3

Financial Literacy

For use after Section 3.5

Making a Theater Production Profitable

You are in charge of the finances for your school's upcoming theater production. The director has given you the following information about the production.

The theater has 400 seats.

Adults are typically charged $8 for admission.

Students are typically charged $5 for admission.

The costs associated with putting on the production amount to $3850.

1. Tickets sold for the first show bring in a total of $2690. Write a system of linear equations for the situation. Graph the system. What do the intercepts represent for each line? What does the intersection of the two lines represent?

2. Solve the system of equations. How many adult tickets and how many student tickets were sold?

3. Suppose you could have raised each ticket price by the same amount. How much would you have had to charge for each type of ticket to equal the cost of putting on the production? Explain how you found your answer.

4. Tickets for the second show also sell out. Can you be sure that the two shows will make enough money to cover the production costs? Explain why or why not. What is the greatest possible profit that could be made from the two shows?

5. A matinee is added to the schedule as a third show, with lower ticket prices. The same ratio of students and adults attend the matinee as did for the first show and the cost of an adult ticket is twice as much as the cost of a student ticket. If $1890 is made in tickets sales, what is the matinee price for each type of ticket? Show your work.

6. It is estimated that your school's next theater production will cost $4200 and that there will be only 2 shows. Devise a plan for selling tickets for this production that will ensure that the production cost is covered. Include equations and a graph to support your reasoning.

Name______________________________ Date__________

Chapter 3 Cumulative Practice

Write the word sentence as an equation.

1. The sum of two times a number h and 3 is 15.
2. The sum of a number p and 48 is equal to 7.
3. The difference between 4 times a number x and 3 is 11.
4. The quotient of a number v and 25 is 13.
5. The product of 18 and a number n is 27.
6. The temperature in a room is $60°$ Fahrenheit. A heater is turned on, and the temperature rises by $2°$ every hour. Write an equation that represents the temperature y after the heater has been on for x hours.

Use the given information to write an equation of the line in slope-intercept form.

7. 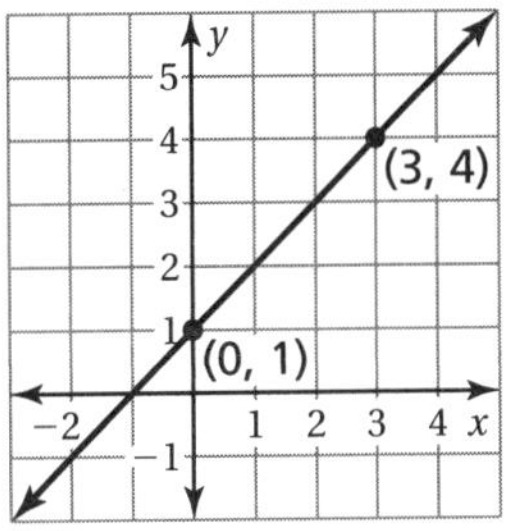

8. 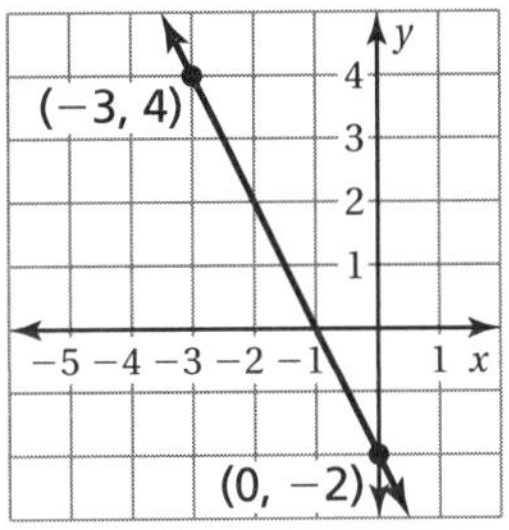

9. 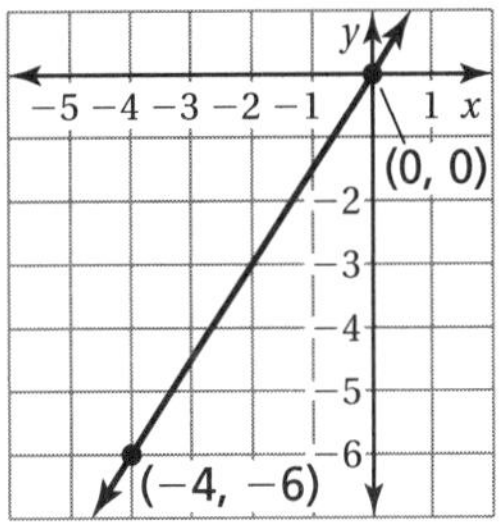

10. The line that passes through $(2, 8)$ and has a slope of $\frac{5}{2}$
11. The line that passes through $(-3, 1)$ and $(3, 7)$
12. The line that passes through $(0, 6)$ and $(4, 0)$
13. The line that passes through $(-5, -4)$ and has a slope of $\frac{1}{5}$
14. There are 28 horses and sheep in a field. There are 6 times as many sheep as horses. Write and solve a system of linear equations to find the number of horses and the number of sheep.
15. There are 20 hardcover and paperback books in a box. There are 10 more hardcover books than paperback books. Write and solve a system of linear equations to find the number of paperback books and the number of hardcover books.

Chapter 4

Name______________________________ Date__________

Chapter 4 Functions

Dear Family,

Many people enjoy putting their photos on their computer and sharing them online with friends and family. Sometimes, when sharing a photo online, you can see jagged edges that don't appear in the actual photo. These jagged edges happen because digital photos are actually created by a grid of colored dots.

The colors in the world we see are an example of a continuous range—any number of colors can exist. However, the colors in our camera are an example of a discrete range—a limited number of colors can exist.

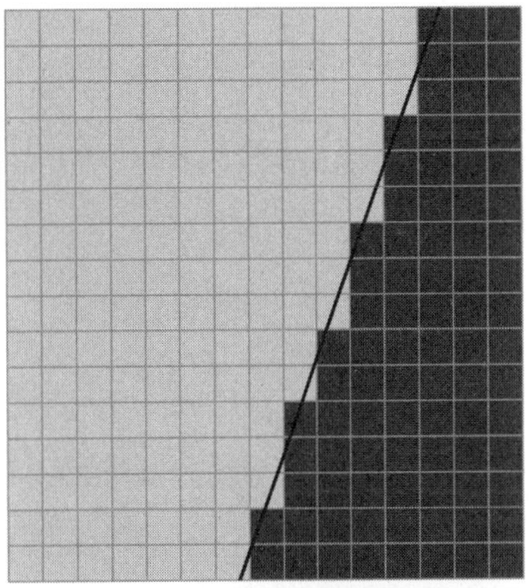

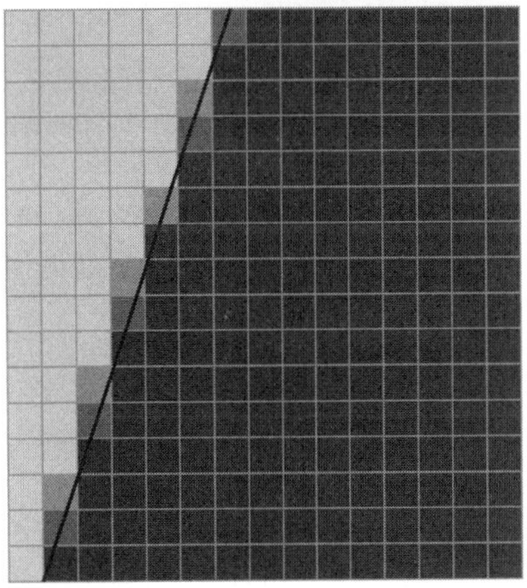

When an edge passes through a dot, the camera has to decide whether to choose the color on one side of the edge or the other. Over a long edge, this can lead to a jagged look as the camera chooses one color or the other.

You and your student can model this using graph paper. Draw a slanted line on the graph paper. Color the blocks on one side of the line with red and the other side with blue. Choose one color or the other for blocks that the line passes through.

- Do you see a jagged pattern?
- What if you change the angle of the line—is the line *more jagged* or *less jagged*?
- What happens if you draw the line using blocks? Does the line look straight or jagged?

To see this, it might help to stand some distance away from the paper so the blocks appear smaller.

Modern computer applications try to eliminate the jagged edges by blending the colors on each side of the line. Try this with your student on a piece of graph paper. Does it improve the appearance of the line—especially from a distance?

Take a picture of your project and share it!

Nombre ______________________________ Fecha __________

Capítulo 4 Funciones

Estimada Familia:

Muchas personas disfrutan colocando sus fotografías en la computadora y compartiéndolas en línea con familiares y amigos. A veces, cuando se comparten fotografías en línea, se pueden ver líneas que no están definidas, las cuales no aparecen en la foto real. Estos bordes irregulares ocurren porque las fotos digitales son creadas mediante una cuadrícula de puntos de colores.

Los colores que vemos en el mundo son un ejemplo de un rango continuo—puede existir un sinfín de colores. Sin embargo, los colores en nuestras cámaras son un ejemplo de un rango definido—existe un número limitado de colores.

Cuando un borde pasa por un punto, la cámara tiene que decidir si escoge entre un color u otro. En un borde largo, esto puede producir una apariencia irregular al elegir la cámara un color u otro.

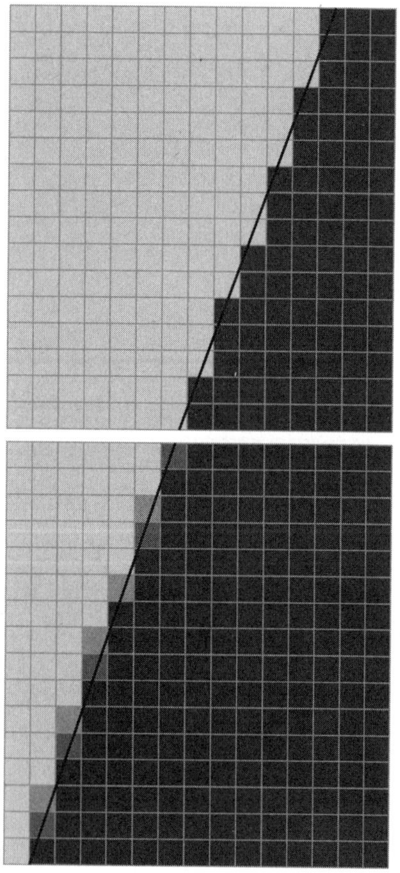

Usted y su estudiante pueden replicar este modelo usando papel cuadriculado. Dibujen una línea oblicua en el papel. Coloreen los cuadrados de un lado con rojo y los del otro lado con azul. Escojan un color u otro para los cuadrados por donde pasa la línea.

- ¿Observan el patrón irregular?
- Y si cambiara el ángulo de inclinación de la línea—¿la línea *es más o menos irregular*?
- ¿Qué pasaría si dibujaran la línea usando cuadrados? La línea aparece recta o irregular?

Para ver esto, ayudaría apreciar el papel a cierta distancia para que los cuadrados aparezcan más pequeños.

Los programas modernos de computadora tratan de eliminar los bordes irregulares mezclando los colores en cada lado de la línea.

Intente hacerlo con su estudiante en un pedazo de papel cuadriculado. ¿Esto mejora la apariencia de la línea—particularmente desde una distancia?

¡Tomen una foto de su proyecto y compártanlo!

Non ______________________________ Dat __________

Chapít 4 Fonksyon

Chè Fanmi:

Anpil moun renmen mete foto yo sou òdinatè yo epi pataje yo sou Entènèt avèk zanmi ak fanmi. Pafwa, lè w'ap pataje yon foto sou Entènèt, ou kapab wè rebò iregilye ki pa parèt sou vre foto a. Rebò iregilye sa yo parèt poutèt foto dijital kreye anfèt avèk yon griy pwen kolore.

Koulè nan monn nou wè a se yon egzanp entèval kontini—nenpòt ki kantite koulè kapab egziste. Sepandan, koulè yo nan kamera nou an se yon egzanp entèval diskrè—yon kantite limite koulè kapab egziste.

Lè yon rebò pase nan yon pwen, kamera a dwe deside si l'ap chwazi koulè ki sou yon bò rebò a oswa sou lòt bò a. Sou yon rebò ki long, sa kapab mennen yon aparans iregilye poutèt kamera a chwazi yon koulè oubyen lòt la.

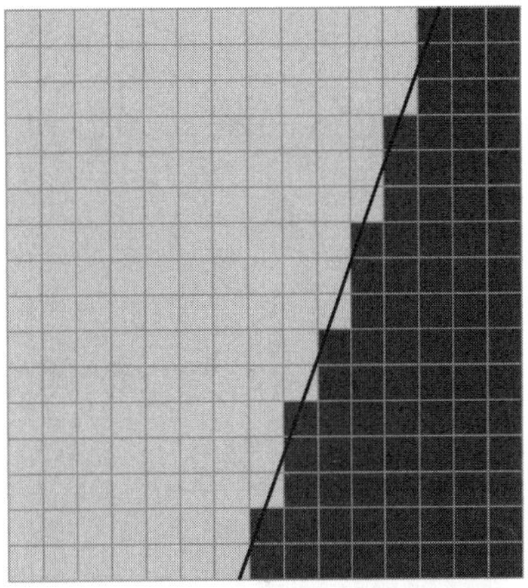

Ou menm ak elèv ou a kapab dekri sa avèk papye kadriye. Desine yon liy oblik sou papye kadriye a. Kolore blòk yo sou yon bò liy nan an wouj epi lòt bò a an ble. Chwazi yon koulè oubyen lòt la pou blòk ke liy nan travèse.

- Èske ou wè yon desen iregilye?
- E si ou chanje ang liy nan—èske liy nan *pi iregilye* oubyen *mwen iregilye*?
- Kisa ki rive lè ou desine liy nan avèk blòk? Èske liy nan parèt dwat oswa iregilye?

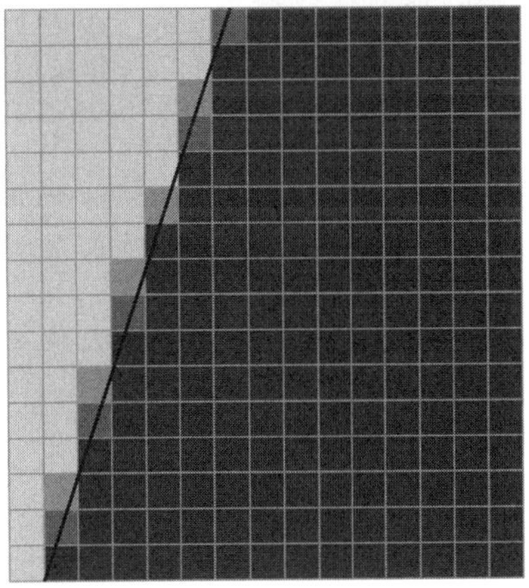

Pou wè sa, sa kapab ede si ou kanpe nan yon sèten distans pa rapò ak papye a pou blòk yo parèt pi piti.

Pou eseye elimine rebò iregilye yo, aplikasyon òdinatè modèn melanje koulè ki sou chak bò liy nan. Eseye fè sa avèk elèv ou a sou yon fèy Papye kadriye. Èske sa amelyore aparans liy nan—espesyalman apati yon sèten distans?

Pran yon foto pwojè ou a epi pataje li!

Activity 4.1 Start Thinking!
For use before Activity 4.1

Study the input-output table below.

Input, x	0	1	2	3	4
Output, y	5	3	1	−1	−3

The **domain** of the function is 0, 1, 2, 3, and 4.

The **range** of the function is 5, 3, 1, −1, and −3.

What do you think the terms **domain** and **range** mean?

Activity 4.1 Warm Up
For use before Activity 4.1

Copy and complete the input-output table for the function.

1. $y = -x + 2$

x	−2	−1	0	1	2
y					

2. $y = 2x - 7$

x	−2	−1	0	1	2
y					

3. $y = \frac{1}{2}x + 3$

x	0	1	2	3	4
y					

4. $y = \frac{1}{3}x - 1$

x	−6	−3	0	3	6
y					

Lesson 4.1 Start Thinking!

For use before Lesson 4.1

The installation and set-up fees for cable internet come to \$150. The monthly cost for internet access is \$40 per month.

Write a function for the cost y of x months of internet service.

What are the domain and range of the function if you only have internet service for 6 months?

Lesson 4.1 Warm Up

For use before Lesson 4.1

1. The number of hard cover books and paper back books you can buy with \$10 at a used book sale is represented by the equation $2x + y = 10$. The table shows the numbers of each type of book you can buy.

Hard cover books, *x*	0	1	2	3	4	5
Paper back books, *y*	10	8	6	4	2	0

   a. Write the equation in function form.

   b. Find the domain and range.

   c. Why is $x = 6$ not in the domain of the function?

Name ____ Date __________

4.1 Practice A

Find the domain and range of the function represented by the graph.

1.

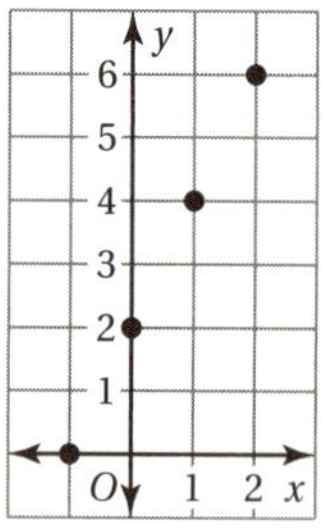

2.

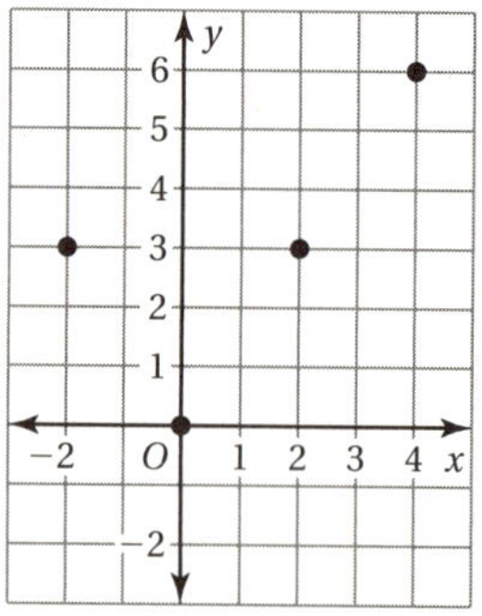

3. The number of girls and boys on the committee is represented by the equation $x + y = 6$. The table shows the number of girls and boys.

Girls, x	0	2	4	6
Boys, y	6	4	2	0

a. Write the equation in function form.

b. Find the domain and range.

c. How many people are on the committee?

d. Why is $y = 8$ not in the range of the function?

Copy and complete the input-output table for the function. Then find the domain and range of the function.

4. $y = 2x - 5$

x	–1	0	1	2
y				

5. $y = \frac{1}{3}x + 1$

x	0	3	6	9
y				

6. In the sport of Catamaran Sailing, boats compete around a triangular race course. For each race, a boat receives points according to its position in crossing the finish line. The table shows the race number x and the place y a boat took for each race.

x	**y**
1	4
2	2
3	1
4	1

a. Find the domain and range of the function represented by the table.

b. Interpret the domain and range.

c. Using the Low Score System, a boat's worst finish is thrown out and the remaining scores are added. The boat with the lowest total score wins. Is this boat guaranteed to be the winner? Explain.

Name______________________________ Date__________

4.1 Practice B

Find the domain and range of the function represented by the graph.

1.

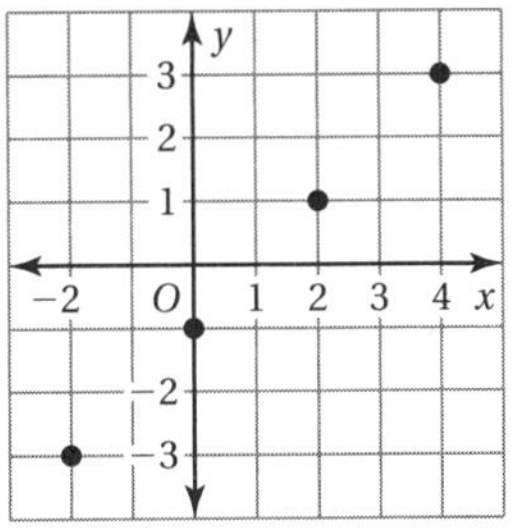

2.

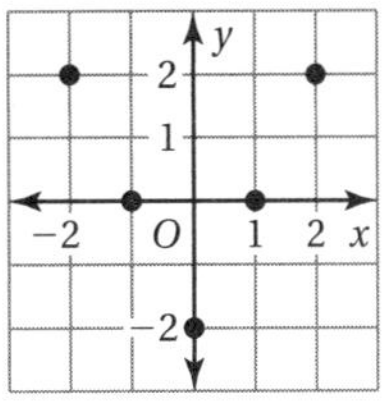

3. The number of packages of limes and lemons you can buy with \$16 is represented by the equation $4x + 2y = 16$. The table shows the number of packages of limes and lemons.

Limes, x	0	1	2	3	4
Lemons, y	8	6	4	2	0

a. Write the equation in function form.

b. Find the domain and range.

c. What is the minimum number of packages that you can purchase for \$16?

d. Which package costs more, *limes* or *lemons*? Explain.

Copy and complete the input-output table for the function. Then find the domain and range of the function.

4. $y = -\frac{2}{3}x + 4$

x	0	3	6	9
y				

5. $y = 1.2x - 3$

x	−1	0	1	2
y				

6. The triceps machine at the gym is set for an individual to lift 60% of his or her body weight with each repetition.

a. Write an equation in function form that represents the amount of weight y (in pounds) an individual lifts with each repetition for his or her weight x.

b. Create an input-output table for the equation in part (a). Use the inputs 125, 150, 175, 200, and 225.

c. Find the domain and range.

d. If the machine were set to lift 75% of his or her body weight, what would be the equation in function form?

Name ______________________________ Date __________

4.1 Enrichment and Extension

Describing Graphs

Use the phrases in the word bank to describe the graphs. List all the phrases that apply. Each item in the word bank will be used at least once.

Is a function	Linear	The domain is $x \geq 0$.
Is non-linear	The range is $y \leq 0$.	The domain is $-1 \leq x \leq 1$.
Is not a function	The range is $-1 \leq y \leq 1$.	The range is $y \geq 0$.

1.

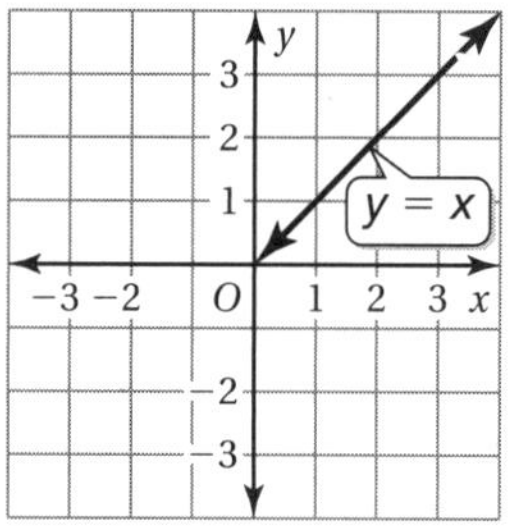

2.

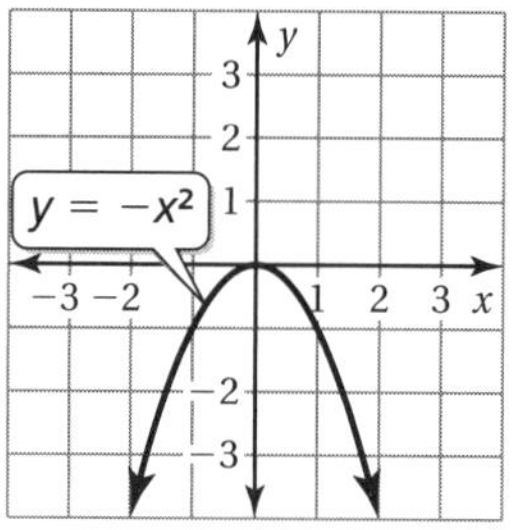

3.

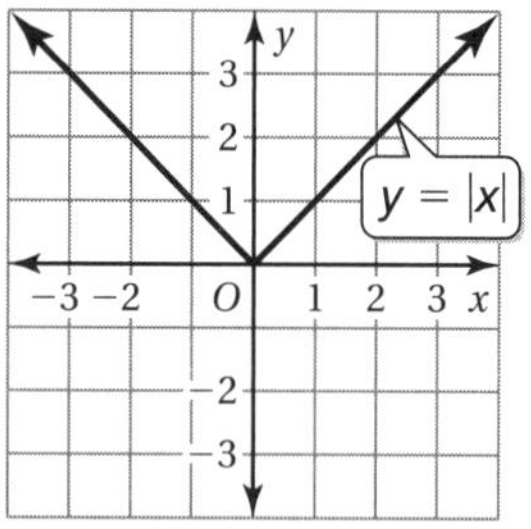

4.

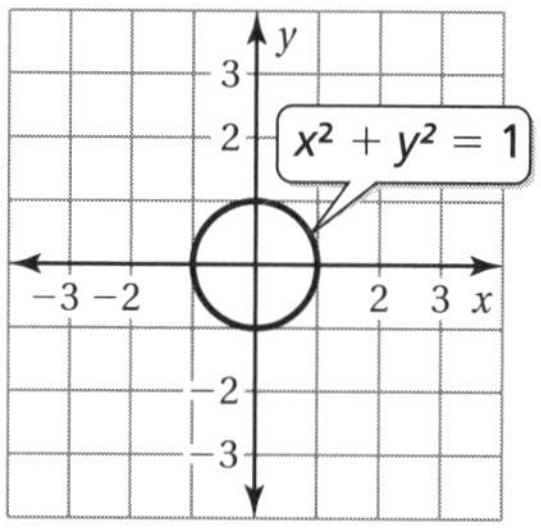

Name____ Date__________

4.1 Puzzle Time

What's Stranger Than Playing The Piano With Your Feet?

Write the letter of each answer in the box containing the exercise number.

Find the range of the function with the given domain.

1. $y = x + 5$, domain: $-1, 0, 1, 2$

2. $y = 3x - 4$, domain: $-2, 0, 2, 4$

3. $y = -7x + 2$, domain: $-2, -1, 0, 1$

4. $2x + y = -5$, domain: $-1, 0, 1, 2$

5. $-x + y = 6$, domain: $1, 2, 3, 4$

Find the domain of the function with the given range.

6. $y = x + 4$, range: $3, 4, 5, 6$

7. $y = x - 8$. range: $-6, -4, -2, 0$

8. $y = -x + 6$, range: $12, 6, 0, -6$

9. $2x + y = 3$, range: $3, 1, -1, -3$

10. $-3x + y = -7$, range: $-13, -7, -1, 5$

11. You are making necklaces and you use 3 beads per inch. The lengths of the necklaces are 8 inches, 10 inches, 12 inches, and 14 inches. Find the range for the function that represents how many beads are needed to make each necklace.

12. Each year a soccer league has teams with 12 players. Shirts are made for all of the players on each team. The number of shirts that had to be made in each of the past four soccer seasons was 60, 72, 84 and 96. Find the domain for the function that represents how many teams are in the league.

Answers

R. $-2, 0, 2, 4$

I. $-3, -5, -7, -9$

N. $4, 5, 6, 7$

Y. $0, 1, 2, 3$

A. $-10, -4, 2, 8$

A. $16, 9, 2, -5$

Y. $-1, 0, 1, 2$

B. $24, 30, 36, 42$

G. $7, 8, 9, 10$

L. $2, 4, 6, 8$

P. $-6, 0, 6, 12$

E. $5, 6, 7, 8$

8	7	3	6	4	1	5		11	9		12	2	10

Activity 4.2

Start Thinking!

For use before Activity 4.2

How much does one apple cost?

How much does one pound of apples cost?

Use the prices you chose to write two functions: one for the cost of x apples, the other for the cost of x pounds of apples.

Graph the functions. When graphing, think about the domain of each function. Can it be any number or is it restricted in any way?

Which graph is a distinct set of points? Which graph is a line segment? Why?

Activity 4.2

Warm Up

For use before Activity 4.2

Find the domain and range of the function represented by the graph.

1.

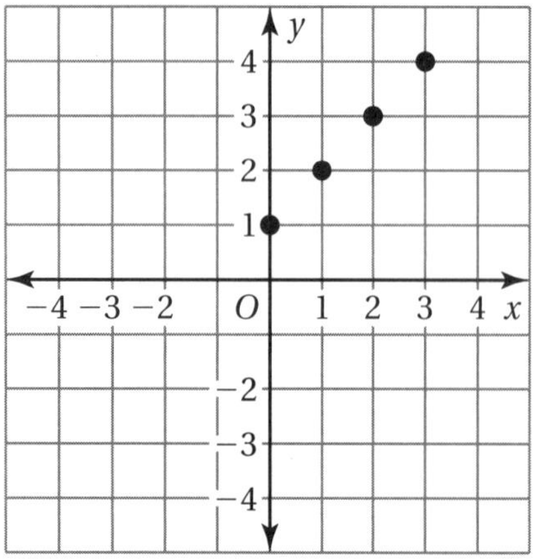

2.

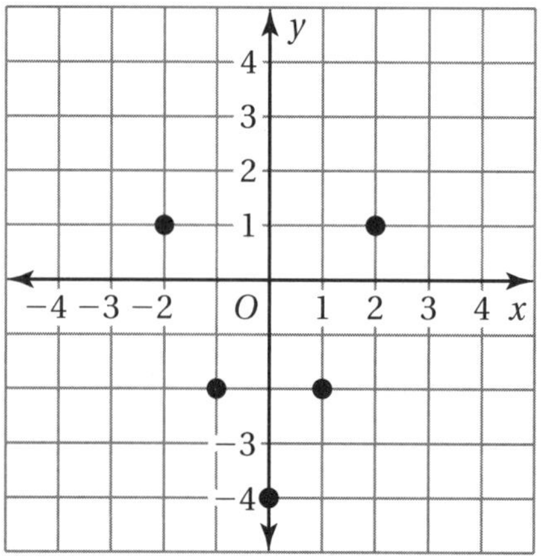

Lesson 4.2 Start Thinking!

For use before Lesson 4.2

Discuss whether the following functions have discrete or continuous domains.

A. The air temperature over the course of a day

B. The cost of hot dogs

C. The distance traveled on a road trip

D. The weight of a baby over his first month

E. The cost of parking for a certain number of hours at a parking garage

Lesson 4.2 Warm Up

For use before Lesson 4.2

Describe the domain and range of the function. Is the function discrete or continuous?

1.

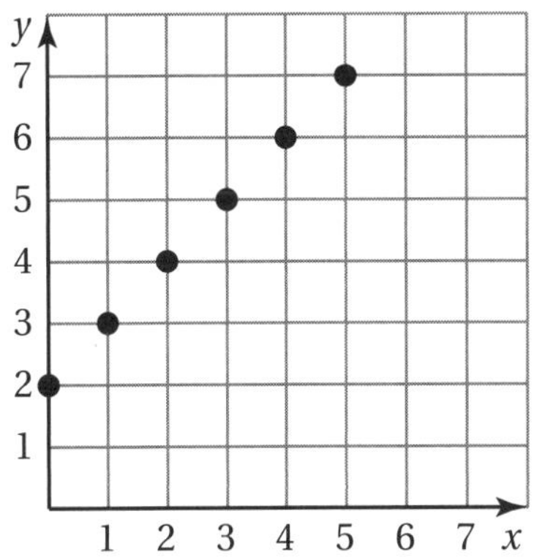

2.

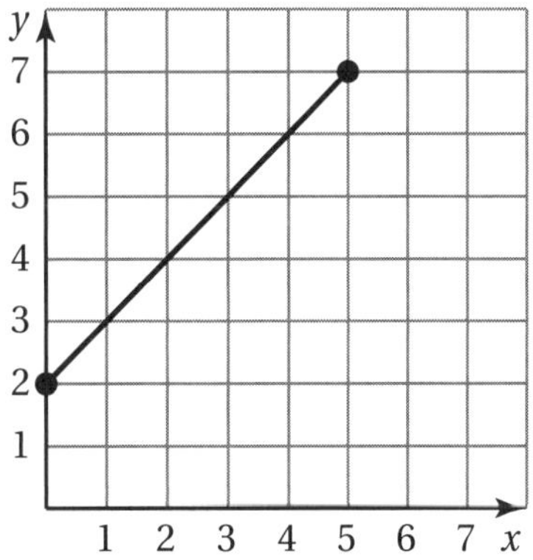

Name ______________________________ Date __________

4.2 Practice A

Describe the domain and range of the function. Is the domain discrete or continuous?

1.

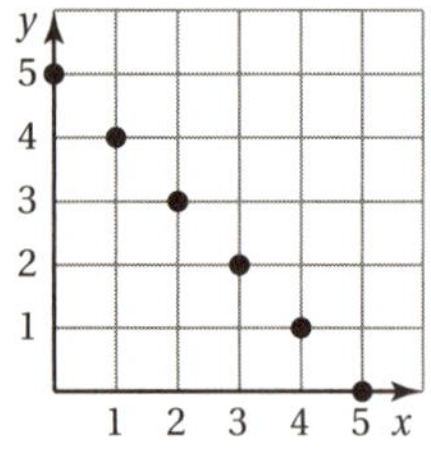

2.

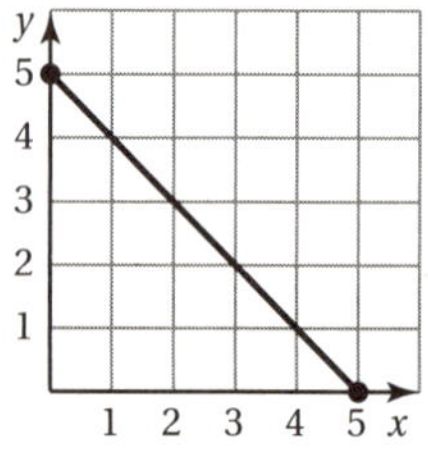

Graph the function. Is the domain discrete or continuous?

3.

Input Years, x	Output Weight of Cat, y
1	5
2	7
3	9

4.

Input Belts, x	Output Cost, y (dollars)
0	0
1	7.50
2	15

5. The function $b = 25 - 2.5n$ represents the balance b (in dollars) on a pre-paid phone card after making n phone calls. Graph the function using a domain of 0, 1, 2, and 3. Is the domain discrete or continuous?

6. The function $y = 1.6x$ converts distance from x miles to y kilometers. Graph the function. Is the domain discrete or continuous?

7. Describe and correct the error in classifying the domain.

✗ The domain is discrete.

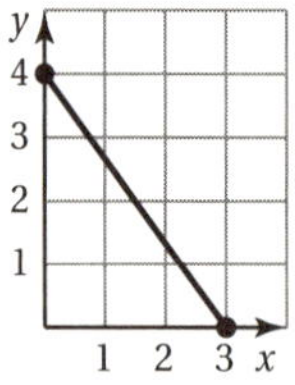

8. You spent $24 on surfboard wax and sunscreen. The function $y = -\frac{1}{3}x + 4$ relates x, the number of containers of surfboard wax you purchased, to y, the number of tubes of sunscreen you purchased.

a. Is 2 in the range? Explain.

b. Is 2 in the domain? Explain.

Name____ Date__________

4.2 Practice B

Describe the domain and range of the function. Is the domain discrete or continuous?

1.

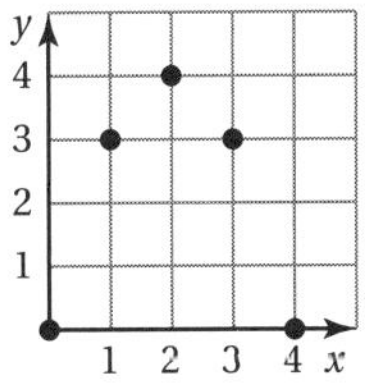

2.

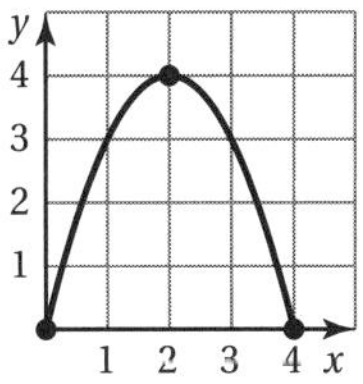

Graph the function. Is the domain discrete or continuous?

3.

Input Radius, *x*	Output Area, *y*
1	3.14
2	12.56
3	28.26

4.

Input Cell Phones, *x*	Output Cost, *y* (dollars)
1	45
2	54.99
3	64.98

5. The input of one function is number of accidents. The input of another function is time. Which function has a discrete domain? Explain.

6. The function $B = 212 - 0.0018x$ represents the boiling point of water B (in degrees Fahrenheit) at an altitude of x feet. Graph the function using a domain of 0, 10,000, 20,000, and 30,000. Is the domain discrete or continuous?

7. The circumference C of a circle is a function of its radius r.

a. Write the equation in function form.

b. Graph the function.

c. Is the domain discrete or continuous?

d. Is -2 in the domain? Explain.

e. Is 6 in the range? Explain.

Name ____ Date __________

4.2 Enrichment and Extension

Classifying Domains

Determine whether the domain of the graph is *continuous*, *discrete*, or *both*. Place the letter of the graph in the corresponding section of the diagram.

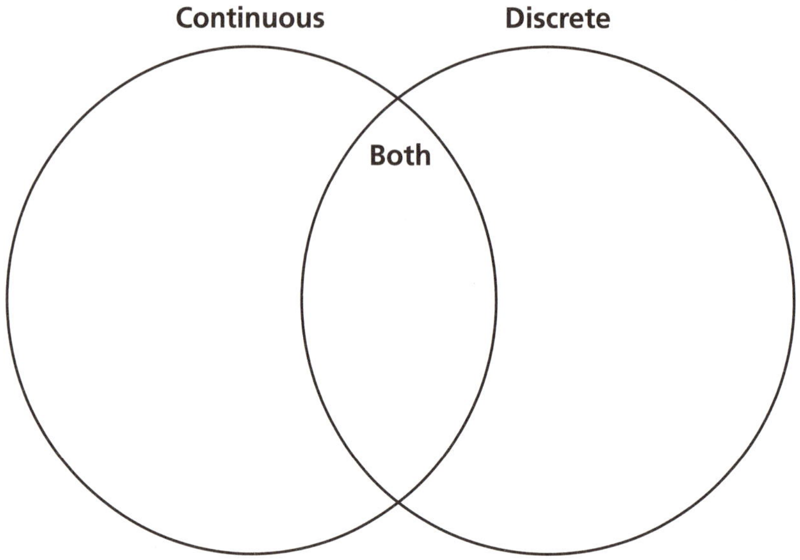

A. 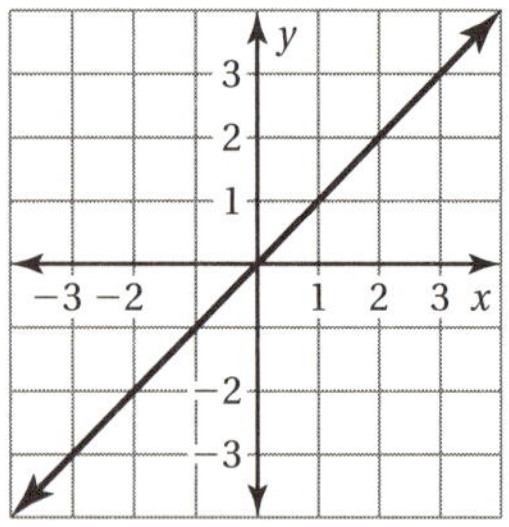

B. 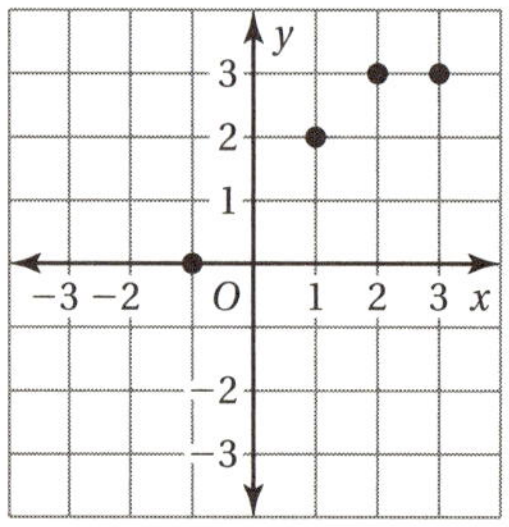

C. 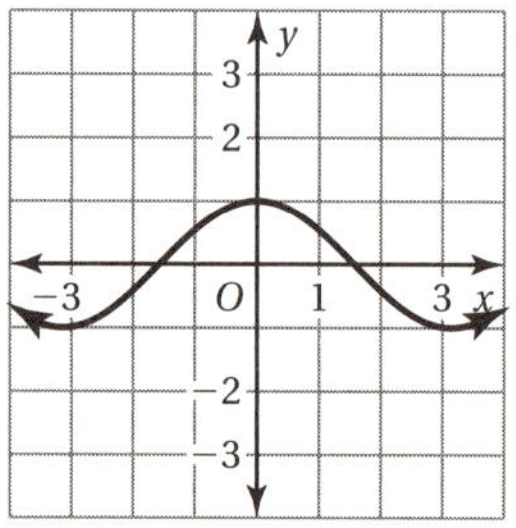

D. 

E. 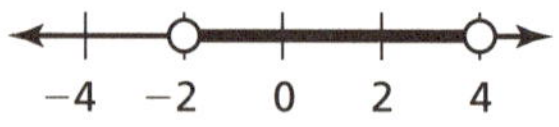

F. 

1. Describe the common characteristics of the graphs in the continuous circle.

2. Describe the common characteristics of the graphs in the discrete circle.

3. Are there any graphs in the intersection of the circles?

4. Can you construct an example of a graph that would appear in the intersection? Why or why not? Explain your reasoning.

Name____ Date__________

4.2 Puzzle Time

How Did the Gardner Feel At The End Of A Long Day?

Write the letter of each answer in the box containing the exercise number.

Determine whether the domain of the function is discrete or continuous.

1. A car traveling 65 miles per hour brakes at a steady rate to a stop. The function $y = -5t + 65$ represents the speed y of the car after t seconds.

   D. Discrete **E.** Continuous

2. A fitness club membership has a \$100 initiation fee plus a \$19 monthly fee. The function $y = 19x + 100$ represents the cost y (in dollars) of a membership after x months.

   S. Discrete **T.** Continuous

3. The function $y = 12x$ represents the cost y (in dollars) to hit x buckets of golf balls at a driving range.

   U. Discrete **V.** Continuous

4. The function $F = \frac{9}{5}C + 32$ converts temperature from Celsius C to Fahrenheit F.

   C. Discrete **D.** Continuous

5. The function $15 = 3x + 2y$ represents how many \$3 packs of colored pens x and \$2 drawing pads y you can buy spending exactly \$15.

   H. Discrete **I.** Continuous

6. The perimeter of a carpet is 32 feet. The function $32 = 2\ell + 2w$ represents the length ℓ and width w of the carpet.

   A. Discrete **B.** Continuous

6	3	2	5	1	4

Activity 4.3 Start Thinking!
For use before Activity 4.3

In the block pattern below, two blocks are added each step to form a pattern.

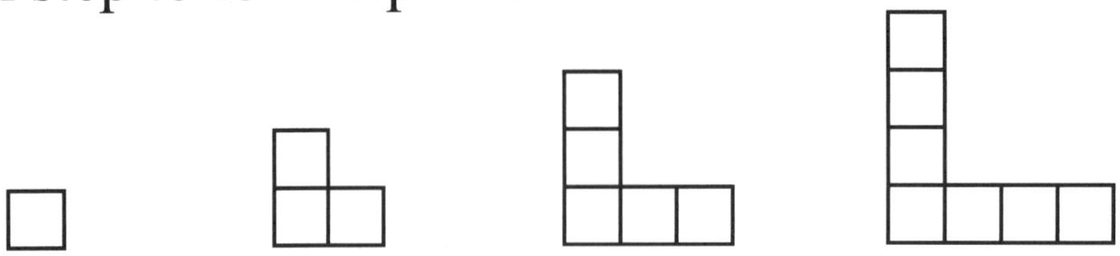

Step 1 **Step 2** **Step 3** **Step 4**

Use blocks to come up with a pattern in which the same number of blocks is added in each step. Sketch the first 4 steps in your pattern.

Complete the table. Then graph the function.

Step number, *x*				
Number of blocks, *y*				

Activity 4.3 Warm Up
For use before Activity 4.3

Write an equation of the line in slope-intercept form.

1.

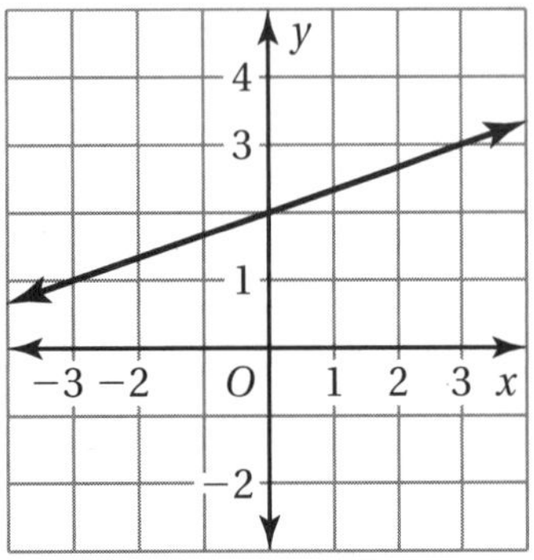

2.

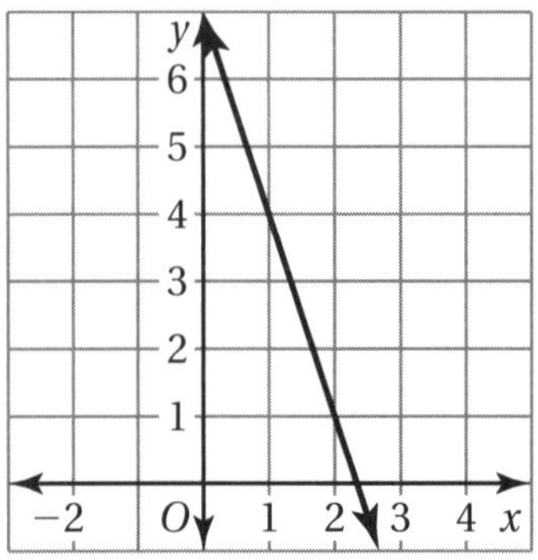

Lesson 4.3 Start Thinking!
For use before Lesson 4.3

Think about the term *linear function*.

What do you think it means?

Give an example.

Lesson 4.3 Warm Up
For use before Lesson 4.3

The table shows a familiar linear pattern from geometry.

x	1	2	3	4	5
y	4	8	12	16	20

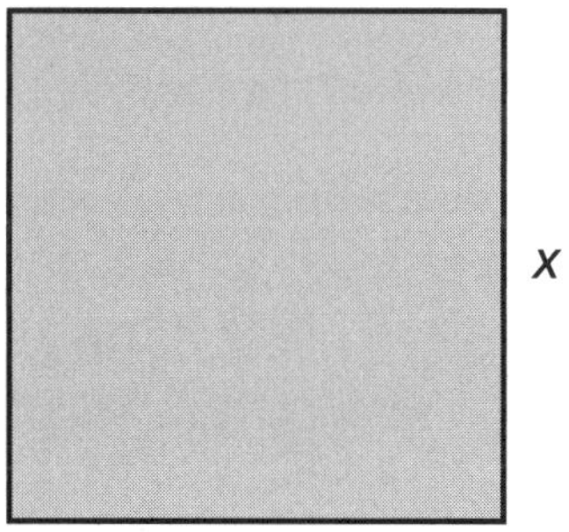

1. Write a linear function that relates y to x.

2. What do the variables x and y represent?

3. Graph the linear function.

Name ____ Date __________

4.3 Practice A

Use the graph or table to write a linear function that relates *y* to *x*.

1.

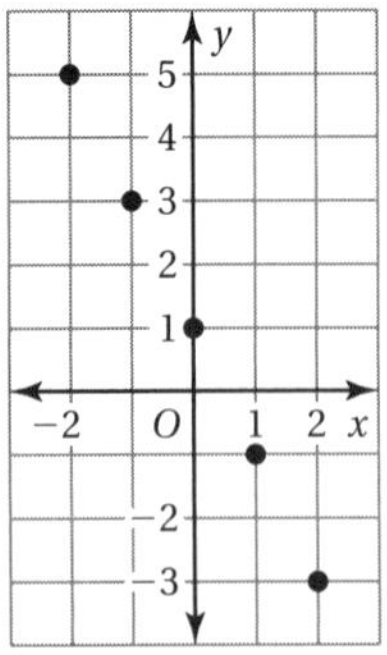

2.

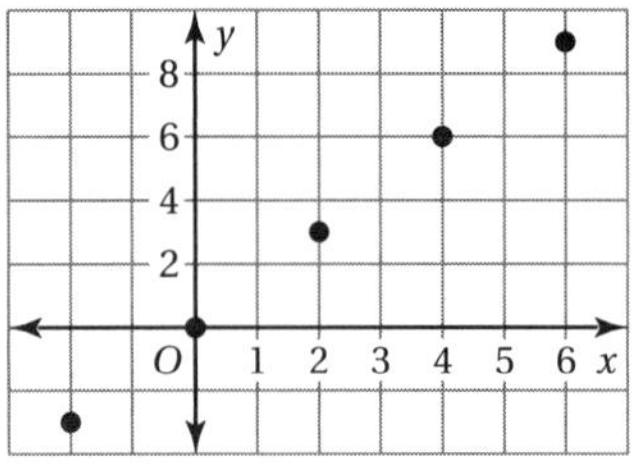

3.

x	–1	0	1	2
y	3	0	–3	–6

4.

x	–10	–5	0	5
y	–2	–1	0	1

5. The table shows the cost y (in dollars) of x ounces of brewed coffee.

Ounces, x	0	8	16	24
Cost, y	0	0.5	1	1.5

a. Graph the data.

b. Is the domain discrete or continuous?

c. Write a linear function that relates y to x.

d. How much does it cost to purchase 32 ounces of brewed coffee?

6. The table shows the area y (in square feet) of a triangle whose height is x feet.

Height, x	0	1	2	3
Area, y	0	3	6	9

a. Graph the data.

b. Is the domain discrete or continuous?

c. Write a linear function that relates the area of the triangle to the height of the triangle.

d. The formula for the area of a triangle is $A = \frac{1}{2}bh$. What is the length (in feet) of the base of the triangle?

Name____ Date__________

4.3 Practice B

Use the graph or table to write a linear function that relates *y* to *x*.

1.

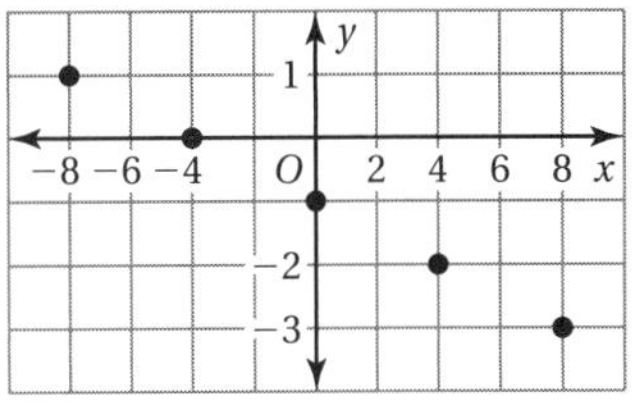

2.

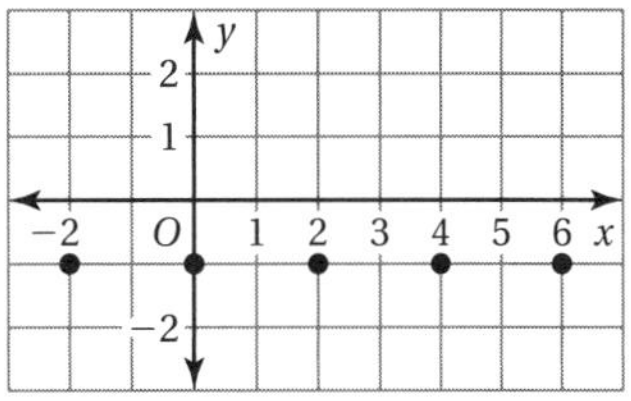

3.

x	−4	−2	0	2
y	8	4	0	−4

4.

x	−5	0	5	10
y	1	3	5	7

5. The table shows the time y (in minutes) it takes to make x burritos.

Burritos, *x*	1	2	3	4
Minutes, *y*	0.75	1.5	2.25	3

a. Graph the data.

b. Is the domain discrete or continuous?

c. Write a linear function that relates y to x.

d. How long does it take to make 7 burritos?

6. The table shows the distance traveled y (in miles) in a car in x hours.

Hours, *x*	0	2	4	6
Miles, *y*	0	128	256	384

a. Graph the data.

b. Is the domain discrete or continuous?

c. Write a linear function that relates distance traveled to hours.

d. What was the distance traveled in 5 hours?

e. The formula $d = rt$ relates distance with time for a given rate. Use the formula to determine the rate at which the car was traveling.

f. How long will it take to travel 400 miles?

Name ______________________________ Date __________

4.3 Enrichment and Extension

Finding and Creating a Pattern

A *fractal* is a geometric object that is self-similar. In other words, if you enlarge any part of the fractal, it will be similar to the original object you started with.

Fractals are produced by repeating the same set of steps multiple times. The number of times you have performed the procedure is call the *iteration number.*

This example is called the *Box Fractal.*

0 1 2 3 4 Iteration Number

1. Describe the procedure used to transform the box at iteration 0 into the box at iteration 1.

2. What is the relationship between the original box and each of the five boxes in iteration 1?

3. Copy and complete the table using the fractal pattern.

Iteration	Number of Squares
0	1
1	5
2	
3	
4	

4. What pattern do you notice in the table? Can you generalize the pattern?

5. Is the pattern in the table linear? Explain your reasoning.

6. How many iterations do you expect this pattern to continue for? Explain your reasoning.

Name______________________________________ Date__________

4.3 Puzzle Time

What Would You Get If You Crossed A Vampire With A Snowman?

Write the letter of each answer in the box containing the exercise number.

Use the graph or table to find a linear function that relates *y* to *x*.

1.

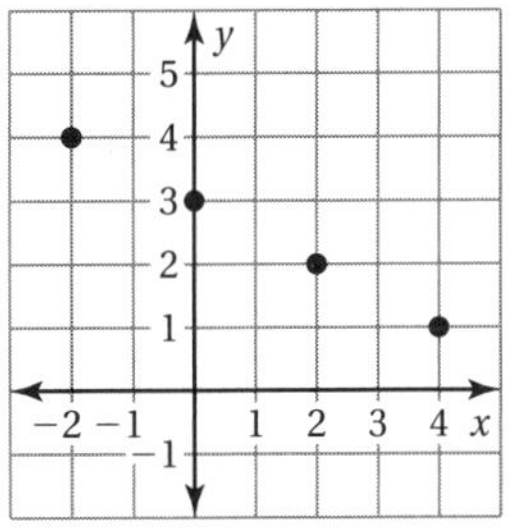

2.

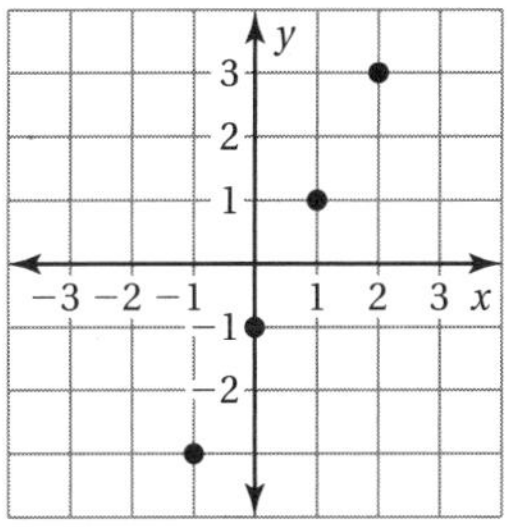

3.

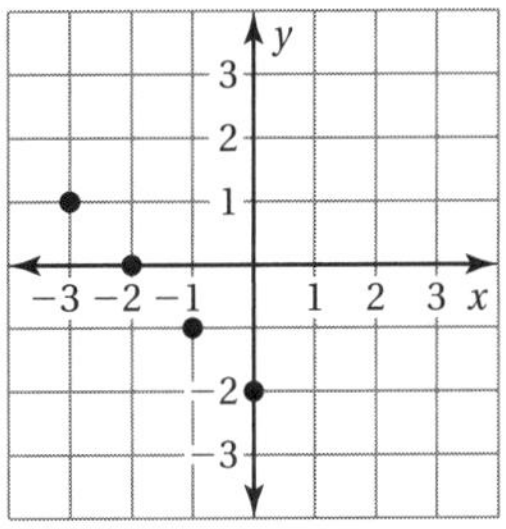

4.

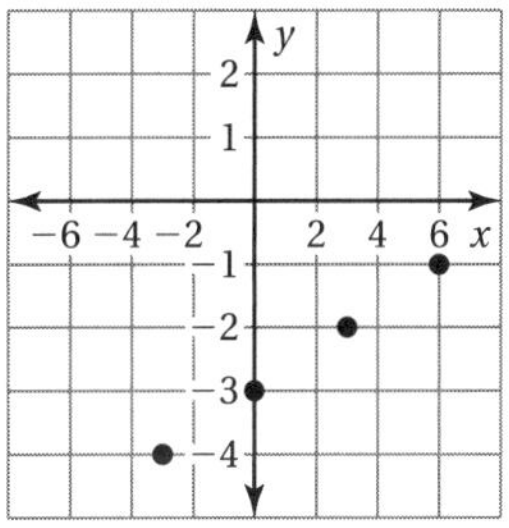

5.

x	−6	−2	2	6
y	2	4	6	8

6.

x	−4	−2	0	2
y	0	−2	−4	−6

7.

x	2	4	6	8
y	−2	1	4	7

8.

x	2	4	6	8
y	0	−2	−4	−6

9. The table shows how many cups of pancake mix x are needed to make y pancakes. Find the linear function that relates the number of pancakes to the number of cups of pancake mix.

Cups of mix, *x*	1	2	3	4
Pancakes, *y*	6	12	18	24

Answers

S. $y = 2x - 1$

R. $y = \frac{1}{2}x + 5$

E. $y = -\frac{1}{2}x + 3$

O. $y = \frac{1}{3}x - 3$

T. $y = \frac{3}{2}x - 5$

I. $y = -x - 2$

B. $y = 6x$

F. $y = -x + 2$

T. $y = -x - 4$

8	5	4	2	7	9	3	6	1

Activity 4.4 Start Thinking!
For use before Activity 4.4

Consider the following:

- The height of a bouncing ball
- The height of a person on an escalator
- The height of a baseball hit over an outfield fence
- The height of a roller coaster car

Draw a graph to represent each situation and then tell whether it is *linear* or *nonlinear*.

Activity 4.4 Warm Up
For use before Activity 4.4

Find the perimeter and area of the rectangle.

1.

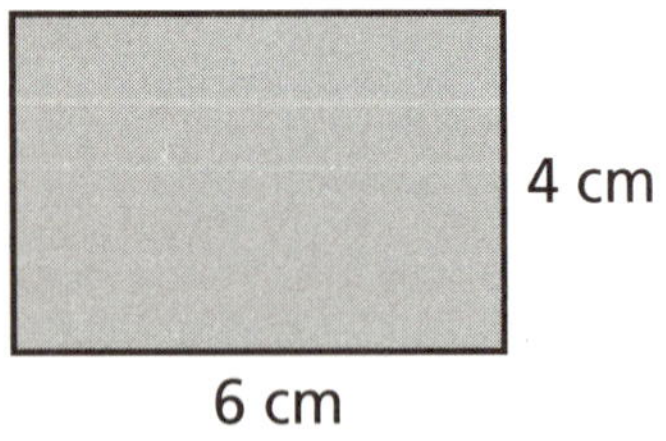

2.

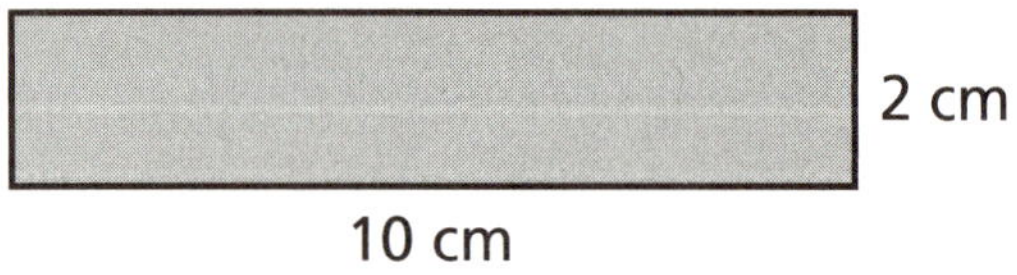

3.

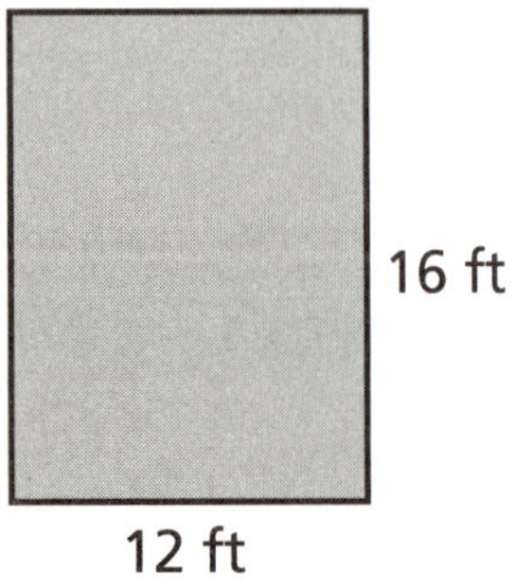

4.

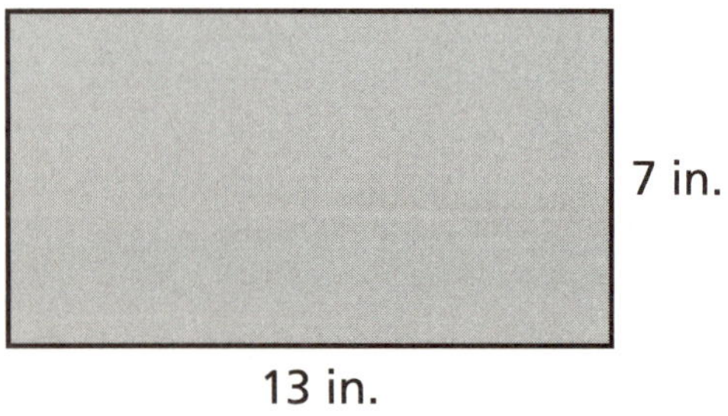

Lesson 4.4 Start Thinking!

For use before Lesson 4.4

Describe two real-life situations at an amusement park: one that can be represented by a linear function and one that cannot.

Lesson 4.4 Warm Up

For use before Lesson 4.4

Graph the data in the table. Decide whether the function is *linear* or *nonlinear.*

1.

x	0	1	2	3
y	6	4	2	0

2.

x	0	1	2	3
y	3	5	8	12

3.

x	0	1	2	3
y	15	25	20	30

4.

x	0	1	2	3
y	–7	–2	3	8

Name ______________________________ Date __________

4.4 Practice A

Graph the data in the table. Decide whether the function is *linear* or *nonlinear*.

1.

x	0	1	2	3
y	5	10	15	20

2.

x	1	2	3	4
y	4	6	9	13

Does the table or graph represent a *linear* or *nonlinear* function? Explain.

3.

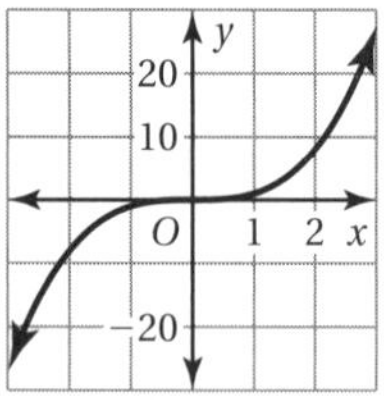

4.

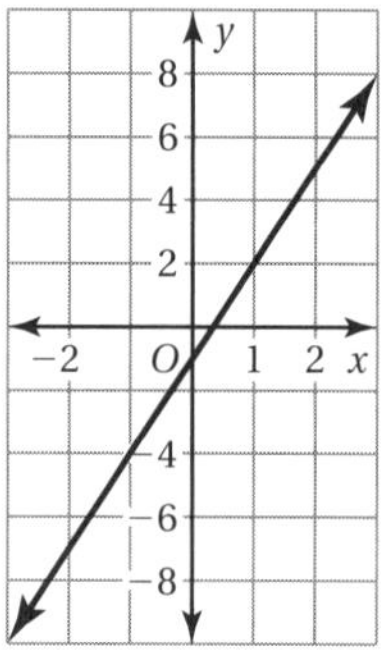

5.

x	3	5	7	9
y	5	3	0	3

6.

x	4	7	10	13
y	−2	0	2	4

7. The table shows the area A (in square centimeters) of a circle with radius r centimeters. Does the table represent a *linear* or *nonlinear* function? Explain.

Radius, *r*	1	2	3	4	5	6	7	8
Area, *A*	π	4π	9π	16π	25π	36π	49π	64π

8. The table shows the cost y (in dollars) of x ounces of cereal.

Ounces, *x*	8	12	16
Cost, *y*	?	2.5	3.5

a. What is a missing y-value that makes the table represent a nonlinear function?

b. What is the missing y-value that makes the table represent a linear function?

c. Write a linear function that relates cost y to ounces x.

Name____ Date__________

4.4 Practice B

Graph the data in the table. Decide whether the function is *linear* or *nonlinear*.

1.

x	4	3	2	1
y	1	3	7	11

2.

x	2	5	8	11
y	3	6	9	12

Does the graph or equation represent a *linear* or *nonlinear* function? Explain.

3.

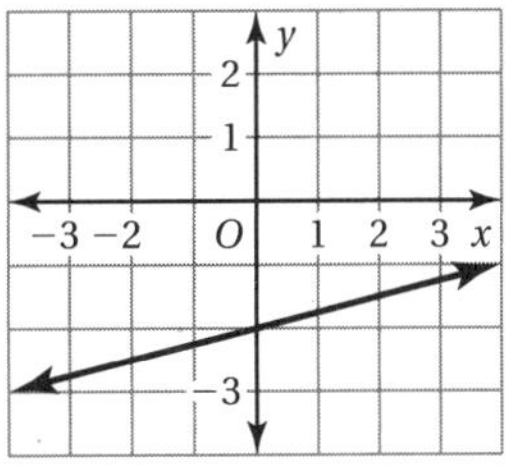

4.

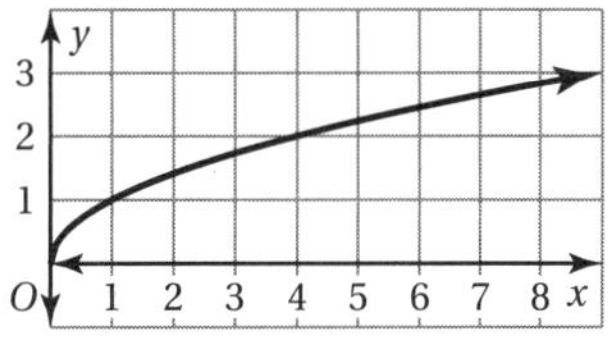

5. $y = \frac{3}{x} - 1$

6. $5x - y = 8$

7. The table shows the profit P (in dollars) of selling x pairs of flip flops. Does the table represent a *linear* or *nonlinear* function? Explain.

Flip Flops, *x*	1	2	3	4	5
Profit, *P*	4	8	12	16	20

8. The table shows the commission y (in dollars) of selling x cell phone plans.

Cell phone plans, *x*	1	2	3	4
Commission, *y*	100	150	250	400

a. Does the table represent a *linear* or *nonlinear* function? Explain.

b. Based on the pattern in the table, what is the commission of selling 5 cell phone plans?

9. The formula for the volume V of a sphere with radius r is $V = \frac{4}{3}\pi r^3$.

Does this formula represent a *linear* or *nonlinear* function? Explain.

Name ______________________________ Date __________

4.4 Enrichment and Extension

Direct and Inverse Variation

Recall that two quantities x and y show direct variation when $y = kx$, where k is a number and $k \neq 0$. Also, two quantities x and y show inverse variation when $y = \frac{k}{x}$, where k is a number and $k \neq 0$.

Air Over a Bottle

When air is blown over a bottle, sound is produced with a frequency F (in hertz) given by the equation $Fr = 775$ where r represents the radius of the bottle base (in centimeters).

1. Solve the equation for F.

2. What is the value of k in the equation?

3. Complete the table using the equation you found in Exercise 1. Then graph the data.

Radius (cm), r	Frequency (hertz), F
5	
10	
15	
20	
25	

4. Describe the equation from Exercise 1. Does the equation show *direct variation* or *inverse variation*? Explain your reasoning.

5. How does the graph of the data support your conclusion in Exercise 4?

6. Are there any radius values that do not make sense to evaluate in the equation? How do you know?

7. What is the domain of the equation? Will the domain be the same for any inverse variation equation?

Name______________________________ Date__________

4.4 Puzzle Time

What Belongs To You, But Is Used More By Other People?

Write the letter of each answer in the box containing the exercise number.

Choose the equation that represents a *nonlinear* function.

1. **N.** $y + x = 2x - 1$ **O.** $xy = 2x - 1$ **P.** $2y + 1 = 2x$

2. **P.** $3y = 4x + 3$ **Q.** $x - y = 4$ **R.** $y = 3x^2 + 4$

3. **A.** $5y = \frac{7}{x}$ **B.** $7y = 5x$ **C.** $5y + x = 7x$

4. **C.** $y = 6\pi$ **D.** $y = 6\pi x$ **E.** $y = 6\pi x^2$

Choose the missing *y*-value that makes the points represent a linear function.

5. (1, 15), (2, 19), (3, ?)

   U. 23 **V.** 24 **W.** 25

6. (−3, 9), (−2, ?), (−1, 7)

   K. 12 **L.** 6 **M.** 8

7. (4, ?), (7, −3), (10, −3)

   X. −2 **Y.** −3 **Z.** −4

8. (25, 360), (40, 320), (55, ?)

   L. 300 **M.** 380 **N.** 280

7	1	5	2		8	3	6	4

Name ______________________________ Date __________

Chapter 4

School-to-Work

For use after Section 4.3

Sales Associate

You are a sales associate at a retail clothing store. The sales manager has announced that today the store is having a clearance event—everything in the store will be discounted by 20% when rung up at the registers.

1. To help promote the clearance event, the sales manager has asked that you make a sign that lists the original prices and corresponding sale prices for different priced items in the store. Make such a sign for the items shown below.

   Jeans: \$39 T-shirts: \$15 Sweaters: \$42 Skirts: \$28

2. Write the sale price s as a function of the original price p. Graph the function on a coordinate plane. Is the function discrete or continuous? Explain.

3. Use your graph to find the sale price for a sweatshirt that originally cost \$35. Check your answer algebraically.

4. A customer decides to buy a pair of jeans. At the register, the customer produces a coupon for \$5 off any clothing item. You subtract the \$5 from the sale price of the jeans. What is the final cost of the jeans? For a customer with a coupon that is applied after the 20% discount, what function describes the final price of an item in terms of the original price?

5. Does it matter whether the \$5 coupon is applied before or after the 20% discount? Explain your reasoning. What function describes the final price of an item when the coupon is applied first?

Name ______________________________ Date ____________

Chapter 4 Study Help

You can use a **comparison chart** to compare new topics.

On Your Own

Make a comparison chart to help you study and compare these topics.

1. discrete data and continuous data

After you complete this chapter, make comparison charts for the following topics.

2. linear functions with positive slopes and linear functions with negative slopes

3. linear functions and nonlinear functions

Name ______________________________ Date __________

Chapter 4 Cumulative Practice

Describe the pattern of inputs *x* and outputs *y*.

1.

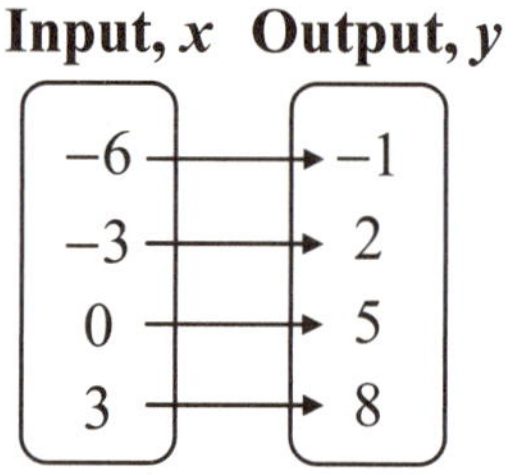

2.

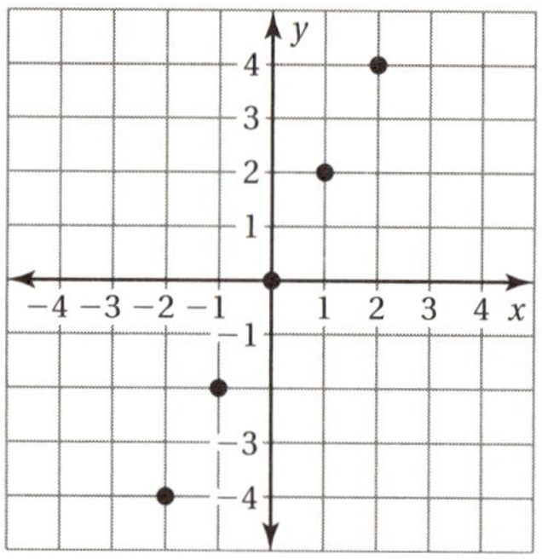

3.

Input, *x*	2	1	0	−1
Output, *y*	−6	−3	0	3

Use the table or graph to write a linear function that relates *y* to *x*. For the graphs, determine if the domain is discrete or continuous.

4.

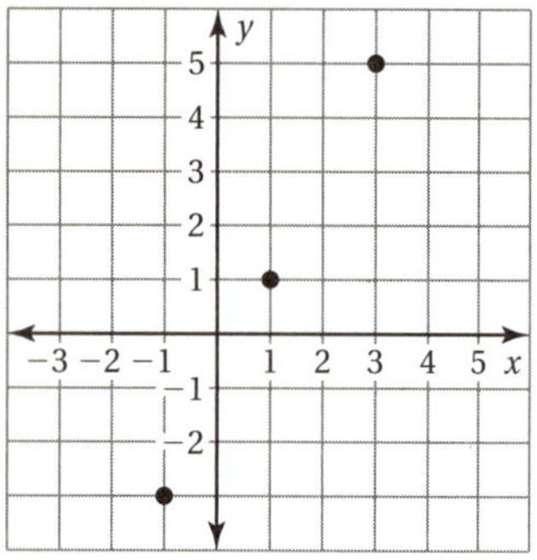

5.

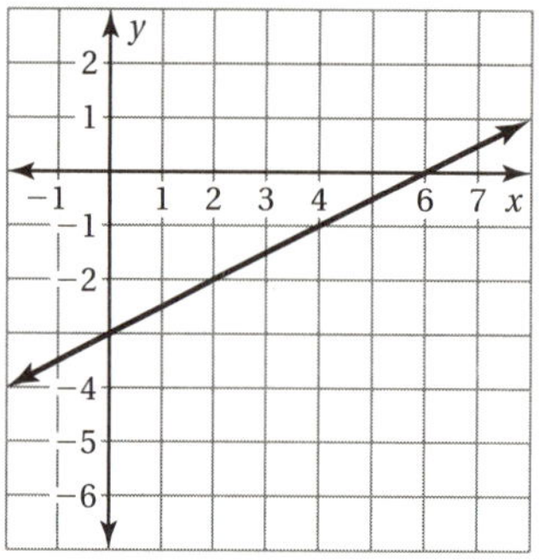

6.

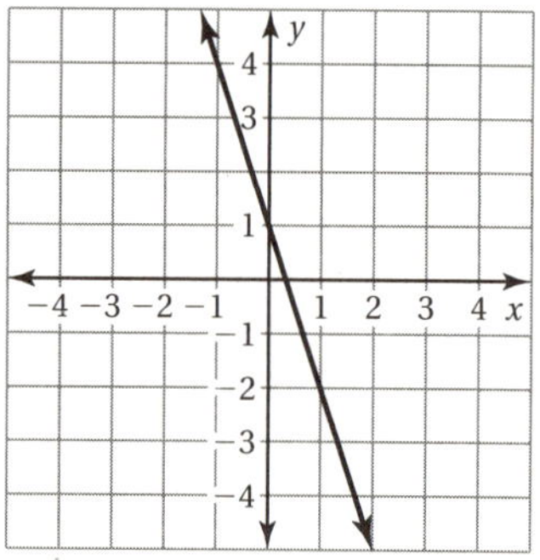

7.

x	1	3	5	7
y	−6	−4	−2	0

8.

x	−4	−2	0	2
y	3	2	1	0

9.

x	0	3	6	9
y	4	6	8	10

10. The equation for the area of a circle with a radius of r is $A = \pi r^2$. Is the area of a circle a linear function of the radius of the circle? Explain.

11. The function $m = 20 - \frac{3}{4}b$ represents the amount m of money you have left after buying b bookmarks. Is the domain of the function discrete or continuous?

Name____ Date __________

Unit 1 Project: Basketball Accuracy

For use after Unit 1

Objective **Analyze and model how your ability to make a basket changes with distance.**

Materials Graph paper, a yardstick, a wastebasket, and paper wadded-up to form a ball

Investigation

1. Work with two other people. One person will shoot baskets, one will measure the distance from the basket, and one should record the number of baskets made.

2. Stand 0 yards from the wastebasket. Shoot your paper ball into the basket 10 times and record the number of shots that go into the basket.

Distance from basket (yd)	Number of shots that go into the basket (out of 10 shots)
0	
1	
2	

3. Step back 1 yard. Shoot your paper ball into the basket 10 times from this distance and record the number of balls that go into the basket.

4. Repeat Step 3 until you reach a distance where none of your shots go into the basket. After one person finishes shooting baskets, switch roles and repeat the experiment until everyone has a chance to shoot baskets. (There should be one table of data for each person.)

Data Analysis

5. Write the ordered pairs represented by your data. Use the distance from the basket as the input.

6. Describe the domain and range of the function represented by your data. Is the domain *continuous* or *discrete*? Explain.

7. Graph the ordered pairs on a graph. What does the y-intercept represent in this situation? What does the x-intercept represent?

8. Use two of the ordered pairs you wrote to write an equation of a line to represent your data. Graph the equation on the same graph as the ordered pairs. Do you think your line represents the data well? Why or why not?

9. Is the rate of change in the number of shots that go into the basket *constant* or *not constant*? Explain.

Written Report Write a report about your experiment that includes your data, graph, models, and explanations. You may also want to use your data to describe how a basketball player could use this experiment to decide whether to try for a three-point shot. Why do you think the three-point rule was created? How far away would you put a "three-point line" from the wastebasket based on your data? Explain.

Name ______________________________ Date __________

Unit 1 Student Grading Rubric

For use after Unit 1

		Student Score	Teacher Score
Cover Page 10 points			
	a. Name (4 points)	______	______
	b. Class (2 points)	______	______
	c. Project Name (2 points)	______	______
	d. Due Date (2 points)	______	______
Investigation 40 points			
	a. Complete the table until the number of shots that go into the basket is zero. (20 points)	______	______
	b. Measure the distances from the basket for another person in your group. (10 points)	______	______
	c. Record the data for another person in your group. (10 points)	______	______
Data Analysis 110 points			
	a. Write the ordered pairs represented by your data. (20 points)	______	______
	b. Describe the domain and range of the function and explain if the domain is continuous or discrete. (20 points)	______	______
	c. Graph the ordered pairs. Describe what the y-intercept and x-intercept represent. (30 points)	______	______
	d. Write an equation of a line to represent the data. Graph the equation. Explain whether the line represents the data well. (30 points)	______	______
	e. Describe the rate of change and provide an explanation of how you arrived at your conclusion. (10 points)	______	______
Written Report 40 points			
	a. Report includes data, graphs, models, and explanations. (30 points)	______	______
	b. Report is neat and clearly written. (5 points)	______	______
	c. Includes an analysis of the three-point shot (5 points)	______	______
FINAL GRADE		______	______

Unit 1

Teacher's Project Notes

For use after Unit 1

Materials Graph paper, a yardstick, enough wastebaskets for every three students, scrap paper to use to wad-up in a ball

Alternatives You could do the investigation with a real basketball and net, but the starting point should be a little further away. The data with a real basketball and net will not be as consistent as with the wastebasket model, as it requires more ability.

After completing the wastebasket investigation, you can ask students to predict their results for a real basketball and net. Then repeat the experiment using a real basketball and net.

Common Errors Students may write the ordered pairs in the opposite order. Make sure they realize the number of baskets is *dependent* on the distance from the wastebasket, so the distance from the wastebasket is the input.

Students may think that the domain of the function is discrete, because they only found integer distances from the wastebasket. Ask them if fractional distances make sense in this situation to help them realize that the function is continuous.

Suggestions You may want to have the students in a group combine their data and analyze the impact this has on the data. Does it change the answers to any of their questions?

You may want to work with the class to find a linear model and an exponential model for a set of data using the regression feature of a graphing calculator. Ask students which model they think best fits the data and why.

Unit 1

Grading Rubric

For use after Unit 1

Scoring Rubric	
A	179-200
B	159-178
C	139-158
D	119-138
F	118 or below

Cover Page 10 points

a. Name (4 points)

b. Class (2 points)

c. Project Name (2 points)

d. Due Date (2 points)

Investigation 40 points

a. Complete the table until the number of shots that go into the basket is zero. (20 points)

b. Measure the distances from the basket for another person in your group. (10 points)

c. Record the data for another person in your group. (10 points)

Data Analysis 110 points

a. Write the ordered pairs represented by your data. (20 points)

b. Describe the domain and range of the function and explain if the domain is continuous or discrete. (20 points)

c. Graph the ordered pairs. Describe what the y-intercept and x-intercept represent. (30 points)

d. Write an equation of a line to represent the data. Graph the equation. Explain whether the line represents the data well. (30 points)

e. Describe the rate of change and provide an explanation of how you arrived at your conclusion. (10 points)

Written Report 40 points

a. Report includes data, graphs, models, and explanations. (30 points)

b. Report is neat and clearly written. (5 points)

c. Includes an analysis of the three-point shot (5 points)

Chapter 5

Name ______________________________ Date ________

Chapter 5 Angles and Similarity

Dear Family,

Take a look at the structures in your neighborhood. What shape do you see over and over again? Triangles can be seen everywhere. Peaked roofs are made with triangular sections; umbrellas use triangular wedges; bridges use triangular shapes to span great distances.

The triangle is used in many structures because it is a stable shape. There is only one way to form a polygon with three equal segments and this makes the shape strong. A square can shift into a parallelogram. A circle can be deformed into an oval shape without changing the perimeter. But a triangle retains its shape.

You may want to use toothpicks and gum drops to create a triangle, square, and rectangle. Test the strength of each shape to reinforce that the triangle provides the most support.

Triangles can be used to form the sides of pyramids or the ends of prisms, making them popular for building roofs. More complicated patterns can be used to approximate nearly any shape. A geodesic dome uses triangles to create a sphere-like shape. Try this with your student. Choose a structure, such as your home or a nearby building or bridge.

- What types of triangles are used in the structure? Are they all similar, or are different triangles used?
- What types of angles are used? Are they acute (less than 90 degrees), right (equal to 90 degrees), or obtuse (greater than 90 degrees)? Why do you think those angles are chosen?
- Are there triangular parts inside of the structure? Try looking at the trusses in an attic or under a bridge. Why do you think the shapes you see were chosen?

You may want to build a model of your structure using toothpicks and gum drops or other similar materials.

Have fun looking at your structure from "every angle"!

Nombre ____ Fecha _________

Capítulo 5 Ángulos y Similitudes

Estimada Familia:

Observe las estructuras en su barrio. ¿Qué formas observa una y otra vez? Los triángulos se pueden ver por todas partes. Los techos puntiagudos están compuestos de secciones triangulares; los paraguas utilizan pedazos triangulares, los puentes usan formas triangulares para cubrir grandes distancias.

El triángulo se utiliza en muchas estructuras porque es una forma estable. Sólo hay un modo para formar un polígono con tres segmentos iguales y esto hace que la forma sea fuerte. Un cuadrado puede convertirse en un paralelogramo. Un círculo puede deformarse en una forma ovalada sin cambiar el perímetro. Pero un triángulo conserva su forma.

Puede querer usar palitos de dientes y gomitas dulces para crear un triángulo, un cuadrado y un rectángulo. Ponga a prueba la resistencia de cada forma para comprobar que el triángulo brinda el mayor soporte.

Los triángulos pueden ser usados para formar los lados de las pirámides o las terminaciones de un prisma, haciéndolos populares para construir techos. Los patrones más complicados pueden ser usados para aproximarse casi a cualquier forma. Un domo geodésico utiliza un triángulo para crear la forma de una esfera. Intente lo siguiente con su estudiante. Escojan una estructura, como su casa o un edificio o puente cercano.

- ¿Qué tipos de triángulos son usados en esta estructura? ¿Son todos similares o se utilizan distintos tipos de triángulos?
- ¿Qué clase de ángulos se utilizan? ¿Son agudos (menores de 90 grados), rectos (iguales a 90 grados) u obtusos (mayores de 90 grados)? ¿Por qué creen que se escogen estos ángulos?
- ¿Hay partes triangulares dentro de la estructura? Observen las estructuras en un ático o bajo un puente. ¿Por qué creen que las formas que observan fueron escogidas?

Querrán construir un modelo de sus estructuras usando palitos de dientes, gomitas dulces o materiales similares.

¡Diviértanse observando su estructura desde "todos los ángulos"!

Non ______________________________ Dat __________

Chapít 5 Ang ak Similarite

Chè Fanmi:

Gade estrikti ki nan katye ou a. Ki fòm ou wè toutan? Ou kapab wè triyang tout kote. Twa pwenti fèt ak seksyon triyangilè; parapli sèvi ak kal triyangilè; pon sèvi ak fòm triyangilè pou janbe gwo distans.

Triyang nan sèvi nan anpil estrikti poutèt se yon fòm estab. Gen sèlman yon sèl jan pou fòme yon poligòn avèk twa segman egal donk sa fè fòm nan solid. Yon kare kapab chanje an paralelogram. Yon sèk kapab defòme an fòm oval san perimèt la pa chanje. Men triyang la kenbe fòm li.

Ou gen dwa vle sèvi avèk kidan ak boul gòm pou kreye yon triyang, yon kare, ak yon rektang. Teste solidite chak fòm pou ranfòse ke triyang nan ofri plis sipò.

Triyang kapab sèvi pou fòme bò piramid oswa pwent pris, sa ki fè yo popilè pou bati. Motif pi konplike kapab sèvi pou pwoche prèske tout fòm. Yon donm jeyodezik sèvi ak triyang pou kreye yon fòm ki sanble yon esfè. Eseye sa avèk elèv ou a. Chwazi yon estrikti, tankou lakay ou oswa yon bilding oswa pon ki toupre a.

- Ki tip triyang yo itilize nan estrikti a? Èske yo tout sanble, oubyen yo itilize diferan triyang?
- Ki tip ang yo itilize? Èske yo egi (mwens pase 90 degre), dwat (egal 90 degre), oubyen obti (pi gran pase 90 degre)? Dapre ou poukisa yo chwazi ang sa yo?
- Èske gen pati triyangilè anndan estrikti a? Eseye gade chapant la nan yon grenye oswa anba yon pon. Dapre ou poukisa yo te chwazi fòm ou wè yo?

Nou gen dwa vle konstwi yon modèl estrikti nou an avèk kidan ak boul gòm oswa lòt materyèl sanblab.

Amize nou nan gade estrikti nou an nan "tout ang"!

Activity 5.1 Start Thinking!

For use before Activity 5.1

What is an acute angle?

What is an obtuse angle?

Use a dictionary to look up the non-math definitions of *acute* and *obtuse*. How do the definitions relate to the math definitions?

Activity 5.1 Warm Up

For use before Activity 5.1

Graph the linear function.

1. $y = -x + 3$

2. $y = 2x - 2$

3. $y = \frac{1}{2}x - 1$

4. $y = \frac{1}{3}x + 2$

5. $y = -\frac{3}{4}x + 2$

6. $y = \frac{2}{3}x + 4$

Lesson 5.1 Start Thinking!

For use before Lesson 5.1

Complete the statement.

Two angles are ___?___ if the sum of their measures is 90°.

Two angles are ___?___ if the sum of their measures is 180°.

People often have trouble remembering which is 90° and which is 180°. Make up your own way to help you remember the definitions.

Lesson 5.1 Warm Up

For use before Lesson 5.1

Tell whether the statement is *always*, *sometimes*, or *never* true. Explain.

1. If x and y are supplementary angles, then x is right.

2. If x and y and complementary angles, then y is acute.

3. If x is a right angle and y is an acute angle, then x and y are supplementary angles.

4. If x is acute and y is obtuse, then x and y are supplementary angles.

Name______________________________________ Date__________

5.1 Practice A

Tell whether the statement is *always*, *sometimes*, or *never* true. Explain.

1. If x and y are supplementary angles, then y is __?__ acute.

2. If x and y are complementary angles, then x is __?__ obtuse.

Tell whether the angles are *complementary*, *supplementary*, or *neither*.

3. 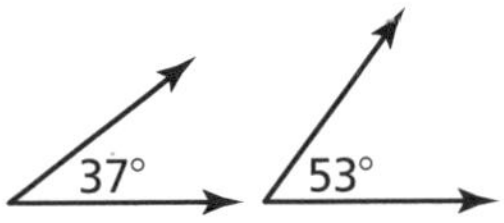

4. 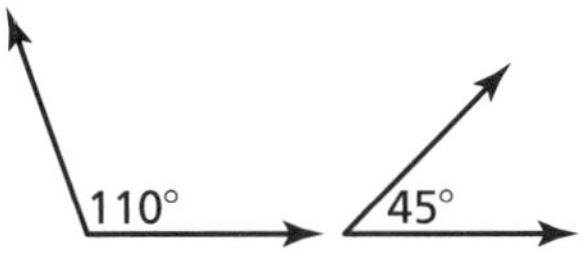

5. 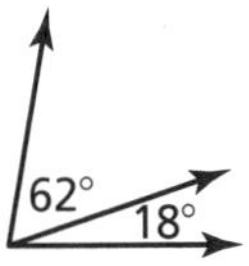

6. 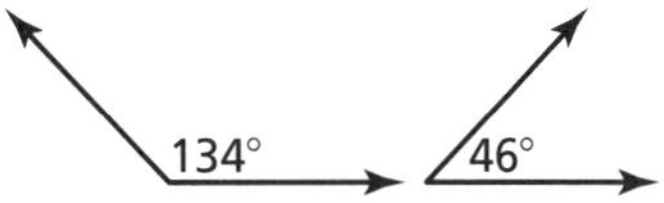

Find the value of *x*.

7. 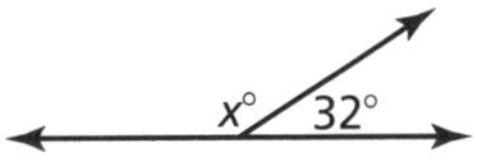

8. 

9. 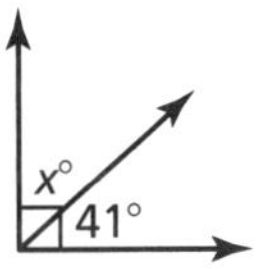

10. 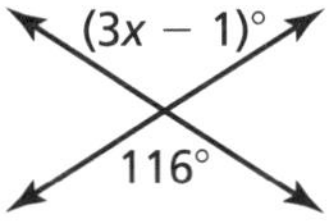

11. Angle x and angle y are complementary. Angle x is supplementary to a 128° angle. What are the measures of angle x and angle y?

Name ______________________________ Date __________

5.1 Practice B

Tell whether the angles are *complementary*, *supplementary*, or *neither*.

1.

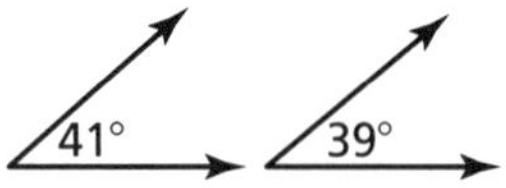

2.

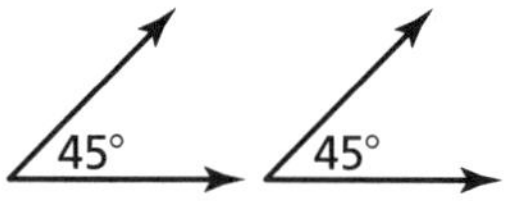

3.

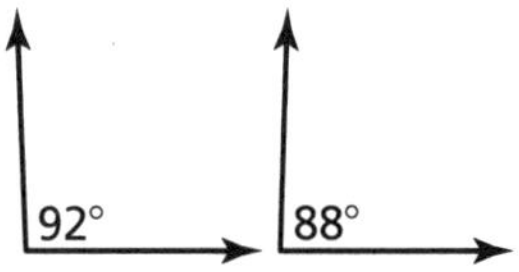

4.

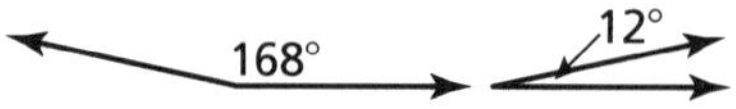

Find the value of *x*.

5.

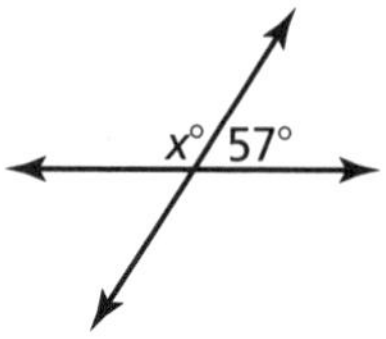

6.

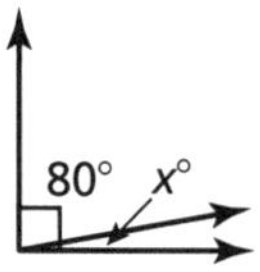

7.

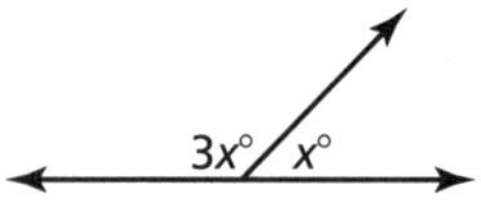

8.

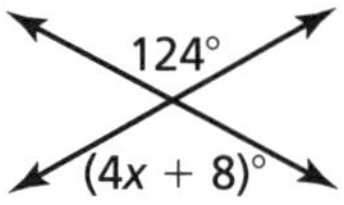

9. The measures of two supplementary angles have a ratio of 2 : 4. What is the measure of the smaller angle?

10. Let x be an angle measure. Let c be the measure of the complement of the angle and let s be the measure of the supplement of the angle. Write equations for c as a function of x, and for s as a function of x. What is the domain of each function? What is the range of each function? Explain.

Name________________________________ Date__________

5.1 Enrichment and Extension

Compass Straightedge Construction

In addition to modern technology, a geometer's most important tools are a compass and a straightedge (ruler without marks). These instruments can be used for geometric drawings called constructions.

Bisecting the Angle

Bisecting the angle is the process of cutting a given angle in half so that both halves are equal. This can be done using only a compass and straightedge.

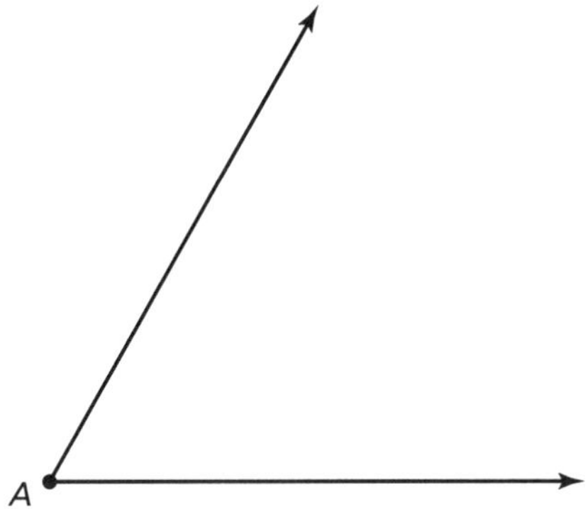

1. Classify angle A.

2. Open the compass. Place the point of the compass at A and draw an arc (a small curved line that is part of a circle) that intersects both sides of the angle. Label the points where the curved line intersects the angle as B and C.

3. Place the point of the compass on B. Draw an arc that is contained within the angle. Without changing the width of the compass, place the point of the compass on C and draw an arc that intersects the one just drawn.

4. Label the intersection point of the two arcs G. Use a straightedge to connect points A and G. The line segment AG bisects angle A.

5. Use a protractor to measure angle A. Divide this measure by two to determine what each smaller angle should measure. Then, use your protractor and measure the smaller angles. How do the numbers compare?

Name ____________________ Date ________

5.1 Puzzle Time

What Runs Around All Day And Lies Under The Bed With Its Tongue Hanging Out?

Write the letter of each answer in the box containing the exercise number.

Tell whether the angles are *complementary*, *supplementary*, or *neither*.

1.

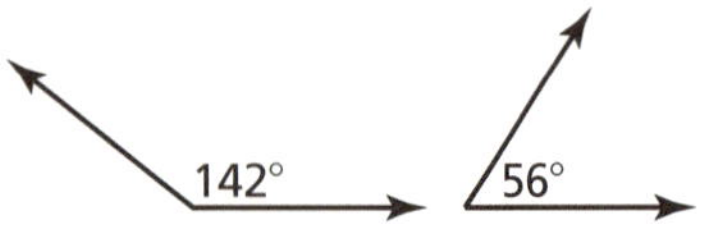

2.

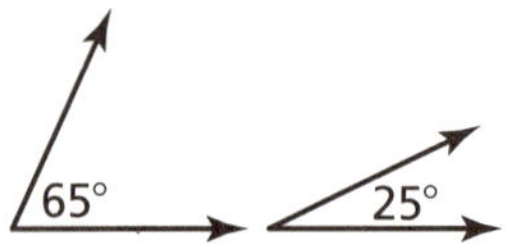

3.

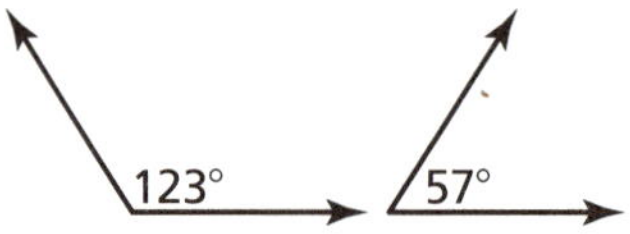

Find the value of *x*.

4.

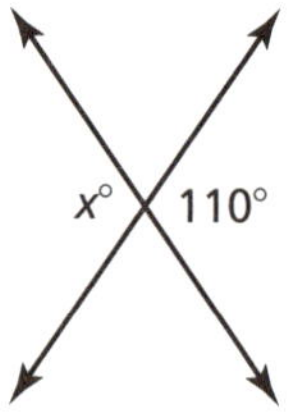

5.

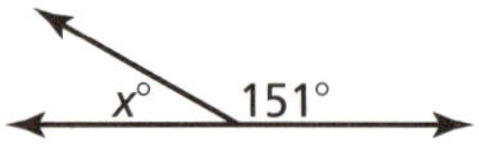

6.

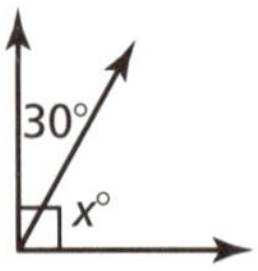

7.

8. A road intersects another road at an angle of 45°. Find the value of x.

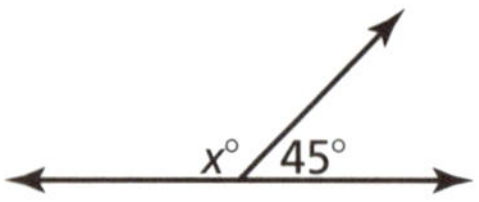

Answers

N. 29

K. complementary

A. 160

S. neither

E. 60

R. supplementary

E. 135

A. 110

7		1	5	8	4	2	6	3

Activity 5.2 Start Thinking!
For use before Activity 5.2

Describe some triangles in real-life. What kind of triangles are they?

Activity 5.2 Warm Up
For use before Activity 5.2

Classify the angle as *acute*, *right*, *obtuse*, or *straight*.

1. 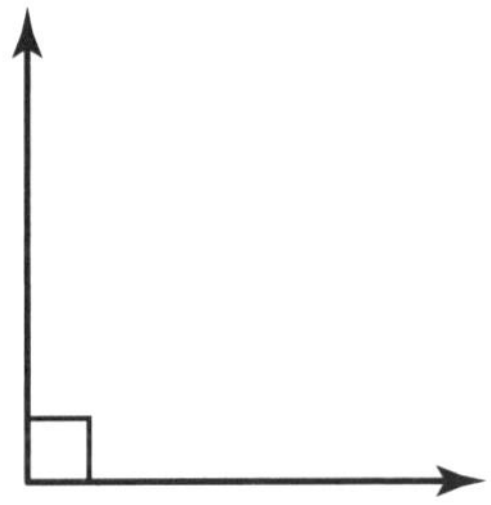

2. 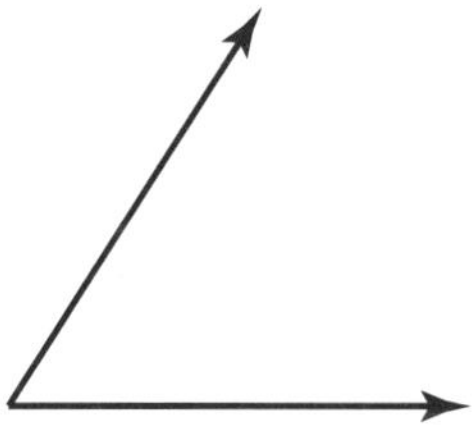

3. 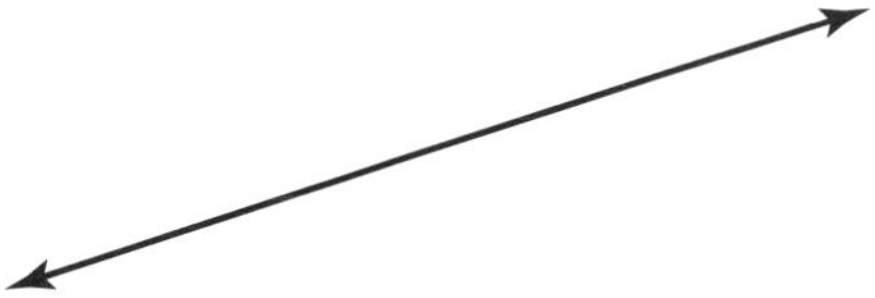

4. 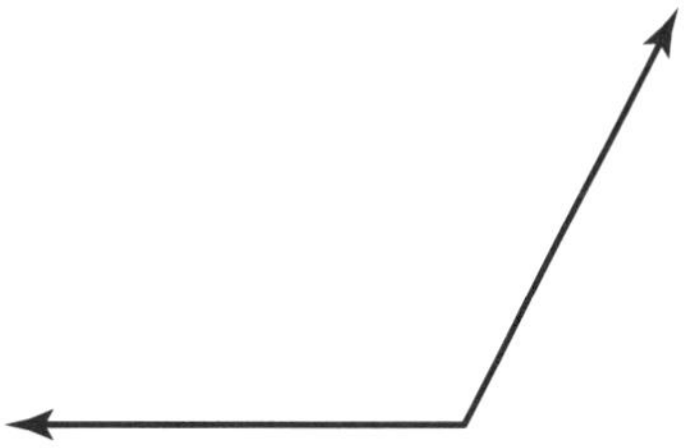

Lesson 5.2 Start Thinking!
For use before Lesson 5.2

In real life, when is it important to know the angle measures of a triangle?

Lesson 5.2 Warm Up
For use before Lesson 5.2

Classify the triangle in as many ways as possible.

1.
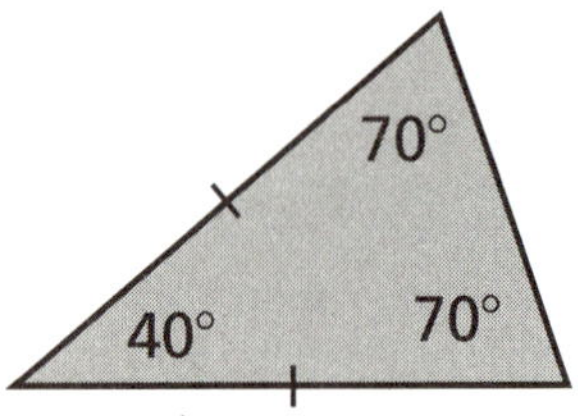

2.
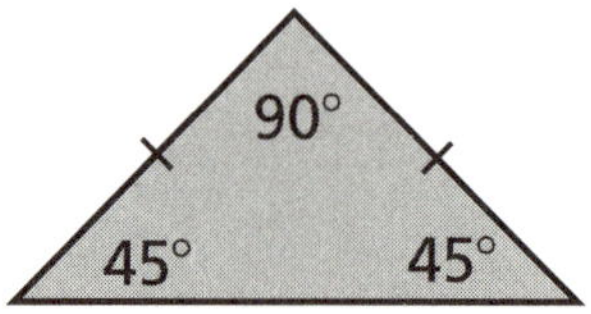

3.
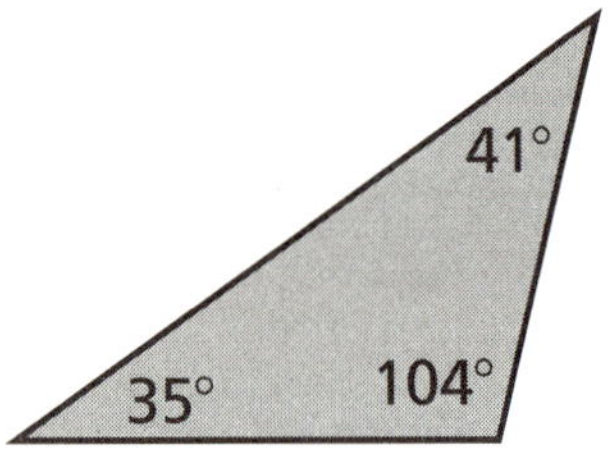

4.

5.
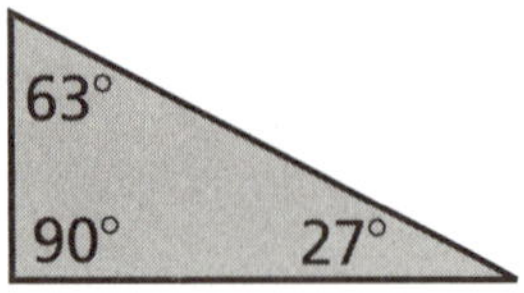

6.
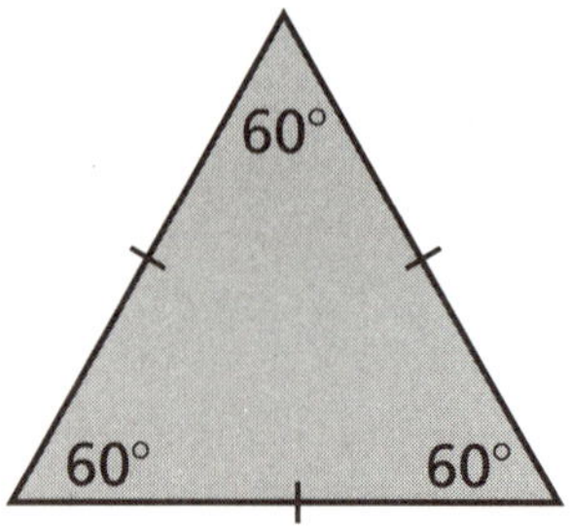

Name____ Date__________

5.2 Practice A

Classify the triangle in as many ways as possible.

1. 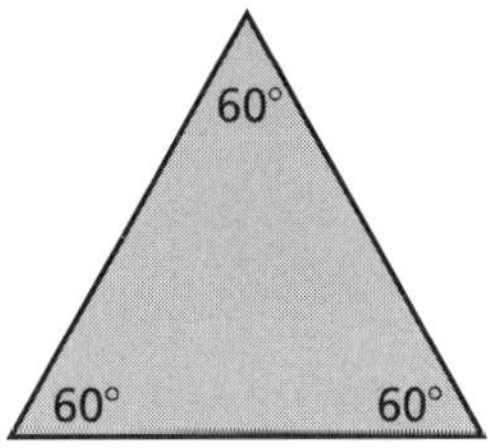

2. 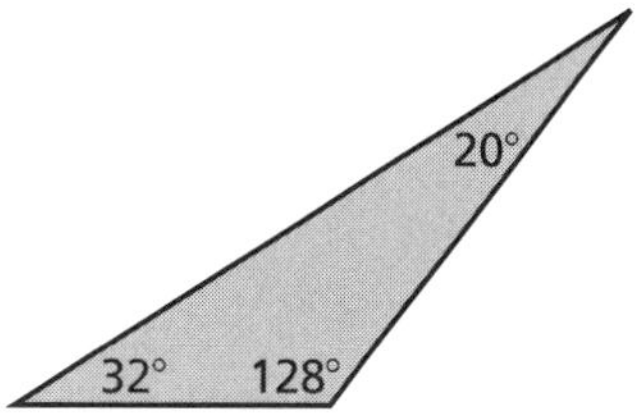

Find the value of *x*. Then classify the triangle in as many ways as possible.

3. 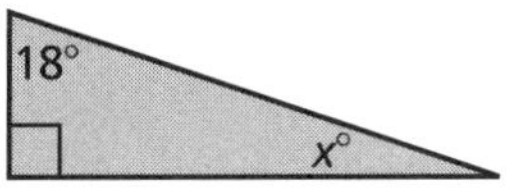

4. 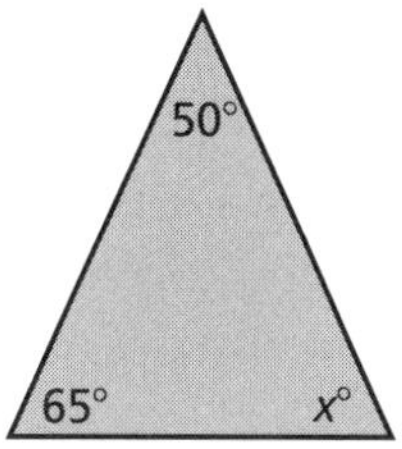

Tell whether a triangle can have the given angle measures. If not, change the first angle measure so that the angle measures form a triangle.

5. $36.9°, 110.4°, 33.7°$

6. $62°, 44\frac{3}{4}°, 73\frac{1}{4}°$

7. Consider the three isosceles right triangles.

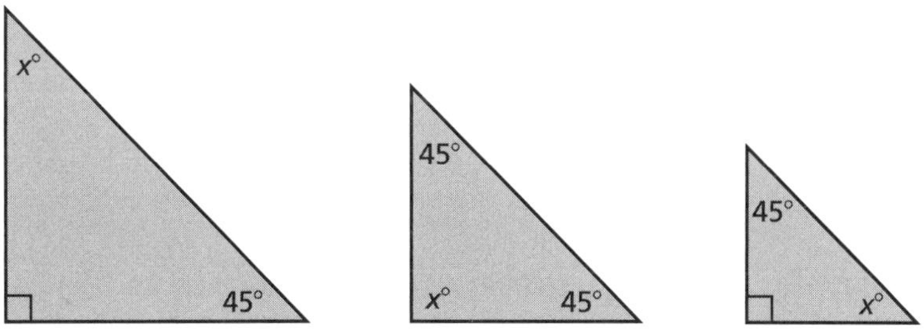

   a. Find the value of x for each triangle.

   b. What do you notice about the angle measures of each triangle?

   c. Write a rule about the angle measures of an isosceles right triangle.

Name ______________________ Date ________

5.2 Practice B

Find the value of *x*. Then classify the triangle in as many ways as possible.

1.

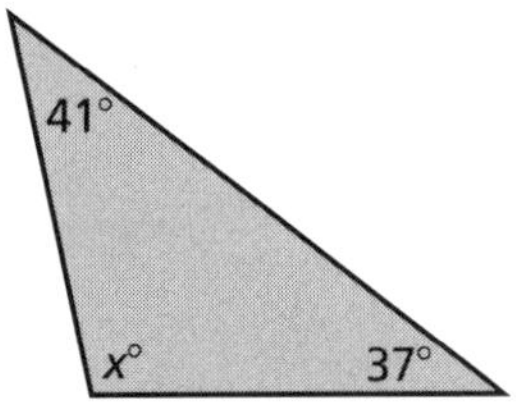

2.

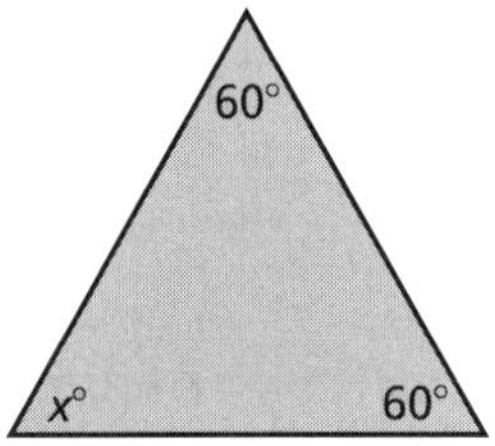

3.

4.

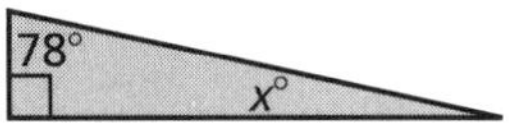

Tell whether a triangle can have the given angle measures. If not, change the first angle measure so that the angle measures form a triangle.

5. $46\frac{1}{3}°, 81\frac{1}{2}°, 52\frac{1}{6}°$

6. 36.9°, 121.4°, 33.7°

7. Using 3 equal-sized craft sticks, put the ends together to make a triangle.

a. Use a protractor to find the measure of each angle.

b. Classify the triangle that you made in as many ways as possible.

c. Replace one of the sticks with either a longer stick or a longer pencil. Use a protractor to find the measure of each angle and classify this new triangle in as many ways as possible.

d. Replace the longest side with a stick or pencil that is shorter than the two other sides. Use a protractor to find the measure of each angle and classify this new triangle in as many ways as possible.

e. What do you notice about the triangle when two of its sides are equal in length?

Name____ Date__________

5.2 Enrichment and Extension

Writing Ratios

The sides of a right triangle can be named by their location with respect to an angle of the triangle.

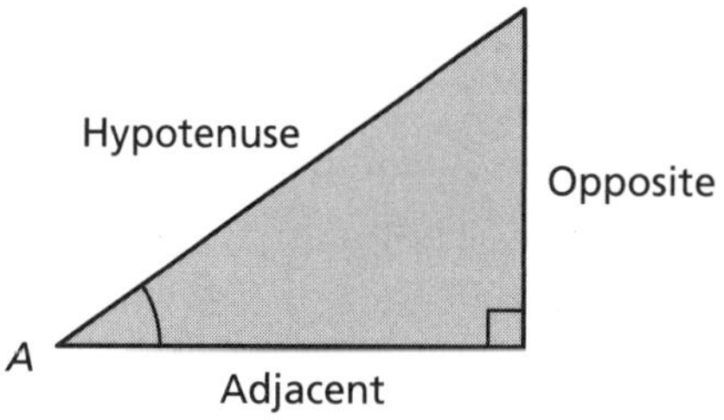

Trigonometry

It is possible to write ratios that compare the lengths of the sides in the triangle using special functions and a given angle. These ratios are called sine (sin), cosine (cos), and tangent (tan) and are studied in depth in a branch of mathematics called trigonometry.

$$\sin A = \frac{\text{Opposite}}{\text{Hypotenuse}} \qquad \cos A = \frac{\text{Adjacent}}{\text{Hypotenuse}} \qquad \tan A = \frac{\text{Opposite}}{\text{Adjacent}}$$

Write the ratios. Use your answers and the color key to shade the mosaic.

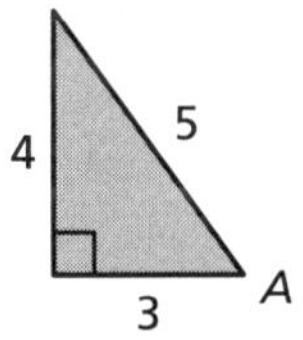

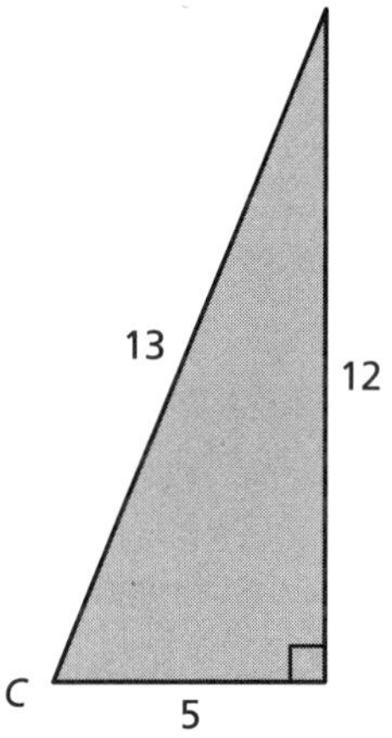

1. $\sin A$ **2.** $\tan C$

3. $\cos A$ **4.** $\tan A$

5. $\sin C$ **6.** $\cos C$

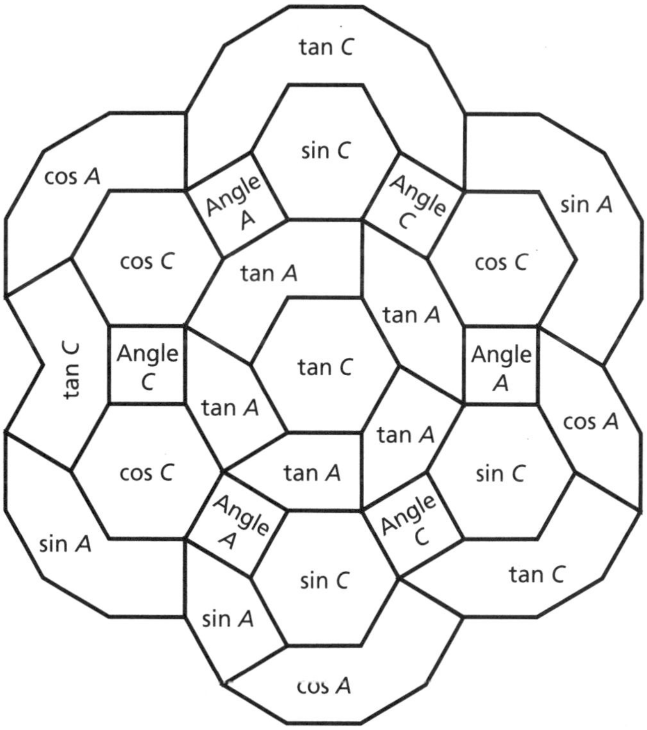

Key:
Angles = Blue
5 in the denominator = Red
13 in the denominator = Yellow
3 in the denominator = Purple

5.2 Puzzle Time

Why Did The Kindergartener Take Her Books To The Zoo?

A	B	C	D	E	F
G	H	I			

Complete each exercise. Find the answer in the answer column. Write the word under the answer in the box containing the exercise letter.

Equilateral Triangle TO
80 LIONS
60 READ
Obtuse Triangle WANTED
46 BETWEEN

Classify the triangle.

A.

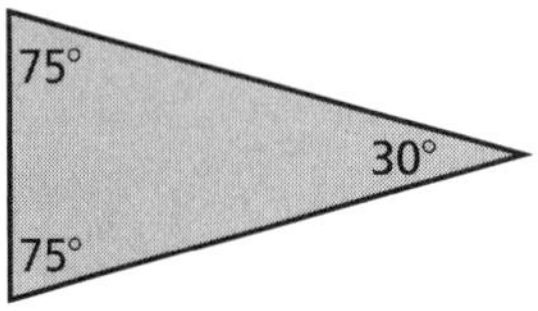

B.

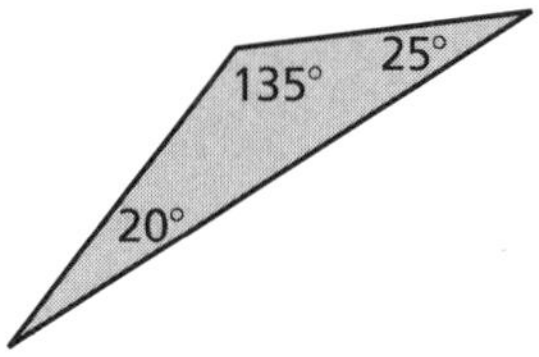

C.

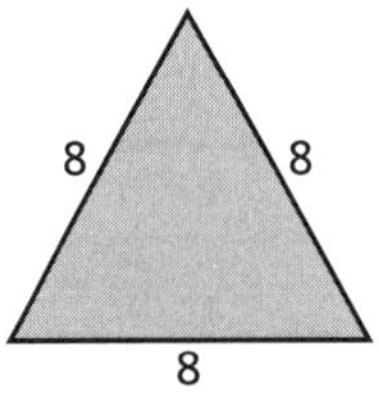

D.

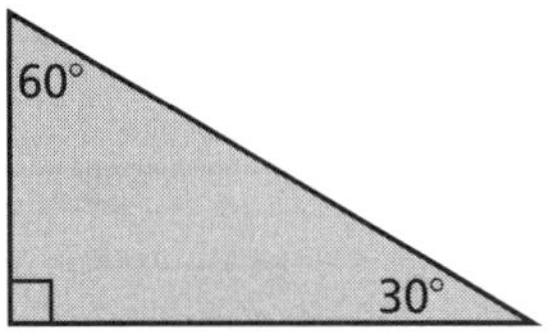

Find the value of *x*.

E.

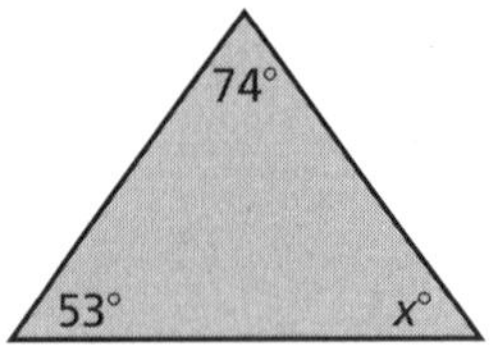

F.

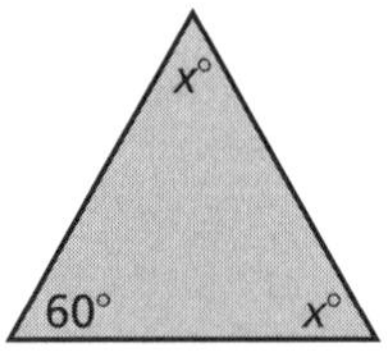

G.

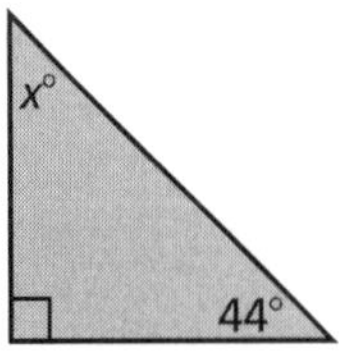

H.

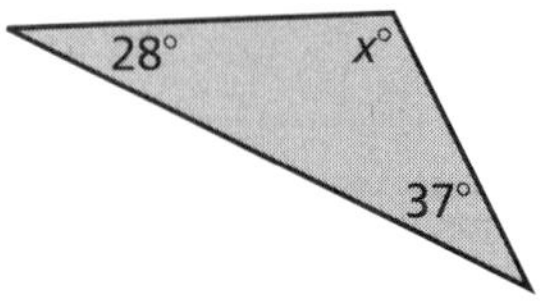

I. A triangle contains angles measuring 25° and 75°. How many degrees is the third angle of the triangle?

53 TO
Isosceles Triangle SHE
Right Triangle LEARN
115 THE

Activity 5.3 Start Thinking!
For use before Activity 5.3

Many road signs are polygons. Try to come up with as many different road signs as you can. For each sign, classify the polygon by the number of sides.

Activity 5.3 Warm Up
For use before Activity 5.3

Find the value of *y* for the given value of *x*.

1. $y = \frac{1}{2}x - 3;\ x = -2$

2. $y = -x + 2;\ x = 15$

3. $y = 10(x - 2);\ x = 10$

4. $y = 13(x - 3);\ x = 1$

5. $y = \frac{3}{2}x + \frac{5}{2};\ x = -2$

6. $y = 3(x + 4);\ x = -5$

Lesson 5.3 Start Thinking!

For use before Lesson 5.3

Explain how you can draw an octagon on graph paper to make it easy to find the angle measures. Use this method to find each angle measure.

What is the sum of the angle measures? Does this agree with the equation you wrote in Activity 2?

Lesson 5.3 Warm Up

For use before Lesson 5.3

Use triangles to find the sum of the angle measures of the polygon.

1. 

2. 

3. 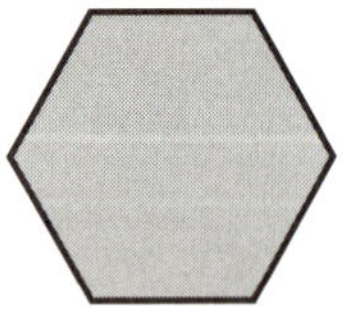

4. 

5. 

6. 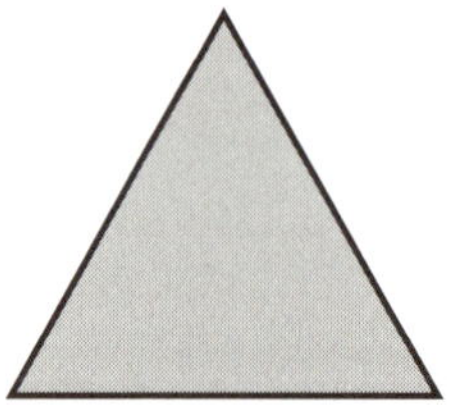

Name____ Date __________

5.3 Practice A

Use triangles to find the sum of the angle measures of the polygon.

1.

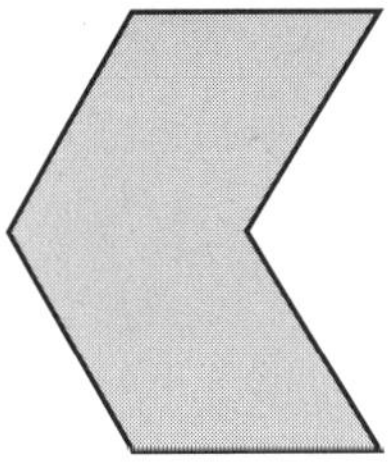

2.

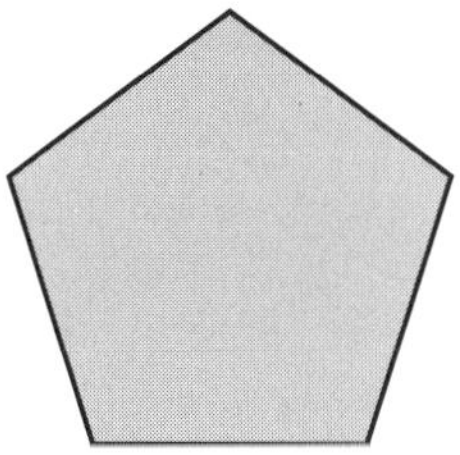

Find the sum of the angle measures of the polygon.

3.

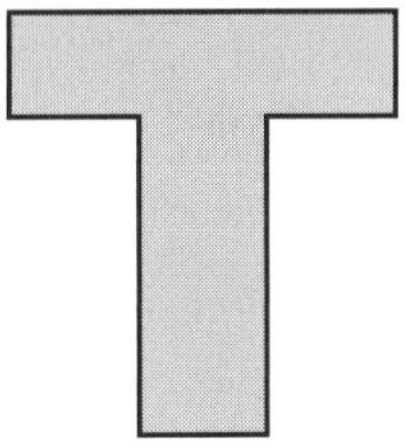

4.

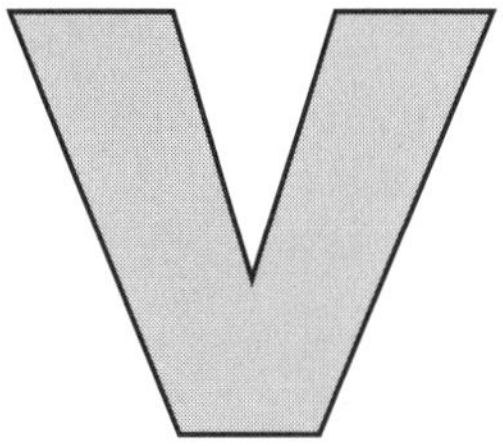

5. Can an octagon have angles that measure 100°, 156°, 125°, 90°, 175°, 134°, 160°, and 140°? Explain.

Find the value of *x*.

6.

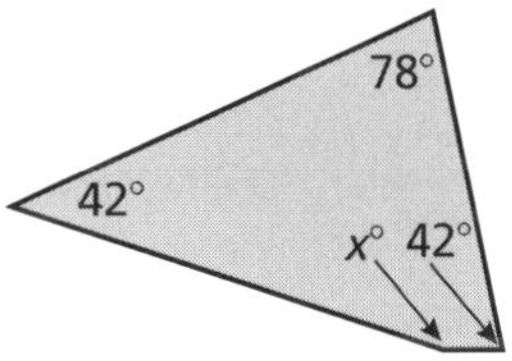

7.

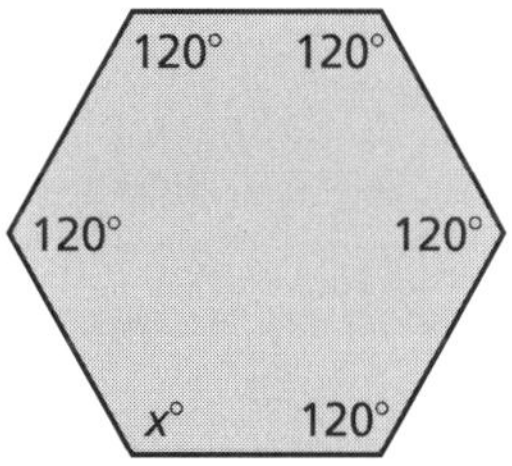

8. A stop sign is in the shape of a regular octagon. What is the measure of each angle?

Tell whether the polygon is *convex* or *concave*. Explain.

9.

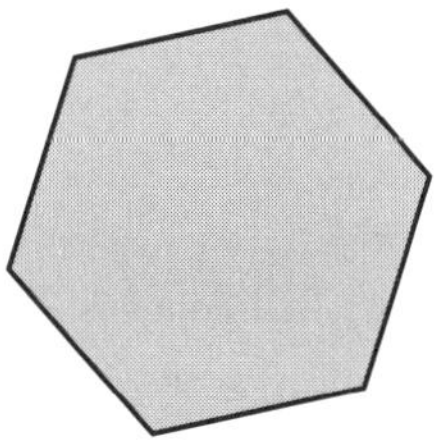

10.

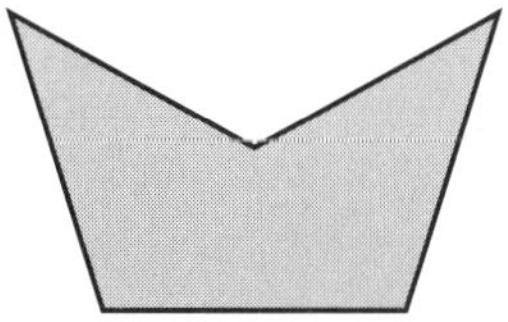

Name ______________________________ Date __________

5.3 Practice B

Use triangles to find the sum of the angle measures of the polygon.

1.

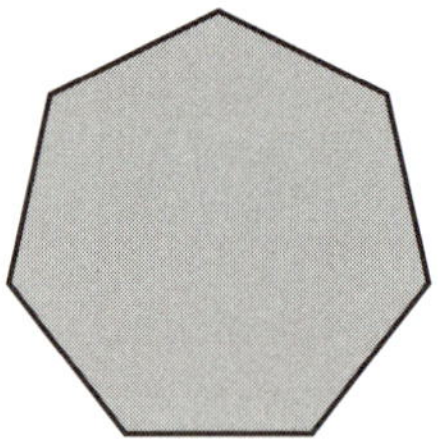

2.

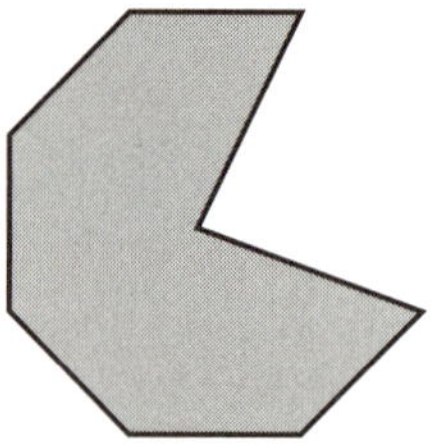

Find the sum of the angle measures of the polygon.

3.

4.

5. Four angles of a pentagon measure 50°, 73°, 146°, and 161°. Find the missing angle measure.

Find the value of *x*.

6.

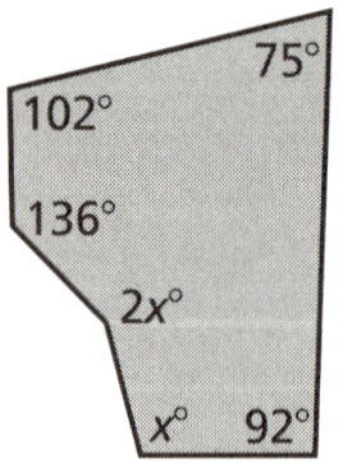

7.

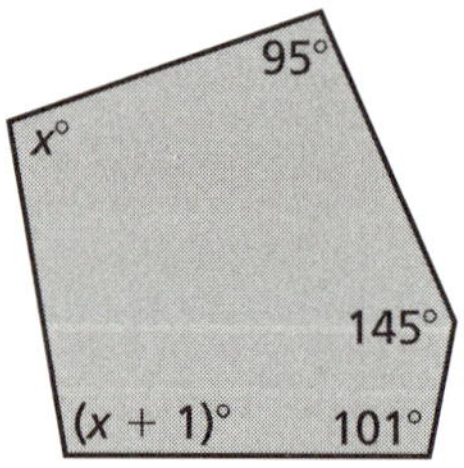

8. The angles of a regular polygon each measure 135°. How many sides does the polygon have?

9. Use the polygon shown.

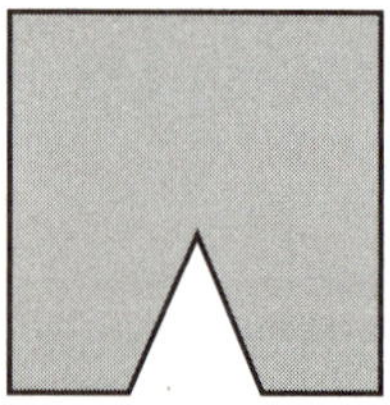

a. Is the polygon *convex* or *concave*?

b. Is the polygon *regular* or *not regular*?

c. What is the name of the polygon?

d. What is the sum of the angle measures in the polygon?

Name____ Date__________

5.3 Enrichment and Extension

Area of Regular Polygons

To find the area of a regular polygon with more than four sides, you need to know the length of the *apothem*. The apothem is the distance from the center to any side of the polygon.

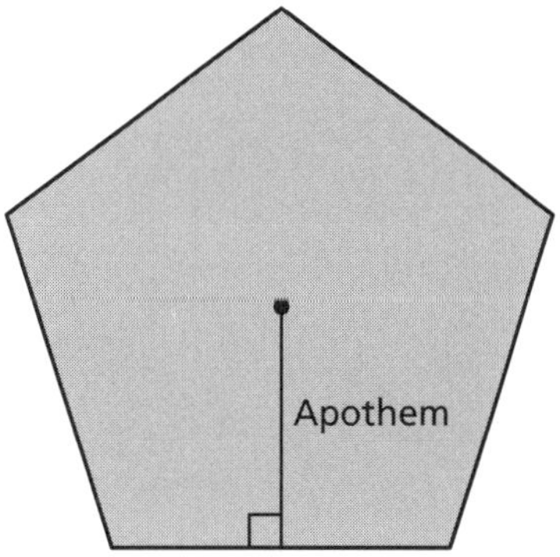

Once you know the length of the apothem, you can calculate the area of a regular polygon using the formula $A = \frac{1}{2}ans$, where a is the length of the apothem, n is the number of sides in the polygon, and s is the length of the sides.

Find the area of the regular polygon.

1.

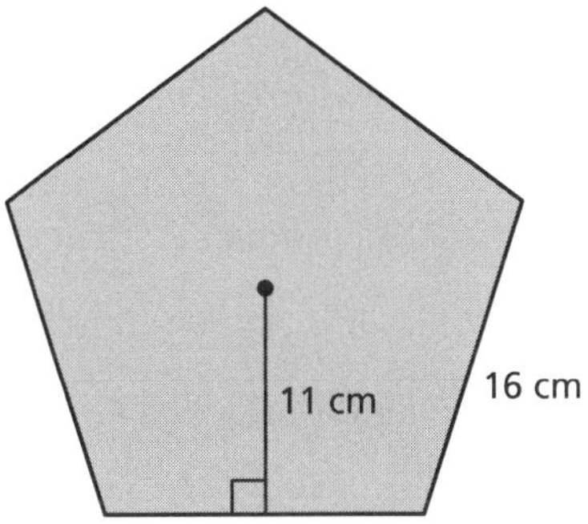

2.

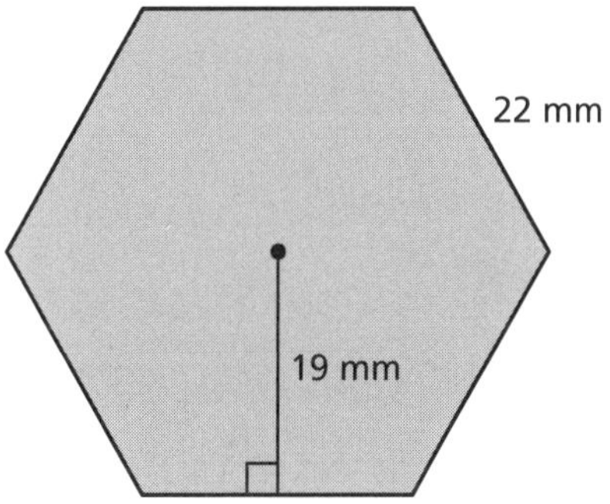

3.

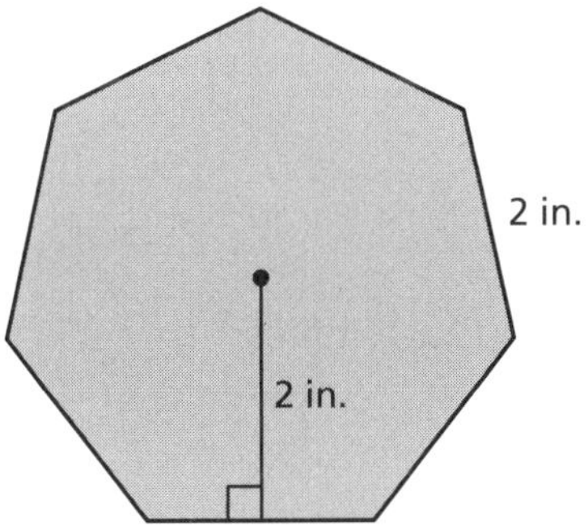

4.

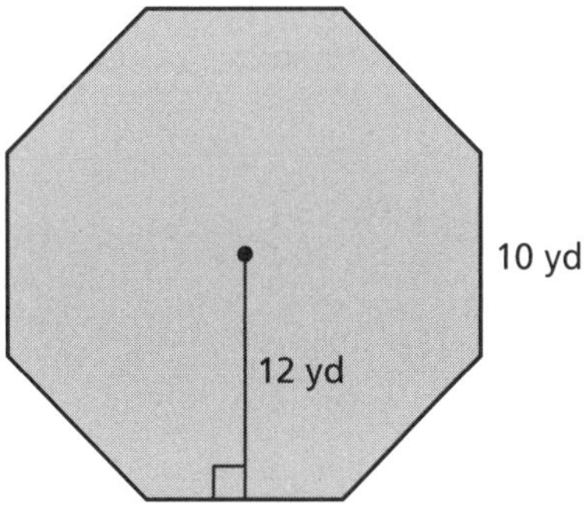

Name ____________________ Date __________

5.3 Puzzle Time

Why Did Old Mother Hubbard Scream When She Went To Fetch Her Poor Dog A Bone?

A	B	C	D	E	F
G	H				

Complete each exercise. Find the answer in the answer column. Write the word under the answer in the box containing the exercise letter.

720°	GOT
135°	WAS
124	THE
convex	SHE

Tell whether the polygon is *convex* or *concave*.

A. 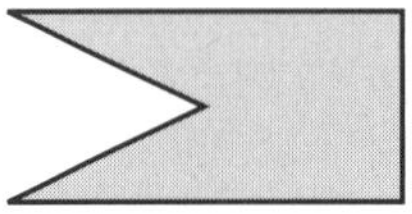

B. 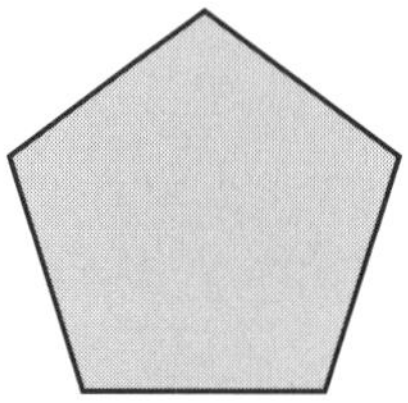

Find the sum of the angle measures of the polygon.

C. 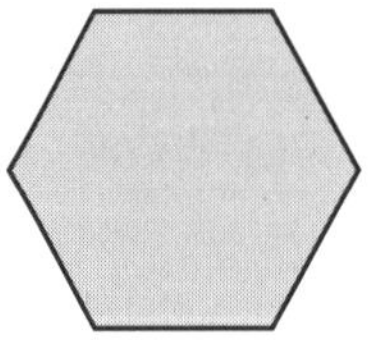

D. 

Find the value of *x*.

E. 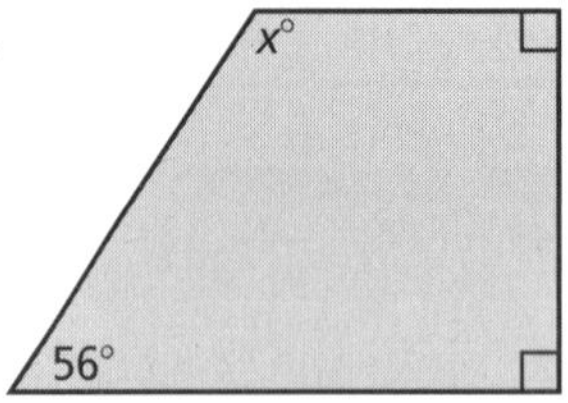

F. 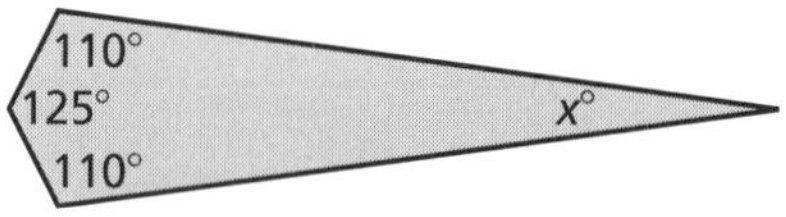

Find the measure of each angle of the regular polygon.

G. 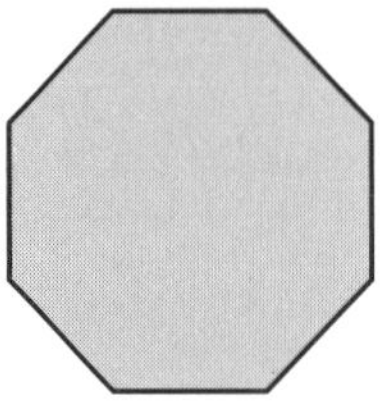

H. 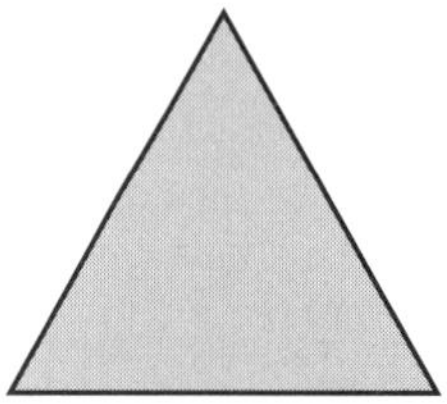

concave	WHEN
15	CUPBOARD
1080°	THERE
60°	BEAR

Activity 5.4 Start Thinking!
For use before Activity 5.4

Which pairs of shapes appear to be *similar*? How can you tell?

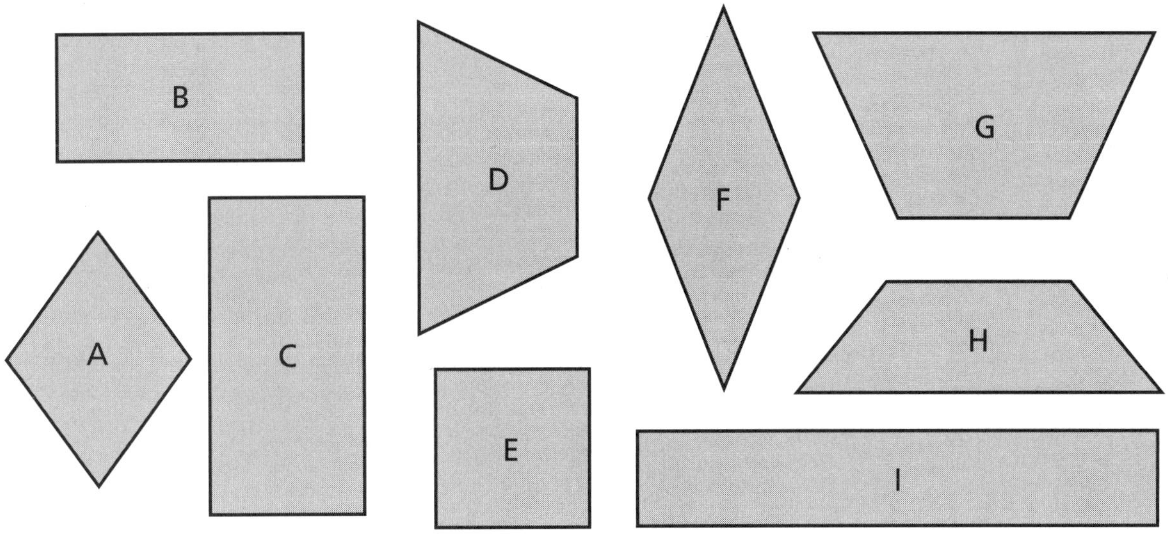

Activity 5.4 Warm Up
For use before Activity 5.4

The triangles are similar. Find the value of *x*.

1.

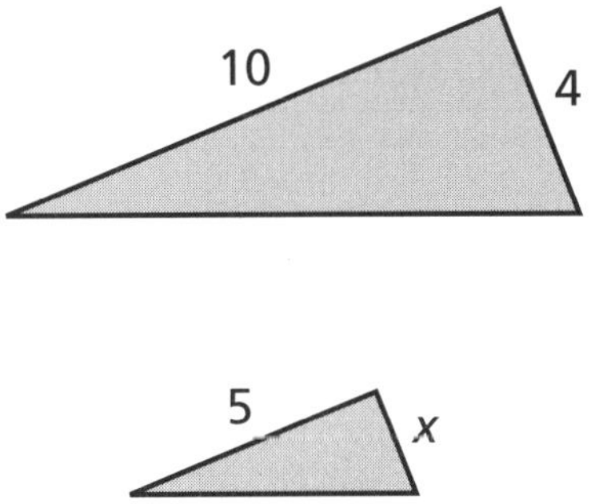

2.

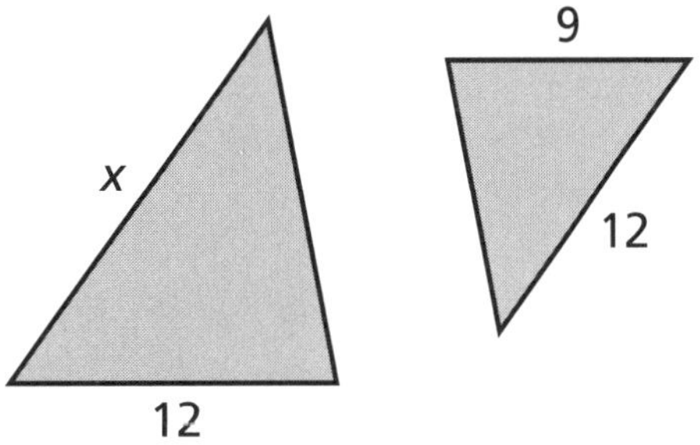

Lesson 5.4 Start Thinking!

For use before Lesson 5.4

Thales was a Greek philosopher and mathematician who lived around 600 B.C. There are several accounts of how he used indirect measurement to find the height of the Great Pyramid in Giza.

According to one account, when his shadow was the same length as his height, he measured the length of the Great Pyramid's shadow.

What does this have to do with similar triangles?

Lesson 5.4 Warm Up

For use before Lesson 5.4

Make a triangle that is larger than the one given and has the same angle measures. Find the ratios of the corresponding side lengths.

1. 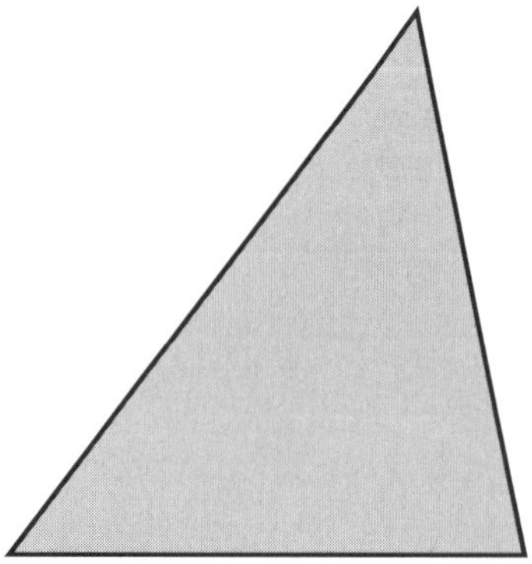

2. 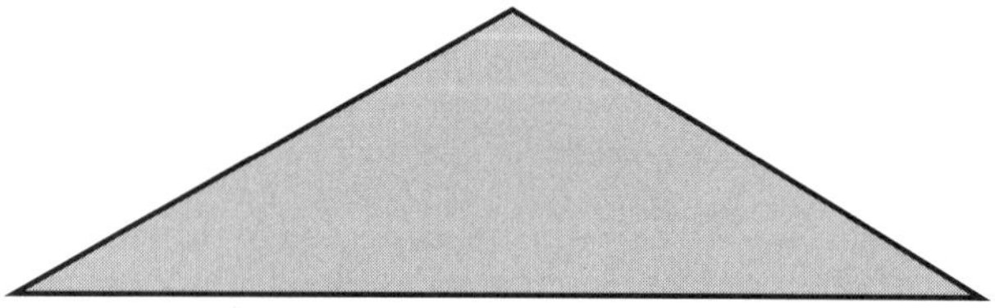

Name____________________________ Date__________

5.4 Practice A

Tell whether the triangles are similar. Explain.

1.

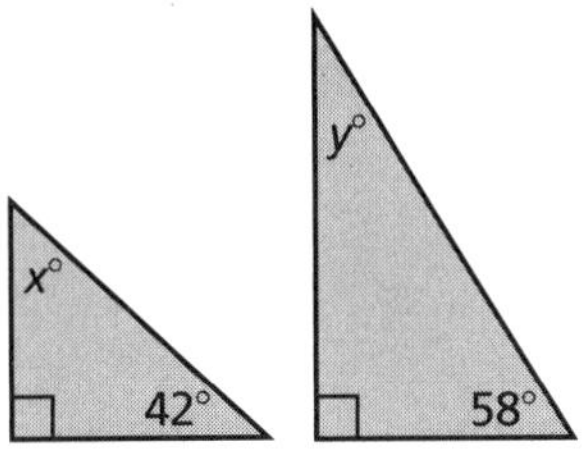

2.

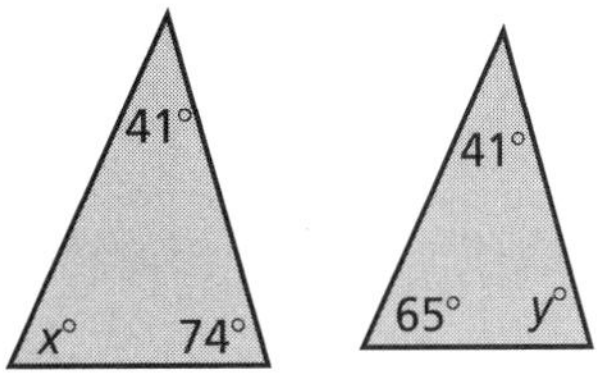

3. The triangles are similar. Find the value of x.

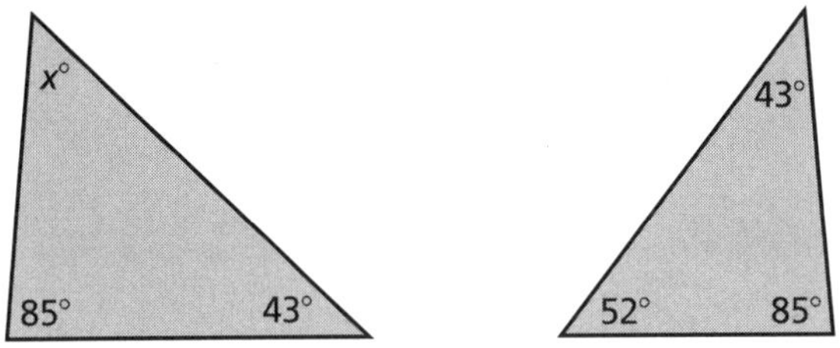

4. You can use indirect measurement to estimate the height of a building. First measure your distance from the base of the building and the distance from the ground to a point on the building that you are looking at. Maintaining the same angle of sight, move back until the top of the building is in your line of sight.

a. Explain why $\triangle ABC$ and $\triangle DBE$ are similar.

b. What is the height of the building?

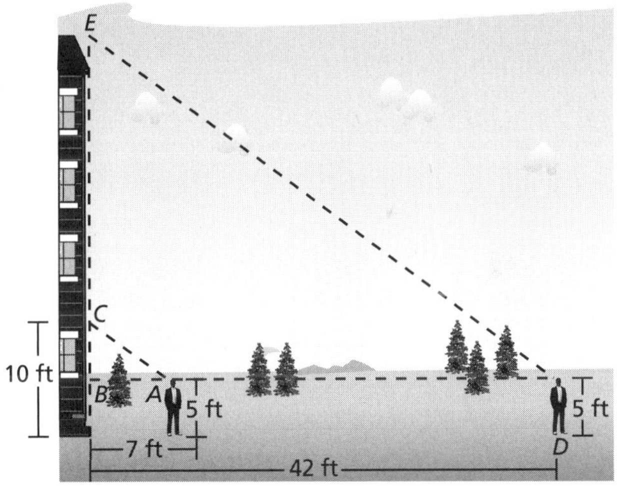

5. You and your friend are practicing for a rowing competition and want to know how far it is to an island in the Indian River Lagoon. You take measurements on your side of the lagoon and make the drawing shown.

a. Explain why $\triangle ABC$ and $\triangle DEC$ are similar.

b. What is the distance to the island?

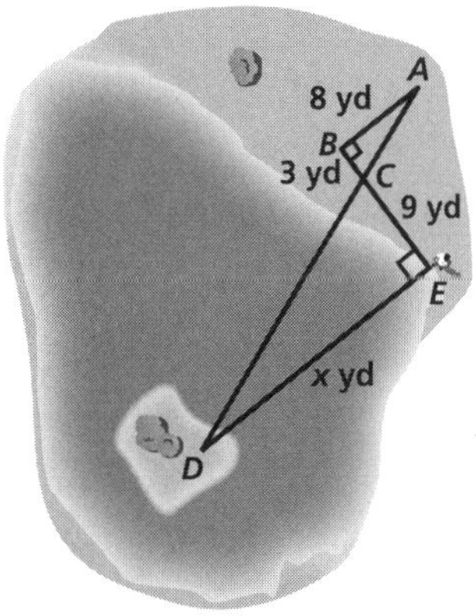

Name ____ Date __________

5.4 Practice B

Tell whether the triangles are similar. Explain.

1.

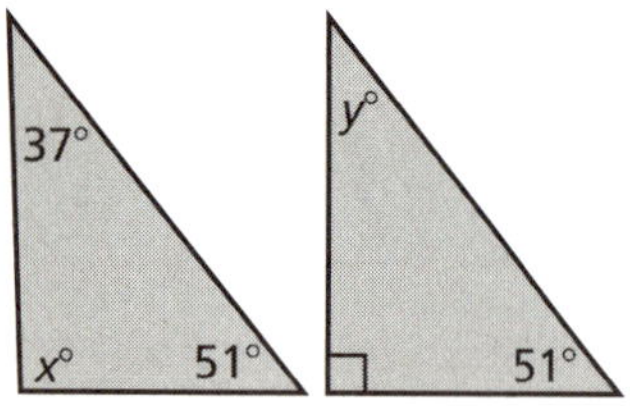

2.

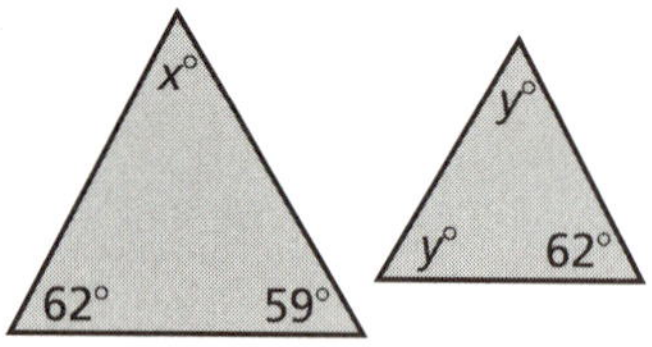

3. The triangles are similar. Find the value of x.

4. You can use indirect measurement to estimate the height of a flag pole. First measure your distance from the base of the flag pole and the distance from the ground to a point on the flag pole that you are looking at. Maintaining the same angle of sight, move back until the top of the flag pole is in your line of sight.

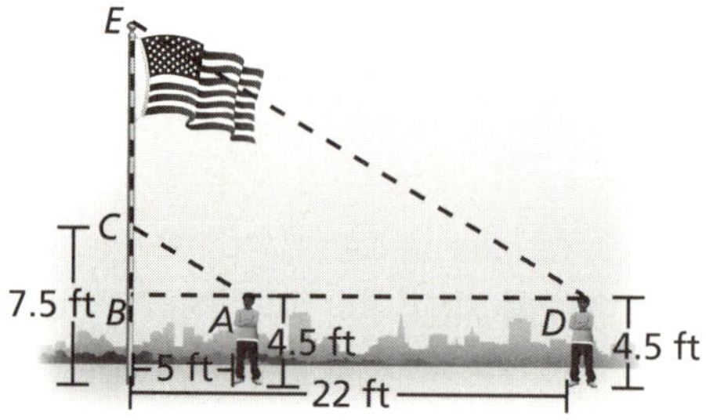

a. Explain why $\triangle ABC$ and $\triangle DBE$ are similar.

b. What is the height of the flag pole?

5. You are on a boat in the ocean, at Point A. You locate a lighthouse at Point D, beyond the line of sight of the marker at point C. You drive 0.2 mile west to Point B and then 0.1 mile south to Point C. You drive 0.3 mile more to arrive at Point E, which is due east of the lighthouse.

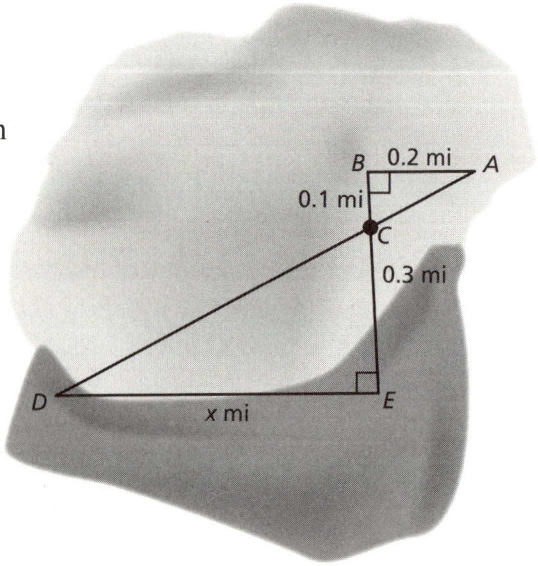

a. Explain why $\triangle ABC$ and $\triangle DEC$ are similar.

b. What is the distance from Point E to the lighthouse?

Name______________________________ Date__________

5.4 Enrichment and Extension

Telescopes

The most basic design for a telescope is a long tube with a lens at each end. The larger lens, called the objective lens, focuses a distant image to a point inside the tube, called the focal point. The smaller lens, called the eyepiece, has the same focal point and works like a magnifying glass. The result is that your eye sees an enlarged view of the image from the objective lens.

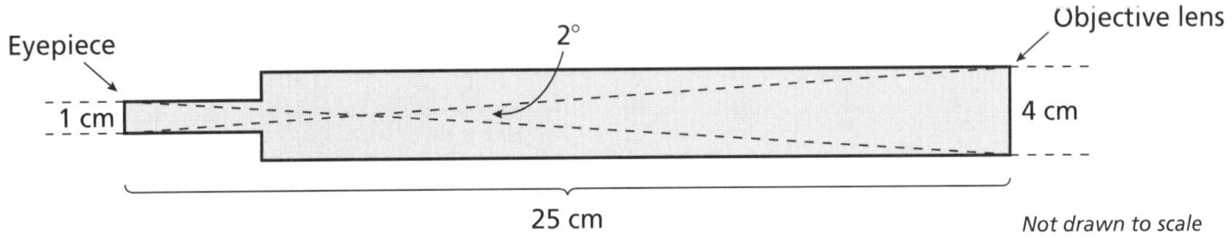

1. A lens and its focal point form an isosceles triangle. What are the angle measures of the triangle formed by the objective lens and the focal point?

2. Does the triangle formed by the eyepiece and the focal point have the same angle measures? Explain your reasoning.

3. What is the ratio of the diameters of the objective lens to the eyepiece?

4. The distance from a lens to its focal point is called the focal length. Use indirect measurement to find the focal length (in centimeters) of each lens.

5. The focal length of the eyepiece determines the magnification of the telescope. The number of times m an image is magnified is represented by the equation $m = \frac{250}{f}$, where f is the focal length (in millimeters). How many times is the image magnified in the telescope above?

6. If the length of the telescope increased, would the magnification of the telescope increase or decrease? Explain.

Name ______________________________ Date __________

5.4 Puzzle Time

What Do You Call A Dandelion Floating In The Ocean?

Write the letter of each answer in the box containing the exercise number.

Choose the correct letter that describes the triangles.

Answers
A. D
E. 58
D. A
S. G
E. 18
W. 70
E. F

1.

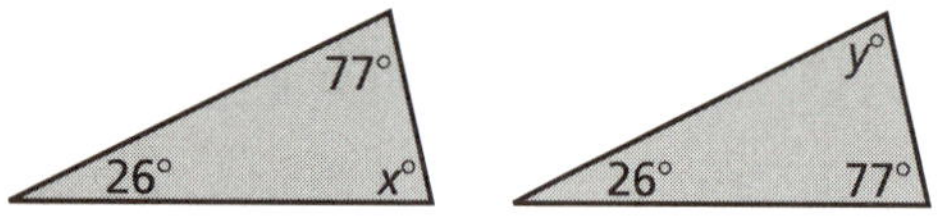

A. Similar **B.** Not Similar

2.

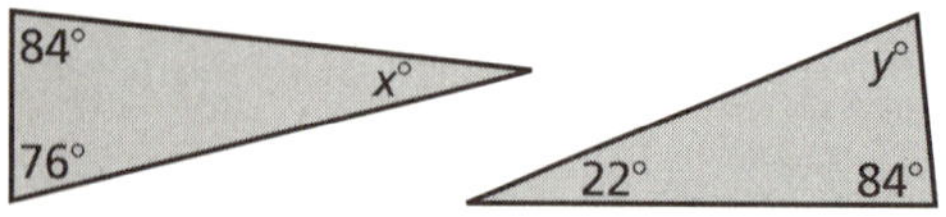

C. Similar **D.** Not Similar

3.

E. Similar **F.** Not Similar

4.

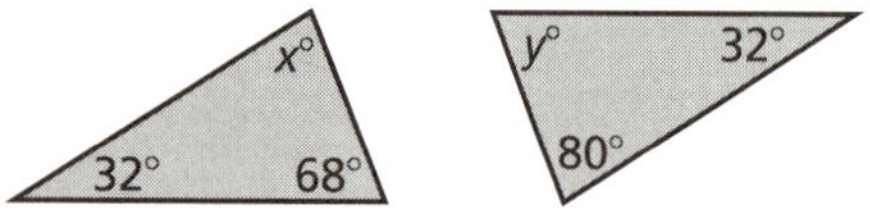

G. Similar **H.** Not Similar

The triangles are similar. Find the value of *x*.

5.

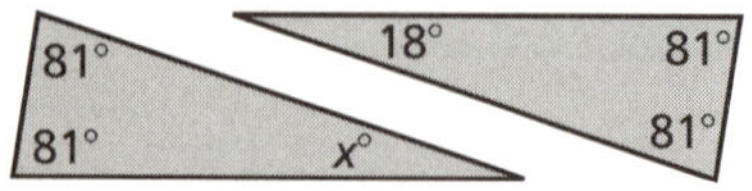

6.

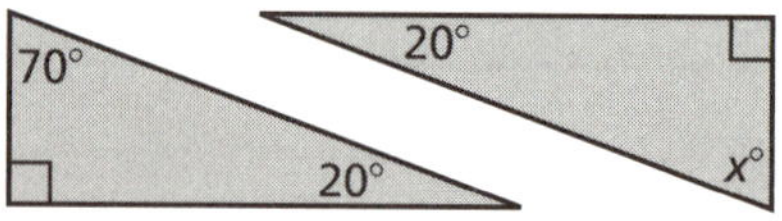

7.

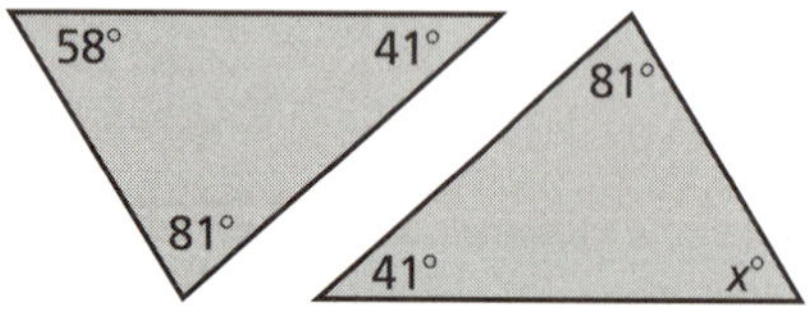

4	7	2		6	5	3	1

Activity 5.5 Start Thinking!
For use before Activity 5.5

Give some examples of parallel lines in your classroom.

Activity 5.5 Warm Up
For use before Activity 5.5

Solve the proportion.

1. $\frac{16}{3} = \frac{x}{9}$

2. $\frac{2}{5} = \frac{5}{x}$

3. $\frac{15}{12} = \frac{x}{8}$

4. $\frac{x}{2} = \frac{11}{4}$

5. $\frac{100}{x} = \frac{25}{8}$

6. $\frac{x}{5} = \frac{3}{8}$

Lesson 5.5 Start Thinking!
For use before Lesson 5.5

If you know the measure of one of the angles formed by two parallel lines and a transversal, can you find the measures of all the other angles without using a protractor? Why or why not?

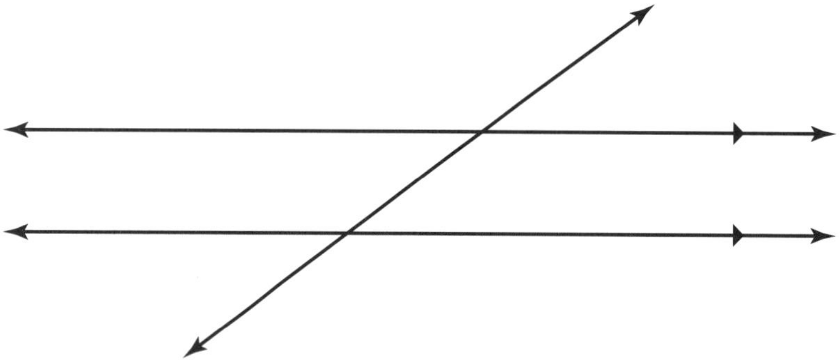

Lesson 5.5 Warm Up
For use before Lesson 5.5

Use the figure.

1. Identify the parallel lines.

2. Identify the transversal.

3. How many angles are formed by the transversal?

4. Which of the angles are congruent?

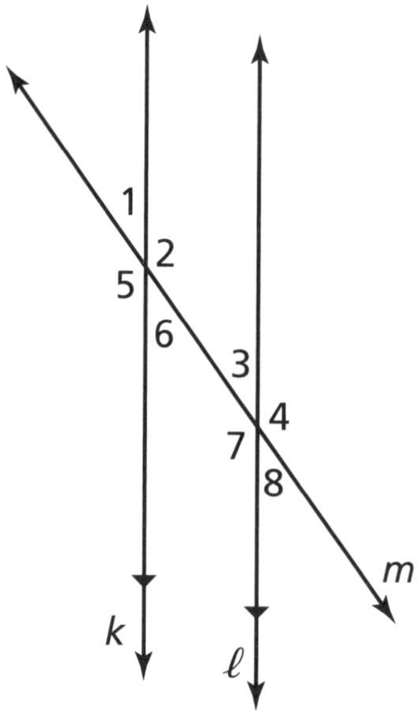

Name______________________________ Date__________

5.5 Practice A

Use the figure to find the measures of the numbered angles.

1.

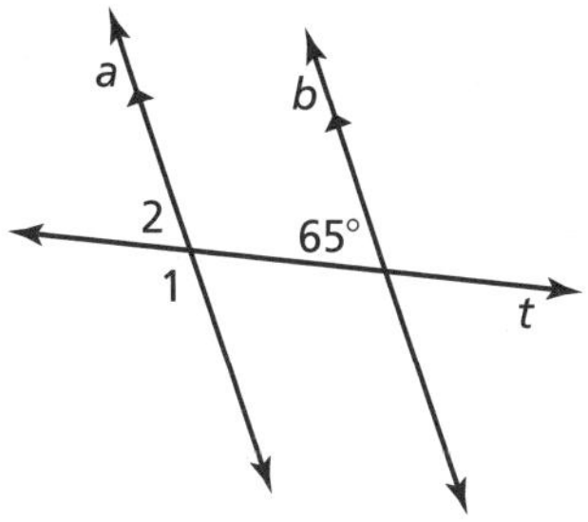

2.

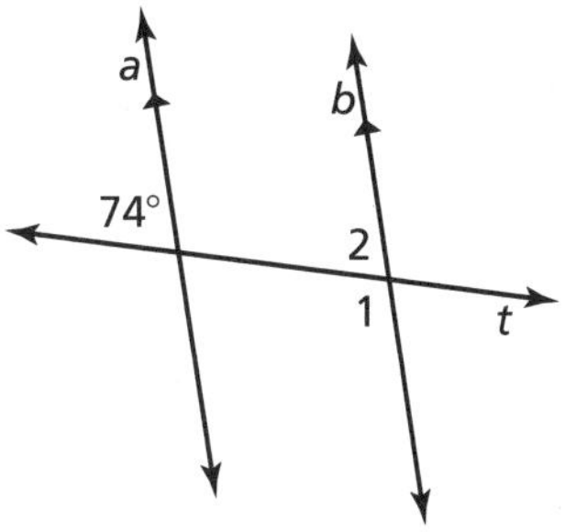

Use the figure to find the measures of the numbered angles. Explain your reasoning.

3.

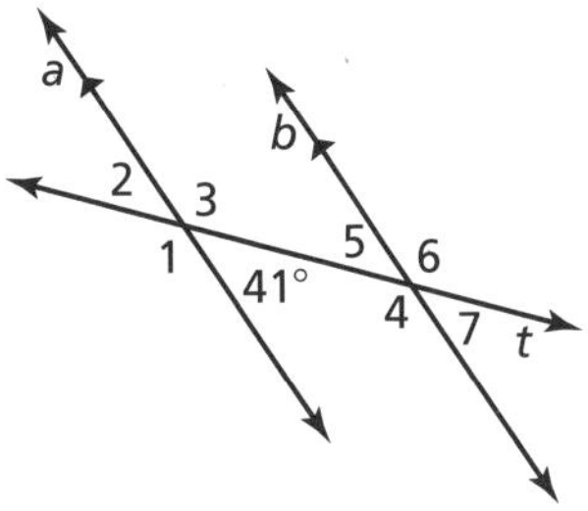

4.

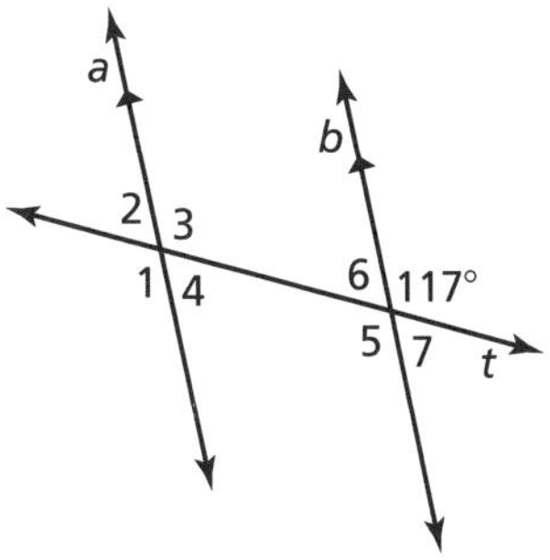

Complete the statement. Explain your reasoning.

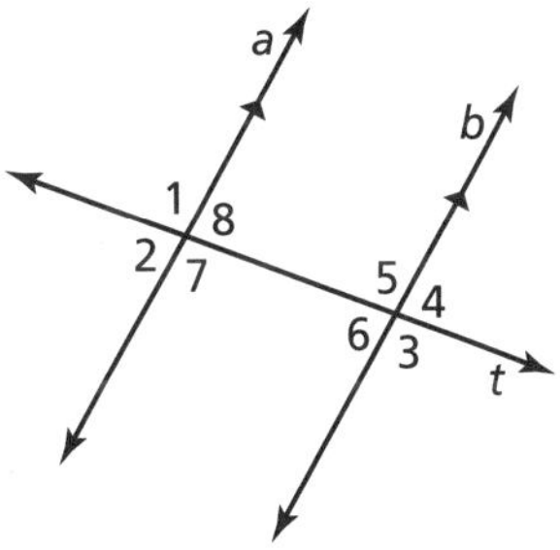

5. If the measure of $\angle 1 = 160°$, then the measure of $\angle 5 = \underline{\ ?\ }$.

6. If the measure of $\angle 6 = 37°$, then the measure of $\angle 4 = \underline{\ ?\ }$.

7. If the measure of $\angle 8 = 82°$, then the measure of $\angle 3 = \underline{\ ?\ }$.

8. If the measure of $\angle 4 = 60°$, then the measure of $\angle 5 = \underline{\ ?\ }$.

Correct the following statements about the numbered angles by replacing the underlined words with the correct words.

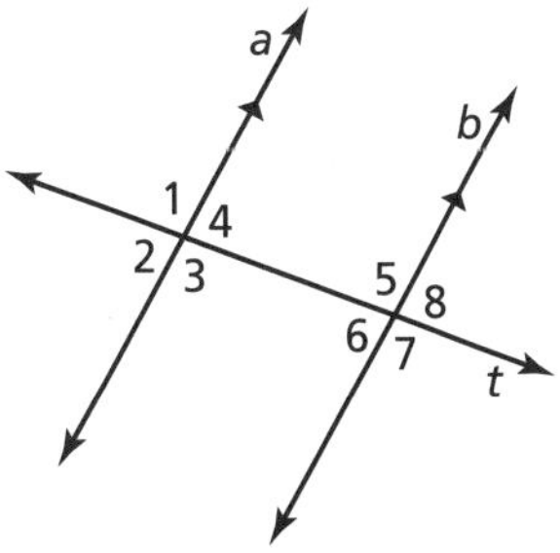

9. $\angle 2$ is $\underline{\text{congruent}}$ to $\angle 4$. $\angle 4$ is $\underline{\text{congruent}}$ to $\angle 8$. So, $\angle 2$ is $\underline{\text{supplementary}}$ to $\angle 8$.

10. $\angle 6$ is $\underline{\text{congruent}}$ to $\angle 3$. $\angle 3$ is $\underline{\text{congruent}}$ to $\angle 1$. So, $\angle 6$ is $\underline{\text{congruent}}$ to $\angle 1$.

Name ______________________________ Date __________

5.5 Practice B

Use the figure to find the measures of the numbered angles. Explain your reasoning.

1.

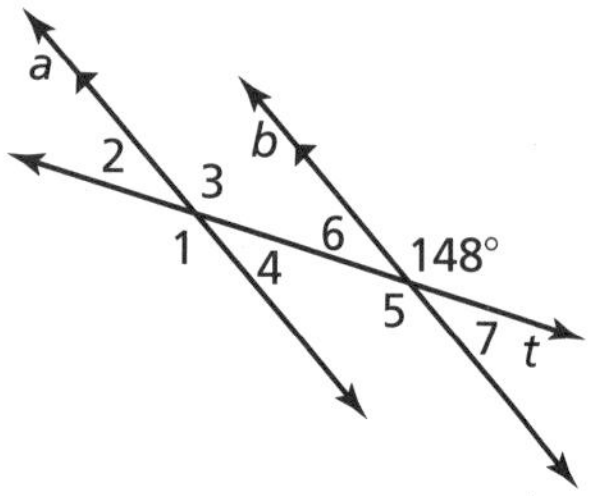

2.

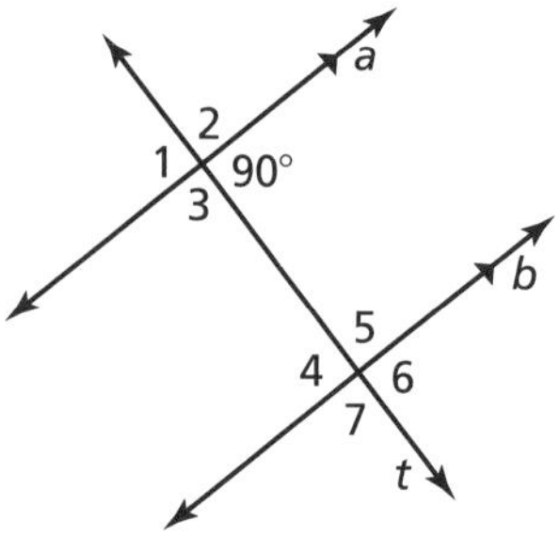

Complete the statement. Explain your reasoning.

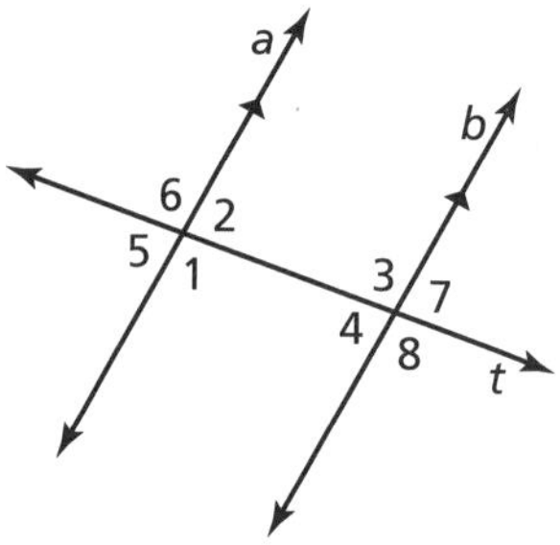

3. If the measure of $\angle 1 = 130°$, then the measure of $\angle 8 =$ _?_.

4. If the measure of $\angle 5 = 53°$, then the measure of $\angle 3 =$ _?_.

5. If the measure of $\angle 7 = 71°$, then the measure of $\angle 3 =$ _?_.

6. If the measure of $\angle 4 = 65°$, then the measure of $\angle 6 =$ _?_.

Using the diagram for angle placement only (the measurement of the angles may change), indicate if the following statements are *always*, *never*, or *sometimes* true. Explain.

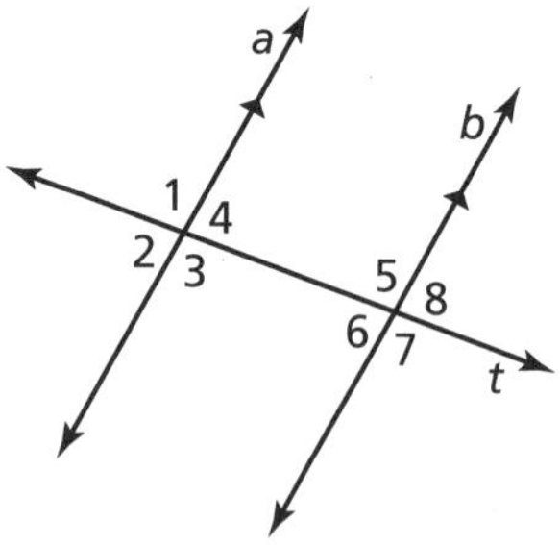

7. $\angle 1$ is congruent to $\angle 3$.

8. $\angle 6$ is supplementary to $\angle 8$.

9. $\angle 2$ is complementary to $\angle 1$.

10. $\angle 8$ and $\angle 5$ are vertical angles.

11. $\angle 2$ is congruent to $\angle 8$.

12. If a transversal intersects two parallel lines, is it possible for all of the angles formed to be acute angles? Explain.

Name____ Date __________

5.5 Enrichment and Extension

Tessellations

A *tessellation* is a collection of figures that cover a plane with no gaps or overlaps.

There are several different types of tessellations. A *semi-regular tessellation* is a pattern formed by using two or more regular polygons. In a regular polygon, all sides have the same measure and all angles have equal measures.

Use the tessellation to answer the questions.

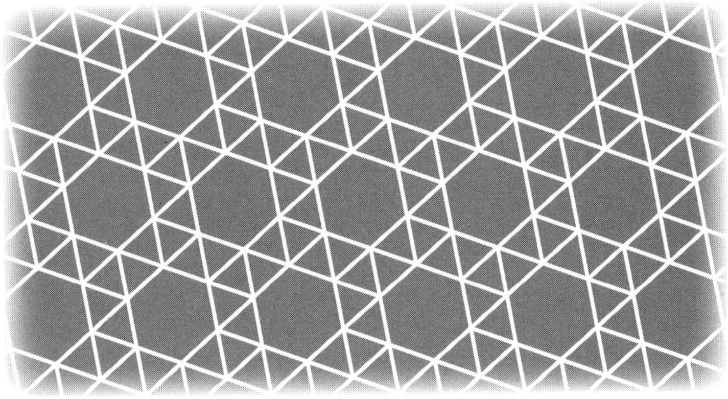

1. Is the tessellation semi-regular? How do you know?
2. Identify the two regular polygons used in the tessellation.
3. What is the measure of one angle of the triangle used in the tessellation?
4. What is the measure of one angle of the hexagon used in the tessellation?
5. A vertex is the point at which several polygons come together. Identify and mark one vertex in the tessellation.
6. Traveling clockwise, name the polygons that meet at your vertex.
7. What is the sum of the angles that meet at the vertex?
8. Give two examples where tessellations are used in real life.

Name ______________________________ Date ________

5.5 Puzzle Time

Why Did The Rabbit Wear A Shower Cap?

A	B	C	D	E	F
G	H				

Complete each exercise. Find the answer in the answer column. Write the word under the answer in the box containing the exercise letter.

147 HARE
44 WANT
129 ITS
c IT

Use the figure to answer the questions.

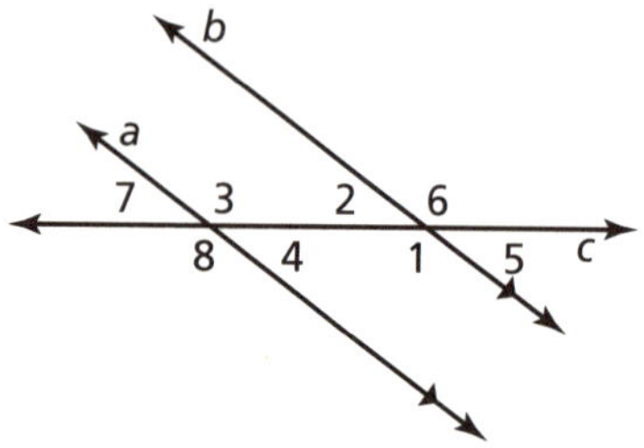

A. Identify the transversal.

B. Identify the parallel lines.

C. If the measure of $\angle 1 = 136°$, then the measure of $\angle 2 =$ ________°.

D. If the measure of $\angle 4 = 44°$, then the measure of $\angle 8 =$ ________°.

E. If the measure of $\angle 8 = 129°$, then the measure of $\angle 4 =$ ________°.

F. If the measure of $\angle 3 = 129°$, then the measure of $\angle 1 =$ ________°.

a, *b* DIDN'T
51 GET
112 WET
136 TO

Find the value of *x*.

G.

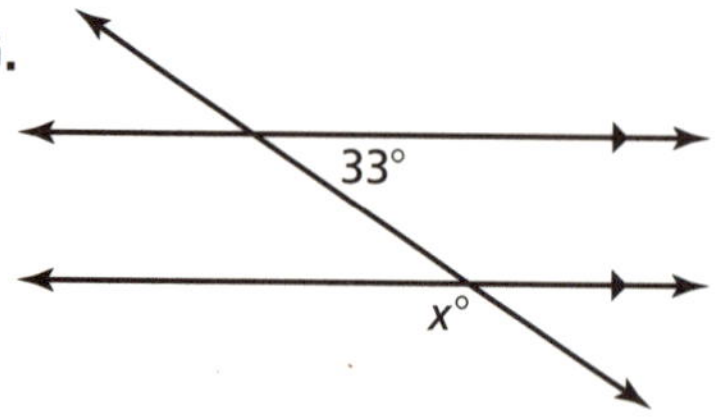

H.

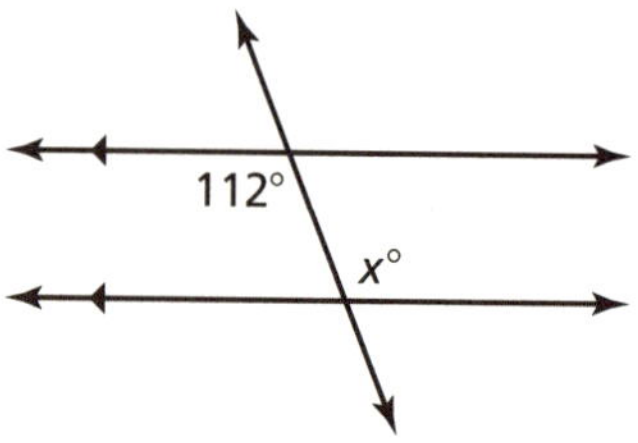

Name____ Date __________

Chapter 5

School-to-Work

For use after Section 5.3

Contractor's Apprentice

You are working as an apprentice to a contractor. The contractor has put you in charge of suggesting tile patterns for a new bathroom floor. The tiles you can choose from are regular hexagons, squares, and equilateral triangles, as shown below. You have an unlimited supply of all tile shapes.

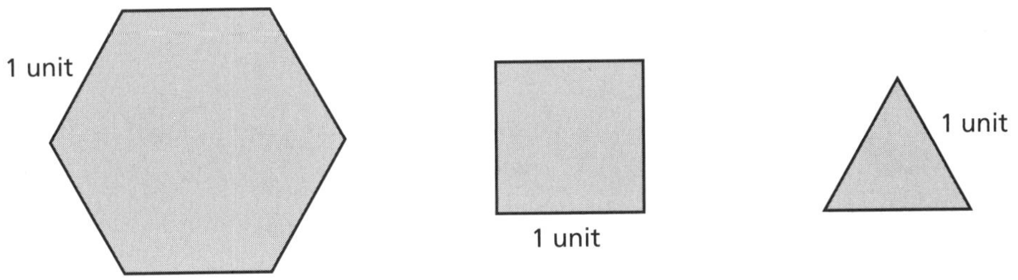

1. What is the measure of the interior angles of each polygon? How do you know?

2. How can you tile the floor using just triangles? using just squares? using just hexagons? Draw sketches to show how.

3. In your sketches of tile designs using just one polygon, what is the sum of the angle measures at any one vertex? How does this explain how the polygons fit together?

4. Sketch a tile design using just triangles and squares. Use angle measures to verify that the polygons fit together with no gaps or overlaps.

5. Sketch a tile design using all three polygons. Use angle measures to verify that the shapes fit together with no gaps or overlaps.

6. The three tile shapes are available in a variety of colors. Sketch a design for the bathroom floor that you will recommend to the contractor. Include the use of at least two different colors. Explain why you recommend this design.

Name ______________________________ Date __________

Chapter 5 Study Help

You can use an **example and non-example chart** to list examples and non-examples of a vocabulary word or item.

Examples	Non-Examples

On Your Own

Make an example and non-example chart to help you study these topics.

1. isosceles triangles
2. equilateral triangles
3. regular polygons
4. convex polygons
5. concave polygons

After you complete this chapter, make example and non-example charts for the following topics.

6. similar triangles
7. transversals
8. interior angles

Name____ Date______________

Chapter 5

Financial Literacy

For use after Section 5.5

Planning a Road Trip

Your family is visiting your aunt, who lives in Tallahassee, Florida. You are planning a road trip from Tallahassee to Orlando, from Orlando to Jacksonville, and then from Jacksonville back to Tallahassee. The route is shown on the map below. You want to estimate the amount you will spend in gasoline, using distance estimates from the map.

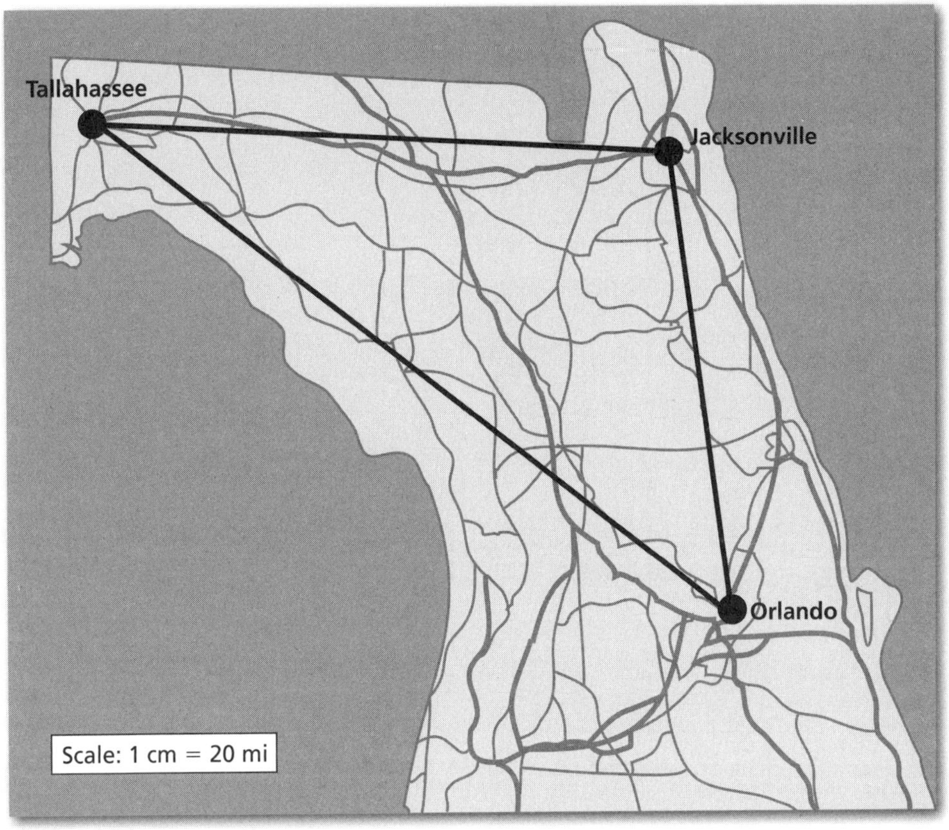

1. The triangular path shown on the map is similar to the actual triangular path among the three cities. How are the distances on the map related to the actual distances? How do you know?

2. Measure the distance between Tallahassee and Orlando in centimeters. What is the actual distance between Tallahassee and Orlando? Show your work.

3. Find the other two actual distances between cities. What is the total distance along the actual triangular path?

4. Your car gets about 24 miles per gallon of gasoline. About how many gallons of gasoline will you need for the trip? If your car's tank holds 16 gallons of gasoline, how many times will you need to fill the tank?

5. Research current gasoline prices. About how much money will you need to spend on gasoline for the trip? Explain why the actual cost of gasoline for the trip is likely to be greater than your estimate.

Name ______________________________ Date __________

Chapter 5 Cumulative Practice

Find the value of *x*. If there are two figures, they are similar.

1. 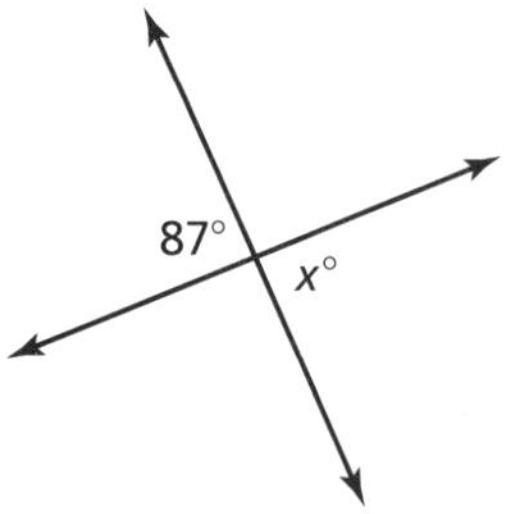

2. 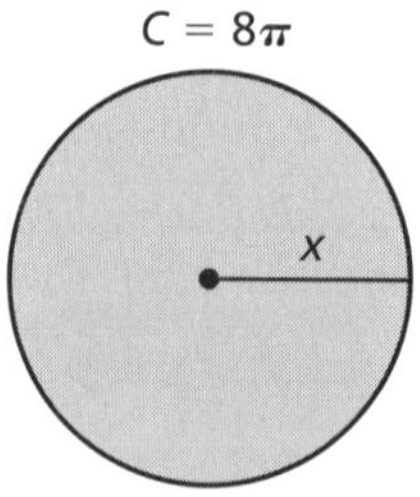

3. 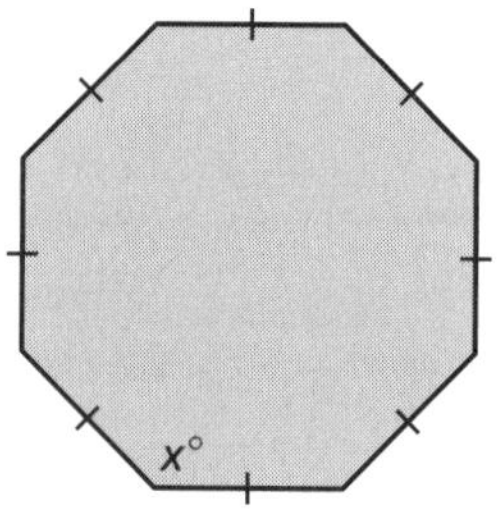

4. 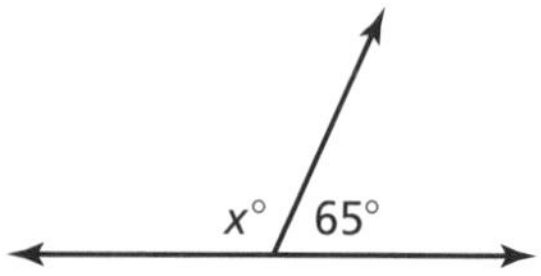

5. 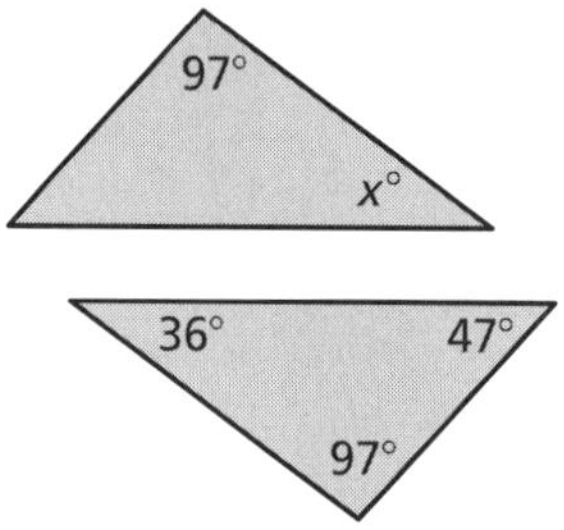

6. 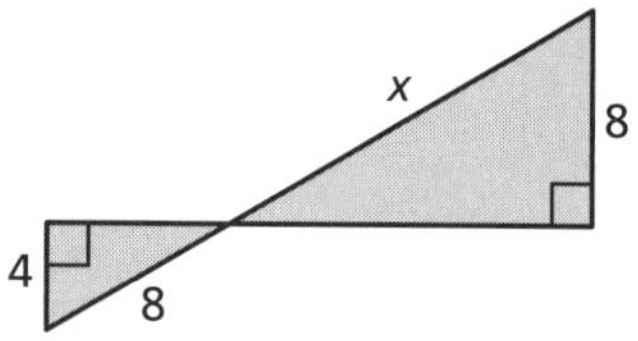

7. 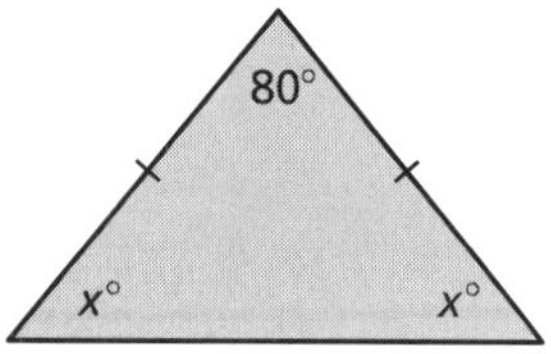

8. 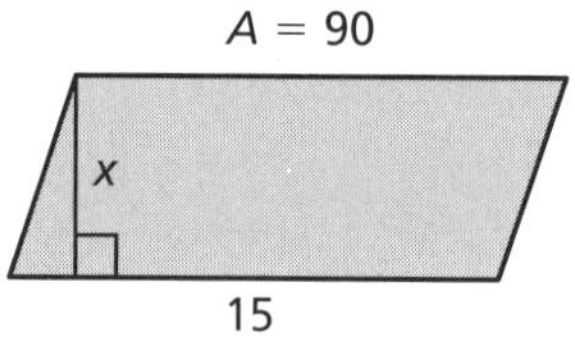

9. 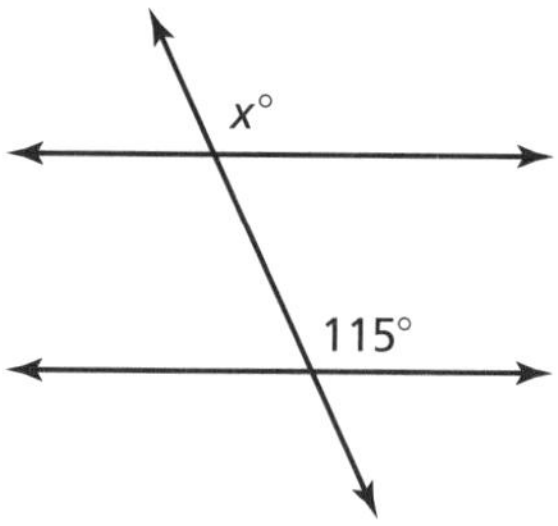

10. The ratio of the side lengths of two rectangles is 2 : 5. What is the ratio of the areas of the two rectangles?

11. Is the polygon convex or concave?

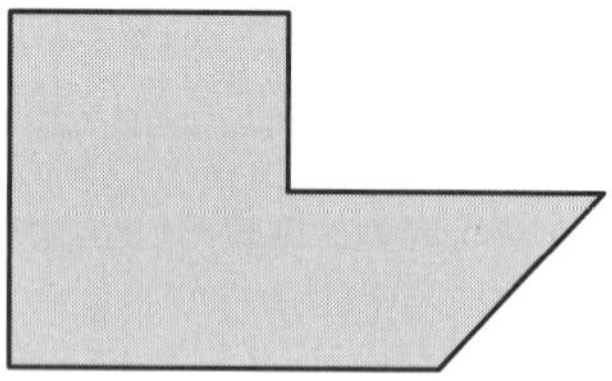

12. The measure of an angle is 43°. What is the measure of an angle that is complementary to that angle?

13. The sum of the angle measures of a polygon is 900°. How many sides does the polygon have?

Chapter 6

Name ______________________________ Date __________

Chapter 6 Square Roots and the Pythagorean Theorem

Dear Family,

When adding or multiplying small numbers, you rely on tables you memorized long ago. For larger numbers, you follow the rules you've learned. For example, when adding large numbers, you line up the place values and start adding from the right, carrying digits to the left.

The "add and carry" method is an example of a rule that follows a strict, predictable procedure. Perhaps surprisingly, not all problems in mathematics have rules that are this straightforward. One of the oldest ways of solving problems is to use the "guess and check" method.

This method requires us to make a reasonable guess about the answer and check how close it is. You then refine your guess and check the new estimate. Each time you do this, you try to get closer to the answer.

Try this with your student to find the square root of a number. For example, to find the square root of 19, you might do the following steps.

- The square root of 16 is 4 $\left(\text{because } 4^2 = 16\right)$ and the square root of 25 is 5 $\left(\text{because } 5^2 = 25\right)$. Because 19 is between 16 and 25, the square root of 19 is greater than 4 and less than 5, so guess 4.5.
- Check: $(4.5)^2 = 20.25$, which is too big, so refine your guess. Try 4.2.
- Check: $(4.2)^2 = 17.64$, which is too small, so refine your guess. Try 4.4.
- Check: $(4.4)^2 = 19.36$, which is getting closer, but still a little too big.

If you continue this method, you will soon find out that $19 \approx (4.36)^2$. You could keep going to get the precision you need.

It may appear that computers and calculators have functions like these memorized, because the answers are shown immediately. However, many types of calculations are done using a process very similar to "guess and check". Because computers and calculators can make millions of guesses per second, the answer simply appears to be memorized.

So don't be afraid to guess the answer—just remember to check it!

Nombre ____ Fecha__________

Capítulo 6 Raíces Cuadradas y el Teorema Pitagórico

Estimada Familia:

Al sumar o multiplicar números pequeños, dependemos de tablas que memorizamos hace muchos años. Para números más grandes, seguimos reglas que hemos aprendido. Por ejemplo, al sumar números grandes, alineamos las posiciones de valores y empezamos a sumar desde el lado derecho, llevando dígitos hacia el lado izquierdo.

El método de "sumar y llevar" es un ejemplo de una regla que sigue un procedimiento estricto y predecible. Quizás, y sorprendentemente, no todos los problemas en matemáticas tienen reglas tan simples como ésta. Una de las formas más antiguas de resolver problemas es usando el método de "predecir y verificar".

Este método requiere que hagamos una predicción razonable sobre la respuesta y que verifiquemos qué tan cerca estamos. Luego refinamos la predicción y verificamos la nueva aproximación. Cada vez que hacemos esto, estamos más cerca de la respuesta.

Intente esto con su estudiante para hallar la raíz cuadrada de un número. Por ejemplo, para encontrar la raíz cuadrada de 19, pueden hacer los siguientes pasos:

- La raíz cuadrada de 16 es 4 $\left(\text{porque } 4^2 = 16\right)$ y la raíz cuadrada de 25 es 5 $\left(\text{porque } 5^2 = 25\right)$. Ya que 19 se encuentra entre 16 y 25, la raíz cuadrada de 19 es mayor que 4 y menor que 5, entonces predecimos 4.5.
- Verifique: $(4.5)^2 = 20.25$, que es demasiado grande, así que refine su predicción. Intente con 4.2.
- Verificar: $(4.2)^2 = 17.64$, que es demasiado pequeño, así que refine su predicción. Intente con 4.4.
- Verificar: $(4.4)^2 = 19.36$, lo cual está más cerca, pero todavía es un poco más grande.

Si continúa con este método, pronto averiguará que $19 \approx (4.36)^2$. Puede continuar para obtener la precisión deseada.

Puede parecer que las computadoras y calculadoras tengan funciones como éstas memorizadas, ya que las respuestas se muestran inmediatamente. Sin embargo, muchos tipos de cálculos se realizan con un proceso muy similar al de "predecir y verificar". Ya que las computadoras y calculadoras pueden hacer millones de predicciones por segundo, la respuesta simplemente aparece como memorizada.

Así que no tema predecir la respuesta—¡sólo recuerde verificarla!

Non ______________________________ Dat __________

Chapít 6 Rasin Kare ak Teyorèm Pitagò a

Chè Fanmi:

Lè w'ap adisyone oswa miltipliye ti chif, ou fye ou ak tab ou te aprann pa kè sa fè lontan. Pou gwo chif, ou swiv règ ou aprann. Paregzanp, lè w'ap adisyone gwo chif, ou aliyen valè pozisyon yo epi ou kòmanse adisyone apatide bò dwat la, retni chif sou bò gòch la.

Metòd "adisyone ak retni" an se yon egzanp règ ki swiv yon pwosedi estrik, san sipriz. Petèt sa ap fè ou sezi, se pa tout pwoblèm nan matematik ki gen règ ki senp konsa. Youn nan mannyè pi ansyen pou rezoud pwoblèm se sèvi avèk metòd "sipoze ak verifye" a.

Metòd sa a egzije pou nou fè yon sipozisyon rezonab sou repons la epi verifye nan ki pwen li pwòch. Apre sa ou rafine sipozisyon ou an epi ou verifye nouvo estimasyon an. Chak fwa ou fè sa, ou eseye vin pi pre repons la.

Eseye sa avèk elèv ou a pou jwenn rasin kare yon chif. Paregzanp, pou jwenn rasin kare 19, ou gen dwa pase pa etap sila yo.

- Rasin kare 16 se 4 $\left(\text{paske } 4^2 = 16\right)$ epi rasin kare 25 se 5 $\left(\text{paske } 5^2 = 25\right)$. Poutèt 19 nan mitan 16 ak 25, rasin kare 19 pi gran pase 4 ak pi piti pase 5, donk sipoze 4.5.
- Verifye: $(4.5)^2 = 20.25$, ki twò gran, donk rafine sipozisyon ou an. Eseye 4.2.
- Verifye: $(4.2)^2 = 17.64$, ki twò piti, donk rafine sipozisyon ou an. Eseye 4.4.
- Verifye: $(4.4)^2 = 19.36$, ki pi pre, men ki toujou yon ti jan twò gran.

Si ou kontinye metòd sa a, w'ap jwenn byento ke $19 \approx (4.36)^2$. Ou ta kapab kontinye ale pou jwenn presizyon ou bezwen an.

Sa gen dwa sanble ke òdinatè ak kalkilatris gen fonksyon tankou sa yo nan memwa yo, poutèt yo montre repons yo imedyatman. Sepandan, anpil tip kalkil fèt avèk yon pwosede ki sanblan anpil ap "sipoze ak verifye." Poutèt òdinatè ak kalkilatris kapab fè plizyè milyon sipozisyon pa segonn, repons la senpleman sanble li nan memwa li.

Donk ou pa bezwen pè sipoze repons la—annik sonje verifye li!

Activity 6.1 Start Thinking!
For use before Activity 6.1

When you know the area of a rectangle, can you determine the lengths of its sides? Why or why not?

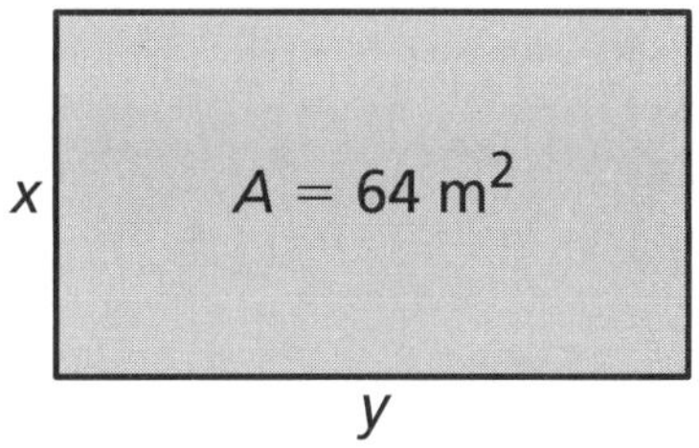

When you know the area of a square, can you determine the lengths of its sides? Why or why not?

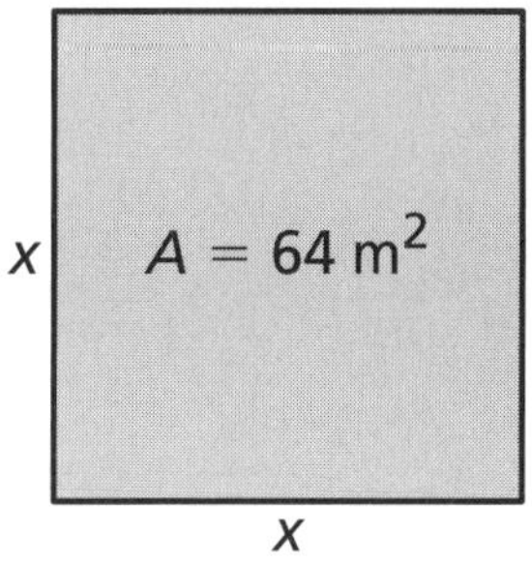

Activity 6.1 Warm Up
For use before Activity 6.1

Find the product.

1. 12×12

2. 9×9

3. 18×18

4. 1.6×1.6

5. 2.5×2.5

6. $\frac{2}{3} \times \frac{2}{3}$

Lesson 6.1

Start Thinking!

For use before Lesson 6.1

Shelley says that there are two solutions to the equation $x^2 = 400$. Gina says that there is only one solution. Who is correct? Explain.

Lesson 6.1

Warm Up

For use before Lesson 6.1

Find the side length of the square. Check your answer by multiplying.

1.

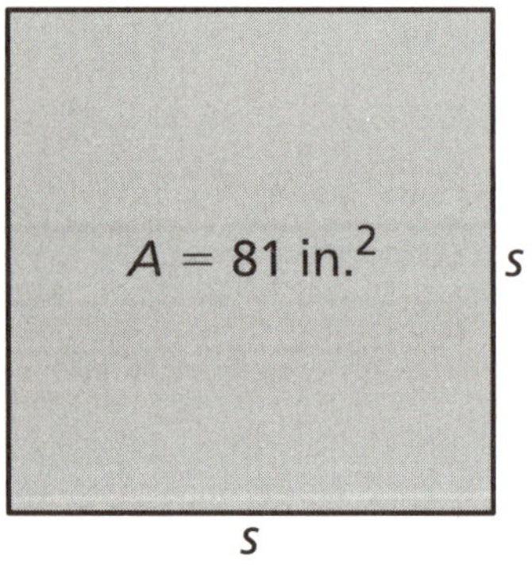

2.

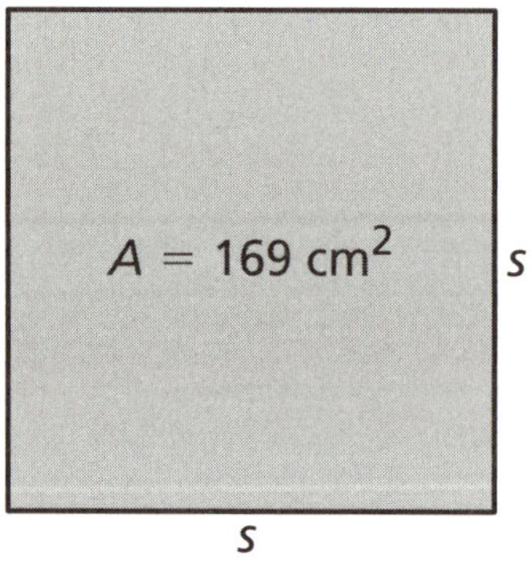

3.

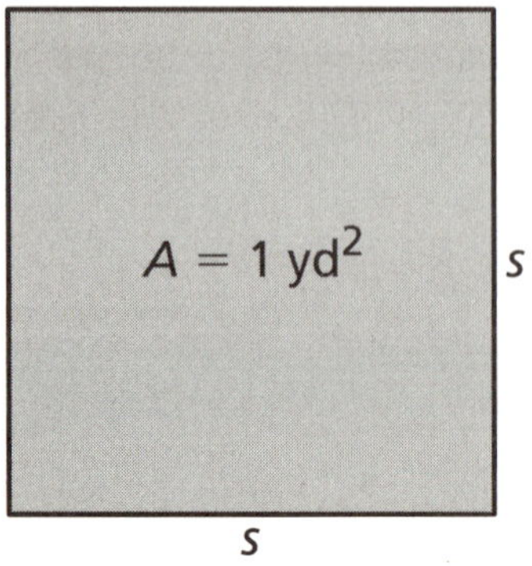

4.

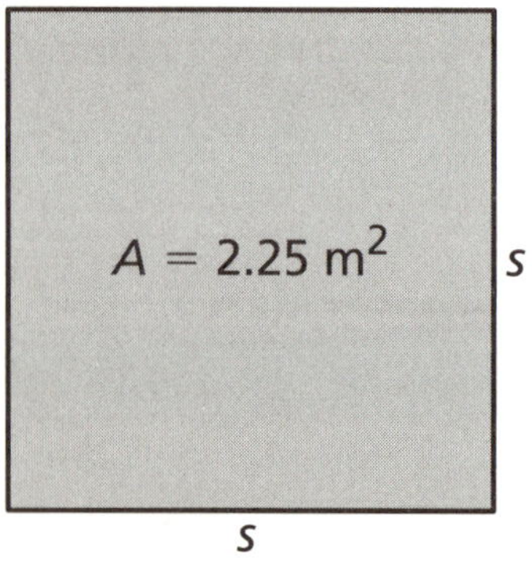

Name____ Date__________

6.1 Practice A

Find the side length of the square. Check your answer by multiplying.

1. Area $= 196 \text{ in.}^2$

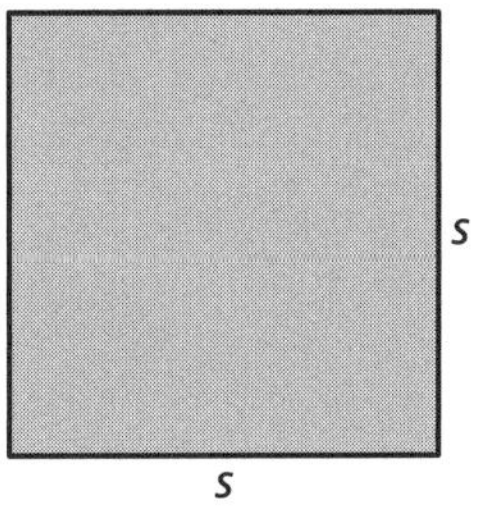

2. Area $= \dfrac{49}{81} \text{ m}^2$

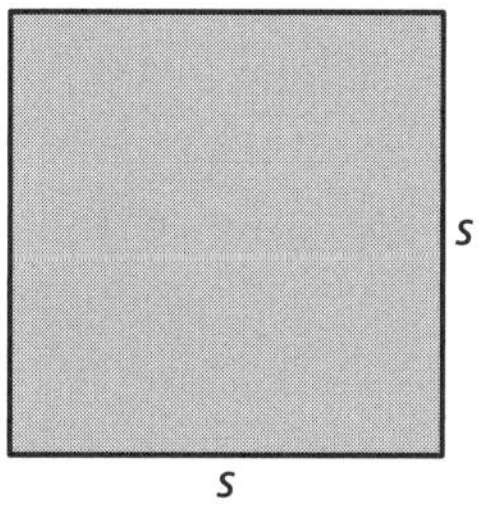

Find the two square roots of the number.

3. 16

4. 0

Find the square root(s).

5. $\sqrt{121}$

6. $-\sqrt{\dfrac{1}{36}}$

7. $\pm\sqrt{\dfrac{289}{49}}$

8. $-\sqrt{0.64}$

Evaluate the expression.

9. $2\sqrt{25} + 3$

10. $7 - 12\sqrt{\dfrac{1}{9}}$

Copy and complete the statement with <, >, or =.

11. $\sqrt{64}$ __?__ 5

12. 0.6 __?__ $\sqrt{0.49}$

13. The volume of a right circular cylinder is represented by $V = \pi r^2 h$, where r is the radius of the base (in feet). What is the radius of a right circular cylinder when the volume is 144π cubic feet and the height is 9 feet?

14. The cost C (in dollars) of producing x widgets is represented by $C = 4.5x^2$. How many widgets are produced if the cost is \$544.50?

15. Two squares are drawn. The larger square has area of 400 square inches. The areas of the two squares have a ratio of 1 : 4. What is the side length s of the smaller square?

Name ______________________________ Date __________

6.1 Practice B

Find the side length of the square. Check your answer by multiplying.

1. Area $= \dfrac{169}{225}\ \text{cm}^2$

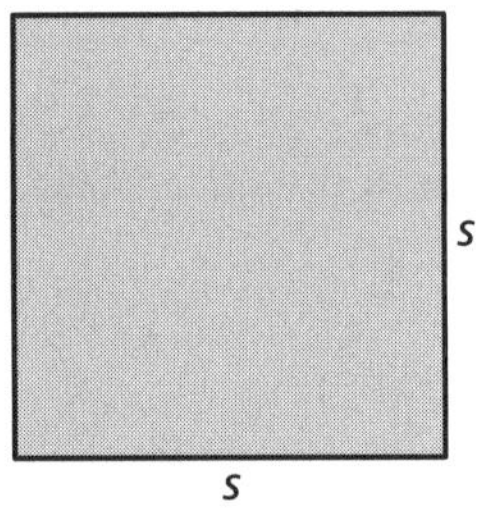

2. Area $= 2.56\ \text{yd}^2$

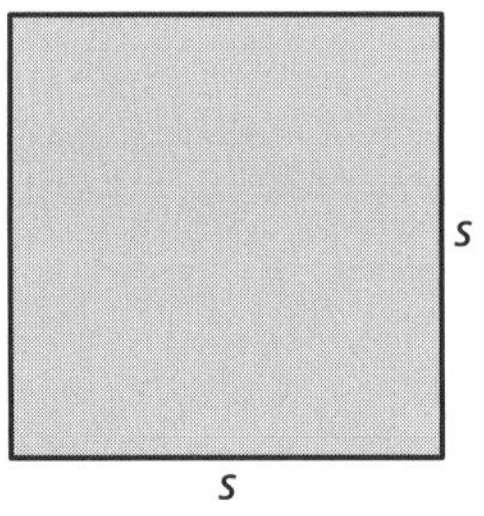

Find the two square roots of the number.

3. 225

4. 400

Find the square root(s).

5. $-\sqrt{484}$

6. $\pm\sqrt{\dfrac{25}{64}}$

7. $\sqrt{6.25}$

8. $\pm\sqrt{1.69}$

Evaluate the expression.

9. $6\sqrt{2.25} - 4.2$

10. $3\left(\sqrt{\dfrac{48}{3}} - 2\right)$

Copy and complete the statement with $<$, $>$, or $=$.

11. $\sqrt{\dfrac{49}{9}}\ \underline{\ \ ?\ \ }\ 2$

12. $\dfrac{2}{5}\ \underline{\ \ ?\ \ }\ \sqrt{\dfrac{12}{75}}$

13. The area of a sector of a circle is represented by $A = \dfrac{5}{18}\pi r^2$, where r is the radius of the circle (in meters). What is the radius when the area is 40π square meters?

14. Is the quotient of two perfect squares always a perfect square? Explain your reasoning.

15. Two squares are drawn. The smaller square has an area of 256 square meters. The areas of the two squares have a ratio of 4 : 9. What is the side length s of the larger square?

Name____ Date__________

6.1 Enrichment and Extension

Finding Cube Roots

A square root of a number is a number that when multiplied by itself, equals the given number. A *cube root* of a number is a number that when used as a factor in a product three times, equals the given number. The notation for the cube root of n is $\sqrt[3]{n}$.

Complete the table.

1.

n	n^2	$\sqrt{(n^2)}$	Check
1	1	1	$1 \bullet 1 = 1$
2			
3			
4			
5			

2.

n	n^3	$\sqrt[3]{(n^3)}$	Check
1	1	1	$1 \bullet 1 \bullet 1 = 1$
2			
3			
4			
5			

Find the cube root of the number.

3. 216

4. -8

5. $-\frac{1}{512}$

6. $\frac{64}{729}$

7. A CD case is in the shape of a cube. The volume is 343 cubic inches. What is the length (in inches) of one side of the CD case?

8. There are three numbers that are their own cube roots. What are these numbers?

Name ______________________________ Date __________

6.1 Puzzle Time

How Did The Man At The Seafood Restaurant Cut His Mouth?

Circle the letter of each correct answer in the boxes below. The circled letters will spell out the answer to the riddle.

Find the side length of the square with the given area.

1. Area $= 169$

2. Area $= 576$

3. Area $= \dfrac{49}{64}$

4. Area $= 2.56$

Find the square root(s).

5. $\sqrt{400}$

6. $-\sqrt{225}$

7. $\pm\sqrt{\dfrac{9}{16}}$

8. $\sqrt{\dfrac{36}{25}}$

9. $\pm\sqrt{7.84}$

10. $-\sqrt{56.25}$

Evaluate the expression.

11. $6 - 2\sqrt{81}$

12. $\sqrt{53.29} + \sqrt{2.89}$

13. $\sqrt{21.16} - \sqrt{1.69}$

14. $7\sqrt{\dfrac{25}{49}} + \sqrt{\dfrac{36}{64}}$

15. The bottom of a circular swimming pool has an area of 200.96 square feet. What is the radius of the swimming pool? Use 3.14 for π.

R	**E**	**L**	**C**	**A**	**F**	**T**	**M**	**I**	**H**	**N**	**U**	**S**	**B**	**G**	**R**	**D**
25	±2.8	−10	7.5	1.6	2.3	$\pm\frac{3}{4}$	$4\frac{3}{4}$	13	28	$\frac{7}{8}$	±3.4	$4\frac{1}{3}$	−5.5	−12	30	±5.2
S	**I**	**T**	**W**	**N**	**O**	**P**	**R**	**G**	**D**	**V**	**F**	**I**	**Y**	**S**	**L**	**H**
−15	$3\frac{1}{4}$	−6.5	$5\frac{3}{4}$	3.4	20	±1.8	$\frac{6}{5}$	12	8	−1.6	3.3	24	−6.1	−7.5	14	9

Activity 6.2 Start Thinking!
For use before Activity 6.2

Cut three narrow strips of paper that are 3 inches, 4 inches, and 5 inches long.

Form a triangle using the three strips. What kind of a triangle is formed?

Notice that $3^2 + 4^2 = 5^2$.

Do you know any other lengths of a triangle that would illustrate a similar equation?

Activity 6.2 Warm Up
For use before Activity 6.2

Find the square root(s).

1. $\sqrt{1.44}$

2. $\pm\sqrt{900}$

3. $\sqrt{\frac{4}{9}}$

4. $-\sqrt{441}$

5. $\pm\sqrt{484}$

6. $-\sqrt{2500}$

Lesson 6.2

Start Thinking!

For use before Lesson 6.2

How can you use the Pythagorean Theorem in sports?

Lesson 6.2

Warm Up

For use before Lesson 6.2

Find the missing length of the triangle.

1.

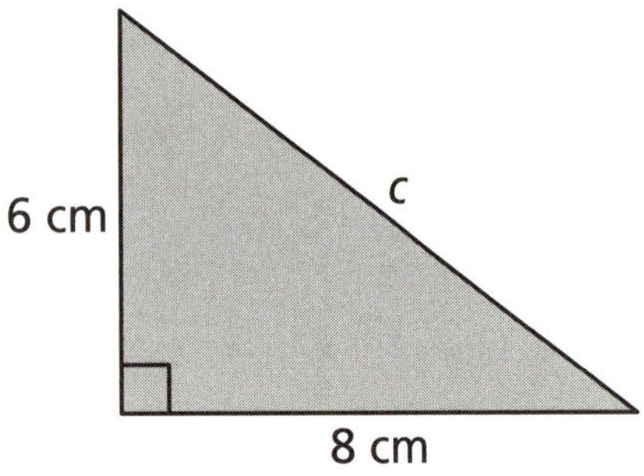

2.

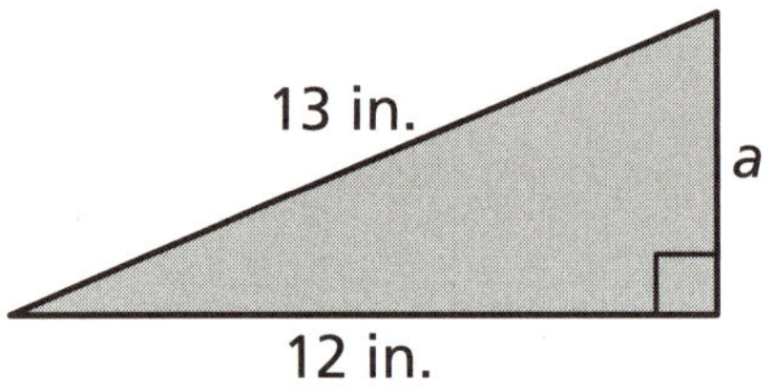

3.

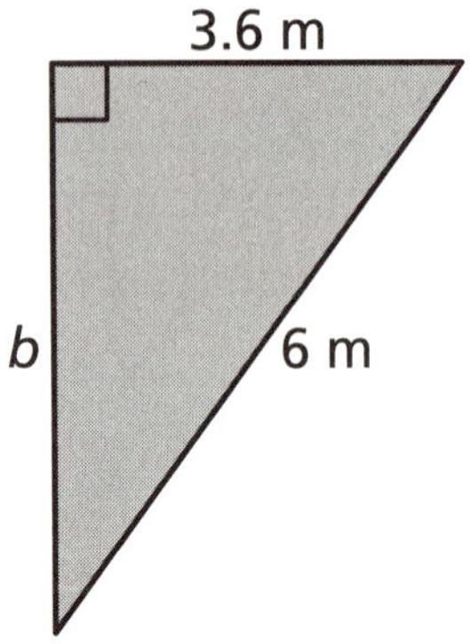

4.

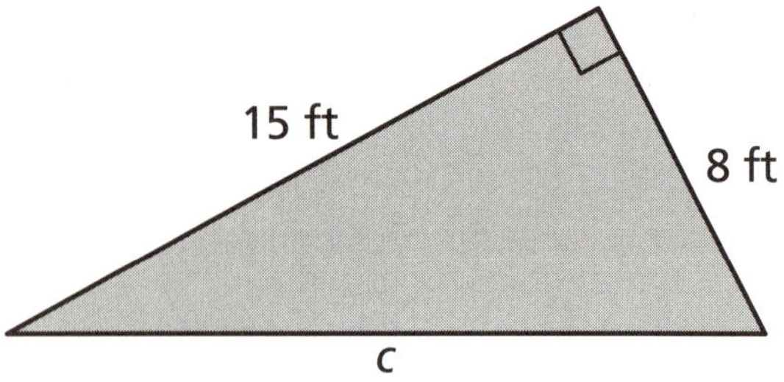

Name____ Date__________

6.2 Practice A

Find the missing length of the triangle.

1.

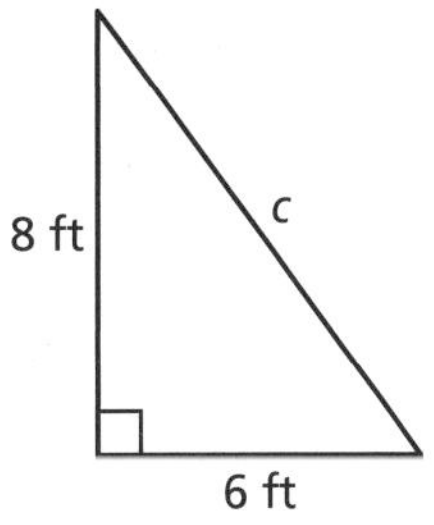

2.

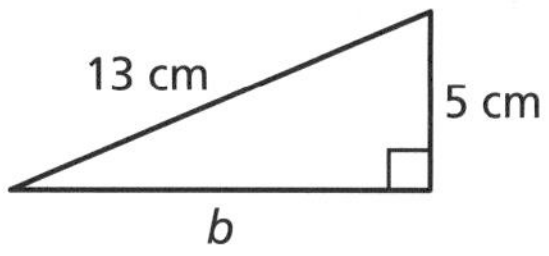

3.

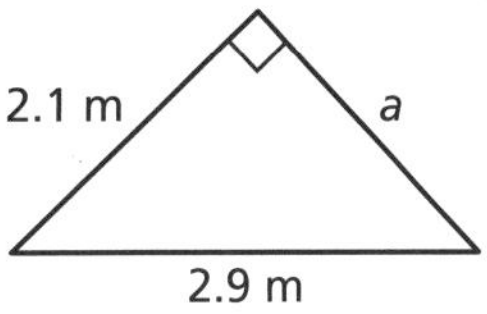

4.

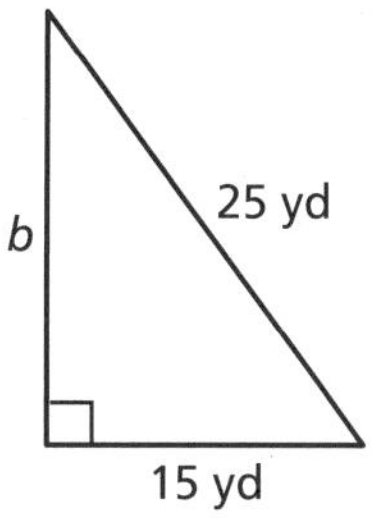

5. A small shelf sits on two braces that are in the shape of a right triangle. The leg (brace) attached to the wall is 4.5 inches and the hypotenuse is 7.5 inches. The leg holding the shelf is the same length as the width of the shelf. What is the width of the shelf?

Find the value of *x*.

6.

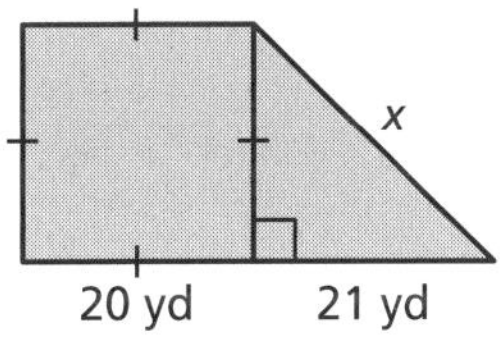

7.

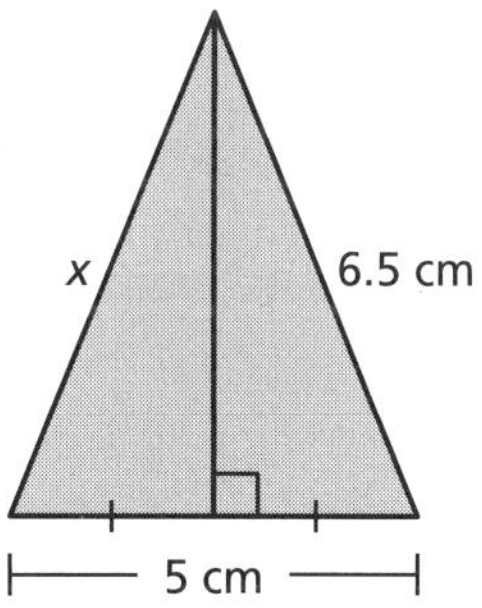

8. Can a right triangle have a leg that is 10 meters long and a hypotenuse that is 10 meters long? Explain.

9. One leg of a right triangular piece of land has a length of 24 yards. The hypotenuse has a length of 74 yards. The other leg has a length of $10x$ yards. What is the value of x?

Name ______________________________ Date __________

6.2 Practice B

Find the missing length of the triangle.

1.

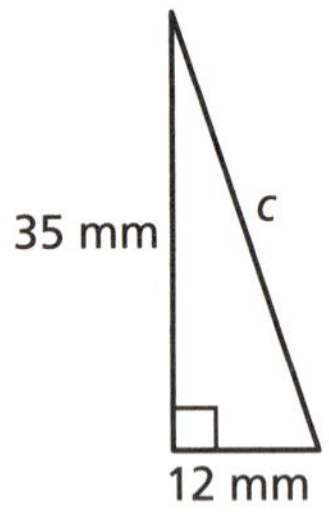

2.

3.

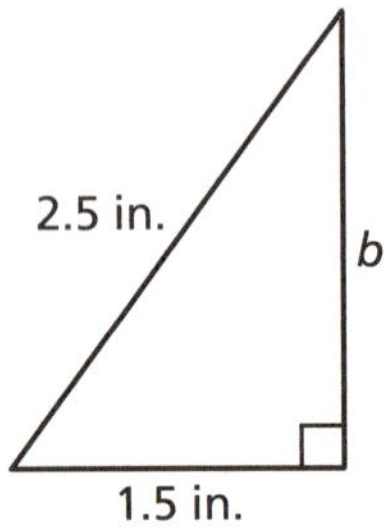

4.

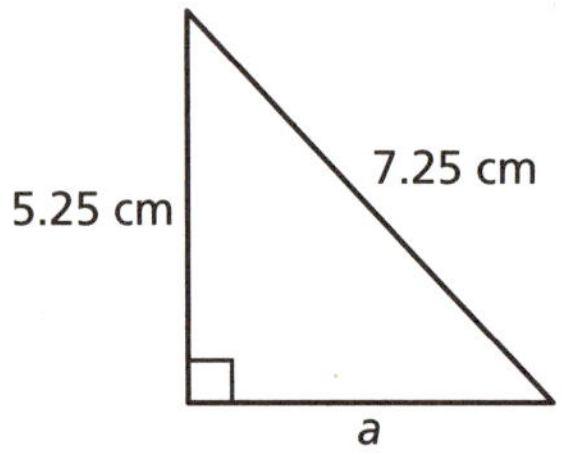

5. You built braces in the shape of a right triangle to hold your surfboard. The leg (brace) attached to the wall is 10 inches and your surfboard sits on a leg that is 24 inches. What is the length of the hypotenuse that completes the right triangle?

6. Laptops are advertised by the lengths of the diagonals of the screen. You purchase a 15-inch laptop and the width of the screen is 12 inches. What is the height of its screen?

7. In a right isosceles triangle, the lengths of both legs are equal. For the given isosceles triangle, what is the value of x?

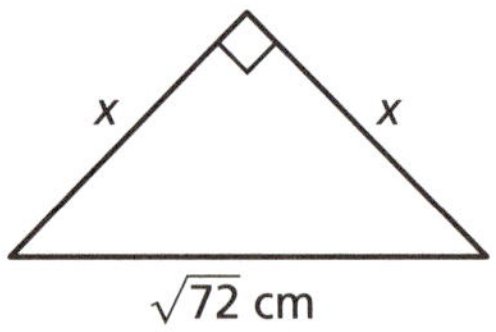

8. To get from your house to your school, you ride your bicycle 6 blocks west and 8 blocks north. A new road is being built that will go directly from your house to your school, creating a right triangle. When you take the new road to school, how many fewer blocks will you be riding to school and back?

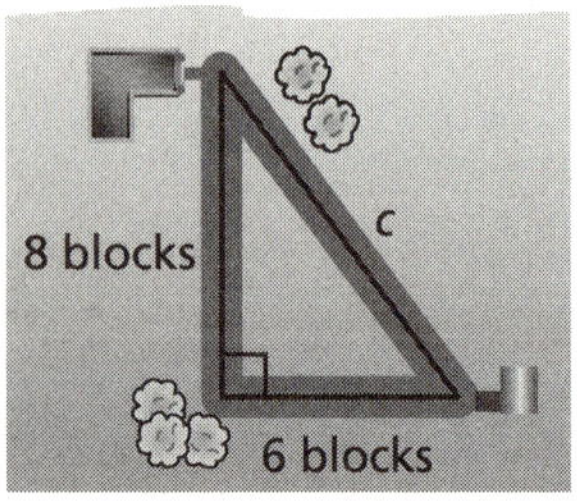

Name______________________________ Date__________

6.2 Enrichment and Extension

The Bermuda Triangle

The Bermuda Triangle is in the Atlantic Ocean between Bermuda, Miami, Florida, and San Juan, Puerto Rico. There are many stories about strange events that occur within the Bermuda Triangle.

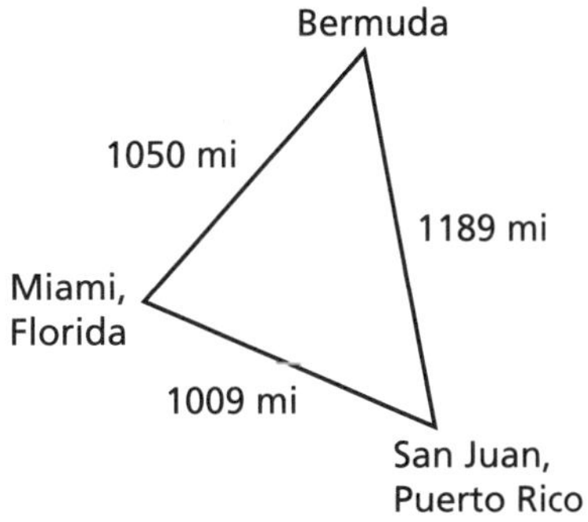

The Bermuda Triangle is not a right triangle. In order to find the area, you need to use a different method.

1. Find the perimeter of the triangle.

2. The semi-perimeter of a triangle is equal to half the perimeter. Find the semi-perimeter s of the triangle.

3. Find the differences between the semi-perimeter and each side of the triangle, $s - a$, $s - b$, and $s - c$.

4. Use the values you found to evaluate the product $R = s(s - a)(s - b)(s - c)$.

5. The area of the triangle is equal to $\sqrt{R}$. What is the area (in square miles) of the Bermuda Triangle?

6. This method of finding the area of a triangle is called Heron's Formula. Use this method to find the area of the triangle below.

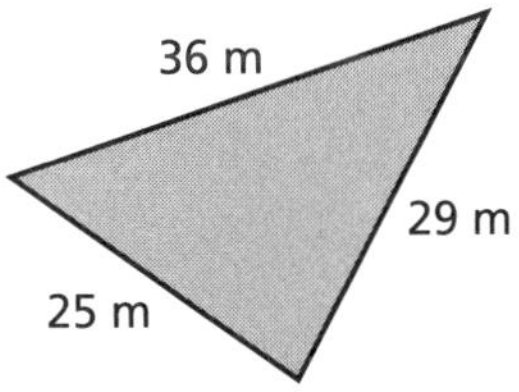

Name ______________________________ Date __________

6.2 Puzzle Time

What Did One Dog Say To The Other Dog?

Write the letter of each answer in the box containing the exercise number.

Find the hypotenuse *c* of the right triangle with the given side lengths *a* and *b*.

1. $a = 15, b = 20$

2. $a = 5, b = 12$

3. $a = 13, b = 84$

4. $a = 65, b = 72$

5. $a = 6, b = 17.5$

6. $a = 6\frac{2}{3}, b = 7$

Find the side length *b* of the right triangle with the given hypotenuse *c* and side length *a*.

7. $c = 61, a = 11$

8. $c = 82, a = 80$

9. $c = 34, a = 16$

10. $c = 65, a = 63$

11. $c = 13, a = 6.6$

12. $c = 10\frac{3}{5}, a = 5\frac{3}{5}$

13. The flap of an envelope has two side lengths that are each 10 centimeters long and meet at a right angle. How long is the envelope? Round your answer to the nearest tenth.

14. A middle school gym is 60 feet wide and 100 feet long. If you stand in one corner of the gym, how many feet away is the corner diagonally across from you? Round your answer to the nearest tenth.

Answers

T. $9\frac{2}{3}$

P. 14.1

E. 18.5

D. 18

N. 25

U. 9

O. 97

H. 116.6

N. 60

G. 30

O. 13

M. 11.2

I. 85

S. 16

10	6	2	13		14	4	12	1	8	3	7	9		11	5

Activity 6.3 Start Thinking!

For use before Activity 6.3

An irrational number is a number that cannot be written as a ratio of integers. Decimals that do not repeat and do not terminate are irrational.

Do you know any examples of irrational numbers?

Activity 6.3 Warm Up

For use before Activity 6.3

Use the Pythagorean Theorem to find the hypotenuse of a right triangle with the given legs.

1. 30, 40

2. 10, 24

3. 16, 30

4. 9, 40

5. 54, 72

6. 2.5, 6

Lesson 6.3

Start Thinking!

For use before Lesson 6.3

How can you find the side length of a square that has the same area as an 8.5-inch by 11-inch piece of paper?

Lesson 6.3

Warm Up

For use before Lesson 6.3

Tell whether the rational number is a reasonable approximation of the square root.

1. $\frac{577}{408}, \sqrt{2}$

2. $\frac{401}{110}, \sqrt{8}$

3. $\frac{271}{330}, \sqrt{21}$

4. $\frac{521}{233}, \sqrt{5}$

5. $\frac{795}{153}, \sqrt{27}$

6. $\frac{441}{150}, \sqrt{12}$

Name____ Date __________

6.3 Practice A

Tell whether the rational number is a reasonable approximation of the square root.

1. $\frac{277}{160}, \sqrt{3}$

2. $\frac{590}{160}, \sqrt{17}$

Tell whether the number is *rational* or *irrational*. Explain.

3. $-\sqrt{14}$

4. $1.\overline{3}$

5. 2.375

6. 4π

7. You are finding the area of a circle with a radius of 2 feet. Is the area a *rational* or *irrational* number? Explain.

Estimate the nearest integer.

8. $\sqrt{33}$

9. $\sqrt{630}$

10. $-\sqrt{8}$

11. $\sqrt{\frac{7}{2}}$

12. A swimming pool is in the shape of a right triangle. One leg has a length of 10 feet and one leg has a length of 15 feet. Estimate the length of the hypotenuse to the nearest integer.

Which number is greater? Explain.

13. $\sqrt{70}, 8$

14. $-\sqrt{16}, 3$

15. $\sqrt{210}, 16\frac{1}{4}$

16. $\sqrt{\frac{4}{25}}, \frac{3}{10}$

17. Find a number a such that $2 < \sqrt{a} < 3$.

18. Is $\sqrt{\frac{1}{9}}$ a rational number? Explain.

19. Is $\sqrt{\frac{5}{9}}$ a rational number? Explain.

20. Is $\sqrt{\frac{2}{18}}$ a rational number? Explain.

Name ________________________________ Date __________

6.3 Practice B

Tell whether the rational number is a reasonable approximation of the square root.

1. $\frac{2999}{490}, \sqrt{41}$

2. $\frac{2298}{490}, \sqrt{22}$

Tell whether the number is *rational* or *irrational*. Explain.

3. $2\frac{2}{9}$

4. $2\pi + 3$

5. $2.\overline{41}$

6. $\sqrt{130}$

7. You are finding the circumference of a circle with a diameter of 10 meters. Is the circumference a *rational* or *irrational* number? Explain.

Estimate the nearest integer.

8. $-\sqrt{\frac{250}{9}}$

9. $\sqrt{395}$

Estimate to the nearest tenth.

10. $\sqrt{0.79}$

11. $\sqrt{1.48}$

12. A patio is in the shape of a square, with a side length of 35 feet. You wish to draw a black line down one diagonal.

    a. Use the Pythagorean Theorem to find the length of the diagonal. Write your answer as a square root.

    b. Find the two perfect squares that the length of the diagonal falls between.

    c. Estimate the length of the diagonal to the nearest tenth.

Which number is greater? Explain.

13. $\sqrt{220}, 14\frac{3}{4}$

14. $-\sqrt{135}, -\sqrt{145}$

15. $\sqrt{\frac{7}{64}}, \frac{3}{8}$

16. $-0.25, -\sqrt{\frac{1}{4}}$

17. Find two numbers a and b such that $7 < \sqrt{a} < \sqrt{b} < 8$.

Name____ Date__________

6.3 Enrichment and Extension

Approximating Square Roots

Before there were calculators and computers, mathematicians developed several methods of approximating square roots by hand. One popular method is sometimes called the divide-and-average method. It uses the following steps.

Use the divide-and-average method to calculate $\sqrt{47}$.

1. What two perfect squares is 47 between?

2. Let $g = \sqrt{47}$. Estimate g to the nearest whole number.

3. Find the quotient $q = 47 \div g$. Round your answer to two decimal places.

4. Find the average of g and q. This gives the approximate value of $\sqrt{47}$. To get a closer approximation, you can repeat this process multiple times by using the average as g.

5. Check the accuracy by squaring the average and comparing it to 47. How close are the numbers?

6. Use this method to estimate $\sqrt{30}$ by repeating the process three times. How close is the square of the estimate and 30?

Name ______________________________ Date ________

6.3 Puzzle Time

Did You Hear About...

A	B	C	D	E	F
G	H	I	J	K	L
M					

Complete each exercise. Find the answer in the answer column. Write the word under the answer in the box containing the exercise letter.

18.5 CLAW
$\sqrt{\frac{4}{9}}$ HE
7 BECAME
−34 SEASHELL
6.8 IN
$\sqrt{55}$ OCEAN
11.7 BELIEVED
14 THE
$-\sqrt{83}$ COURT
12 A

Estimate to the nearest integer.

A. $\sqrt{195}$

B. $-\sqrt{1220}$

C. $-\sqrt{306}$

D. $\sqrt{\frac{315}{6}}$

Which number is greater?

E. $\sqrt{55}$, 12

F. $-\sqrt{83}$, −9

G. $-\sqrt{0.75}$, $-\sqrt{0.85}$

H. $\sqrt{\frac{4}{9}}$, $\frac{1}{2}$

Estimate to the nearest tenth.

I. $\sqrt{137}$

J. $\sqrt{45.9}$

K. $\sqrt{342.5}$

L. $\sqrt{\frac{38}{7}}$

M. You are standing 15 feet from a 25-foot tall tree. Estimate the distance from where you are standing to the top of the tree? Round your answer to the nearest tenth.

−9 POLICEMAN
13 DUTY
−18 SAND
−35 LOBSTER
$-\sqrt{0.85}$ CLAM
2.3 AND
29.2 ORDER
$-\sqrt{0.75}$ BECAUSE
$\frac{1}{2}$ LAWYER
−17 THAT

Activity 6.4

Start Thinking!

For use before Activity 6.4

Use a ruler and a protractor to draw a regular pentagon with side lengths 1 inch long. (*Hint*: First find the measure of an interior angle of a regular pentagon.)

Use a ruler to verify that the length of a diagonal of a regular pentagon with 1-inch sides is equal to the golden ratio, $\dfrac{1 + \sqrt{5}}{2}$ inch.

Activity 6.4

Warm Up

For use before Activity 6.4

Use a calculator to find a decimal approximation of the expression. Round your answer to the nearest thousandth.

1. $\dfrac{\sqrt{7}}{7}$

2. $\dfrac{\sqrt{3}}{2}$

3. $\dfrac{1 + \sqrt{3}}{2}$

4. $\dfrac{\sqrt{3} - 1}{3}$

5. $\dfrac{2 + \sqrt{2}}{3}$

6. $\dfrac{2 - \sqrt{2}}{4}$

Lesson 6.4 Start Thinking!
For use before Lesson 6.4

In previous courses, you have learned how to simplify fractions. When is a fraction simplified?

Square roots can also be simplified.

A square root is simplified when the number under the radical sign has no perfect square factors other than 1.

Which of the following expressions are simplified? Explain why.

$$\sqrt{2}, \sqrt{4}, \sqrt{10}, \sqrt{50}, 3\sqrt{5}, 3\sqrt{8}$$

Lesson 6.4 Warm Up
For use before Lesson 6.4

Find the ratio of the side lengths. Is the ratio close to the golden ratio?

1.

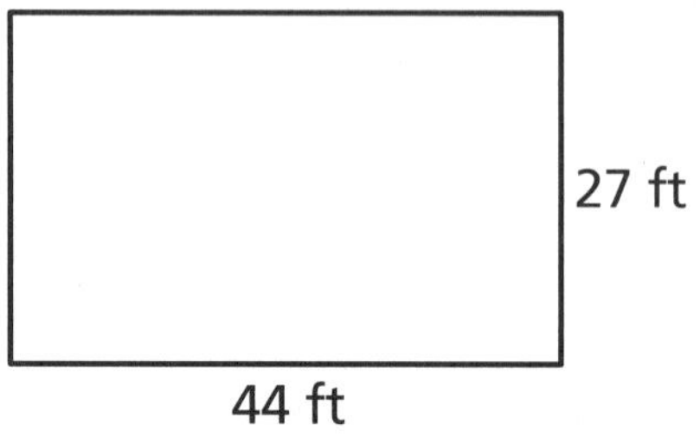

2.

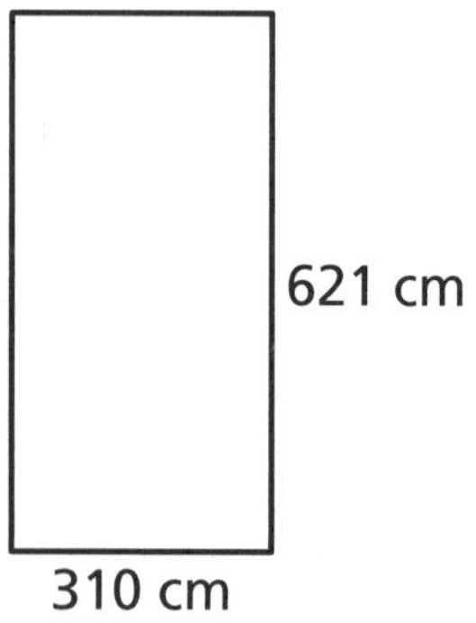

Name______________________________ Date__________

6.4 Practice A

Simplify the expression.

1. $5\sqrt{2} + 4\sqrt{2}$

2. $9\sqrt{5} - 4\sqrt{5}$

3. $\frac{1}{3}\sqrt{7} + \frac{2}{3}\sqrt{7}$

4. $\sqrt{10} - 8\sqrt{10}$

5. $\frac{1}{4} + \frac{\sqrt{5}}{4}$

6. $\frac{\sqrt{3}}{6} + \frac{1}{6}$

7. The side lengths of a triangle are $4\sqrt{2}$, $\sqrt{2}$, and 5. What is the perimeter of the triangle?

Simplify the expression.

8. $\sqrt{20}$

9. $\sqrt{32}$

10. $\sqrt{50}$

11. $\sqrt{\frac{7}{16}}$

12. $\sqrt{\frac{11}{25}}$

13. $\sqrt{\frac{33}{144}}$

14. The area of a square is 24 square centimeters. Find the side length s of the square.

Simplify the expression.

15. $4\sqrt{3} + \sqrt{27}$

16. $\sqrt{50} - 4\sqrt{18}$

17. The ratio $7 : x$ is equivalent to the ratio $x : 5$. What are the possible values for x?

18. You are designing a table in the shape of a right triangle. The side lengths are 20 inches and 10 inches.

    a. What is the length of the hypotenuse?

    b. You reduce the side lengths by half, resulting in side lengths of 10 inches and 5 inches. What is the length of the hypotenuse?

    c. What happened to the length of the hypotenuse when the side lengths were reduced by half?

Name ______________________________ Date __________

6.4 Practice B

Simplify the expression.

1. $9\sqrt{11} - 4\sqrt{11}$

2. $\frac{3}{5}\sqrt{10} - \frac{4}{5}\sqrt{10}$

3. $\frac{5}{6}\sqrt{15} + \frac{3}{6}\sqrt{15}$

4. $\sqrt{7} - \sqrt{7}$

5. $\frac{1}{5} + \frac{\sqrt{11}}{5}$

6. $\frac{\sqrt{3}}{2} - \frac{1}{2}$

7. The length of a rectangle is $5\sqrt{3}$ inches and the width is $2\sqrt{3}$. What is the perimeter of the rectangle?

Simplify the expression.

8. $\sqrt{98}$

9. $\sqrt{300}$

10. $\sqrt{80}$

11. $\sqrt{\frac{14}{169}}$

12. $\sqrt{\frac{7}{625}}$

13. $\sqrt{\frac{67}{100}}$

14. The area of a circle is 40π square meters. What is its radius?

Simplify the expression.

15. $\sqrt{8} - \frac{9}{4}\sqrt{2}$

16. $\sqrt{128} + 3\sqrt{200}$

17. The ratio $6 : x$ is equivalent to the ratio $x : 10$. What are the possible values for x?

18. You are designing an orange right circular cone to block off a parking space. It has a height of 60 centimeters and a volume of 240π cubic centimeters.

a. What is the radius of the cone?

b. You double the height of the cone to 120 centimeters and the volume of the cone to 480π cubic centimeters. What is the radius of the cone?

c. When the height and volume were doubled, what happened to the radius?

Name____ Date__________

6.4 Enrichment and Extension

Simplifying Square Roots

Simplify the expression. Find the answer in the grid below and write the number of the exercise next to the appropriate dot. When you have completed all twelve exercises, connect the dots in order according to the exercise numbers and connect the last point to the first point. What polygon is formed in the grid?

1. $\sqrt{3} + 8\sqrt{3}$
2. $13\sqrt{2} + \sqrt{32}$
3. $9\sqrt{11} + \sqrt{99}$
4. $4\sqrt{7} - \sqrt{112}$
5. $15\sqrt{5} - \sqrt{80}$
6. $3\sqrt{6} - \sqrt{216}$
7. $\sqrt{\dfrac{17}{36}}$
8. $\sqrt{\dfrac{35}{729}}$
9. $\sqrt{\dfrac{13}{100}}$
10. $\sqrt{(15)(60)}$
11. $\sqrt{(13)(52)}$
12. $\sqrt{(16)(12)(27)}$

• $\dfrac{17}{6}$	• $9\sqrt{3}$	• $-4\sqrt{21}$	• $12\sqrt{11}$	• $\dfrac{\sqrt{42}}{81}$
• 72	• $9\sqrt{6}$	• $17\sqrt{2}$	• $\dfrac{35}{\sqrt{27}}$	• 0
• $29\sqrt{2}$	• 26	• $3\sqrt{6}$	• $11\sqrt{5}$	• $5\sqrt{5}$
• 30	• $9\sqrt{110}$	• $\dfrac{\sqrt{35}}{27}$	• $4\sqrt{13}$	• $-3\sqrt{6}$
• 900	• $\dfrac{\sqrt{13}}{10}$	• $\sqrt{0.13}$	• $\dfrac{\sqrt{17}}{6}$	• $8\sqrt{27}$

Name ______________________ Date __________

6.4 Puzzle Time

Why Shouldn't You Give A Little Girl Spaghetti Late At Night?

Circle the letter of each correct answer in the boxes below. The circled letters will spell out the answer to the riddle.

Simplify the expression.

1. $\frac{\sqrt{5}}{2} + \frac{\sqrt{5}}{2}$

2. $8\sqrt{11} - 5\sqrt{11}$

3. $\frac{1}{3}\sqrt{6} + \frac{2}{3}\sqrt{6}$

4. $\frac{1}{5}\sqrt{2} - \frac{6}{5}\sqrt{2}$

5. $3.7\sqrt{3} - 1.7\sqrt{3}$

6. $4.8\sqrt{2} + 2.2\sqrt{2}$

7. $\sqrt{325}$

8. $\sqrt{192}$

9. $\sqrt{40}$

10. $\sqrt{63}$

11. $\sqrt{\frac{13}{144}}$

12. $\sqrt{\frac{27}{100}}$

13. $2\sqrt{3} + \sqrt{48}$

14. $\sqrt{54} - 5\sqrt{6}$

15. $\frac{3}{4}\sqrt{2} + \frac{1}{4}\sqrt{18}$

16. $(\sqrt{15})(\sqrt{21})(\sqrt{35})$

I	R	T	M	I	A	S	G	P	D	A	L	S	T	A
$8\sqrt{3}$	$8\sqrt{11}$	$5\sqrt{13}$	$2\sqrt{5}$	$-\sqrt{2}$	$-\sqrt{6}$	$3\sqrt{7}$	$9\sqrt{3}$	$\frac{3\sqrt{2}}{2}$	45	$7\sqrt{2}$	$3\sqrt{8}$	$\sqrt{5}$	$\frac{\sqrt{13}}{12}$	105
S	**B**	**U**	**E**	**N**	**D**	**O**	**H**	**T**	**F**	**I**	**C**	**M**	**E**	**R**
$-\sqrt{3}$	$-2\sqrt{6}$	25	$3\sqrt{11}$	$7\sqrt{13}$	$\frac{3\sqrt{3}}{10}$	$\sqrt{2}$	305	$2\sqrt{10}$	$3\sqrt{6}$	$\sqrt{6}$	$2\sqrt{13}$	$6\sqrt{3}$	$2\sqrt{3}$	$5\sqrt{2}$

Activity 6.5 Start Thinking!
For use before Activity 6.5

How can you use the Pythagorean Theorem to find the height of a kite?

Activity 6.5 Warm Up
For use before Activity 6.5

Find the missing length of the triangle. Round your answer to the nearest tenth.

1.

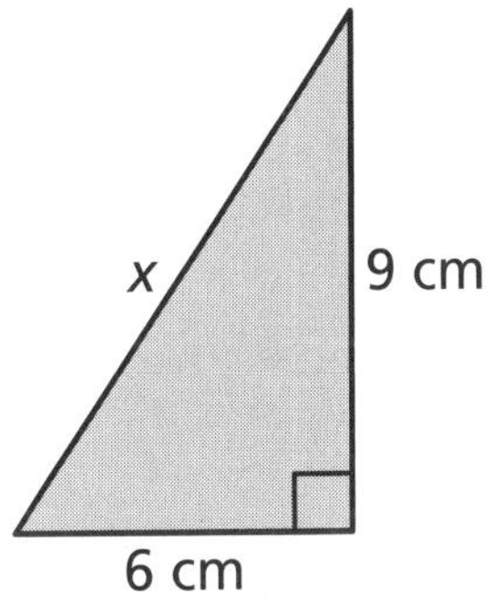

2.

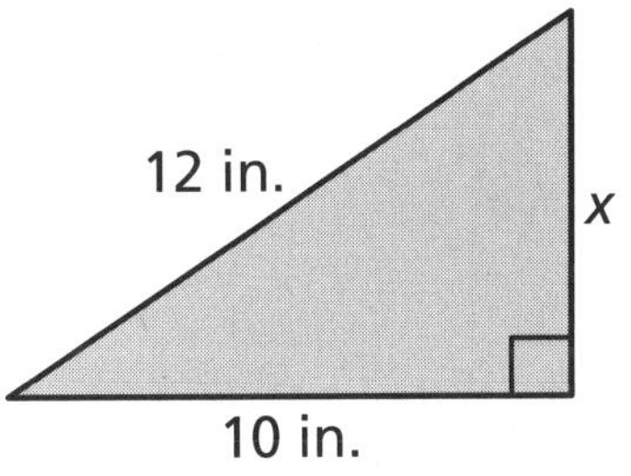

3.

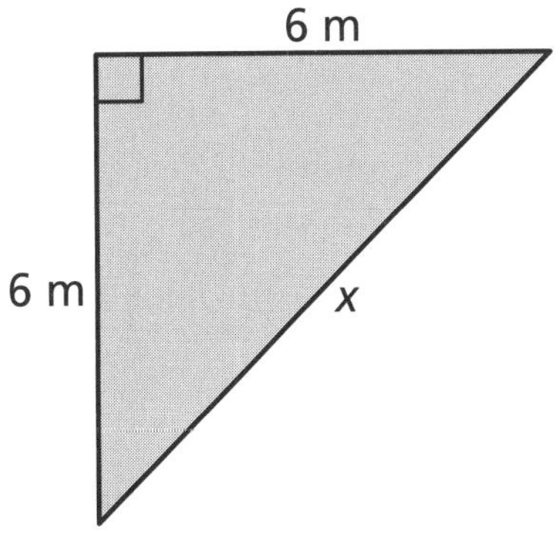

4.

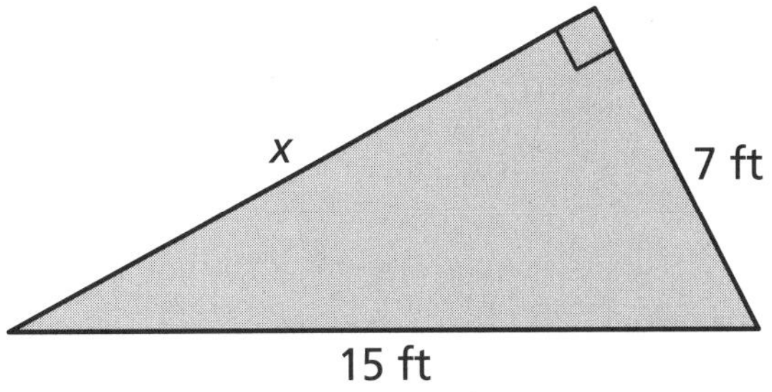

Lesson 6.5

Start Thinking!

For use before Lesson 6.5

Write a word problem that can be solved using the Pythagorean Theorem. Be sure to include a sketch of the situation.

Lesson 6.5

Warm Up

For use before Lesson 6.5

Find the perimeter of the figure. Round your answer to the nearest tenth.

1. Right triangle

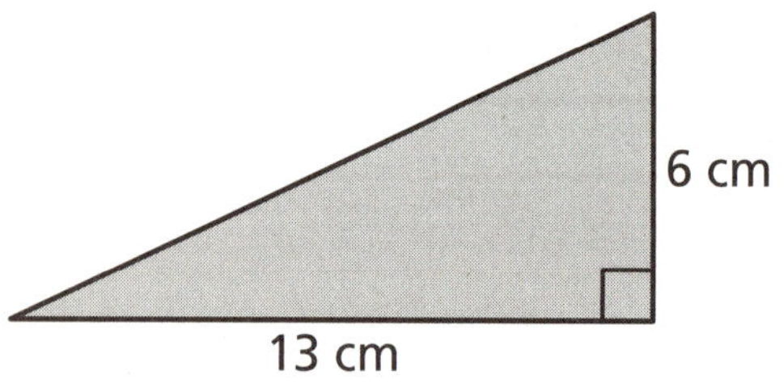

2. Right triangle

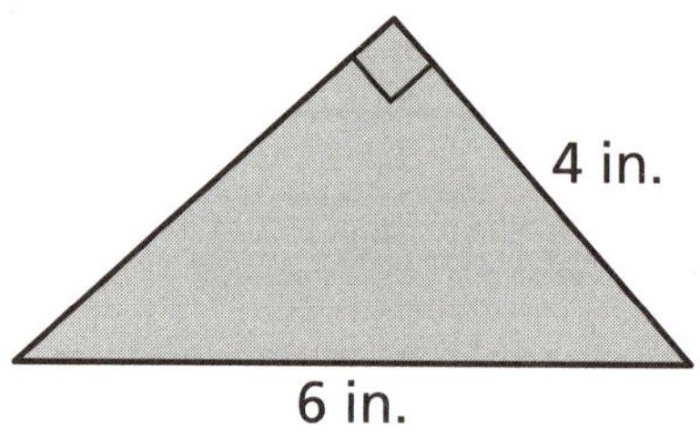

3. Square

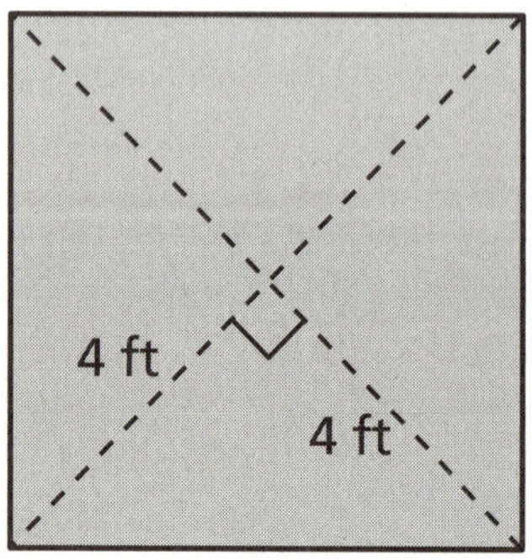

4. Parallelogram

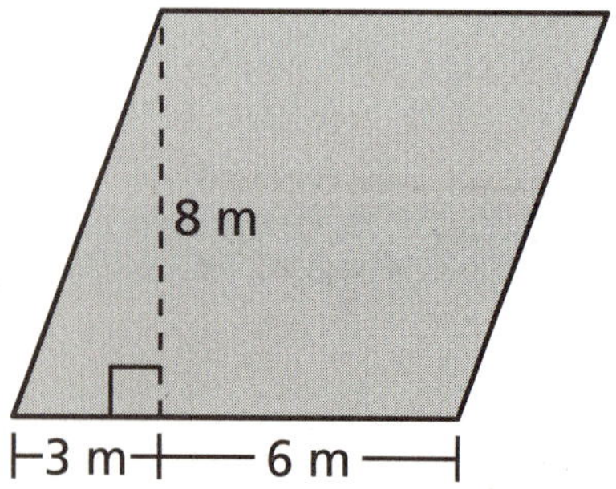

Name____ Date__________

6.5 Practice A

Find the perimeter of the figure. Round your answer to the nearest tenth.

1. Right Triangle

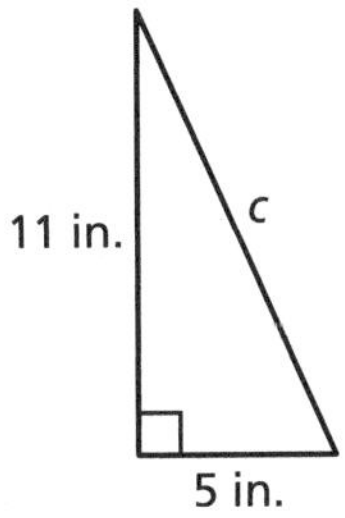

2. Parallelogram

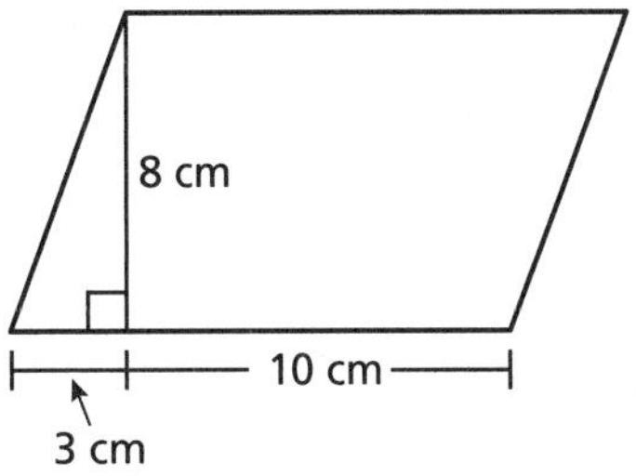

Find the distance *d*. Round your answer to the nearest tenth.

3.

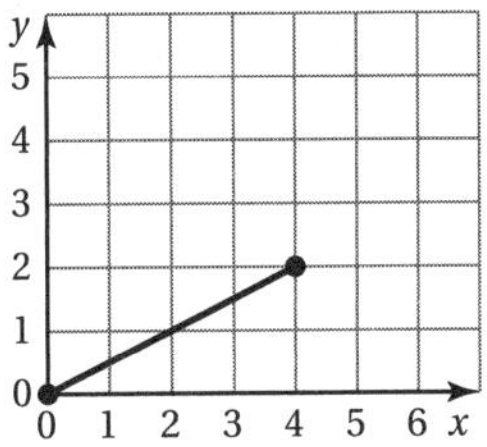

4.

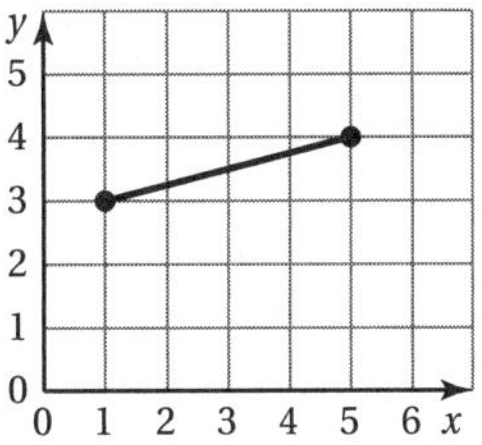

Estimate the height. Round your answer to the nearest tenth.

5.

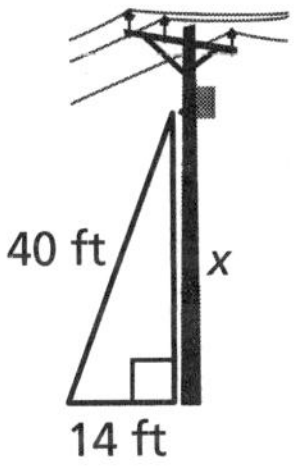

6.

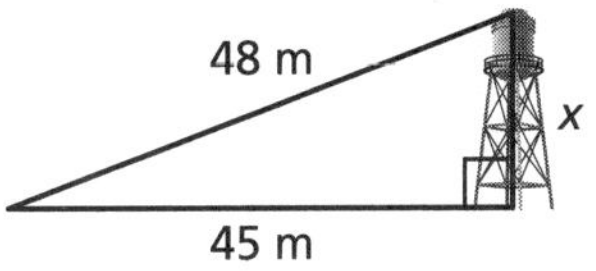

Tell whether the triangle with the given side lengths is a right triangle.

7. 20 ft, 21 ft, 29 ft

8. $\frac{3}{5}$ m, 1 m, $\frac{6}{5}$ m

9. On the Junior League baseball field, you run 60 feet to first base and then 60 feet to second base. You are out at second base and then run directly along the diagonal to home plate. Find the distance that you ran. Round your answer to the nearest tenth.

Name ______________________________ Date __________

6.5 Practice B

Find the perimeter of the figure. Round your answer to the nearest tenth.

1. Parallelogram

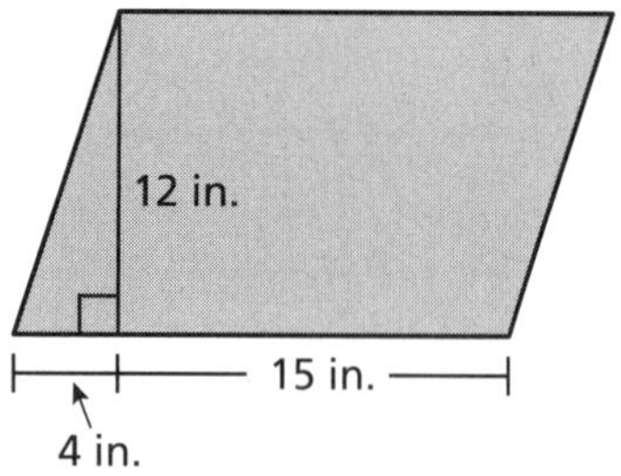

2. Square

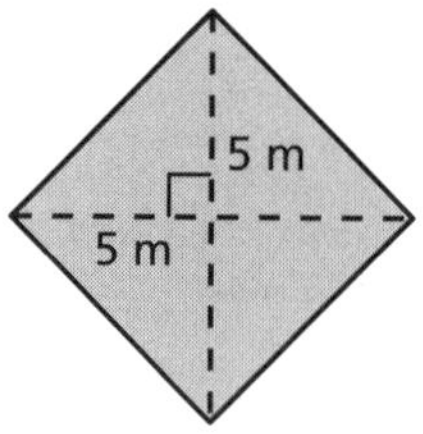

Find the distance *d*. Round your answer to the nearest tenth.

3.

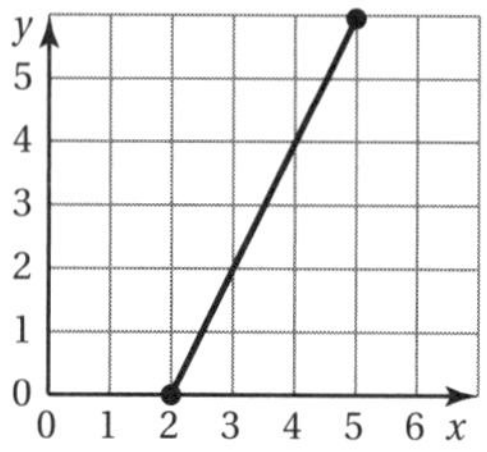

4.

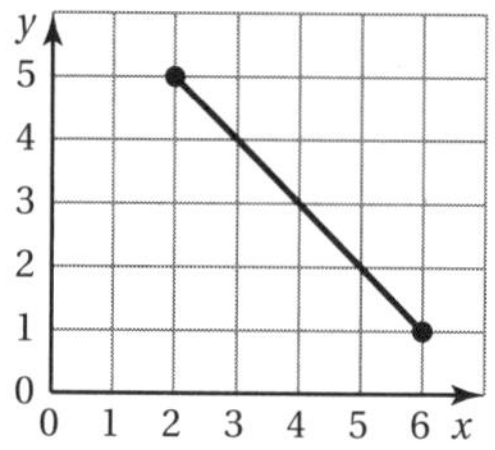

Estimate the height. Round your answer to the nearest tenth.

5.

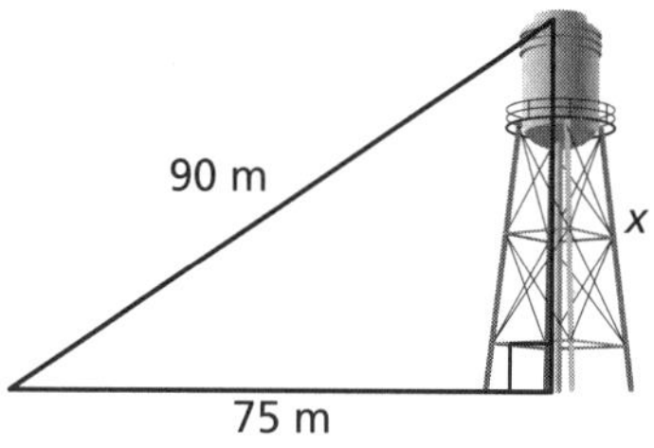

6.

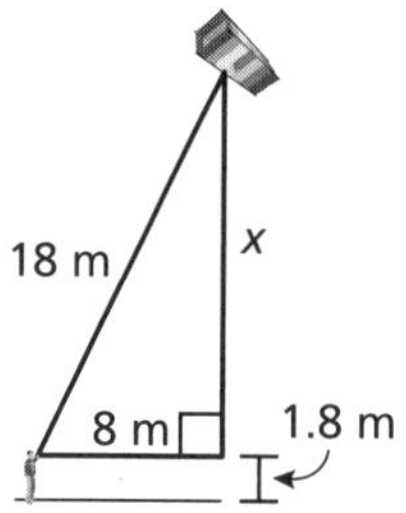

Tell whether the triangle with the given side lengths is a right triangle.

7. $\frac{3}{20}$ cm, $\frac{1}{5}$ cm, $\frac{1}{4}$ cm

8. 4 ft, 9.6 ft, 10.4 ft

9. You are creating a flower garden in the triangular shape shown. You purchase edging to go around the flower garden. The edging costs $1.50 per foot. What is the cost of the edging? Round your lengths to the nearest whole number.

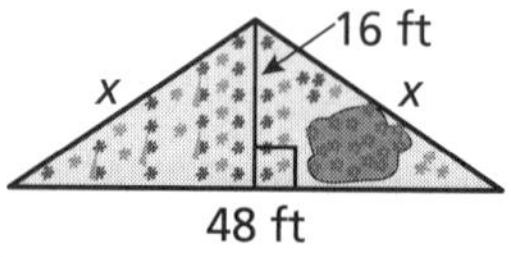

Name____ Date__________

6.5 Enrichment and Extension

Making Pythagorean Triples

You can generate a Pythagorean triple by picking a value for b and using a system of equations to find a and c.

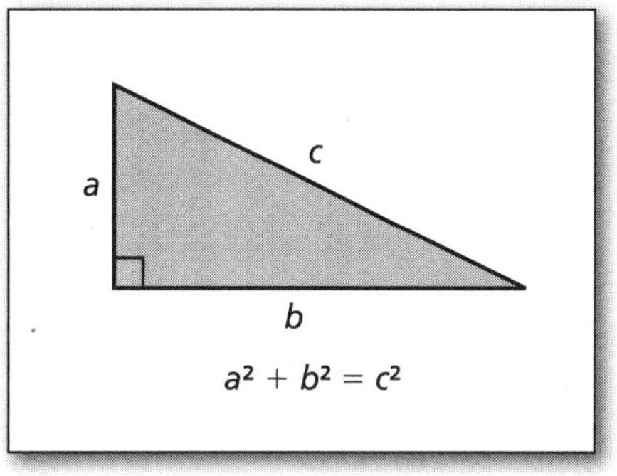

Let $b = 20$.

1. Find b^2.

2. Factor b^2 into the product of 8 and a number.

3. Write a system of linear equations. Set $c + a$ equal to the larger factor and $c - a$ equal to the smaller factor.

4. Solve the system of linear equations.

5. Now you have values for a, b, and c. Use the Converse of the Pythagorean Theorem to check that a triangle with these side lengths is a right triangle.

6. Use the same method to generate a Pythagorean triple using

   a. $b = 24$ and 18 as a factor of b^2.

   b. $b = 15$ and 9 as a factor of b^2.

Name ______________________________ Date ________

6.5 Puzzle Time

What Are Twins' Favorite Kind Of Fruit?

Write the letter of each answer in the box containing the exercise number.

1. Your friend sits 4 desks in front of you. The center of each desk is five feet away from the center of the next desk in a row. Your other friend sits 3 seats to your right. The desks going this direction are 4 feet apart from center to center. About how far away from each other are your two friends?

D. 16 feet **E.** 23.3 feet **F.** 30.3 feet

2. A ramp used by a moving van has a base that is 8 feet long. The height of the ramp is 5 feet. What is the approximate length of the ramp?

Q. 6.2 feet **R.** 7.6 feet **S.** 9.4 feet

3. The shopping mall is 4.6 miles south of your house. Your favorite restaurant is 7.4 miles east of your house. What is the approximate distance between the shopping mall and your favorite restaurant?

A. 8.7 miles **B.** 9.5 miles **C.** 10.2 miles

4. A basketball hoop is 10 feet high. The horizontal distance from the free throw line to directly below the backboard is 15 feet. What is the approximate distance from the free throw line to the backboard?

R. 18 feet **S.** 20 feet **T.** 22 feet

5. A backyard tool shed has a roof that forms a right angle. The two sides of the roof have the same length. The distance between the lower parts of the two sides of the roof is about 12.8 feet. What is the length of each side of the roof?

N. 7 feet **O.** 8 feet **P.** 9 feet

5	1	3	4	2

Name____ Date __________

Chapter 6

School-to-Work

For use after Section 6.5

Carpenter

You are working as a carpenter, building the frame for the roof of a house. The frame consists of several sections called *trusses*. The plan for one truss is shown below. It is important that you construct the truss so that the posts meet the ridge beam at right angles.

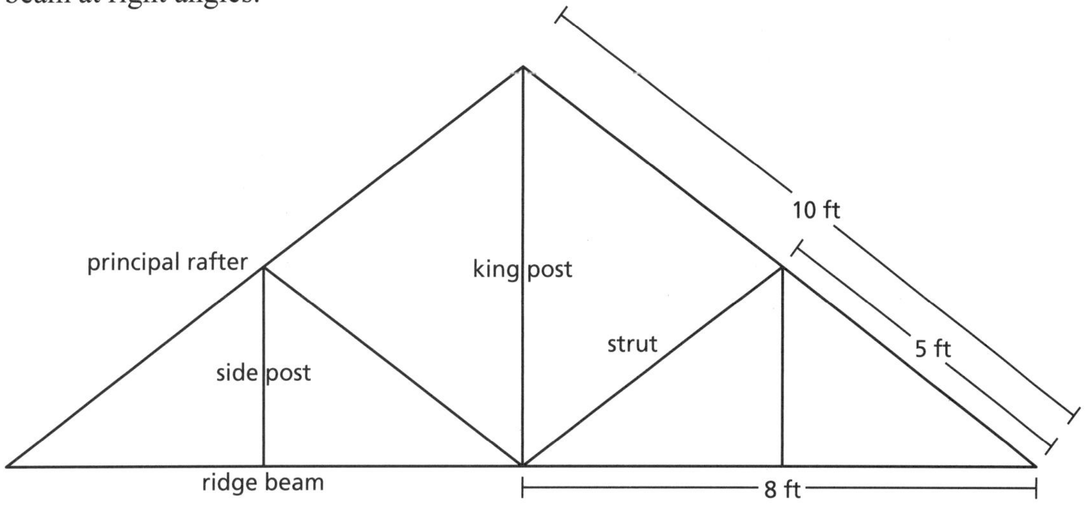

1. What is the relationship between the principal rafter, the king post, and half the length of the ridge beam? Show how you can use this relationship to find the length of the king post.

2. The vertical posts are to be evenly spaced along the ridge beam. What is the distance between the king post and each side post? What is the distance between each side post and each lower corner of the truss?

3. You cut a piece of wood for a side post and nail it in place. You then measure and determine that it is 3.5 feet long. Does this side post meet the ridge beam at a right angle? How do you know?

4. What is the length of each strut? Explain how you know.

5. Is the triangle formed by the principal rafters and the ridge beam a right triangle? Justify your answer.

6. Is the triangle formed by the king post, one strut, and half the principal rafter a right triangle? If not, what kind of triangle is it?

Name ______________________________ Date __________

Chapter 6 Study Help

You can use a **summary triangle** to explain a concept.

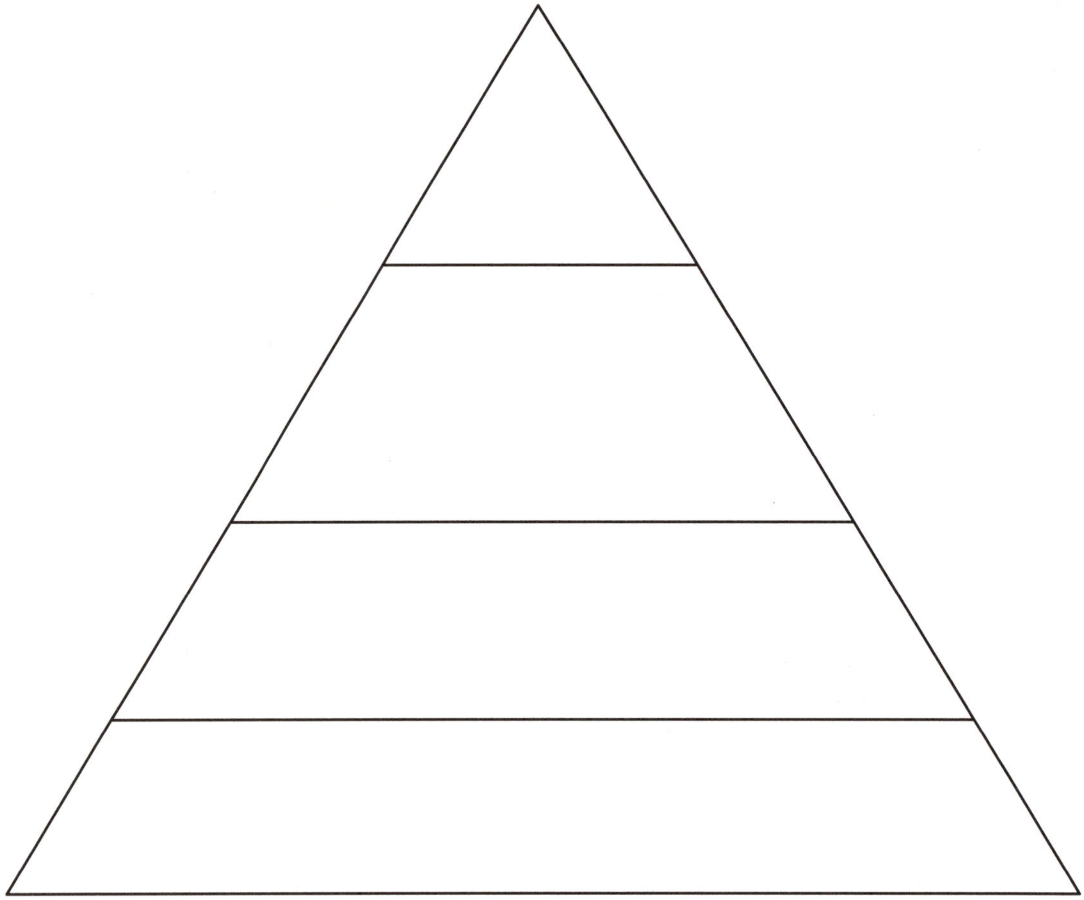

On Your Own

Make a summary triangle to help you study these topics.

1. finding square roots
2. evaluating expressions involving square roots
3. finding the length of a leg of a right triangle

After you complete this chapter, make summary triangles for the following topics.

4. approximating square roots
5. simplifying square roots

Name ______________________________ Date __________

Chapter 6

Financial Literacy

For use after Section 6.3

Organizing a School Dance

As a member of the student council, you are responsible for organizing a spring dance for your school. The dance is to be held outside, so it will be necessary to rent a dance floor and tent in addition to hiring a DJ. All other rentals are optional. A list of rental options is given below.

Dance floor (allow 4 ft^2 per dancer)			Tents	
192 ft^2	\$320		120 ft^2	\$480
320 ft^2	\$530		350 ft^2	\$890
480 ft^2	\$800		600 ft^2	\$1230
672 ft^2	\$1120		950 ft^2	\$1450
DJ (3 hours)	**\$500**		**Table (5 ft round)**	**\$25**
			Chair	**\$1**

1. How many students are enrolled at your school? Of these students, how many do you predict will attend the dance?

2. What is the greatest number of dance attendees you think will be on the dance floor at any one time? What size dance floor should you rent? Explain your reasoning.

3. If you have the dance floor set up as a square, what would be the approximate side length? Give your answer to the nearest tenth of a foot. Show your work.

4. What size tent should you rent? If the tents all cover a square area, what is the approximate side length of the square area? Give your answer to the nearest tenth of a foot. Show your work.

5. What, if any, other equipment do you think should be rented? Draw a diagram of how the equipment should be set up for the dance.

6. What is the total cost of putting on the dance? Assume that refreshments will be donated.

7. Based on the cost of putting on the dance, how much should you charge each student for admission? Explain your reasoning.

Name ______________________________ Date __________

Chapter 6 Cumulative Practice

Simplify the expression.

1. $\sqrt{\dfrac{16}{25}}$

2. $9 - 5 \bullet 3$

3. $\sqrt{49} + 22$

4. $7(8.2 - 14.9)$

5. $\dfrac{2}{5} + \dfrac{\sqrt{6}}{5}$

6. $\sqrt{8} - 3\sqrt{2}$

7. $\sqrt{68}$

8. $\pm\sqrt{529}$

9. $\sqrt{\dfrac{19}{121}}$

10. $\dfrac{5}{9} \bullet 3 + \dfrac{15}{8} \div 5$

11. $8 \div 4 - 2$

12. $-\sqrt{\dfrac{1}{16}} + \dfrac{1}{2}$

13. A rectangular prism has side lengths of $\sqrt{15}$ centimeters, $\sqrt{45}$ centimeters, and $\sqrt{27}$ centimeters. What is the volume of the rectangular prism?

Find the missing length of the triangle. Round your answer to the nearest tenth, if necessary.

14.

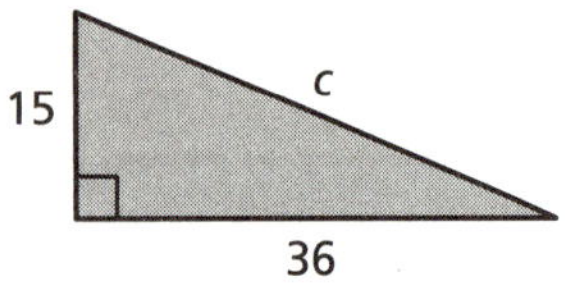

15.

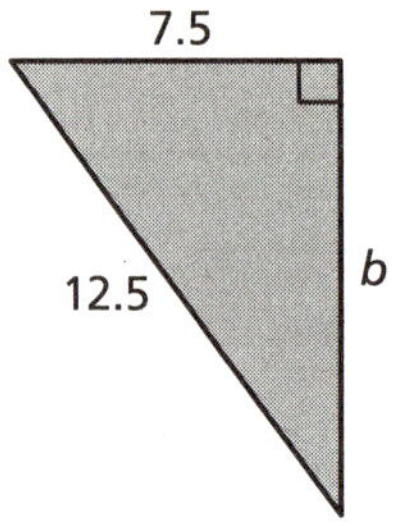

16.

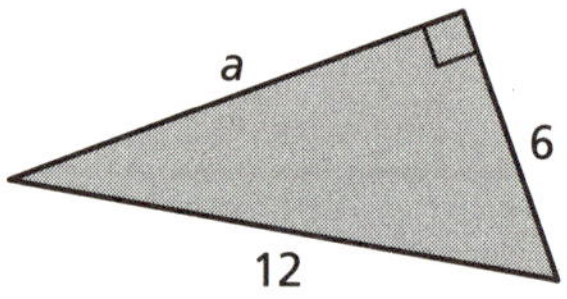

17. You are 18 feet away from a building that is 45 feet tall. What is the distance from where you stand to the roof of the building?

Copy and complete the statement with <, >, or =.

18. -1.6 __?__ $-\sqrt{2.56}$

19. $\sqrt{\dfrac{7}{4}}$ __?__ 2.1

20. $1.6\overline{1}$ __?__ $\dfrac{\pi}{2}$

21. The distance between your school and the library is 3 miles. The distance between your home and your school is $\sqrt{10}$ miles. Is your school closer to your home or the library?

Name______________________________ Date__________

Unit 2 Project: Analyzing Stride Length

For use after Unit 2

Objective **Draw and analyze similar triangles to predict your height from your stride.**

Materials Yardstick, ruler, protractor, calculator

Investigation

1. Work in a group of 3. Measure the length of your leg from the ground to your hip.
2. Take one normal step and "freeze." Have a member of your group measure the length from the toe on the back foot to the toe on the front foot. This is your *walking stride length.*
3. Take one running stride and have someone measure your running stride length.
4. Record the measurements for each person in the group.

Data Analysis

5. Sketch an isosceles triangle to represent the length of your legs and your walking stride. Choose a scale, such as 10 inches (actual stride length) to 2 centimeters (stride length in drawing.)
6. Use the Pythagorean Theorem to calculate h. Then find the measure of $\angle A$.
7. Repeat Steps 5 and 6 for your running stride length.

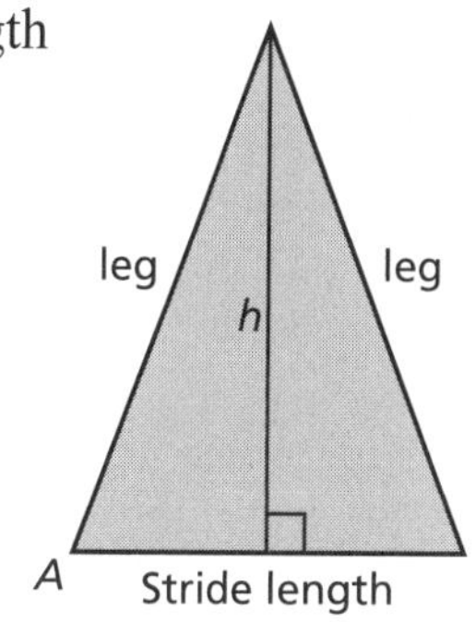

8. Calculate the ratio $\dfrac{\text{height}}{\text{leg length}}$. Share your ratio with the members of your group. Find the mean of your group ratios, rounded to the nearest tenth.
9. Gather the measures of $\angle A$ for the walking stride length from the members of your group. Find the mean of these angle measures. Repeat this for the measures of $\angle A$ for the running stride length.
10. You will receive two sets of footprints. Measure the stride lengths. Use the stride lengths, the mean of $\angle A$, and the mean of your group's height : leg length ratio to make a scale drawing for these stride lengths.
11. Measure the leg length on the drawing with a ruler and use your scale to find the actual leg length for these footprints. Calculate the approximate height of the person who left the footprints.

Make a Poster Explain the Investigation. Display your data, scale drawings, and calculations. Describe how you determined the height of the person who made the "mystery" footprints.

Name ______________________________ Date __________

Unit 2 Student Grading Rubric

For use after Unit 2

		Student Score	Teacher Score
Cover Page 10 points			
	a. Name (4 points)	______	______
	b. Class (2 points)	______	______
	c. Project Name (2 points)	______	______
	d. Due Date (2 points)	______	______
Investigation 20 points			
	a. Measurements for leg length, walking stride length, and running stride length are shown. (20 points)	______	______
Data Analysis 120 points			
	a. Includes all scale drawings. Drawing are labeled correctly and drawn to scale. (30 points)	______	______
	b. Shows calculations to find h and the measure of $\angle A$. (15 points)	______	______
	c. Shows height : leg length ratios and means. (15 points)	______	______
	d. Finds the mean of $\angle A$ for walking stride lengths and running stride lengths. (15 points)	______	______
	e. Accurately measures stride length of footprints and makes accurate scale drawings. (30 points)	______	______
	f. Calculates a reasonable height for the person who made the footprints. (15 points)	______	______
Poster 50 points			
	a. Includes a description of the investigation, all data, all scale drawings, and calculations. (25 points)	______	______
	b. Describes the process for estimating the height of the person who made the footprints. (15 points)	______	______
	c. Poster is neat and well laid out. (10 points)	______	______
FINAL GRADE		______	______

Unit 2

Teacher's Project Notes

For use after Unit 2

Materials Yardstick, ruler, protractor, calculator; You will need to provide the students with a running and walking stride length for another person, such as yourself. Prepare these ahead of time and have enough for each group. For added interest, you can draw actual footprints representing the stride length on newsprint or poster board.

Alternatives Students who are on crutches or unable to walk could measure the strides of others. They might also measure the strides of a jointed doll or a cooperative pet. Are the triangles formed by the stride lengths for a dog the same as for a small human, or different? Could you use them to estimate the size of a bear from its tracks?

Determining size from footprints is used in both forensic medicine and paleontology. The class project might focus on one of these, e.g. "Who left the footprints running away from the crime scene?" or "How tall was the bipedal dinosaur who left these walking footprints?"

Common Errors Students may need help finding a formula for h:

$$h = \sqrt{(\text{leg length})^2 - (\text{half of stride})^2}.$$

Small children have shorter legs and arms proportionate to their size than adults and adolescents. If students are looking for a shorter stride to use in their measurements, they should not use very small children. You can illustrate this by drawing two stick figures on the board who are the same height, but one with a larger head, longer torso, and shorter legs, and ask which represents an adult and which a toddler. This could spark a discussion about using proportion in drawing.

Note that the actual height of the mystery strider may be more or less than the height students calculate using their model.

Suggestions Explain to students that the footprints of walkers and runners vary: runners, for example, have a deeper imprint at the ball of the foot. You can illustrate this if you have access to sand or soft dirt that two students can cross.

Students use the mean angle measures and ratios to create a model triangle. Then, given a stride length and information as to whether the strider is walking or running, they assume that the unknown strider's triangle is similar to their model triangle. A class discussion prior to the project about how models are similar (in the mathematical sense) to what they represent will help students grasp the different ways similar figures are (and are not) used in this application.

Unit 2

Grading Rubric

For use after Unit 2

Scoring Rubric	
A	179-200
B	159-178
C	139-158
D	119-138
F	118 or below

Cover Page 10 points

a. Name (4 points)

b. Class (2 points)

c. Project Name (2 points)

d. Due Date (2 points)

Investigation 20 points

a. Measurements for leg length, walking stride length, and running stride length are shown. (20 points)

Data Analysis 120 points

a. Includes all scale drawings. Drawing are labeled correctly and drawn to scale. (30 points)

b. Shows calculations to find h and the measure of $\angle A$. (15 points)

c. Shows height : leg length ratios and means. (15 points)

d. Finds the mean of $\angle A$ for walking stride lengths and running stride lengths. (15 points)

e. Accurately measures stride length of footprints and makes accurate scale drawings. (30 points)

f. Calculates a reasonable height for the person who made the footprints. (15 points)

Poster 50 points

a. Includes a description of the investigation, all data, all scale drawings, and calculations. (25 points)

b. Describes the process for estimating the height of the person who made the footprints. (15 points)

c. Poster is neat and well laid out. (10 points)

Chapter 7

Name ______________________________ Date ________

Chapter 7 Data Analysis and Displays

Dear Family,

Volunteering is a rewarding way to spend time with friends and family while helping your community.

Many charitable and non-profit organizations require a lot of management—as much as, or more than some businesses that operate for profit. Managers must raise money through donations and recruit volunteers—both of which involve convincing people of the value of supporting the organization. Organizations often use marketing campaigns and community presentations to promote their cause.

With your student, decide on a volunteer opportunity to explore. Take a look at a number of organizations. Do research on the Internet, attend community presentations, and view the organization's marketing material to learn as much as you can. Have your student analyze the data presented.

- Did the organization use measures of central tendency to represent their data? Are they the best measurements for the data?
- Did the organization use data displays? Did they use the best types of displays to make their point? If not, what would be a better display? Why?
- Does the volunteer opportunity fit your abilities and schedule?

If your analysis leaves you with questions, you and your student should contact the volunteer coordinator at the organization. After analyzing each organization, compare each opportunity to find one that is the best fit for you.

Not every volunteer effort requires a long-term commitment. Often, a community group will organize a neighborhood beautification or cleanup project. Talk with your student about how they would get such a project started. What kind of data would your student present to convince people to volunteer? Would your family like to put the plan into motion?

May you have a cause to celebrate!

Nombre ____ Fecha__________

Capítulo 7 Análisis de Datos y Representaciones

Estimada Familia:

El voluntariado es una manera gratificante de pasar tiempo con sus amigos y familia mientras ayuda a su comunidad.

Muchas organizaciones de caridad y sin fines de lucro requieren muchos recursos administrativos—tanto o más que los negocios en sí. Los administradores deben recaudar fondos a través de donaciones y reclutar voluntarios—ambos requieren convencer a las personas sobre la importancia de apoyar a la organización. Las organizaciones generalmente utilizan campañas de mercadeo y presentaciones a la comunidad para promover su causa.

Con su estudiante, decidan explorar una oportunidad de voluntariado. Examinen un número de organizaciones. Hagan investigaciones en Internet, vayan a presentaciones en la comunidad y revisen el material de mercadeo de la organización para aprender lo más que puedan. Haga que su estudiante analice los datos presentados.

- ¿La organización utilizó medidas de tendencia central para representar los datos? ¿Estas son las mejores medidas para los datos?
- ¿La organización utilizó representaciones de datos? ¿Usaron los mejores tipos de representaciones para explicar su mensaje? Si no, ¿qué representación sería mejor? ¿Por qué?
- ¿La oportunidad de voluntariado se ajusta a sus habilidades y horarios?

Si su análisis le deja interrogantes, usted y su estudiante deben contactar a la coordinador de voluntarios de la organización. Luego de analizar cada organización, comparen cada oportunidad para encontrar aquélla que mejor se adapte a ustedes.

No todos los voluntariados requieren un compromiso a largo plazo. A menudo un grupo en la comunidad organiza el embellecimiento del barrio o inicia un proyecto de limpieza. Hable con su alumno acerca de cómo ellos pueden iniciar un proyecto de esa naturaleza. ¿Qué tipo de información presentaría su estudiante para convencer a las personas para hacer un voluntariado? ¿A su familia le gustaría llevar a cabo esta iniciativa?

¡Que tengan una causa para celebrar!

Non ______________________________ Dat __________

Chapít 7 Analiz ak Reprezantasyon Done

Chè Fanmi:

Ede kòm volontè se yon mannyè ki gen gratifikasyon ladann pou pase tan avèk zanmi ak fanmi pandan w'ap ede kominote ou a.

Anpil òganizayon charitab ak sa ki pa gen pou bi fè lajan mande anpil jesyon— menm kantite ak, oswa plis pase kèk biznis ki fonksyone pou pwofi. Manadjè yo dwe kolekte lajan atravè don ak rekritman benevòl—toulède enplike konvenk moun ke soutni òganizayon an vo lapèn. Òganizasyon souvan sèvi ak kanpay maketin ak prezantasyon nan kominote a pou fè reklam kòz yo a.

Avèk elèv ou a, deside ki opòtinite benevòl nou pwal esplore. Gade yon sèten kantite òganizayon. Fè rechèch sou Entènèt, al nan prezantasyon pou kominote a, epi gade materyèl maketin òganizasyon an pou aprann tout sa nou kapab. Fè elèv ou a analize done yo prezante yo.

- Èske òganizasyon an te sèvi ak mezi tandans santral pou reprezante done yo? Èske se pi bon mezi pou done yo?
- Èske òganizasyon an te sèvi avèk reprezantasyon done? Èske yo te sèvi avèk pi bon tip reprezantasyon yo pou jistifye pwendvi yo? Si se pa sa, kisa ki t'ap yon pi bon reprezantasyon? Poukisa?
- Èske opòtinite benevòl la ale avèk kapasite ak orè nou?

Si analiz nou an kite nou avèk keksyon, ou menm ak elèv ou a ta dwe kontakte kowòdinatè benevòl la nan òganizasyon an. Apre nou fin analize chak òganizasyon, konpare chak opòtinite pou jwenn youn ki pi bon pou nou.

Se pa tout efò benevòl ki egzije angajman alontèm. Souvan, yon gwoup kominotè ap òganize yon pwojè anbeli oswa netwaye katye. Pale avèk elèv ou a sou kijan yo t'ap kòmanse ak yon pwojè konsa. Ki kalite done elèv ou a t'ap prezante pou konvenk moun pou yo fè benevòl? Èske fanmi ou ta renmen mete plan an an aksyon?

Mwen swete pou ou gen yon kòz pou selebre!

Activity 7.1 Start Thinking!

For use before Activity 7.1

The mean, median, and mode are called *measures of central tendency*. Why do you think they are called that?

Give an example of a measure of central tendency used in real life.

Activity 7.1 Warm Up

For use before Activity 7.1

Find the mean of the data set.

1. 1, 4, 6, 7, 8, 8, 8, 12

2. 1, 1, 3, 3, 5, 5, 7, 7, 9, 9

3. 17, 18, 19, 19, 26, 27, 28

4. 5, 5, 5, 5, 5, 5, 5, 34

5. 11, 38, 39, 40, 44, 44

6. 1, 5, 7, 3, 8, 5, 3, 2, 0, 9, 1

Lesson 7.1 Start Thinking!
For use before Lesson 7.1

Record the number of siblings of each person in your class.

Find the mean, median, and mode of the data.

If you had instead recorded the number of children in each person's family, how would the mean, median, and mode have changed? Explain.

Lesson 7.1 Warm Up
For use before Lesson 7.1

Draw a line plot of the data. Then find the mean, median, and mode of the data.

1. The weights (in pounds) of nine watermelons: 12, 14, 13, 13, 14, 11, 12, 13, 15

2. The number of email accounts 10 people have: 1, 4, 2, 4, 3, 1, 0, 2, 4, 3

Name____ Date__________

7.1 Practice A

Find the mean, median, and mode of the data.

1.

Bananas Per Bunch		
5	8	7
9	8	4
6	4	3

2.

Changes in Gas Prices			
0.25	0.30	−0.10	0.05
−0.20	0.20	0.05	−0.15
0.30	0.05	−0.10	−0.05

3. Pages per Chapter

Stem	Leaf
1	0 4
2	5 7
3	1 3 3 9
4	1 5

Key: 1 | 4 = 14 pages

4.

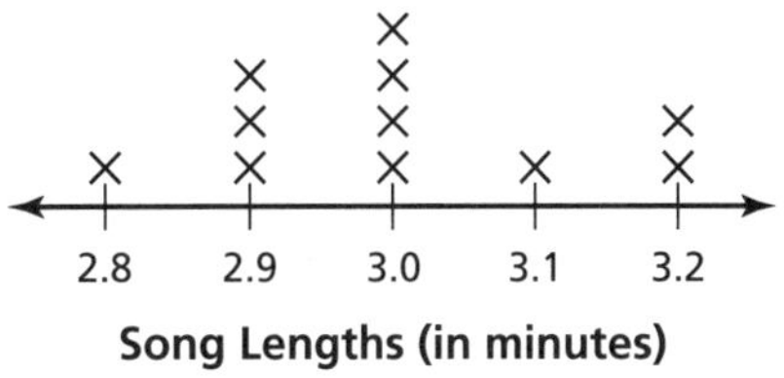

Find the value of *x*.

5. Mean is 4; 5, 1, 6, 3, 4, x

6. Median is 28; 21, 24, x, 30, 30, 32

7. A nutritionist records the amount of sodium in 6 different brands of turkey breast. The amounts of sodium (in milligrams) are 460, 280, 320, 850, 290, and 350.

a. Find the mean, median, and mode of the milligrams.

b. Identify the outlier.

c. How does omitting the outlier affect the mean?

8. You record the scores per game for your favorite baseball team. The scores in the games so far are 3, 7, 6, 8, and 10.

a. Find the mean of the scores.

b. Multiply each score by 2. How does this affect the mean?

Name ______________________________ Date __________

7.1 Practice B

Find the mean, median, and mode of the data.

1.

ACT Scores		
29	25	27
31	29	28
24	30	29

2.

Cost To Download a Song			
0.99	0.99	1.25	0.99
1.05	0.95	0.99	1.05
0.99	1.05	0.99	1.19

3. Points on a Quiz

Stem	Leaf
1	5
2	4 8 8
3	2 2 4
4	5

Key: 1 | 5 = 15 points

4.

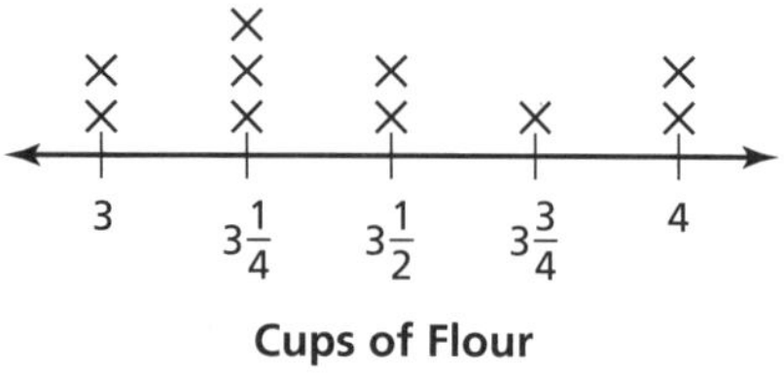

Find the value of *x*.

5. Mean is -0.35; 2.25, -3.75, -1.25, 3.5, x

6. Median is 72; 35, 61, x, 120

7. The circle graph shows the favorite sport of 180 students.

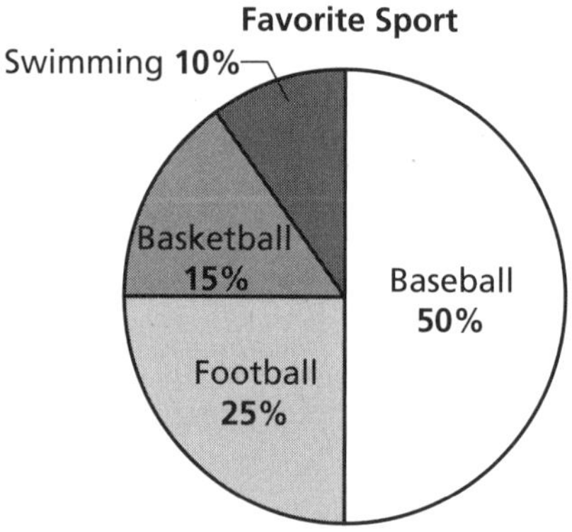

   a. Find the mean, median, and mode of the numbers of students who favor each sport.

   b. Find the mean number of students who favor each sport if only 90 students had been surveyed (half as many). How does this mean compare with the mean in part (a)?

8. Create a data list of 8 numbers that has the given number of modes.

   a. 3 modes

   b. no mode

Name ______________________________ Date __________

7.1 Enrichment and Extension

Expected Value

To find the mean of a data set, you average the numbers. To find the average outcome or expected value of a spinner, you must use a weighted average. Instead of adding the values and dividing by the number of values, you multiply each value by the probability of landing on that value and then add.

1. What is $P(2)$, the probability of the spinner landing on 2?

2. What is $P(3)$?

3. What is $P(5)$?

4. The expected value E can be found using the formula $E = 2 \bullet P(2) + 3 \bullet P(3) + 5 \bullet P(5)$. What is the expected value of the spinner?

5. Is the expected value a possible outcome of the spinner?

6. You spin the spinner 10 times and the outcomes are 3, 3, 5, 3, 2, 5, 3, 3, 5, and 3.

   a. What is the mean of the outcomes?

   b. How does the mean compare to the expected value?

   c. Do you think the mean would be closer to the expected value if you spun the spinner 100 times? Explain.

7. Use the same process to find the expected value of the spinner below. The sections for 4, 9, and 8 are equal.

Name ______________________________ Date __________

7.1 Puzzle Time

How Did The Parrot Get To The Vet?

Write the letter of each answer in the box containing the exercise number.

Answers

U. mean = 23.62, median = 24.5, mode = 20.2

L. mean = 49, median = 50, no mode

H. mean = 29, median = 25.1, mode = 20.2

F. mean = 3, median = 3, mode = 3

E. mean = 3, median = 4, mode = 4

Find the mean, median, and mode of the data.

1.

Gains and Losses of Yards		
9	−7	4
15	8	−4
−9	4	7

2.

Stem	Leaf
3	5
4	0 5 9
5	1 2 5
6	5

Key: 3 | 5 = 35

3.

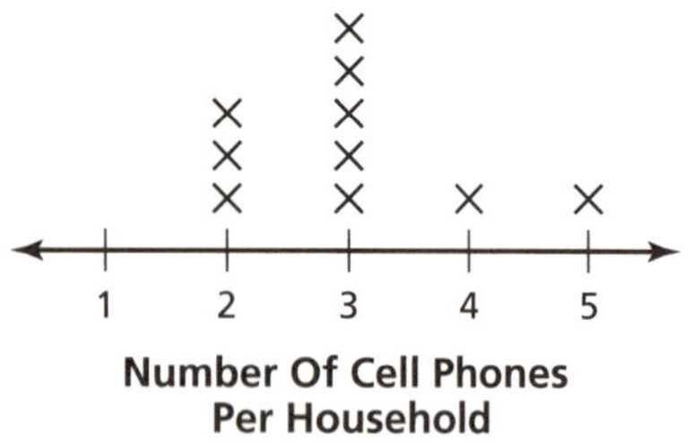

4.

Number of Minutes Spent on Leisure Reading Each Night		
24.5	20.2	55.9
20.2	27.5	25.7

5. Identify the outlier in Exercise 4. Find the mean, median, and mode without the outlier.

4	1		3	2	5

Activity 7.2 Start Thinking!

For use before Activity 7.2

The ages of the Presidents of the United States at the time of their inauguration are shown in a list at the right and in a box-and-whisker plot below. Each section of the plot represents one quarter of the data values.

57	64	55	62
61	50	55	43
57	48	54	55
57	65	42	56
58	52	51	61
57	56	56	52
61	46	55	69
54	54	51	64
68	49	54	46
51	50	51	54
49	47	60	47

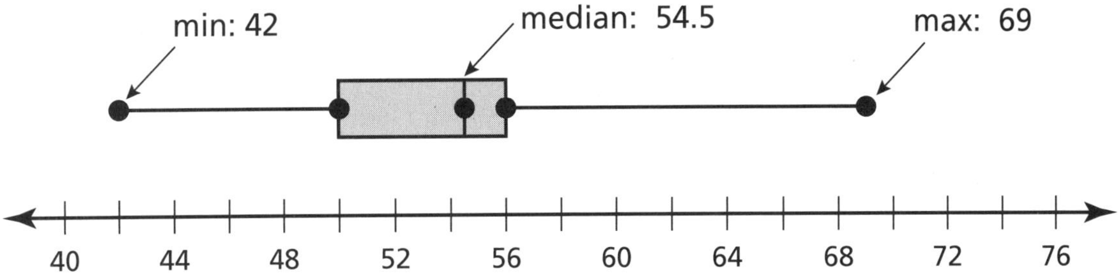

Is it easier to get a feel for the data from the list or from the box-and-whisker plot? Why?

Activity 7.2 Warm Up

For use before Activity 7.2

Find the median, least value, and greatest value in the data set.

1. 1, 3, 4, 5, 5, 6, 8, 10, 12

2. 0.5, 0.6, 0.8, 1.1, 1.2, 1.7

3. 45, 55, 55, 57, 59, 89

4. 5, 14, 54, 56, 78, 79, 100, 120, 222

Lesson 7.2 Start Thinking!
For use before Lesson 7.2

Do you think it costs more to see a professional baseball game or basketball game?

The box-and-whisker plots show the average ticket price for each team in Major League Baseball (MLB) and the National Basketball Association (NBA).

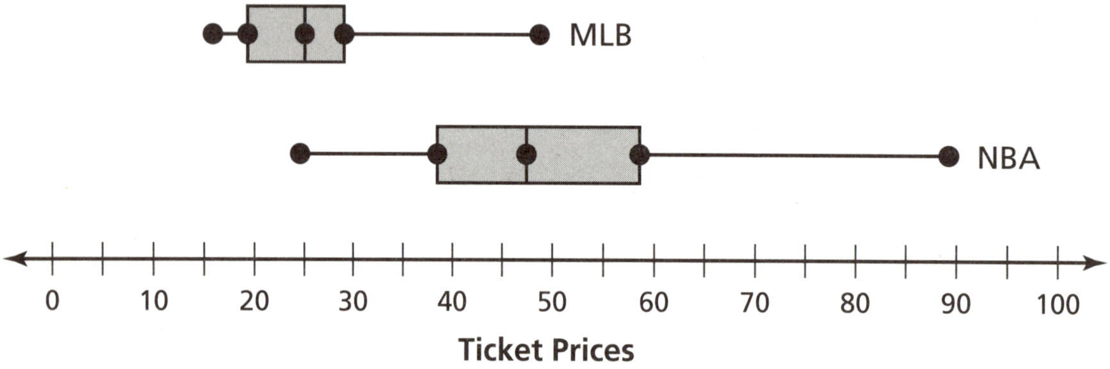

Analyze the box-and-whisker plots to compare the prices in MLB and the NBA.

Lesson 7.2 Warm Up
For use before Lesson 7.2

1. The box-and-whisker plots show the time (in minutes) two friends spent doing homework each day for a month. Compare and contrast the time spent doing homework by the two friends.

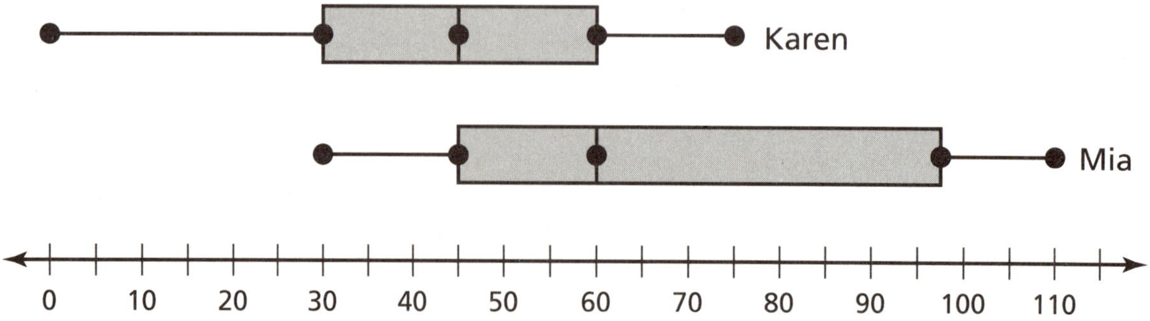

Name____ Date__________

7.2 Practice A

1. The box-and-whisker plots show the minutes it took for the members of two cross country teams to run three miles.

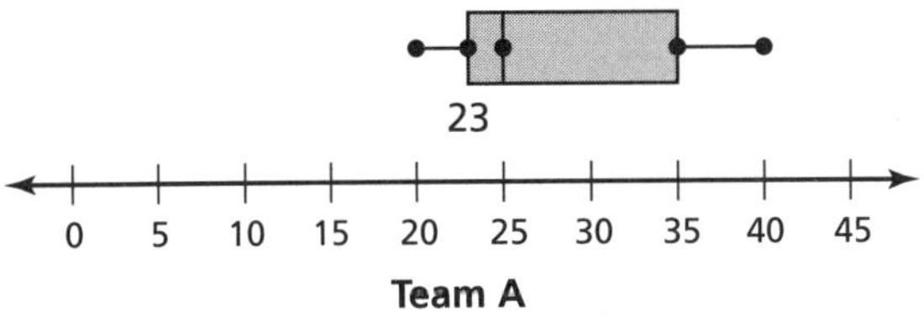

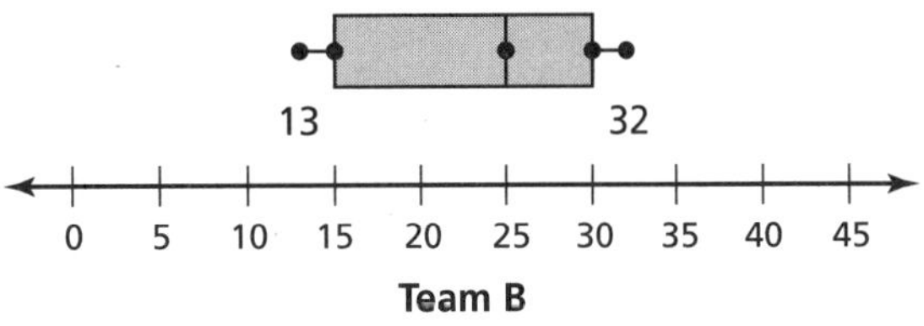

   a. Compare and contrast the minutes of the two teams.

   b. If these two teams were to compete in a race, which team would you expect to win, *Team A* or *Team B*? Explain.

Make a box-and-whisker plot for the data.

2. Service calls per day:

   5, 8, 2, 4, 6, 5, 7, 4

3. Blocks from school:

   1, 5, 3, 9, 5, 7, 1, 4, 10, 12

4. The laps swam during warm-up by members of the swim team are 12, 8, 11, 15, 6, 14, 8, 9, 10, 14, 11, and 20.

   a. Make a box-and-whisker plot for the data.

   b. Does any swimmer stand out? Explain.

5. The square footages of the houses in the neighborhood are 2100, 2450, 2800, 1800, 2500, 3000, 5000, and 3200.

   a. Make a box-and-whisker plot for the data.

   b. Identify the outlier.

   c. Make another box-and-whisker plot without the outlier.

   d. Describe how the outlier affects the whiskers, the box, and the quartiles of the box-and-whisker plot.

6. A box-and-whisker plot consists of the *left whisker*, the *box,* and the *right whisker.*

   a. Which part of the box-and-whisker plot represents the upper 25% of the data?

   b. Which part of the box-and-whisker plot represents the middle 50% of the data?

Name ______________________________ Date __________

7.2 Practice B

1. The box-and-whisker plots show the number of moves made in order to win the chess match for two members of the chess team.

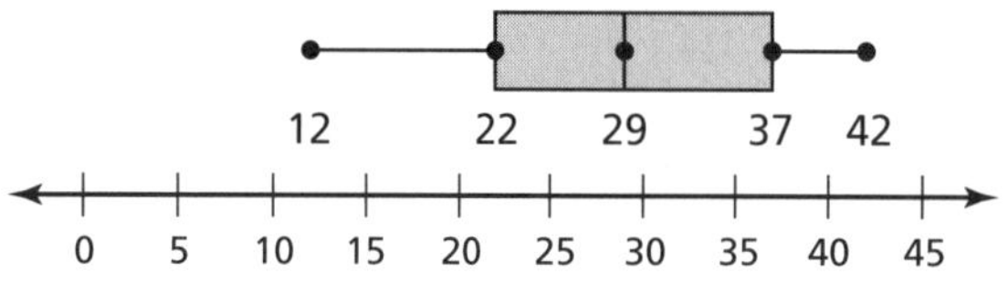

Competitor A

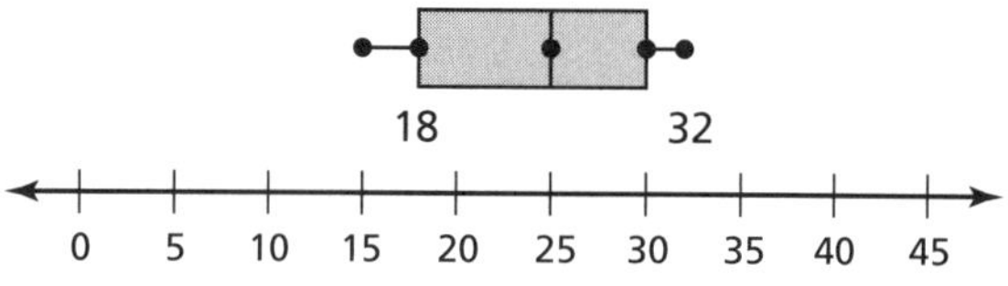

Competitor B

   a. Compare and contrast the number of moves made by the two members of the chess team.

   b. If these two members of the chess team were to compete in a chess match, who would you expect to win, *Competitor A* or *Competitor B*? Explain.

Make a box-and-whisker plot for the data.

2. Desks in classrooms:

   25, 20, 22, 18, 25, 24, 28, 24, 19, 21

3. Golf scores:

   1, −3, −2, 1, 0, 3, 2, −1, −1, 0, 4

4. The box-and-whisker plots compare the pulse rates (in beats per minute) of two different patients.

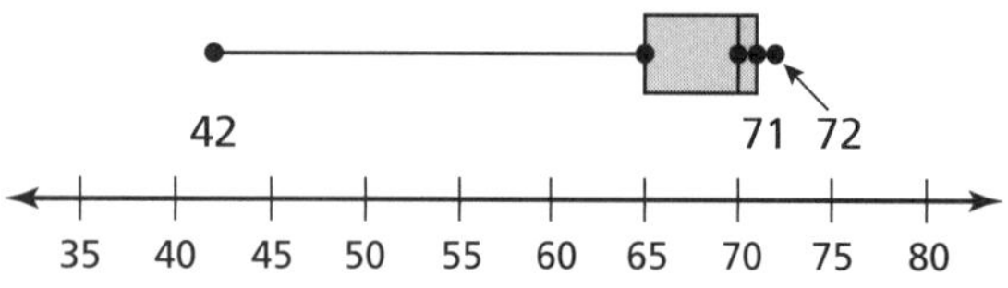

Patient A

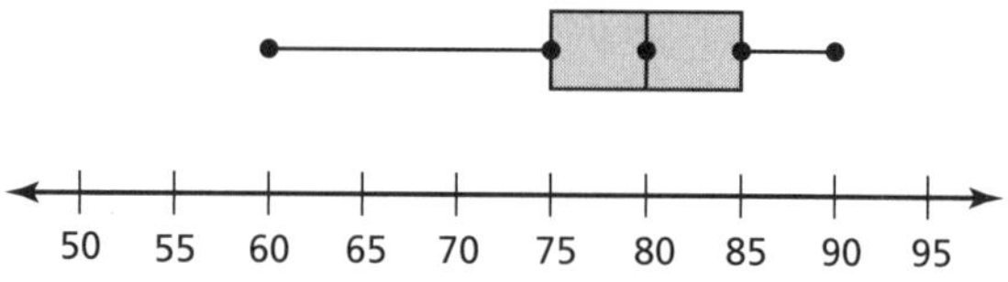

Patient B

   a. Which patient has an overall lower pulse rate?

   b. Compare the ranges of the pulse rates of the two patients.

   c. For both graphs, the longest part is the left whisker. What does this tell you about the patients' pulse rates? During what time of day might this be occurring?

   d. Compare the relationships between the median, the first quartile, and the third quartile for Patient A.

5. The number of fluid ounces of coffee poured in cups for a meeting are 8, 8.2, 7.9, 8.1, 7.8, 8, 7.9, 8, 6.5, and 8.1

   a. Make a box-and-whisker plot for the data.

   b. Identify the outlier.

   c. Make another box-and-whisker plot without the outlier.

   d. Describe how the outlier affects the whiskers, the box, and the quartiles of the box-and-whisker plot.

Name____ Date__________

7.2 Enrichment and Extension

Outliers

Outliers are values that are much higher or lower than other data values. Because outliers are not always easy to identify, statisticians developed a numerical test to determine if a value is actually an outlier.

Botanical gardens are large areas of greenhouses and gardens that usually contain a wide variety of plants. A botanical garden records the heights (in feet) of trees that were over 100 years old for a study. The heights are 1, 20, 50, 50, 60, 60, 70, 80, 80, 80, 100, and 120. The data is arranged in a box-and-whiskers plot below.

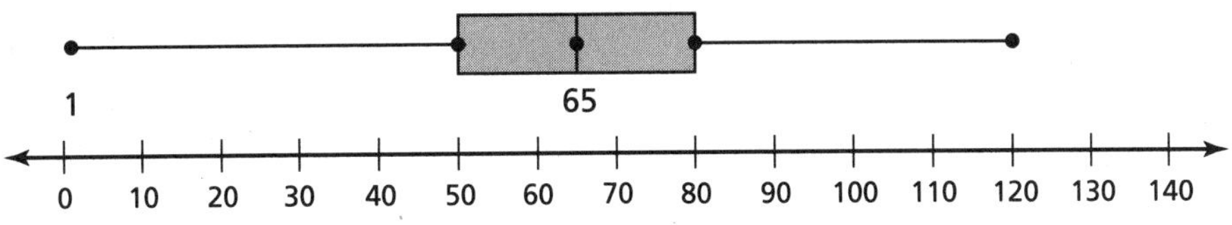

1. What values appear to be outliers?

2. Identify the first and third quartiles.

3. Identify the data values that lie outside the box.

4. What is the difference between the third quartile and the first quartile?

5. Multiply the difference by 1.5 to get a value d.

6. If a data value outside the box is more than d units away from the box, the value is statistically considered an outlier.

   a. Which data value(s) are outliers?

   b. Why might the botanical garden have those outlier(s)?

Name ______________________________ Date __________

7.2 Puzzle Time

Why Couldn't The Egg Lend Money To The Troll?

Write the letter of each answer in the box containing the exercise number.

In Exercises 1–6, use the box-and-whisker plot.

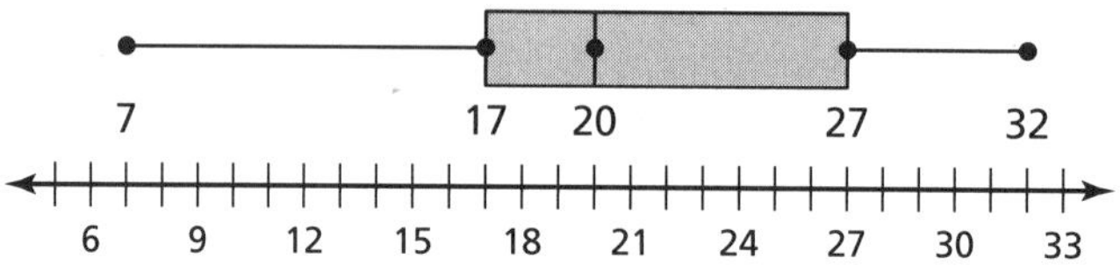

1. What is the least value?
2. What is the greatest value?
3. What is the range of the data?
4. What is the value representing the first quartile?
5. What is the median?
6. What is the value representing the third quartile?

Make a box-and-whisker plot for the data.

7. Test scores in your class: 70, 75, 70, 65, 80, 60, 65, 75
8. Hours spent on the Internet per week: 8, 9, 10, 11, 9, 10, 12, 11
9. Lengths (in inches) of small dogs: 26, 28, 30, 34, 27, 32, 32, 31
10. Temperature changes (in Fahrenheit): 5, 8, 12, –10, 8, –9, 9, 8, –4, –9, 0, 6

Answers for Exercises 1–6

R. 25 **T.** 7

W. 27 **O.** 17

E. 32 **S.** 20

Answers for Exercises 7–10

K.

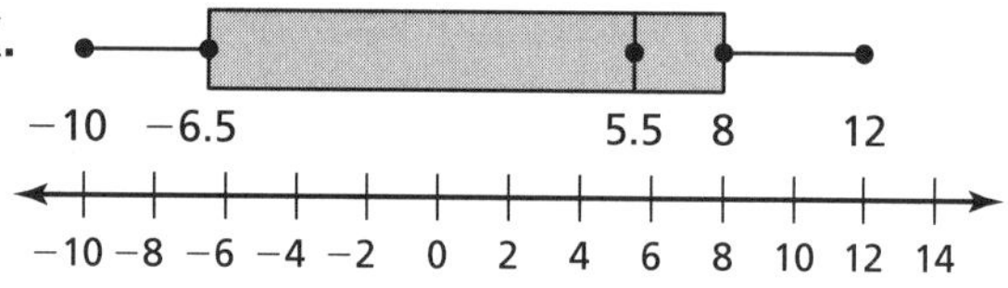

A.

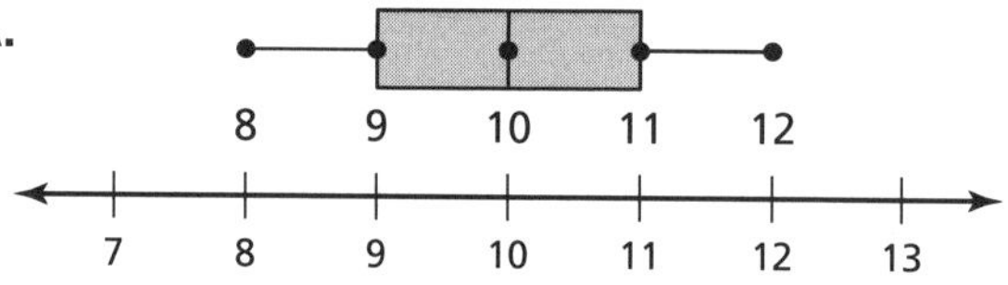

I.

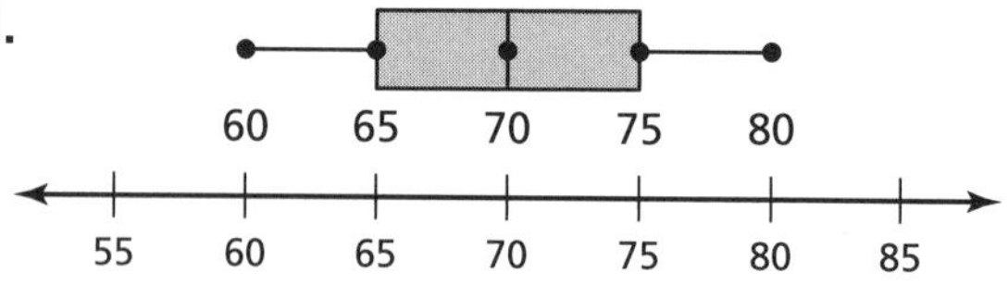

B.

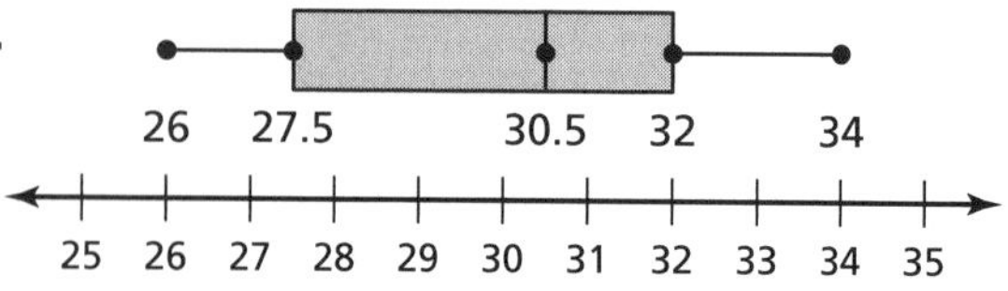

7	1		6	8	5		9	3	4	10	2

Activity 7.3 Start Thinking!

For use before Activity 7.3

Research the population of your state each year for the past several years. You can find the information at *www.census.gov*. Record the results in a table.

Graph the data. Decide what units would be best and keep in mind that you may have to round to make graphing easier.

Is the graph linear? Can you use the graph to predict the population in future years?

Activity 7.3 Warm Up

For use before Activity 7.3

Write an equation of the line that passes through the two points.

1. $(0, 4)$ and $(5, 3)$

2. $(0, 6)$ and $(2, 0)$

3. $(8, 3)$ and $(2, 6)$

4. $(1, 2)$ and $(5, 6)$

5. $(9, 3)$ and $(3, 1)$

6. $(4, 16)$ and $(2, 12)$

Lesson 7.3 Start Thinking!
For use before Lesson 7.3

In which graph is the line shown most representative of the data? Explain.

A

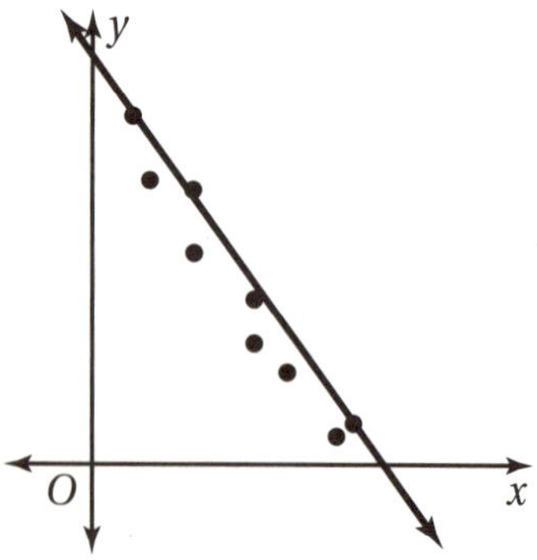

B

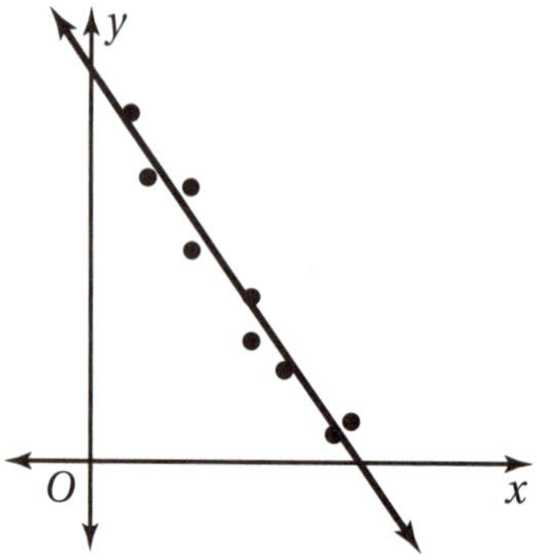

C

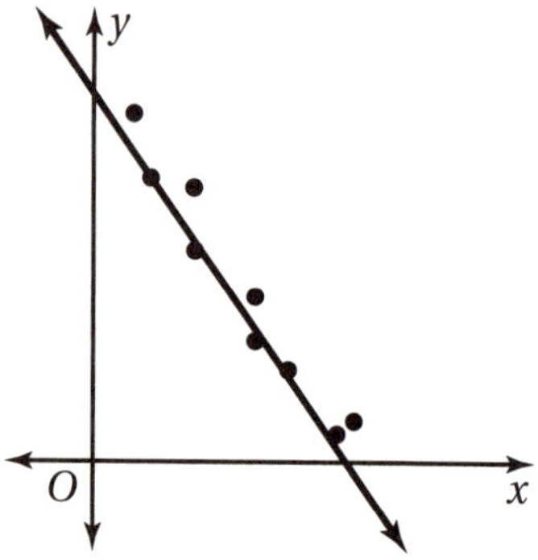

Lesson 7.3 Warm Up
For use before Lesson 7.3

The table shows the weight of *x* bananas.

Number of Bananas	0	1	2	3	4	5
Weight (ounces)	0	5	8	14	17	20

1. Graph the data in the table.

2. Draw the straight line that you think best approximates the points.

3. Write an equation of the line you drew.

4. Use the equation to predict the weight of 10 bananas.

Name____ Date__________

7.3 Practice A

1. The scatter plot shows the weights y of an infant from birth through x months.

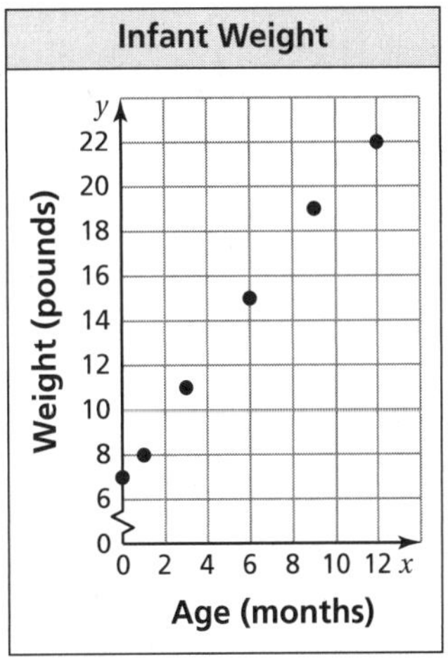

   a. At what age did the infant weigh 11 pounds?

   b. What was the infant's weight at birth?

   c. Draw the straight line that you think best approximates the points.

   d. Write an equation of the line you drew.

   e. Use the equation to predict the weight of the infant at 18 months.

   f. Does the data show a *positive*, a *negative*, or *no* relationship?

2. The table shows the number of losses y at a video game during week x.

Week, x	1	2	3	4	5	6	7
Losses, y	15	12	10	7	6	3	1

   a. Make a scatter plot of the data.

   b. Draw a line of best fit.

   c. Write an equation of the line of best fit.

   d. Does the data show a *positive*, a *negative*, or *no* relationship?

   e. Interpret the relationship.

3. The scatter plot shows the relationship between the number of girls and the number of boys in 10 different classrooms.

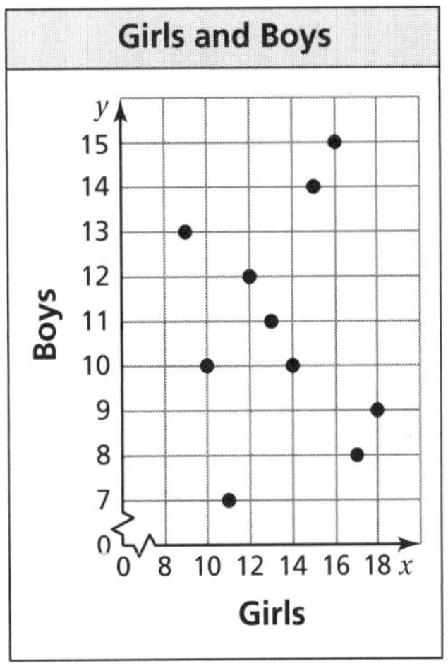

   a. What type of relationship, if any, does the data show?

   b. Is it possible to find the line of best fit for the data? Explain.

   c. Is it reasonable to use this scatter plot to predict the number of boys in the classroom based on the number of girls? Explain.

Name ______________________________ Date __________

7.3 Practice B

1. The scatter plot shows the costs y of bottles containing x fluid ounces of juice.

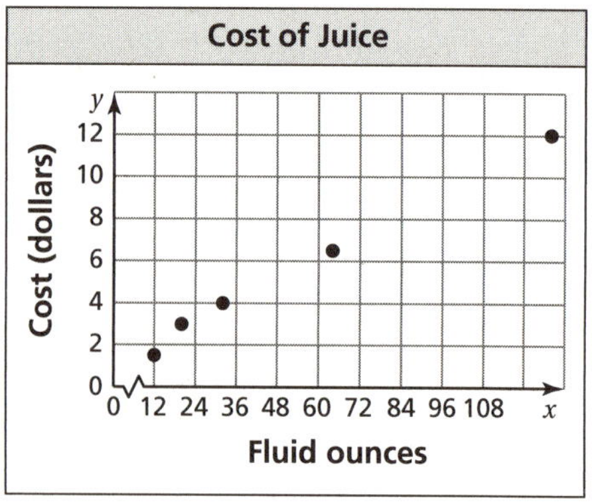

   a. How much does a gallon of juice cost?

   b. How many fluid ounces of juice can you purchase for $3?

   c. Draw the straight line that you think best approximates the points.

   d. Write an equation of the line you drew.

   e. Use the equation to predict the cost of a 256-fluid ounce container of juice.

   f. Does the data show a *positive*, a *negative*, or *no* relationship?

2. The table shows the mortgage interest rates y at a local bank for the years 2000 through 2008.

Year since 2000, x	0	1	2	3	4	5	6	7	8
Rate (%), y	7.6	6.8	6.2	6.0	5.2	5.8	6.1	5.9	5.5

   a. Make a scatter plot of the data.

   b. Draw a line of best fit.

   c. Write an equation of the line of best fit.

   d. Use the equation to predict the mortgage interest rate for the year 2010.

   e. Does the data show a *positive*, a *negative*, or *no* relationship?

   f. Interpret the relationship.

3. The scatter plot shows the relationship between the age of an individual x and the cost of admission y to a show.

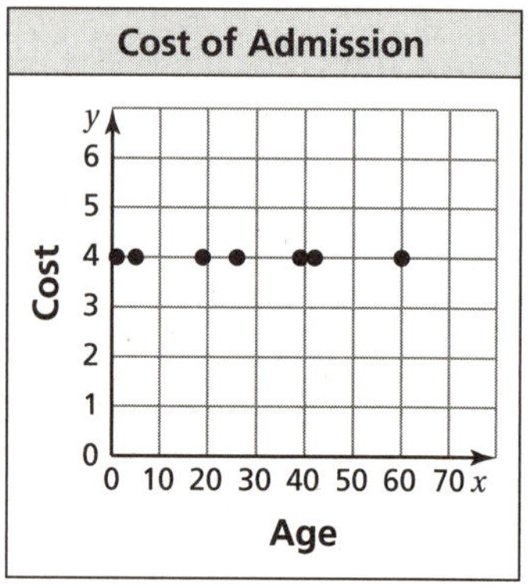

   a. What type of relationship does the data show?

   b. Draw a line of best fit.

   c. Write an equation of the line of best fit.

   d. Interpret the relationship.

Name_______________________________ Date__________

7.3 Enrichment and Extension

Correlation Coefficient

A scatter plot can reveal the relationship that exists between two sets of data. Are all relationships between data the same? Are some relationships stronger than others?

Calculate the correlation coefficient r to help determine the strength of the relationship between data. The correlation coefficient is a value between -1 and 1. When r is close to 1, the data share a strong positive relationship. When r is close to -1, the data share a strong negative relationship. If r is close to 0, the data do not share a relationship.

Analyzing the Relationship

The data in the table show the number of cubic meters x excavated during an archeological dig and the number y of artifacts found.

	x	y	xy	x^2	y^2
	1	1			
	2	3			
	3	4			
	4	5			
	5	6			
Sum	$A = 15$	$B =$	$C =$	$D =$	$F =$

1. Construct a scatter plot for the data. Draw the line of best fit.
2. How many pairs of data are contained in the table? Use this number as your value for n.
3. Copy and complete the table.
4. Use the values in the table and a calculator to evaluate the formula to calculate the correlation coefficient. Round your answer to three decimal places.

$$r = \frac{n(C)-(A)(B)}{\sqrt{\left[n(D)-(A)^2\right]\left[n(F)-(B)^2\right]}}$$

5. What does the value of the correlation coefficient tell you about the relationship between the data? Does your scatter plot support your conclusion?

Name ______________________________ Date __________

7.3 Puzzle Time

What Do You Call A Grouchy Person At The Beach?

Write the letter of each answer in the box containing the exercise number.

In Exercises 1–5, use the scatter plot.

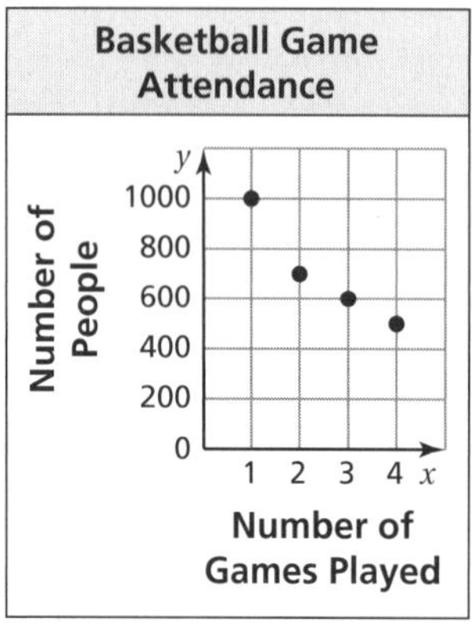

1. Which game did 500 people attend?
2. How many people attended Game 2?
3. Did more people attend Game 2 or Game 3?
4. How many more people attended Game 1 than Game 4?
5. Describe the relationship shown by the data.

Answers

N. 700 **E.** Game 1

R. Game 2 **T.** 600

A. Game 4 **M.** Game 3

C. 500 **I.** 1000

B. Positive relationship

A. Negative relationship

S. No relationship

D. $y = \frac{1}{2}x + \frac{1}{2}$

H. $y = 2x - 1$

Tell whether the data show a *positive*, a *negative*, or *no* relationship.

6.

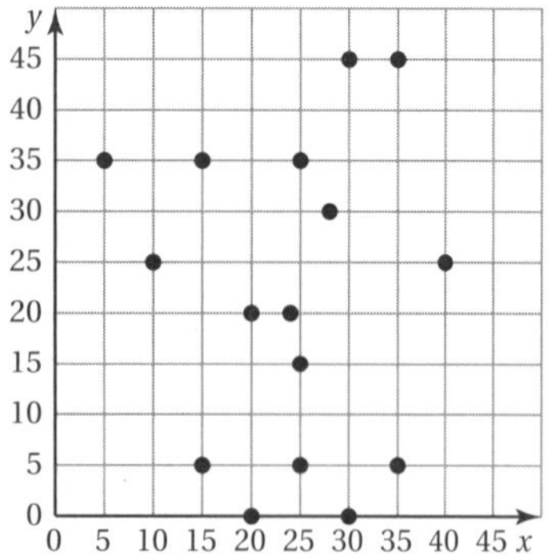

7.

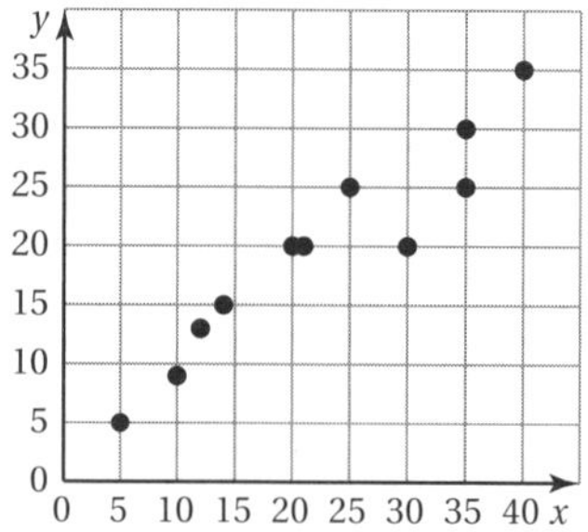

8. Write an equation for the line of best fit.

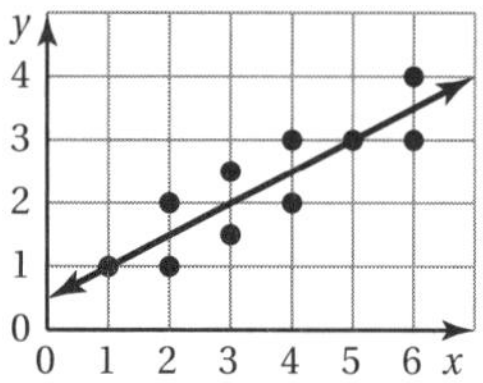

6	5	2	8	4	3	1	7

Activity 7.4 Start Thinking!
For use before Activity 7.4

List all of the different ways you know to display data.

Which can be used to display a list of values?

Which can be used to display information divided into categories?

Which can be used to display the relationship between two sets of data?

Activity 7.4 Warm Up
For use before Activity 7.4

Make a circle graph of the data.

1. A middle school has 390 sixth graders, 310 seventh graders, and 300 eighth graders.

2. The results of a survey in which 20 people were asked to name their favorite color are shown in the table.

Color	blue	red	green	pink	other
People	7	4	3	2	4

Lesson 7.4 Start Thinking!
For use before Lesson 7.4

How are a bar graph and a histogram similar? How are they different?

How are a line graph and a line plot similar? How are they different?

How are a stem-and-leaf plot and box-and-whisker plot similar? How are they different?

Lesson 7.4 Warm Up
For use before Lesson 7.4

1. Analyze and display the data in a way that best describes the data. Explain your choice of display.

Student's Quiz Scores						
Quiz	1	2	3	4	5	6
Grade	82	77	91	88	100	84

Name ______________________________ Date __________

7.4 Practice A

Choose an appropriate data display for the situation. Explain your reasoning.

1. the price of a stock over the last 5 years

2. the numbers of breads, rolls, muffins, and cookies baked this week

3. the number of runners in each 10-year age bracket

4. the comparison of city population and the number of fire stations

Explain why the data set is misleading.

5. 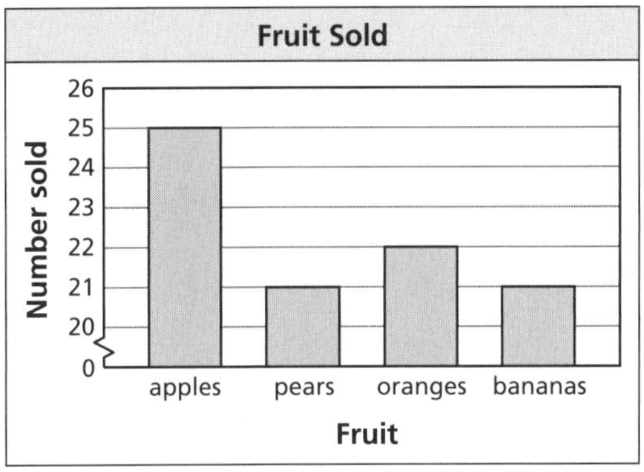

6. 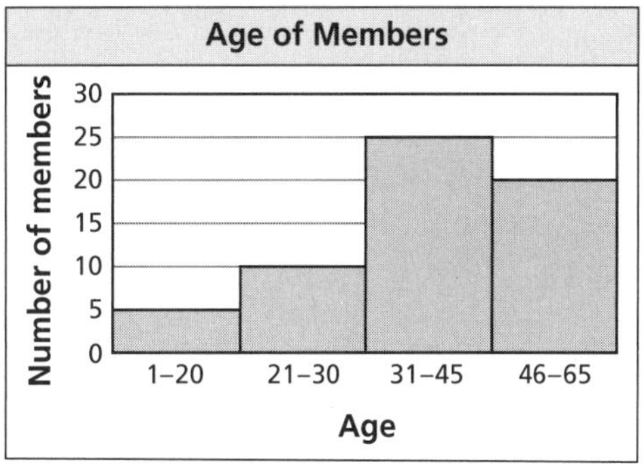

7. An EMT wants to use a data display to show students the relationship between the speed of the vehicle and the number of accidents. Choose an appropriate data display for the situation. Explain your reasoning.

8. What type of data display is appropriate for showing the averages of data over a period of time?

9. The home insurance industry wants to use a data display to show the variability in the costs of insurance for a $200,000 home. Choose an appropriate data display for the situation. Explain your reasoning.

10. A nutritionist wants to use a data display to show kindergarteners the recommended amounts of fruits, vegetables, breads, meats, and dairy that should be consumed each day. Choose an appropriate data display for the situation. Explain your reasoning.

11. The Smithsonian has been collecting data over the last 60 days. They want to use a data display to both put the data in order and to display how the data is distributed. Choose an appropriate data display for the situation. Explain your reasoning.

Name ____ Date __________

7.4 Practice B

Choose an appropriate data display for the situation. Explain your reasoning.

1. the heights of girls in grades 6 through 12

2. the numbers of computers offered within $100 price ranges

3. the comparison of the number of students and the number of office staff

4. the percentages of income budgeted for food, utilities, housing, gas, and education

Explain why the data set is misleading.

5. 

6. 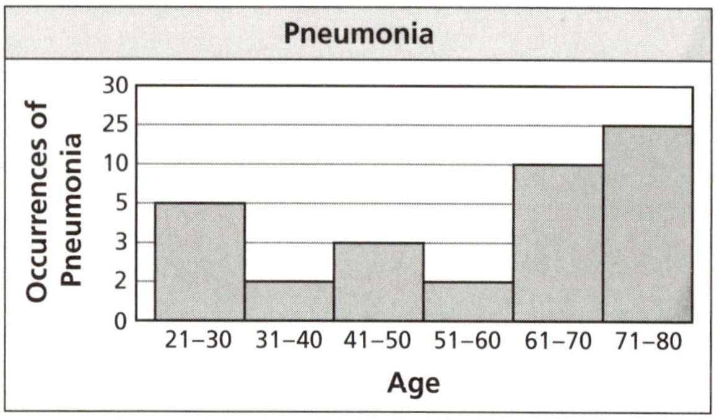

7. You spin the spinner 20 times and want to use a data display to show the number of times each of the numbers 1 through 5 occurred. Choose an appropriate data display for the situation. Explain your reasoning.

8. What type of data display is appropriate for showing the median of the data values?

9. A professor wants to use a data display to show the relationship between class sizes and passing rates for college students. Choose an appropriate data display for the situation. Explain your reasoning.

10. A dentist wants to use a data display to show the percentages of clients using different types of toothbrushes. Choose an appropriate data display for the situation. Explain your reasoning.

11. The new executive was making a presentation to the Board of Directors. He used a pictograph to show the weekly profits made by his department during the last 3 months.

    a. Explain why this would be an inappropriate use of a data display?

    b. Choose an appropriate data display for his situation.

Name______________________________ Date__________

7.4 Enrichment and Extension

Misleading Statistics

When someone quotes a statistic and says "The average number of …," the word "average" can refer to any of the measures of central tendency, not necessarily the mean of the data.

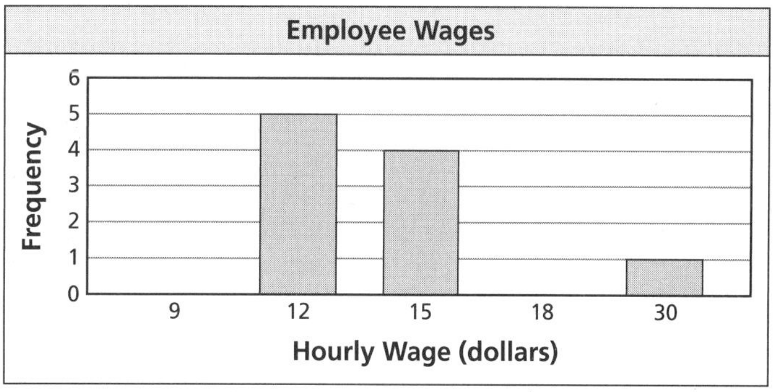

1. The graph shows the hourly wages of 10 employees in a company. Find the mean, median, and mode of the data set.

2. The company advertises that the average pay rate for their employees is $15 per hour.

a. Which measure of central tendency is the company using?

b. Why do you think the company used this measure of central tendency?

3. A competitor claims that the average pay rate is only $12 per hour and a newspaper reports that the average pay rate is $13.50 per hour.

a. Which measure of central tendency is each one using?

b. Why do you think each one chose the measure of central tendency that was used?

4. Which measure of central tendency best represents the data set?

5. The company has 100 employees. The sample used to gather data was the 10 employees who had been working at the company the longest. Is the sample a good representation of the population as a whole? Explain.

Name ______________________________________ Date __________

7.4 Puzzle Time

What Do You Call A Boomerang That Doesn't Come Back?

Write the letter of each answer in the box containing the exercise number.

Choose an appropriate display for the situation.

1. The outcomes of rolling a number cube with sides labeled 1, 2, 3, and 4.

2. The percent of votes that each soda received during a taste testing contest.

3. The number of students on the cross country team each year.

Explain why the data display is misleading.

4. 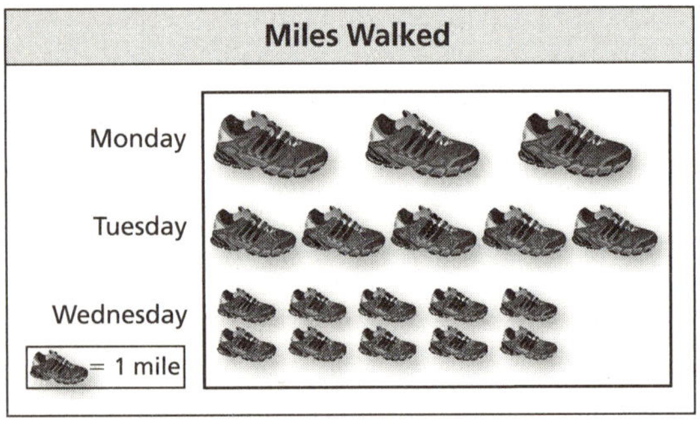

5. 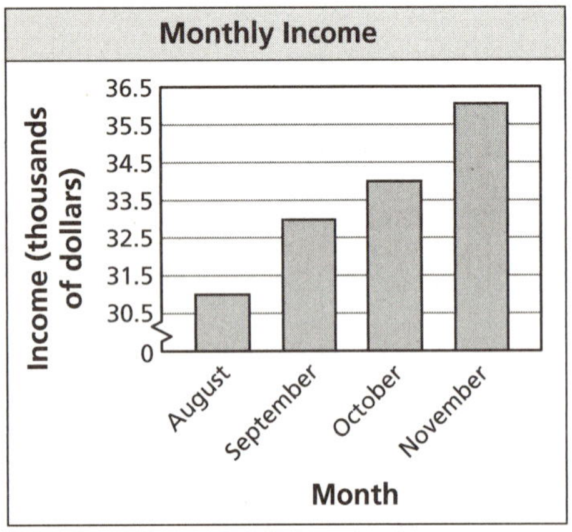

Answers

T. Line graph

K. Circle graph

I. Vertical axis is labeled with small increments

S. Pictures are of different sizes

C. Line plot

4	3	5	1	2

Name____ Date__________

Chapter 7

School-to-Work

For use after Section 7.4

Basketball Coach

You are the coach of a high school basketball team. Halfway into the season, the athletic director asks you to prepare an overview of the team's statistics so he can get an idea of how the team is doing. The key team statistics for each of the first eight games are given below.

Game number	Field goal (make/attempt)	3-point (make/attempt)	Free throw (make/attempt)	Total rebounds	Points scored
1	19/60	2/6	10/18	28	54
2	22/64	1/10	12/20	36	59
3	26/57	5/16	6/14	39	73
4	25/71	3/11	3/10	30	62
5	16/68	2/10	9/16	41	47
6	28/59	5/12	10/16	26	78
7	31/75	1/8	8/18	37	73
8	20/62	2/13	6/8	40	52

1. Find the mean, median, and mode of the points scored for the first half of the season. Which average would give the athletic director the most favorable impression of the basketball team? Which average do you think best describes the data? Explain.

2. You want to show the athletic director how the total of the points for the games were scored. What kind of data display could you use? Make the display. (Field goals are worth 2 points and free throws are worth 1 point.)

3. Make a scatter plot of the game number versus the points scored. Does the data show a *positive*, a *negative*, or *no* relationship? Could you use this relationship to make a prediction for the points scored in the next game? Explain.

4. You want to make a data display to show the athletic director the relationship between the field goals made and field goals attempted for each of the eight games. What kind of data display could you use? Make the display.

5. Discuss how you might present the data for the total rebounds per game. Represent this data in two different ways.

Name ______________________________ Date __________

Chapter 7 Study Help

You can use a **word magnet** to organize information associated with a vocabulary word.

On Your Own

Make a word magnet to help you study these topics.

1. outliers
2. box-and-whisker plots

After you complete this chapter, make word magnets for the following topics.

3. scatter plots
4. lines of best fit
5. data displays

Name____ Date__________

Chapter 7 Cumulative Practice

Organize the data in the given type of display.

1. Circle graph

Favorite Pattern	
Plain	50
Stripes	50
Polka Dot	80
Tartan	70

2. Stem-and-leaf plot

Enrollment	
80	72
76	84
52	83
68	91

3. Bar graph

Favorite Halloween Costume	
Ghost	38
Magician	42
Superhero	28
Zombie	32

4. Box-and-whisker plot

Tempo of a song (beats per minute)			
210	154	178	146
190	166	96	87
160	62	118	130

5. Line plot

Month	1	2	3	4	5	6	7	8
Number of school days	21	22	17	14	18	18	21	19

6. Scatter plot

Number of files, *x*	10	11	12	13	14	15	16	17
Kilobytes of memory, *y*	908	704	856	1024	952	988	1051	1100

Find the mean, median, and mode of the data set.

7. 41, 42, 38, 37, 42, 40, 47

8. 2.68, 1.71, 3.02, 2.45, 1.82, 2.30

Choose an appropriate data display for the situation. Explain your reasoning.

9. the types of flowers in a floral arrangement

10. the amount of time spent playing a video game each day

11. comparison of the weights and prices of bags of pet food

Name ____________________ Date __________

Unit 3 Project: Data Driven

For use after Unit 3

Objective **To collect, display, and analyze data on the distance a toy car can travel when sent down a ramp.**

Materials A small toy car, a ruler or tape measure, and materials to build a ramp (example: wood, cardboard, books, 3-ring binder, tape)

Investigation Have you ever raced a toy car with a friend? In this project, you will construct a ramp for your car and collect data by measuring the distance your car travels when it is released to go down a ramp.

1. Find an open area with a smooth surface to race the car. Construct a ramp using common materials. Make a ramp by stacking books and using cardboard or a three-ring binder. Write 2–3 sentences describing how you constructed your ramp.
2. Place the car at the top of the ramp and let it go down the ramp.
3. Measure and record the distance the car traveled in inches. Repeat the experiment 20 times.

Data Analysis

4. Find the mean, median, and mode of your data.
5. What is the range of the data?
6. Draw a box-and-whisker plot to display your data. What does the box-and-whisker plot tell you about the data?
7. Construct one other appropriate display for your data. Explain your reasoning in choosing this display.
8. Are there any outliers? How does an outlier affect the mean? the median? the mode? Explain.
9. Would a scatter plot be an appropriate display for the data? Explain why or why not.

Write a Report Write a description of the investigation. Include a sketch of your ramp, your data, your data displays, and the answers to the questions above. Think about what factors influence how far the car went. If you were to rebuild your ramp, how would you change it? Why?

Name____ Date__________

Unit 3

Student Grading Rubric

For use after Unit 3

		Student Score	Teacher Score
Cover Page 10 points			
	a. Name (4 points)	_____	_____
	b. Class (2 points)	_____	_____
	c. Project Name (2 points)	_____	_____
	d. Due Date (2 points)	_____	_____
Investigation 40 points			
	a. Construct a ramp and write 2–3 sentences describing how you constructed your ramp. (15 points)	_____	_____
	b. Measure and record the distance the car traveled in inches. Repeat 20 times. (25 points)	_____	_____
Data Analysis 105 points			
	a. Find the mean, median, mode, and range of the data. (40 points)	_____	_____
	b. Draw a box-and-whisker plot to display your data. (15 points)	_____	_____
	c. Construct one other appropriate display for your data. Explain your reasoning for choosing this display. (20 points)	_____	_____
	d. Describe the affect of outliers on measures of central tendency. (10 points)	_____	_____
	e. Explain why a scatter plot would or would not be an appropriate display for the data. (20 points)	_____	_____
Written Report 45 points			
	a. Includes a description of the investigation, sketches of the ramp, the data, the calculations of the measures of central tendency, the data displays, and an analysis of the data. (30 points)	_____	_____
	b. Report is neat, well organized, and clearly written. (15 points)	_____	_____
FINAL GRADE		_____	_____

Unit 3

Teacher's Project Notes

For use after Unit 3

Materials A small toy car, a ruler or tape measure, and materials to build a ramp (example: wood, cardboard, books, 3-ring binder, tape)

Alternatives Students can construct a double box-and-whisker plot to compare their data to other groups or classmates.

Students could enter their data into a spreadsheet and create computer generated displays.

To incorporate scatter plots into the project, change the angle of elevation on the ramp and continue to measure and adjust the ramp's steepness. Students could compare the distance traveled to the angle of elevation and find a line of best fit.

Common Errors To ensure consistency in the results, students should use the same type of car and ramp to collect all data.

Make sure students understand why a scatter plot would not be an appropriate display.

Suggestions Once the project is completed, have students share the additional display they chose with the class and have them explain why they chose the display.

Have students construct their displays on a piece of poster board so they can be displayed in the classroom.

Unit 3

Grading Rubric

For use after Unit 3

Scoring Rubric	
A	179-200
B	159-178
C	139-158
D	119-138
F	118 or below

Cover Page 10 points

a. Name (4 points)

b. Class (2 points)

c. Project Name (2 points)

d. Due Date (2 points)

Investigation 40 points

a. Construct a ramp and write 2–3 sentences describing how you constructed your ramp. (15 points)

b. Measure and record the distance the car traveled in inches. Repeat 20 times. (25 points)

Data Analysis 105 points

a. Find the mean, median, mode, and range of the data. (40 points)

b. Draw a box-and-whisker plot to display your data. (15 points)

c. Construct one other appropriate display for your data. Explain your reasoning for choosing this display. (20 points)

d. Describe the affect of outliers on measures of central tendency. (10 points)

e. Explain why a scatter plot would or would not be an appropriate display for the data. (20 points)

Written Report 45 points

a. Includes a description of the investigation, sketches of the ramp, the data, the calculations of the measures of central tendency, the data displays, and an analysis of the data. (30 points)

b. Report is neat, well organized, and clearly written. (15 points)

Chapter 8

Name______________________________ Date__________

Chapter 8 Linear Inequalities

Dear Family,

Fishing is one of our country's favorite pastimes. Enthusiasts may spend hundreds of dollars on equipment—but a slender pole, some string, and a hook still works as well as it did centuries ago.

Most public fishing spots require a fishing license and have rules that must be followed. Many regulations are given as some form of linear inequality. For example, some states do not require a fishing license for residents age 65 and older. So, you can say "age $<$ 65" requires a fishing license.

Have your student research a family fishing trip. Have them gather information from a Fish and Wildlife Conservation Commission and represent the results using mathematical inequalities.

- Will you be fishing in fresh- or salt-water? Will a license be required for each person going?
- What type of fish are you allowed to keep? How many of each species are you allowed?
- What size of fish are you allowed to keep? Does this depend on the species?
- Are there limits to the time of day or time of year when you can fish at a particular location? Does this depend on the species of fish?

Some limits are open-ended—stating only a minimum weight, for example, or a maximum number of fish. For example, the 65 and older rule stated above is open-ended. On the other hand, some limits are bounded—they have a minimum and a maximum. Have your student write a short summary using his or her inequalities.

With good research you won't find yourself fishing for answers.

Nombre ______________________________ Fecha ________

Capítulo 8 Desigualdades Lineales

Estimada Familia:

Pescar es uno de los pasatiempos favoritos del país. Los entusiastas pueden gastar cientos de dólares en equipos—pero un palo delgado, algo de cuerda y un garfio todavía funcionan tan bien como lo hicieron hace siglos.

La mayoría de lugares públicos para pescar requieren una licencia de pescar y tienen reglas que deben cumplirse. Muchas regulaciones se dan bajo la forma de desigualdades lineales. Por ejemplo, algunos estados no exigen una licencia de pescar para residentes de 65 años a más. Así que se puede decir "edad $<$ 65" requiere una licencia de pescar.

Haga que su estudiante investigue sobre un paseo de pesca familiar. Haga que recoja información de alguna Comisión para la Conservación de los Peces y la Vida Silvestre y que represente los resultados empleando desigualdades matemáticas.

- ¿Pescarán en agua dulce o salada? ¿Se requerirá una licencia para todas las personas que asistan?
- ¿Qué tipo de peces les está permitido conservar? ¿Cuántas de cada una de las especies pueden pescar?
- ¿Qué tamaño de peces les está permitido conservar? ¿Esto depende de la especie?
- ¿Existen límites para la hora del día o época del año cuando pueden pescar en un lugar en particular? ¿Esto depende de la especie del pez?

Algunos límites son abiertos—sólo indicando un peso mínimo, por ejemplo, o un número máximo de peces. Por ejemplo, la regla de los 65 años o más indicada anteriormente es abierta. Por otro lado, algunos límites son cerrados—tienen un mínimo y un máximo. Haga que su estudiante escriba un corto resumen usando sus desigualdades.

Con una buena investigación, no tendrá que "pescar" las respuestas.

Non ______________________________ Dat __________

Chapit 8 Inegalite Lineyè

Chè Fanmi:

Lapèch se youn nan pastan prefere peyi nou an. Moun ki pasyone yo gen dwa depanse plizyè santèn dola nan ekipman—men yon baton fen, yon fil ak yon zen toujou mache osi byen li te mache sa fè plizyè syèk de sa.

Pifò kote piblik pou fè lapèch yo egzije yon pèmi pou lapèch epi yo gen règ ou dwe swiv. Yo bay anpil règleman sou yon fòm inegalite lineyè. Paregzanp, gen eta ki pa egzije yon pèmi pou lapèch pou rezidan ki gen 65 an oswa ki pi gran. Donk, ou kapab di "laj < 65" egzije yon pèmi pou lapèch.

Fè elèv ou a fè rechèch sou yon pwonmnad lapèch familyal. Fè li rasanble enfòmasyon nan men yon Komisyon pou Konsèvasyon Pwason ak Lanati epi reprezante rezilta yo avèk inegalite matematik.

- Èske ou pwal peche nan dlo dous - oubyen dlo sale? Èske chak moun ki pwale ap bezwen yon pèmi?
- Ki tip pwason ou gen dwa kenbe? Konbyen nan chak espès ou gen dwa pran?
- Ki dimansyon pwason ou gen dwa kenbe? Èske sa depann de espès yo?
- Èske gen limit nan lè jounen an oswa epòk ane an kote ou kapab peche nan yon lye espesifik? Èske sa depann de espès pwason yo?

Gen limit ki ouvri—yo bay sèlman yon pwa minimòm, paregzanp, oswa yon kantite pwason maksimòm. Paregzanp, règ 65 an oswa pi gran an ke nou bay pi wo a ouvri. Pakont, gen limit ki bòne—yo gen yon minimòm ak yon maksimòm. Fè elèv ou a ekri yon rezime kout avèk inegalite li yo.

Avèk bon rechèch ou pap gen pou peche repons.

Activity 8.1

Start Thinking!

For use before Activity 8.1

How are inequalities related to height or age restrictions on amusement park rides?

Research the ride restrictions at an amusement park and write inequalities to describe them.

Activity 8.1

Warm Up

For use before Activity 8.1

Measure the line segment to the nearest tenth of a centimeter.

1. ______________________________

2. ____________

3. ____________________

4. ____

5. ____________________________________

6. __________________

Lesson 8.1 Start Thinking!
For use before Lesson 8.1

Write a sentence involving a real-life situation that can be modeled using an inequality.

Which inequality symbol applies: <, ≤, >, or ≥?

Lesson 8.1 Warm Up
For use before Lesson 8.1

Write an inequality for the graph. Then, in words, describe all the values of *x* that make the inequality true.

1. 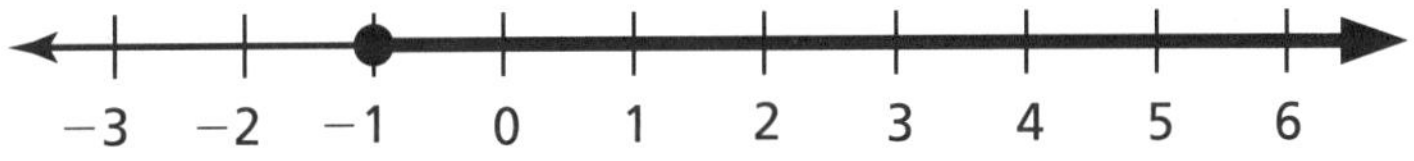

2. 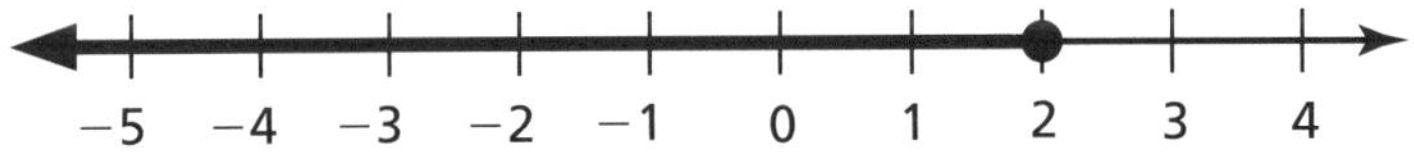

3. 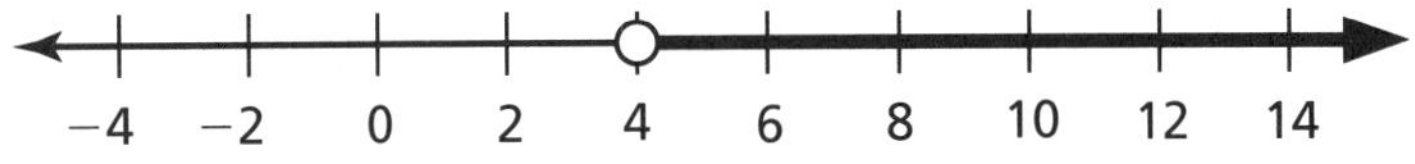

4. 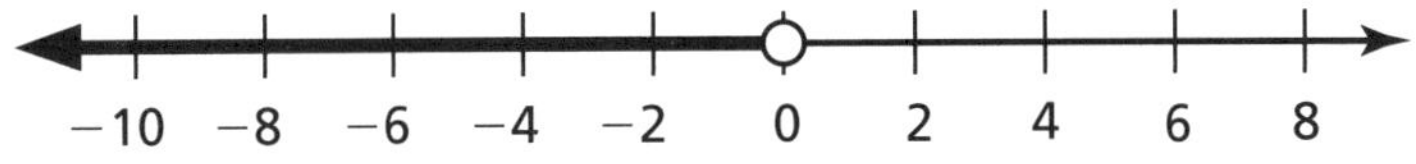

Name ______________________________ Date __________

8.1 Practice A

Write an inequality for the graph. Then, in words, describe all the values of *x* that make the inequality true.

1. 

2. 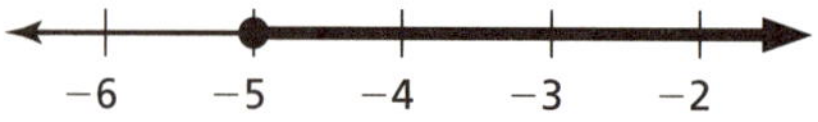

Write the word sentence as an inequality.

3. A number x is at most 3.

4. A number y added to 2 is greater than 7.

5. A number c multiplied by 3 is less than −12.

6. A number m minus 1.5 is no less than 2.

Tell whether the given value is a solution of the inequality.

7. $t - 3 \geq 2;\ t = 10$

8. $6w < -2;\ w = 1$

9. $p + 1.6 \leq 4;\ p = 5$

10. $\frac{1}{2}d > -3;\ d = 0$

Graph the inequality on a number line.

11. $k > 1$

12. $n \leq -2.5$

13. In order to try out for one of the parts in a play at the local theater, you must be at most 12 years old. Write an inequality that represents this situation.

Tell whether the given value is a solution of the inequality.

14. $3h - 7 < h;\ h = 2$

15. $q + 8 \geq \frac{q}{4};\ q = -12$

16. Consider the inequalities $-2x < 10$ and $-6 < -2x$.

a. Is $x = 0$ a solution to both inequalities?

b. Is $x = 4$ a solution to both inequalities?

c. Find another value of x that is a solution to both inequalities.

Name______________________________ Date__________

8.1 Practice B

Write an inequality for the graph. Then, in words, describe all the values of *x* that make the inequality true.

1. 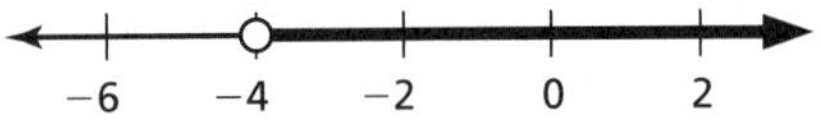

2. 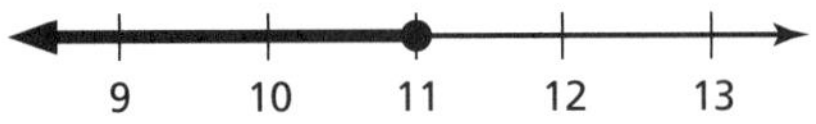

Write the word sentence as an inequality.

3. A number x is at least 15.

4. A number r added to 3.7 is less than 1.2.

5. A number h divided by 2 is more than -5.

6. A number a minus 8.2 is no greater than 12.

Tell whether the given value is a solution of the inequality.

7. $p + 1.7 \geq -4;\ p = -9$

8. $-3y < -5;\ y = 1$

9. $1.5g \leq 6;\ g = 0$

10. $\frac{3}{4} - d > \frac{1}{3};\ d = \frac{1}{2}$

Graph the inequality on a number line.

11. $\ell \leq 3.5$

12. $m > -15$

13. To get a job at the local restaurant, you must be at least 16 years old. Write an inequality that represents this situation.

Tell whether the given value is a solution of the inequality.

14. $5t < 4 - t;\ t = -3$

15. $\frac{q}{5} < q - 20;\ q = 15$

16. In order to qualify for a college scholarship, you must have acceptable scores in either the SAT or the ACT along with the following requirements: a minimum GPA of 3.5; at least 12 credits of college preparatory academic courses; and at least 75 hours of community service.

a. Write and graph three inequalities that represent the requirements.

b. Your cousin has a GPA of 3.6, 15 credits of college preparatory class, and 65 hours of community service. Other than the test scores, does your cousin satisfy the requirements? Explain.

Name ______________________________ Date __________

8.1 Enrichment and Extension

Compound Inequalities

Little League is a commercially sponsored baseball league for boys and girls.

A *compound inequality* is a special type of inequality that places both an upper and lower boundary on a variable. Write a compound inequality that describes the little league rule.

Example: The maximum number of innings in a little league game is 6. Each player must play at least 2 innings. Write a compound inequality that represents the number of innings a player plays.

Let n represent the number of innings a player plays. Because 2 is the minimum number of innings and 6 is the maximum number of innings, the compound inequality that represents the number of innings a player plays is $2 \leq n \leq 6$.

1. To be eligible to play Little League, a player must be at least 9 years old and at most 12 years old. Let a represent the player's age.

a. Write an inequality that represents the minimum age a player must be to participate in Little League.

b. Write an inequality that represents the maximum age a player can be to participate in Little League.

c. Use the inequalities from parts (a) and (b) to write a compound inequality that represents the age restrictions of Little League players.

2. For health and safety reasons, the number of pitches p a player can make per game is limited based on his or her age. A 12-year old may pitch a maximum of 85 pitches in a game day.

a. Write an inequality that represents the minimum number of pitches a player could make during a game.

b. Write an inequality that represents the maximum number of pitches a player could make during a game.

c. Use the inequalities from parts (a) and (b) to write a compound inequality that represents the number of pitches that a player can throw per game.

3. A Little League game lasts for at least 3.5 innings and at most 6 innings. Write a compound inequality that represents the number of innings n that a Little League game lasts.

Name____ Date__________

8.1 Puzzle Time

What Do You Call A Bull That's Sleeping?

Write the letter of each answer in the box containing the exercise number.

Write the word sentence as an inequality.

1. A number x is greater than 25.8.
2. Twice a number x is at most $-\frac{3}{5}$.
3. A number x minus 9.3 is more than 4.6.
4. A number x added to 11.7 is less than 14.

Tell whether the given value is a solution of the inequality.

5. $x - 3.6 \le 2.8;\ x = 6.7$
6. $\frac{5}{6}x > -10;\ x = -6$

Match each inequality with its graph.

7. $x \le -7$
8. $x > 3.2$
9. $x < 3\frac{1}{4}$
10. $x \ge -\sqrt{121}$

Answers

U. $11.7 + x < 14$

L. $x > 25.8$

A.

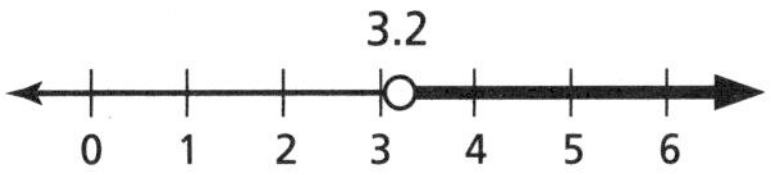

D. yes

E. $2x \le -\frac{3}{5}$

R.

L. $x - 9.3 > 4.6$

Z.

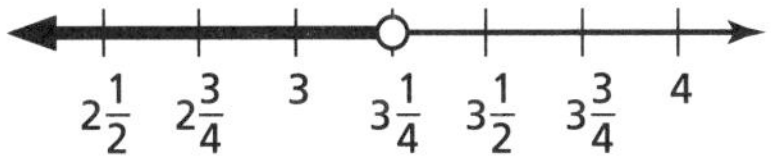

B. no

O.

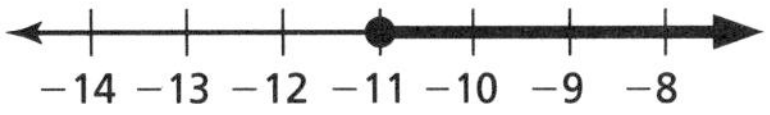

8		5	4	1	3	6	10	9	2	7

Activity 8.2 Start Thinking!

For use before Activity 8.2

Describe two different real-life situations that can represented by the inequality $x + 5 \leq 20$.

Activity 8.2 Warm Up

For use before Activity 8.2

Evaluate the expression when $x = 2$, $y = -6$, and $z = 8$.

1. $xy + 3z$

2. $1.5x - y + 5z$

3. $5x + 3y - 8z$

4. $x(8y - z)$

5. $\dfrac{-x + y + 2z}{z}$

6. $\dfrac{4x + 2y - z}{2x}$

Lesson 8.2 Start Thinking!

For use before Lesson 8.2

Students are selling magazine subscriptions for a fundraiser. The student who sells the most magazines wins a prize. You have sold 26 subscriptions. Your friend is currently in first place, having sold 41. How many subscriptions do you have to sell in order to move into first place?

What does this have to do with solving inequalities using addition or subtraction?

Lesson 8.2 Warm Up

For use before Lesson 8.2

Use the formula below to create a passing record that makes the inequality true.

$$P = \frac{8.4Y + 100C + 330T - 200N}{A}$$

1. $P \geq 120$

Attempts	Completions	Yards	Touchdowns	Interceptions

2. $P + 20 < 100$

Attempts	Completions	Yards	Touchdowns	Interceptions

Name ______________________________________ Date __________

8.2 Practice A

Solve the inequality. Graph the solution.

1. $p - 4 < 2$

2. $s + 1 \geq -5$

3. $k - 14 \leq -10$

4. $n + \frac{3}{2} > 2$

5. $z - \frac{2}{3} \geq \frac{1}{3}$

6. $-\frac{1}{6} + t < -\frac{1}{2}$

7. $d - 2.4 \leq -5.1$

8. $-4.5 + q > 2.5$

9. To stay within your budget, the area of the house and the garage combined is at most 3000 square feet. The area of the garage is 528 square feet. Write and solve an inequality that represents the area of the house.

10. You have $137.26 in a bank account. The bank requires you to have at least $50 in your account or else you are charged a fee. Write and solve an inequality that represents the amount you can write your next check for without being charged a fee.

Write and solve an inequality that represents the value of *x*.

11. The perimeter is less than 20 meters.

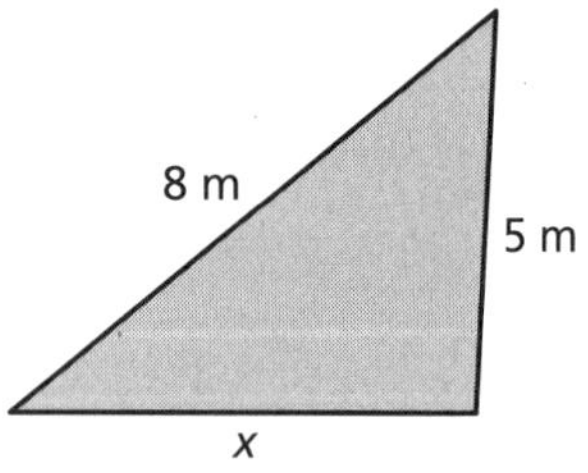

12. The perimeter is at least 18 feet.

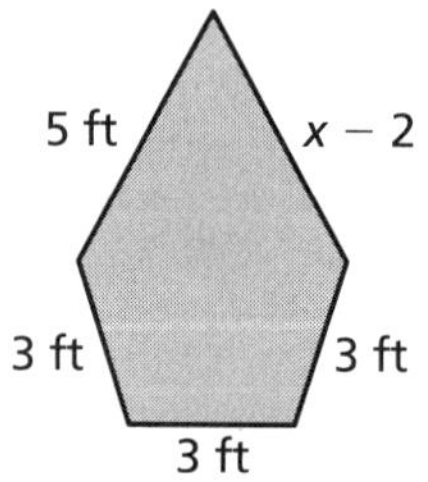

13. You need at least 5000 points to earn a gift card from your bank. You currently have 2700 points.

a. Write and solve an inequality that represents the number of points you need to earn a gift card.

b. You deposit money in your savings account and earn an additional 400 points. How does this change the inequality?

Name______________________________ Date__________

8.2 Practice B

Solve the inequality. Graph the solution.

1. $-12 \le y - 17$

2. $w - 1.8 < 2.5$

3. $v + \frac{1}{3} > 8$

4. $\frac{2}{5} < \frac{4}{5} + k$

5. $q + \frac{3}{4} \ge -\frac{1}{4}$

6. $-\frac{3}{2} + r < \frac{1}{2}$

7. $7.4 > c + 3.9$

8. $p - 10.2 > 3.5$

9. You and two friends are diving for lobster. The maximum number of lobsters you may have on your boat is 18. You currently have 7 lobsters.

   a. Write and solve an inequality that represents the additional lobsters that you may catch.

   b. Another friend comes on your boat and he has 3 lobsters. You may now have 24 lobsters on your boat. Write and solve an inequality that represents the additional lobsters that you may catch.

   c. How many lobsters is each person allowed to catch?

Write and solve an inequality that represents the value of *x*.

10. The length is greater than the width.

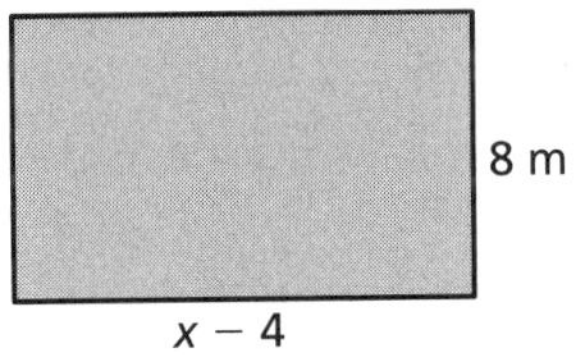

11. The perimeter is less than or equal to 50 inches.

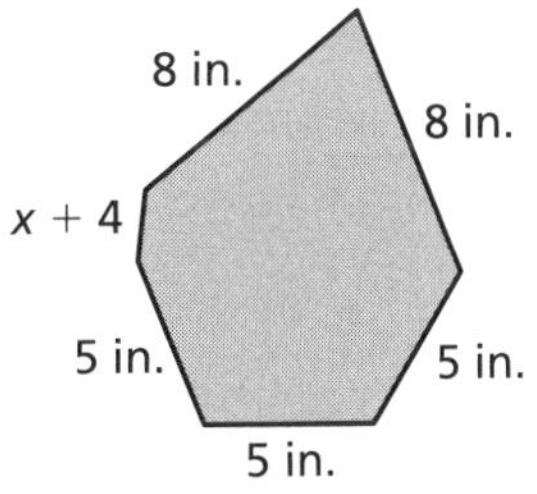

12. The solution of $w - c > -3.4$ is $w > -1.4$. What is the value of c?

13. Describe all numbers that are solutions to $|x| < 5$.

14. The *triangle inequality theorem* states that the sum of the lengths of any two sides of a triangle is greater than the length of the third side. A triangle has side lengths of 6 inches and 17 inches. What are the possible values for the length of the third side? Explain how you found your answer.

Name ______________________________ Date __________

8.2 Enrichment and Extension

Airplanes

The cabin (interior) of an airplane is partitioned into 3 distinct sections, or classes. The cost of a seat in each of the classes is different and the amenities in each class vary.

Seats	Width of seat	Pitch of seat
First class	51 cm	p cm
Business class	x cm	140 cm
Economy class	43 cm	81 cm

1. The width of a seat in Business Class is the average of the widths of the seats in First Class and Economy Class. Find the width of a seat in Business Class.

2. The cabin has a minimum width requirement so that each passenger on the plane has comfortable accommodations.

a. The width of four Business class seats and an aisle y must be at least 304.75 centimeters. Write an inequality that represents this situation.

b. What is the width of the aisle?

c. An airplane row contains 8 Business Class seats and 2 aisles. What is the minimum width w of the cabin?

3. The pitch of an airplane seat refers to the distance between the backs of two consecutive seats.

a. Thirty-eight less than the pitch p between First Class seats is at least as big as 2 times the pitch between Economy Class seats. Write an inequality that models the pitch between seats in First Class.

b. What is the pitch between seats in First Class?

4. Which class contains the greatest number of seats? the least number of seats? Explain your reasoning.

Name________________________________ Date__________

8.2 Puzzle Time

How Do You Make A Hamburger Roll?

A	B	C	D	E	F
G	H	I	J		

Complete each exercise. Find the answer in the answer column. Write the word under the answer in the box containing the exercise letter.

$x \geq 9$ TO
$x \leq -\frac{1}{7}$ GIVE
$x > 11.6$ PUSH
$x \leq 1.4$ A
$x \leq 4$ TAKE

Solve the inequality.

A. $x + 4 \leq 8$

B. $x + 9 < -5$

C. $x + 12 \geq 21$

D. $x + 8 < -2$

E. $6 > x + \frac{1}{3}$

F. $\frac{11}{8} < x + \frac{3}{8}$

G. $-\frac{4}{7} \geq x - \frac{3}{7}$

H. $x - \frac{8}{9} \geq \frac{10}{9}$

I. $10.1 \geq x + 8.7$

J. $6 < x - 5.6$

$x > 1$ AND
$x \geq 2$ IT
$x < \frac{17}{3}$ HILL
$x < -10$ A
$x < -14$ IT

Activity 8.3 Start Thinking!
For use before Activity 8.3

Are the solutions to the following inequalities the same? Explain why or why not.

$2x < -12$ $\qquad$ $-2x < 12$

Activity 8.3 Warm Up
For use before Activity 8.3

Complete the statement with < or >.

1. -7 ___?___ -5

2. 2 ___?___ -2

3. 7 ___?___ -10

4. -13 ___?___ -11

5. -1 ___?___ -2

6. -8 ___?___ 8

Lesson 8.3 Start Thinking!

For use before Lesson 8.3

In football, a team has four attempts to advance the ball at least 10 yards, which earns them a first down.

Write an inequality to describe the average yards per down for a team who has earned a first down in a set of 4 downs.

Lesson 8.3 Warm Up

For use before Lesson 8.3

Use a table to solve the inequality.

1. $-2x > 4$

2. $-4x < -8$

3. $-2x \leq -12$

4. $-3x \geq 18$

5. $\frac{x}{-3} < 21$

6. $\frac{x}{-5} > \frac{1}{2}$

Name ____ Date __________

8.3 Practice A

Solve the inequality. Graph the solution.

1. $8x > 8$

2. $\frac{r}{5} \le 2$

3. $1.6h < -32$

4. $\frac{u}{8} \ge 2.1$

5. $1.5j < -6.6$

6. $3x > -\frac{3}{2}$

Write the word sentence as an inequality. Then solve the inequality.

7. Five times a number is not less than 15.

8. The quotient of a number and 4 is less than -1.

9. An SUV averages 16.5 miles per gallon. The maximum average number of miles that can be driven on a full tank of gas is 363 miles. Write and solve an inequality that represents the number of gallons in a full tank.

Solve the inequality. Graph the solution.

10. $-2p \ge 10$

11. $-2 > \frac{v}{-3}$

12. $\frac{g}{-3.2} > 4$

13. $-\frac{y}{3} \le 1.4$

14. $-12 > -9h$

15. $\frac{a}{-3.5} \le -1.7$

16. You are creating a decorative rope that is at least 20 feet long.

    a. To create the rope you are using beads that are 6 inches long. Write and solve an inequality that represents the number of beads that you can use.

    b. You do not have enough 6-inch beads to make the rope, so you will use 10-inch beads instead. Write and solve an inequality that represents the number of 10-inch beads that you can use.

Name____ Date__________

8.3 Practice B

Solve the inequality. Graph the solution.

1. $3y \le \frac{3}{4}$

2. $-3.2 < \frac{p}{10}$

3. $1.6g \ge 0.48$

4. $2.5k < -100$

5. $\frac{s}{3.1} \ge 4.5$

6. $-\frac{4}{5} < 2x$

Write the word sentence as an inequality. Then solve the inequality.

7. A number divided by 5 is at least 4.

8. The product of 2 and a number is at most –6.

9. The solution of $cx \ge -4$ is $x \ge -8$. What is the value of c?

Solve the inequality. Graph the solution.

10. $-6t < 24$

11. $-\frac{2}{5} \le \frac{u}{-1}$

12. $\frac{q}{-0.4} \le 1.9$

13. $-\frac{d}{2} > \frac{3}{8}$

14. $-1.2 \le -0.8r$

15. $\frac{j}{-5.2} \le -1.5$

16. The height of a room is 10 feet. You are building shelving from the floor to the ceiling.

    a. Each shelf requires 8 inches. Write and solve an inequality that represents the number of shelves that can be made.

    b. You forgot to include the thickness of each shelf in your measurements. The amount of space needed for each shelf is actually 10 inches. Write and solve an inequality that represents the number of shelves that can be made.

Describe all numbers that satisfy both inequalities. Include a graph with your description.

17. $3x < 12$ and $-3x < -3$

18. $\frac{y}{5} \le -2$ and $-\frac{y}{4} \ge 1$

Name ____ Date __________

8.3 Enrichment and Extension

Hockey Pucks

A standard hockey puck has the given dimensions. Any company that manufactures hockey pucks must make sure their product meets the specifications. Hockey pucks are made using machines. No machine is perfect and so there will always be a slight variation of the size of hockey pucks.

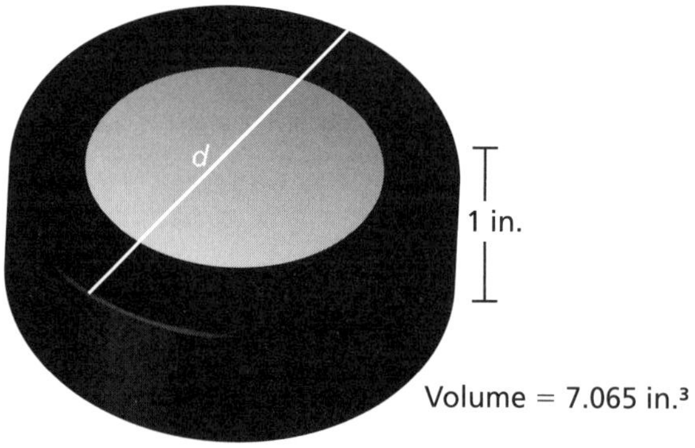

The allowable difference between the required dimensions of the hockey puck and its actual dimensions is called the *margin of error*.

1. What is the ideal radius of a standard hockey puck? Use 3.14 for π.

2. The radius of the puck has a margin of error of 0.005 inch. Write an inequality that models the margin of error e of the puck's radius.

3. A margin of error in the radius will produce a margin of error in the volume of the hockey puck.

   a. What is the smallest radius allowed by the margin of error? What is the volume of a puck with this radius? Use 3.14 for π. Round your answer to the nearest thousandth.

   b. What is the greatest radius allowed by the margin of error? What is the volume of a puck with this radius? Use 3.14 for π. Round your answer to the nearest thousandth.

   c. Write an inequality that models the range of acceptable radius lengths r of a hockey puck.

   d. Write an inequality that models the range of acceptable volumes V of a hockey puck.

4. The radius of a puck is the value you calculated in Exercise 1 and the height is 1.005 inches.

   a. Find the volume of the puck. Use 3.14 for π. Round your answer to the nearest thousandth.

   b. Which margin of error has a greater impact on the volume, the radius or the height? Why? Explain your reasoning.

Name______________________________ Date__________

8.3 Puzzle Time

What Do You Do When Your Smoke Alarm Goes Off?

Write the letter of each answer in the box containing the exercise number.

Solve the inequality.

1. $4x < 24$

2. $\frac{x}{6} \geq -3$

3. $-2.3x > 23$

4. $-15 \geq \frac{x}{3}$

5. $\frac{x}{4} > -4.1$

6. $9 \leq -1.5x$

7. $-6x > -\frac{1}{4}$

8. $4.2x \geq -12.6$

9. Three times a number x is at least -18.

10. The quotient of -7 and a number x is less than 8.

Answers

N. $x < -10$

U. $x \leq -6$

R. $x < \frac{1}{24}$

A. $x > -16.4$

I. $x \geq -6$

T. $x < 6$

F. $x \geq -3$

E. $x > -56$

T. $x \leq -45$

R. $x \geq -18$

7	6	3		5	8	1	10	2		9	4

Activity 8.4 Start Thinking!
For use before Activity 8.4

Write a word problem involving using a gift card to make an online purchase. The situation must be able to be solved using a multi-step inequality. Exchange problems with a classmate and solve your classmate's problem.

Activity 8.4 Warm Up
For use before Activity 8.4

Find the area of the shaded region. Round to the nearest tenth, if necessary.

1.

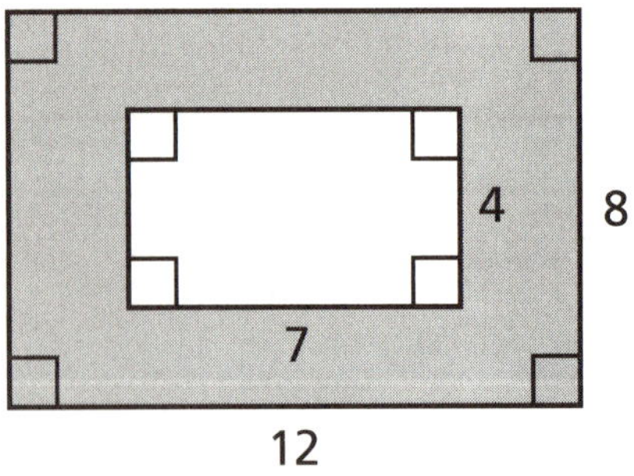

2.

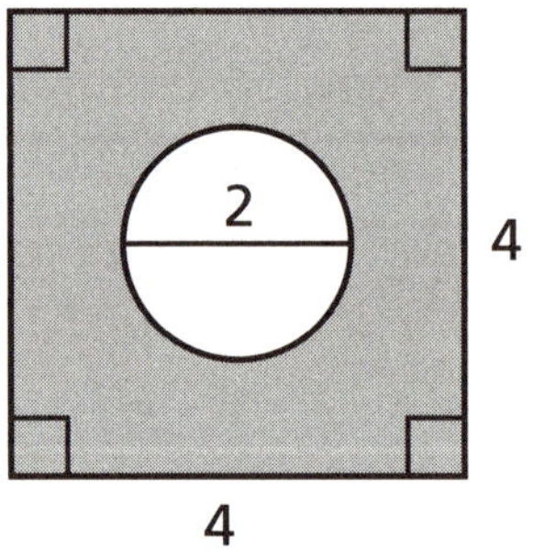

3.

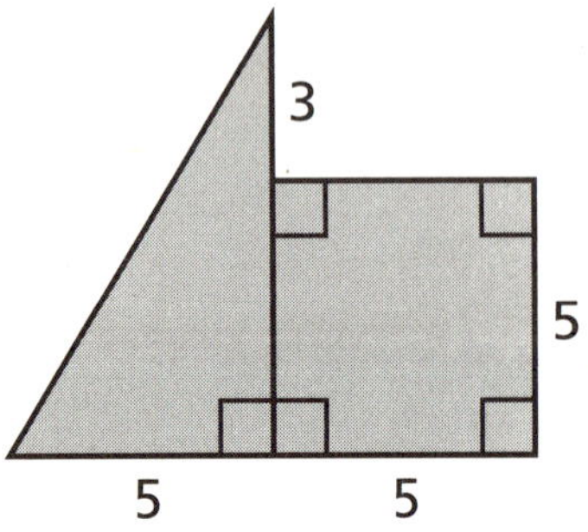

4.

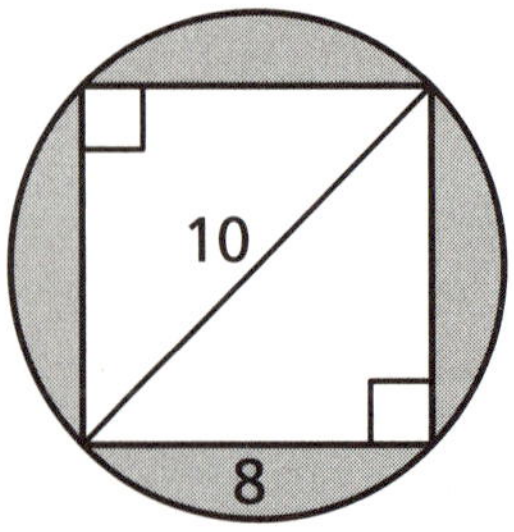

Lesson 8.4

Start Thinking!

For use before Lesson 8.4

What are your math quiz grades so far this grading period?

What must you earn on the next quiz in order to have at least a C quiz average? a B? an A?

Lesson 8.4

Warm Up

For use before Lesson 8.4

1. For what values of s will the perimeter of the hexagon be greater than 72 units?

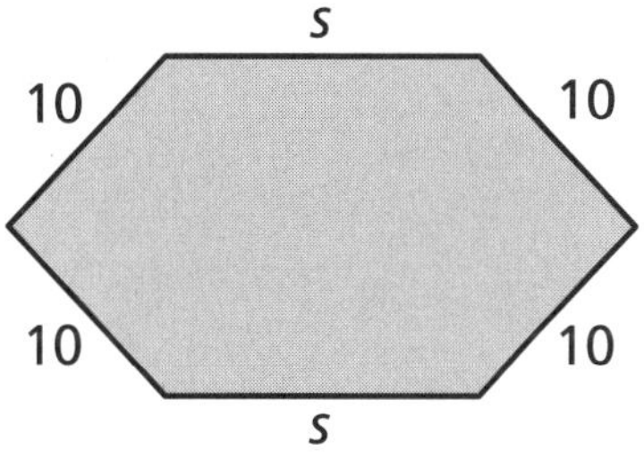

2. For what values of h will the surface area of the solid be less than or equal to 52 square units?

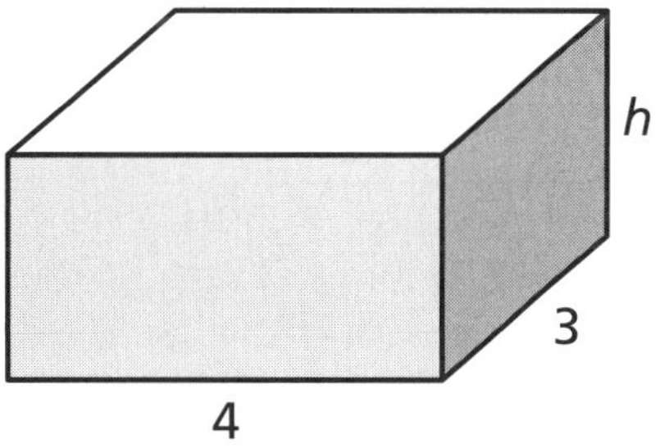

Name ______________________________ Date __________

8.4 Practice A

Solve the inequality. Graph the solution.

1. $3m - 7 < 2$

2. $-13 \leq -5r + 2$

3. $2k + \frac{1}{3} > 1$

4. $4.3 - 1.5c \leq 10$

5. You are renting a moving truck for a day. There is a daily fee of \$20 and a charge of \$0.75 per mile. Your budget allows a maximum total cost of \$65. Write and solve an inequality that represents the number of miles you can drive the truck.

Solve the inequality. Graph the solution.

6. $2(b - 4) > -6$

7. $-8(p + 3) \leq 16$

8. $15 \geq \frac{5}{3}(d - 6)$

9. $3.4 < 0.4(a + 12)$

10. Write and solve an inequality that represents the values of x for which the area of the rectangle will be at least 35 square feet.

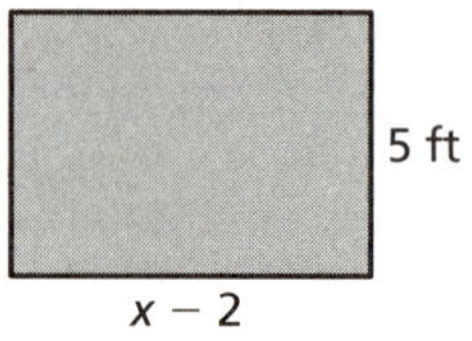

Solve the inequality. Graph the solution.

11. $3x - 7x + 2 < 10 - 12$

12. $14w - 8w - 5.4 \geq 7.3 - 10$

13. Your weekly base salary is \$150. You earn \$20 for each cell phone that you sell.

a. What is the minimum amount you can earn in a week?

b. Write and solve an inequality to find the number of cell phones you must sell to make at least \$630 a week.

c. Write and solve an inequality to find the number of cell phones you must sell to make at least \$750 a week.

d. The company policy is that as a part-time employee, the maximum you can earn each week is \$950. Write and solve an inequality to find the number of cell phones you can sell each week.

Name________________________________ Date__________

8.4 Practice B

Solve the inequality. Graph the solution.

1. $2 - \frac{q}{3} > 6$

2. $7 \le 0.5v + 10$

3. $-\frac{1}{4} \le 4k + \frac{7}{4}$

4. $3.6 - 0.24n < 1.2$

5. An RV park receives \$300 per month from each residential site that is occupied as well as \$2000 per month from their overnight sites. Write and solve an inequality to find the number of residential sites that must be occupied to make at least \$14,000 in revenue each month.

Solve the inequality. Graph the solution.

6. $-5(m - 2) > 30$

7. $-\frac{3}{2}(f + 6) \le -6$

8. $10.5 < 1.5(p - 3)$

9. $30 \ge -7.5(w - 4.2)$

10. Write and solve an inequality that represents the values of x for which the area of the rectangle will be at most 45 square meters.

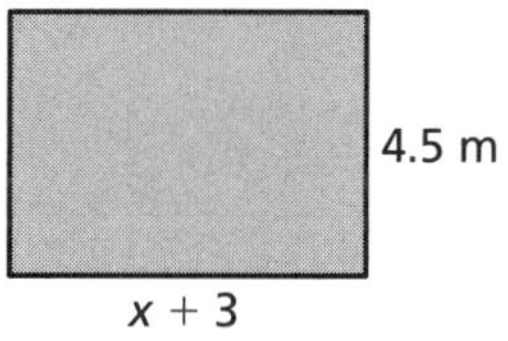

Solve the inequality. Graph the solution.

11. $12x - 5x - 4 \ge 60 - 8$

12. $4v + 6v + 3.2 < 6.8 - 9.2$

13. An animal shelter has fixed weekly expenses of \$750. Each animal in the shelter costs an additional \$6 a week.

a. During the summer months, the weekly expenses are at least \$1170. Write and solve an inequality to find the number of animals at the shelter for expenses to be at least \$1170 a week.

b. During the winter months, the weekly expenses are at most \$900. Write and solve an inequality to find the number of animals at the shelter for expenses to be at most \$900 a week.

c. The cost for each animal has increased by \$2. What will be the maximum weekly expenses during the winter months?

Name ______________________________ Date __________

8.4 Enrichment and Extension

Graphing Linear Inequalities

An inequality that contains only one variable is graphed on a number line. An inequality that contains two variables is graphed in the coordinate plane. Graphing a linear inequality is very similar to graphing a linear equation.

Use the linear inequality $2y - 3x < 6$.

1. Solve the linear inequality for y.

2. What is the y-intercept of the inequality? Plot this point on the graph.

3. What is the slope of the inequality? Use this slope to plot two additional points on the graph.

4. When graphing an inequality on a number line, you must decide whether to use a closed circle or an open circle. When graphing a linear inequality, you must decide whether to use a solid line or a dashed line to connect the points.

   a. In words, identify which inequality symbol is being used.

   b. If you were graphing a problem with this symbol on a number line, would you use a closed circle or an open circle?

   c. Use your answer from part (b) to determine whether to connect the points using a solid line or a dashed line. Then, draw the line.

5. The last step to graphing a linear inequality is determining on which side of the line to shade.

   a. Identify one point above and one point below the line.

   b. Evaluate the inequality using each of these points.

   c. Which of the inequalities from part (b) is true? Is that point above or below the line? Where should you shade? Explain.

   d. Graph the inequality.

6. Graph the linear inequality $4y + 3x \geq 8$.

Name____ Date__________

8.4 Puzzle Time

What Did Ernie Say When Bert Asked Him If He Wanted Ice Cream?

Write the letter of each answer in the box containing the exercise number.

Solve the inequality.

1. $8x - 11 < 13$

2. $3x - 5 \geq 16$

3. $2 - \frac{x}{4} \geq 4$

4. $\frac{6}{7} > -2x - \frac{8}{7}$

5. $4.6 > 1.2 + 1.7x$

6. $8(x - 4) \geq 40$

7. $-30 \leq -\frac{3}{4}(x + 4)$

8. $-6.8 \geq 0.8(x + 1)$

Answers

R. $x < 2$

T. $x \leq 36$

E. $x \geq 7$

S. $x \leq -9.5$

B. $x > -1$

U. $x < 3$

R. $x \leq -8$

E. $x \geq 9$

8	1	3	6		4	2	5	7

Name ______________________________ Date __________

Chapter 8

School-to-Work

For use after Section 8.4

Car Salesperson

You are negotiating the sale of a used car as a salesperson at a car dealership. The buyer is trying to get the best price that he can. As the salesperson, you have to make sure that the selling price is greater than the amount paid for the car by the dealership plus the cost of holding the car at the dealership. Otherwise, the dealership won't make a profit. The costs associated with the car are given below.

Car #58: 4-door sedan

Purchase price: $2850

Cost per day to keep on lot: $25

1. Write an inequality that gives the possible selling price in terms of the number of days the used car is on the lot at the dealership in order for the dealership to make a profit. Include the meaning of each variable.

2. Graph your inequality on a coordinate plane. What does a point in the shaded region represent? What does a point on the line represent?

3. If the car is on the lot for 35 days, describe the profitable selling prices.

4. Would the car dealership experience a profit or a loss if the car is sold for $4200 after 63 days? How much of a profit or a loss? Explain your reasoning.

5. If the car is sold for $5000 and the dealership makes a profit, describe the possible number of days the car could have been on the lot.

6. You earn 15% of the profit made by the dealership when you sell a car. If the sedan is sold for $5150 after 12 days, how much do you earn? Explain how you found your answer.

Name______________________________ Date__________

Chapter 8 Study Help

You can use a **four square** to organize information about a topic. Each of the four squares can be a category, such as *definition*, *vocabulary*, *example*, *non-example*, *words*, *algebra*, *table*, *numbers*, *visual*, *graph*, or *equation*.

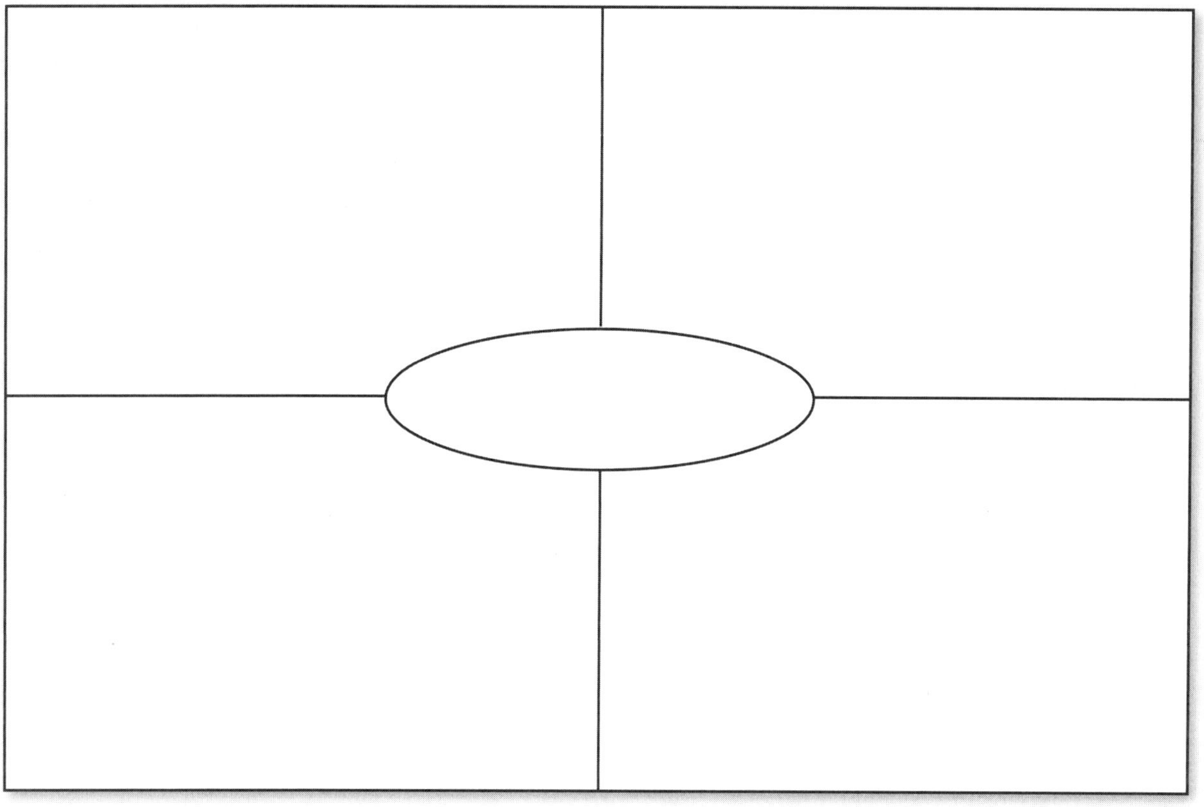

On Your Own

Make a four square to help you study these topics.

1. inequality ($>$)

2. inequality ($\leq$)

3. inequality ($\geq$)

4. solving an inequality using addition

5. solving an inequality using subtraction

After you complete this chapter, make four squares for the following topics.

6. solving an inequality using multiplication

7. solving an inequality using division

Name ______________________________ Date __________

Chapter 8 Cumulative Practice

Write the word sentence as an equation or inequality.

1. A number p multiplied by -4 is no more than 12.

2. The difference of 5 and a number g is greater than 2.

3. A number x divided by 18 is equal to 4.6.

4. The difference of a number c and -2 is 10.

5. The product of a number b and -1.6 is less than -9.6.

6. A gym membership costs \$60 per month. A class at the gym costs \$10 per month. Write an equation that represents the total cost y of taking x classes at the gym each month.

Write an inequality and a word sentence that represent the graph.

7. 

8. 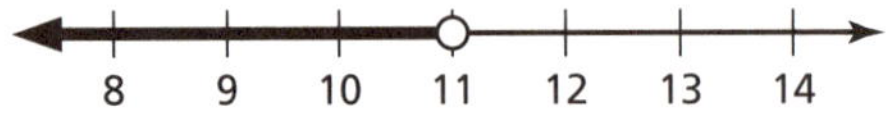

9. A laser tag center has a sign posted that states "Must be 6 years or older to play." Write an inequality that represents the possible ages of laser tag players.

Graph the inequality on a number line.

10. $y \le -3$

11. $n > 2.4$

12. $a \ge \frac{3}{2}$

13. The record for the 50-yard dash at your school is 9.6 seconds. Write an inequality that represents the possible speeds (in feet per second) that would break the record.

Solve the equation or inequality. Graph the solution of each inequality.

14. $-5k \ge -10$

15. $6h - 3 = 21$

16. $\frac{r}{3} \ge 11$

17. $m + 12 > 5$

18. $\frac{3}{4} - d = 2$

19. $2u = 4$

20. An arcade uses tokens for its games. You can buy 75 tokens for \$15. Write and solve an equation to find the price of one token.

Chapter 9

Name ______________________________ Date __________

Chapter 9 Exponents and Scientific Notation

Dear Family,

People shop on the Internet, talk to one another using email, and keep digital photo and video albums. Computers represent all this information using numbers.

The number system that computers use is based on powers of 2 and is called the *binary system*. For example, the number 45 is represented as $32 + 8 + 4 + 1 = 2^5 + 2^3 + 2^2 + 2^0$. Try this with your student.

- Find the first eight powers of two, starting with $2^0 = 1$. Write these values on the blanks under the powers of two in the table.
-

Power	$2^7 =$ ____	$2^6 =$ ____	$2^5 =$ ____	$2^4 =$ ____	$2^3 =$ ____	$2^2 =$ ____	$2^1 =$ ____	$2^0 =$ 1
Digit (0 or 1)								

- Pick any number from 0 to 255. Find the sum of the powers of two that equal your number.
- Put a 1 in the "Digit" row if the power of two is in your number; put a 0 in the "Digit" row if the power of two is not in your number. For example, the digits for 45 are shown below because $45 = 2^5 + 2^3 + 2^2 + 2^0$.

Power	2^7	2^6	2^5	2^4	2^3	2^2	2^1	2^0
Digit	0	0	1	0	1	1	0	1

- Write your number as a binary number. For example, 45 is 00101101.

Computer storage and file sizes are often recorded in kilobytes. A kilobyte is equal to 2^{10} bytes = 1024 bytes. People often round this to 1000 bytes. Ask your student to research what powers of two correspond to megabyte, gigabyte, and terabyte. Find the following information on the Internet or in the library.

- How many kilobytes does a typical page of text use?
- How many megabytes does a typical digital photo use?
- How many gigabytes does one minute of video typically use? How does video compare to audio?

Understanding powers helps make computer terms a *bit* more familiar!

Nombre ______________________________ Fecha__________

Capítulo 9 Exponentes y Notación Científica

Estimada Familia:

Las personas compran por Internet, conversan entre sí usando el correo electrónico y tienen álbumes de videos y fotos digitales. Las computadoras representan toda esta información usando números.

El sistema numérico que las computadoras usan se basa en potencias de 2 y se llama *sistema binario*. Por ejemplo, el número 45 se representa como $32 + 8 + 4 + 1 = 2^5 + 2^3 + 2^2 + 2^0$. Intente hacer esto con su estudiante.

- Encuentren las primeras ocho potencias de dos, empezando con $2^0 = 1$. Escriban estos valores en los espacios en blanco bajo las potencias de dos en la tabla
-

Potencia	$2^7 =$ ____	$2^6 =$ ____	$2^5 =$ ____	$2^4 =$ ____	$2^3 =$ ____	$2^2 =$ ____	$2^1 =$ ____	$2^0 =$ 1
Dígito (0 ó 1)								

- Escojan cualquier número entre 0 y 255. Encuentren la suma de las potencias de dos que equivalen a su número.
- Coloquen un 1 en la fila de "Dígito" si la potencia de dos se encuentra en su número; coloquen un 0 en la fila de "Dígito" si la potencia de dos no se encuentra en su número. Por ejemplo, los dígitos de 45 se muestran a continuación porque $45 = 2^5 + 2^3 + 2^2 + 2^0$.

Potencia	2^7	2^6	2^5	2^4	2^3	2^2	2^1	2^0
Dígito	0	0	1	0	1	1	0	1

- Escriban su número como número binario. Por ejemplo, 45 es 00101101.

La capacidad de almacenamiento de una computadora y los tamaños de los archivos a menudo se registran en kilobytes. Un kilobyte equivale a 2^{10} bytes $= 1024$ bytes. A menudo, la gente redondea esto a 1000 bytes. Pida a su estudiante que investigue qué potencias de dos corresponden a megabyte, gigabyte y terabyte. Encuentren la siguiente información en la Internet o en la biblioteca.

- ¿Cuántos kilobytes usa una hoja de texto común?
- ¿Cuántos megabytes usa una fotografía digital común?
- ¿Cuántos gigabytes usa un minuto de video común? ¿Cómo se puede comparar el video con el audio?

¡Comprender las potencias permitirá que se familiaricen un *poco* más con los términos de computación!

Non ______________________________ Dat __________

Chapít 9 Espozan ak Notasyon Syantifik

Chè Fanmi:

Moun achte sou Entènèt, pale youn ak lòt avèk imel, epi kenbe albòm foto ak videyo dijital. Òdinatè reprezante tout enfòmasyon sa yo avèk chif.

Sistèm nimerik ke òdinatè yo itilize a baze sou pwisans 2 donk li rele *sistèm binè*. Paregzanp, yo reprezante chif 45 la konsa $32 + 8 + 4 + 1 = 2^5 + 2^3 + 2^2 + 2^0$. Eseye sa avèk elèv ou a.

- Jwenn wit premye pwisans de yo, kòmanse avèk $2^0 = 1$. Ekri valè sa yo nan espas ki anba pwisans de yo nan tablo a.
-

Pwisans	$2^7 =$ ____	$2^6 =$ ____	$2^5 =$ ____	$2^4 =$ ____	$2^3 =$ ____	$2^2 =$ ____	$2^1 =$ ____	$2^0 =$ 1
Chif (0 oubyen 1)								

- Chwazi nenpòt ki chif ant 0 ak 255. Jwenn sòm pwisans de ki egal chif ou a.
- Mete yon 1 nan liy "Chif " la si pwisans de a nan chif ou a; mete yon 0 nan liy "Chif" la si pwisans de a pa nan chif ou a. Paregzanp, nou montre chif sa yo pi ba a pou 45 poutèt $45 = 2^5 + 2^3 + 2^2 + 2^0$.

Pwisans	2^7	2^6	2^5	2^4	2^3	2^2	2^1	2^0
Chif	0	0	1	0	1	1	0	1

- Ekri chif ou a sou fòm chif binè. Paregzanp, 45 se 00101101.

Dimansyon memwa ak fichye òdinatè souvan anrejistre an kilo-oktè. Yon kilo-oktè egal 2^{10} oktè $= 1024$ oktè. Moun souvan awondi sa 1000 oktè. Mande elèv ou a pou li fè rechèch sou ki pwisans de ki koresponn avèk mega-oktè, jiga-oktè ak tera-oktè. Jwenn enfòmasyon sila yo sou Entènèt oswa nan bibliyotèk.

- Konbyen kilo-oktè yon paj tèks abityèl itilize?
- Konbyen mega-oktè yon foto dijital abityèl itilize?
- Konbyen jiga-oktè yon minit videyo abityèlman itilize? Kijan videyo ye pa rapò ak odyo?

Konprann pwisans ede fè nou vin yon ti jan pi abitye ak tèm òdinatè!

Activity 9.1 Start Thinking!
For use before Activity 9.1

When have you used exponents before?

What is another way to say "x to the second power?" Why do you think it is called that?

What is another way to say "x to the third power?" Why do you think it is called that?

Activity 9.1 Warm Up
For use before Activity 9.1

Find the product.

1. $5 \times 5 \times 5$
2. $10 \times 10 \times 10$
3. $(-3) \times (-3) \times (-3)$
4. $10 \times 10 \times 10 \times 10 \times 10$
5. $4 \times 4 \times 4 \times 4$
6. $(-2) \times (-2) \times (-2) \times (-2)$

Lesson 9.1 Start Thinking!

For use before Lesson 9.1

Use the nursery rhyme on page 351 of your textbook to guide you in writing your own "power poem."

Lesson 9.1 Warm Up

For use before Lesson 9.1

Write the product using exponents.

1. $2 \bullet 2 \bullet 2$

2. $(-7) \bullet (-7)$

3. $\frac{2}{3} \bullet \frac{2}{3} \bullet \frac{2}{3} \bullet \frac{2}{3} \bullet \frac{2}{3}$

4. $\left(-\frac{1}{6}\right) \bullet \left(-\frac{1}{6}\right) \bullet \left(-\frac{1}{6}\right)$

5. $11 \bullet 11 \bullet 11 \bullet 11 \bullet 11 \bullet 11 \bullet 11$

6. $\left(-\frac{1}{4}\right) \bullet \left(-\frac{1}{4}\right) \bullet \left(-\frac{1}{4}\right) \bullet \left(-\frac{1}{4}\right)$

Name________________________________ Date__________

9.1 Practice A

Write the product using exponents.

1. $6 \bullet 6 \bullet 6 \bullet 6 \bullet 6$

2. $(-2) \bullet (-2) \bullet (-2)$

3. $\frac{2}{3} \bullet \frac{2}{3} \bullet \frac{2}{3} \bullet \frac{2}{3}$

4. $(-1.2) \bullet (-1.2) \bullet (-1.2)$

5. $\frac{1}{5} \bullet \frac{1}{5} \bullet x \bullet x \bullet x$

6. $10 \bullet 10 \bullet (-n) \bullet (-n) \bullet (-n)$

7. $(-5) \bullet (-5) \bullet (-5) \bullet (-5) \bullet y \bullet y \bullet y \bullet y \bullet y$

Evaluate the expression.

8. 9^2

9. -5^4

10. $(-3)^4$

11. $\left(\frac{1}{4}\right)^3$

12. Write the prime factorization of 500 using exponents.

Evaluate the expressions.

13. $7 + (-2) \bullet 3^2$

14. $(15^2 - 5 \bullet 4^2) \div 5$

15. $\left|\frac{1}{3}\left(2^3 - \frac{10^2}{5}\right)\right|$

16. $\frac{3}{2}(4^3 - 2^2 \bullet 3^2)$

17. There are 5 posts supporting the guard rail for the steps to your home. The tallest post is 3 feet tall. The height of each of the other posts is $\frac{5}{6}$ the height of the next larger post.

a. Write an expression for the height of the shortest post.

b. What is the height of the shortest post?

18. You ran 4 miles. Rob ran half as far as you. Tim ran half as far as Rob. Nicole ran half as far as Tim.

a. Write an expression for how far Nicole ran.

b. How far did Nicole run?

Name ______________________________ Date __________

9.1 Practice B

Write the product using exponents.

1. $4 \bullet 4 \bullet 4 \bullet 4 \bullet 4 \bullet 4$

2. $(-12) \bullet (-12) \bullet (-12) \bullet (-12) \bullet (-12)$

3. $-\frac{3}{7} \bullet \frac{3}{7} \bullet \frac{3}{7}$

4. $\left(-\frac{3}{7}\right) \bullet \left(-\frac{3}{7}\right) \bullet \left(-\frac{3}{7}\right)$

5. $(-9) \bullet (-9) \bullet (-9) \bullet (-9) \bullet x \bullet x \bullet x$

6. $25 \bullet 25 \bullet 25 \bullet 25 \bullet (-p) \bullet (-p) \bullet (-p) \bullet (-p) \bullet (-p)$

7. $(-2) \bullet (-2) \bullet x \bullet x \bullet x \bullet y \bullet y \bullet y \bullet y$

Evaluate the expression.

8. 7^3

9. -4^4

10. $(-4)^4$

11. $\left(\frac{2}{5}\right)^3$

12. Write the prime factorization of 1323 using exponents.

Evaluate the expressions.

13. $280 - (-3) \bullet (-5)^3$

14. $\left(20^2 - 3^3 \bullet 8^2\right) \div 16$

15. $\frac{2}{3}\left(16^2 - 17^2\right)$

16. $\left|\frac{1}{5}\left(\frac{6^3}{3^3} - 2^3\right)\right|$

17. Bed A is 7 feet long. Bed B is $\frac{7}{8}$ as long as bed A. Bed C is $\frac{7}{8}$ as long as bed B. Bed D is $\frac{7}{8}$ as long as bed C.

   a. Write an expression for the length of bed D.

   b. What is the length of bed D?

Name____ Date __________

9.1 Enrichment and Extension

The Sierpinski Triangle

A geometric pattern is formed by multiplying by the same number at each step. For example, the numbers 1, 2, 4, 8, 16, … form a geometric pattern because each number is twice the previous number.

The Sierpinski Triangle is constructed using the pattern below.

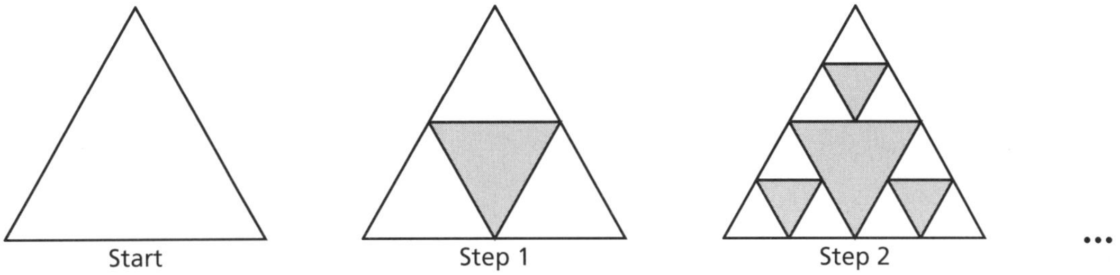

1. Describe the pattern between steps.
2. Draw the next step in the pattern.
3. Complete the table.

Step	White Triangles	Gray Triangles
0	1	0
1	3	1
2		
3		
4		

4. Describe the pattern in the number of white triangles at each step.
5. Describe the pattern in the number of gray triangles at each step.
6. How many white triangles will there be at the ninth step? Write your answer using exponents.

Name ______________________________ Date __________

9.1 Puzzle Time

How Does A Bee Part His Hair?

Write the letter of each answer in the box containing the exercise number.

Evaluate the expression.

1. 3^3
2. -4^5
3. $(-2)^5$
4. $(-7)^4$
5. $\left(\frac{1}{5}\right)^3$
6. $\left(-\frac{1}{3}\right)^5$
7. $-\left(\frac{1}{4}\right)^3$
8. $\left|-\left(\frac{1}{6}\right)^3\right|$
9. $7^3 - 5^4$
10. $4 \bullet 3^4 + 6$
11. $2(11^3 - 10^3)$
12. $\frac{1}{8}(4^6 - 4^3)$
13. $|6^3 - 7^3|$
14. $|9^2 + (-9)^3|$
15. You are sending a package in the mail. The box is a cube measuring 14 inches on all sides. How many cubic inches is the package?
16. You sent an email to 8 friends. They each forwarded the email to 8 friends. And those friends each forwarded the email to 8 friends. The chain of emails is represented by the expression $8 + 8^2 + 8^3$. How many people were sent your email?

Answers

S. $-\frac{1}{243}$

T. -282

C. 584

I. -1024

E. 662

O. $\frac{1}{125}$

I. 27

O. 504

H. 330

B. 2744

W. $\frac{1}{216}$

H. 2401

H. 127

Y. 648

M. $-\frac{1}{64}$

N. -32

8	2	9	4		10	1	6		13	5	3	11	14	16	12	7	15

Activity 9.2 Start Thinking!

For use before Activity 9.2

The Product of Powers Property is a rule for multiplying powers that have the same base, for example $a^m \bullet a^n$. Do you think that Product of Powers Property is a good name for the property? Why or why not?

Activity 9.2 Warm Up

For use before Activity 9.2

Evaluate the power.

1. 3^5

2. 5^4

3. 10^6

4. $(-4)^3$

5. $(-3)^2$

6. $(-2)^5$

Lesson 9.2 Start Thinking!

For use before Lesson 9.2

Think of an example in geometry where you would need to use the Product of Powers Property.

Lesson 9.2 Warm Up

For use before Lesson 9.2

Simplify the expression. Write your answer as a power.

1. $2^4 \bullet 2^3$

2. $7^5 \bullet 7^7$

3. $\left(\frac{1}{2}\right)^2 \bullet \left(\frac{1}{2}\right)^5$

4. $\left(-\frac{3}{5}\right)^3 \bullet \left(-\frac{3}{5}\right)^3$

5. $x^5 \bullet x^{11}$

6. $y^3 \bullet y$

Name____ Date__________

9.2 Practice A

Simplify the expression. Write your answer as a power.

1. $2^3 \bullet 2^2$

2. $9^6 \bullet 9^8$

3. $(-7)^3 \bullet (-7)^5$

4. $\left(\frac{5}{8}\right)^{10} \bullet \left(\frac{5}{8}\right)^2$

5. $c \bullet c^5$

6. $q^4 \bullet q^4$

7. $\left(-\frac{4}{9}\right)^2 \bullet \left(-\frac{4}{9}\right)^5$

8. $(4.7)^3 \bullet (4.7)^2$

9. $(3^2)^3$

10. $(k^5)^{10}$

11. $\left(\left(\frac{1}{2}\right)^4\right)^3$

12. $\left((9.2)^3\right)^6$

Simplify the expression.

13. $(4n)^2$

14. $(-2w)^5$

15. $\left(\frac{1}{3}p\right)^4$

16. $(2.5j)^3$

17. $(ab)^{18}$

18. $3^2(3 \bullet 3^4)$

19. Is $3^2 \bullet 4^2 = 12^4$? Evaluate each side of the equation to explain your answer.

20. The volume of a sphere is $V = \frac{4}{3}\pi r^3$ and the relationship between the radius r and the diameter d is $r = \frac{d}{2}$.

a. Find the volume of the sphere in terms of the diameter d and simplify the expressions.

b. What is the volume of the sphere when the diameter is $\frac{2}{3}$ centimeter?

Name ______________________________ Date ________

9.2 Practice B

Simplify the expression. Write your answer as a power.

1. $8^3 \bullet 8^7$

2. $(-16)^5 \bullet (-16)^{21}$

3. $\left(-\frac{5}{9}\right)^5 \bullet \left(-\frac{5}{9}\right)^5$

4. $\left(\frac{1}{15}\right)^{12} \bullet \left(\frac{1}{15}\right)$

5. $q^7 \bullet q^9$

6. $(13.2)^6 \bullet (13.2)^2$

7. $(-7.4)^9 \bullet (-7.4)^{12}$

8. $\left(9^3\right)^3$

9. $\left(d^2\right)^6$

10. $\left(2.9^3\right)^6$

11. $\left(\left(\frac{5}{8}\right)^2\right)^3$

12. $\left(\left(-\frac{2}{9}\right)^3\right)^5$

Simplify the expression.

13. $(-2p)^4$

14. $\left(\frac{1}{5}k\right)^3$

15. $(1.4c)^3$

16. $(mn)^8$

17. $\left(3^2\right)^4 - 3^5 \bullet 3$

18. $10\left(\frac{1}{5}v\right)^3$

19. The volume of a right circular cylinder is $V = \pi r^2 h$. The relationship between the height h of a given right circular cylinder and the radius r is $r = \frac{2}{3}h.$

a. Find the volume of the right circular cylinder in terms of the height h and simplify the expression.

b. What is the volume of the right circular cylinder when the height is $\frac{3}{4}$ inch?

Find the value of *x* in the equation without evaluating the power.

20. $3^2 \bullet 3^x = 3^{12}$

21. $\left(5^x\right)^4 = 5^{24}$

Name____ Date__________

9.2 Enrichment and Extension

Compound Exponents

Compound exponents have the form a^{m^n}. They are evaluated from the top down.

Example: Simplify 4^{3^2}.

$4^{3^2} = 4^{\left(3^2\right)}$ Rewrite.

$= 4^9$ Simplify.

$= 262{,}144$ Evaluate.

Simplify the expression.

1. x^{4^3} **2.** y^{3^4} **3.** a^{11^2} **4.** b^{9^3}

Compound exponents are used in Fermat numbers. Fermat numbers have the form $2^{2^n} + 1$ and are used in number theory and computational algorithms.

5. Complete the table.

n	2^n	2^{2^n}	Fermat number
1	2	4	5
2			
3			
4			

6. What pattern do you notice in the 2^{2^n} column?

7. Would 3^{2^n} have the same pattern? Explain.

8. Which do you think would grow faster, 2^{2^n} or 2^{n^2}?

Name ____ Date __________

9.2 Puzzle Time

Did You Hear About...

A	B	C	D	E	F
G	H	I	J	K	L
M	N				

Complete each exercise. Find the answer in the answer column. Write the word under the answer in the box containing the exercise letter.

$-4c^5$	NOTES
2058	KEYS
$-1024c^5$	BECAUSE
3^{24}	HOW
6^{10}	PIANO
$8a^3$	DOOR
$\left(\frac{4}{5}\right)^8$	WHO
$\frac{16}{81}m^4$	OF
$6 \bullet 10^7$	STUCK
$(-8)^{30}$	THE

Simplify the expression. Write your answer as a power.

A. $4 \bullet 4^5$

B. $6^3 \bullet 6^7$

C. $(-3)^8 \bullet (-3)^3$

D. $\left(\frac{4}{5}\right)^2 \bullet \left(\frac{4}{5}\right)^6$

E. $\left(7^3\right)^5$

F. $\left(-8^{10}\right)^3$

Simplify the expression.

G. $(8a)^3$

H. $(-4c)^5$

I. $(1.5r)^2$

J. $\left(\frac{2}{3}m\right)^4$

K. $3^2 \bullet 3^4 - 4^4$

L. $7 \bullet \left(7^3 - 7^2\right)$

M. $\left(\frac{1}{3}\right)^4 \bullet \left(9^3\right)^2$

N. The budget for a remake of a movie is 100 times the budget of the original movie. The original movie's budget was $6 \bullet 10^5$ dollars. What is the budget of the remake? Express your answer as a power.

$2.25r^2$	ONE
-3^{11}	TEACHER
$1.5r^2$	CLASS
7^{15}	CALLED
6561	GOT
$512a^3$	LOCKSMITH
$\frac{2}{3}m^4$	SHE
$\left(\frac{4}{5}\right)^{12}$	ARE
4^6	THE
473	HER

Activity 9.3 Start Thinking!
For use before Activity 9.3

You have learned about the Product of Powers Property. What does the property state?

There is a related property called the Quotient of Powers Property. What do you think it might state?

Activity 9.3 Warm Up
For use before Activity 9.3

Write the power as repeated multiplication.

1. 5^4

2. 7^3

3. 6^7

4. $(-4)^3$

5. $(-3)^5$

6. $(-1)^3$

Lesson 9.3

Start Thinking!

For use before Lesson 9.3

Scott learned about the Quotient of Powers Property in math class, but he is not convinced that it is helpful. For example, he thinks that it is just as easy to simplify $\frac{2^5}{2^2}$ by calculating $2^5 = 32$ and dividing by $2^2 = 4$ to get 8. Do you agree or disagree with Scott? Give reasons to support your answer.

Lesson 9.3

Warm Up

For use before Lesson 9.3

Simplify the expression. Write your answer as a power.

1. $\frac{5^9}{5^6}$

2. $\frac{4^8}{4^4}$

3. $\frac{2.5^5}{2.5^2}$

4. $\frac{10.1^7}{10.1^3}$

5. $\frac{(-5)^{12}}{(-5)^{10}}$

6. $\frac{(-2)^7}{(-2)^6}$

Name____ Date__________

9.3 Practice A

Simplify the expression. Write your answer as a power.

1. $\dfrac{3^8}{3^6}$

2. $\dfrac{10^{11}}{10^3}$

3. $\dfrac{(-4)^5}{(-4)^4}$

4. $\dfrac{(5.6)^{15}}{(5.6)^9}$

5. $\dfrac{p^{13}}{p^{11}}$

6. $\dfrac{(-0.7)^{25}}{(-0.7)^{12}}$

7. $\dfrac{s^{28}}{s^7}$

8. $\dfrac{\pi^6}{\pi}$

9. A personal computer developed in the 1980s had approximately 2^{18} bytes of memory. Today a laptop has 1 gigabyte $= 2^{30}$ bytes of memory. How many times more memory does today's laptop have than the personal computer from the 1980s?

Simplify the expression. Write your answer as a power.

10. $\dfrac{6^3 \bullet 6^7}{6^4}$

11. $\dfrac{3^4 \bullet 3^5}{3 \bullet 3^2}$

12. $\dfrac{(-0.5)^8 \bullet (-0.5)^5}{(-0.5)^6 \bullet (-0.5)^2}$

13. $\dfrac{m^{14}}{m^{10}} \bullet \dfrac{m^5}{m^2}$

Simplify the expression.

14. $\dfrac{5^4 \bullet n^4}{5^2}$

15. $\dfrac{x^5 \bullet z^4}{x^2 \bullet z^2}$

16. $\dfrac{c^6 \bullet d^{10} \bullet 2^6}{d^5 \bullet 2^3}$

17. $\dfrac{a^{12}b^8}{a^{10}b^5}$

Find the value of *x* in the equation without evaluating the power.

18. $\dfrac{5^9}{5^x} = 625$

19. $\dfrac{3^7 \bullet 3^x}{3^6} = 9$

Name ______________________________ Date __________

9.3 Practice B

Simplify the expression. Write your answer as a power.

1. $\dfrac{12^{20}}{12^{9}}$

2. $\dfrac{7.6^{13}}{7.6^{3}}$

3. $\dfrac{(-9)^{15}}{(-9)^{3}}$

4. $\dfrac{(-8.5)^{11}}{(-8.5)^{10}}$

5. $\dfrac{u^{33}}{u^{11}}$

6. $\dfrac{\pi^{9}}{\pi^{4}}$

7. $\dfrac{(-1000)^{13}}{(-1000)^{8}}$

8. $\dfrac{t^{21}}{t^{19}}$

9. One kilometer equals 10^3 meters. One terameter equals 10^{12} meters. How many times larger is a terameter than a kilometer?

Simplify the expression. Write your answer as a power.

10. $\dfrac{11^7 \bullet 11^{10}}{11^4 \bullet 11^2}$

11. $\dfrac{2.5^8 \bullet 2.5^3}{2.5 \bullet 2.5^4}$

12. $\dfrac{(-7.9)^{15} \bullet (-7.9)^9}{(-7.9)^{12} \bullet (-7.9)^7}$

13. $\dfrac{b^{35}}{b^{20}} \bullet \dfrac{b^{15}}{b^{10}}$

Simplify the expression.

14. $\dfrac{4^8 \bullet m^7 \bullet n^4}{4^5 \bullet m^2}$

15. $\dfrac{r^{12} \bullet s^7 \bullet t^9}{r^9 \bullet s^3}$

16. $\dfrac{p^{18}q^{11}}{p^{10}q^8}$

17. $\dfrac{3^5a^{17}b^{21}}{3^4a^{15}b^{12}}$

Find the value of *x* in the equation without evaluating the power.

18. $\dfrac{9^7}{9^x} = 729$

19. $\dfrac{2^{12} \bullet 2^x}{2^{10}} = 32$

Name ______________________ Date __________

9.3 Enrichment and Extension

Number Search

Émile Borel was a French mathematician who studied probability theory and randomness. He developed a theory that if monkeys randomly hit the keys of typewriters, over a long period of time, one would eventually by chance type the play *Hamlet* by William Shakespeare.

Evaluate each expression. Circle the corresponding answer in the grid. Answers may read vertically, horizontally, diagonally, backwards, or forwards, but always in a straight line. When you have finished, put the unused numbers in the blanks in order, reading left to right and top to bottom.

5	7	8	1	9	4	3
3	2	6	0	7	6	2
1	4	9	3	5	8	1
5	0	6	1	9	0	4
2	7	7	0	4	8	1
8	0	4	3	6	7	9
7	9	1	2	4	0	5

Question: What is the probability of one monkey randomly typing *Hamlet* on the first try?

Answer: About __ in (__ __)------

1. $2^3 + 3^2 + 753$
2. $(2^2)^2 \bullet 5^2 + 6^2$
3. $95^2 + 10^2 - 1^2$
4. $31^2 - 3^2 \bullet 2$
5. $26^2 + 27^2 + 3^2 - 2$
6. $3^4 + 2^2$
7. $7^3 + 3$
8. $5(5)^3 + 2(5^3)$
9. $\dfrac{9^4 + 19 - (-7)^3}{7}$
10. $\dfrac{1}{2}(6^4 - 9^2 - 1)$
11. $(7 \bullet 2^2)^2 \bullet 10 - 15$
12. $\dfrac{10^5 + 10^4}{2} + 2 \bullet 10^3 + 21^2 - 3 \bullet 2^2 - 2^2$

Name ____________________ Date ________

9.3 Puzzle Time

What Do You Give A Dog That Loves Computers?

Write the letter of each answer in the box containing the exercise number.

Simplify the expression. Write the answer as a power.

1. $\dfrac{8^{12}}{8^6}$

2. $\dfrac{3^{24}}{3^{15}}$

3. $\dfrac{(-7)^{14}}{(-7)^4}$

4. $\dfrac{2.8^7}{2.8^4}$

5. $\dfrac{\pi^{12}}{\pi^5}$

6. $\dfrac{x^8}{x^3}$

7. $\dfrac{3^4 \bullet 3^5}{3^2 \bullet 3^2}$

8. $\dfrac{8^{15}}{8^7 \bullet 8}$

9. $\dfrac{\pi^5 \bullet \pi^9}{\pi^3 \bullet \pi^3}$

10. $\dfrac{x^{14} \bullet x^5}{x^7}$

11. $\dfrac{(-7)^3 \bullet (-7)^9}{(-7)^2 \bullet (-7)}$

12. $\dfrac{2.8^{15} \bullet 2.8^8}{2.8^9 \bullet 2.8^6}$

13. There are about 4×10^5 known species of beetles. The number of known species of caddis flies is about 10^4. How many times more species of beetles are there than caddis flies?

14. The area of the Pacific Ocean is approximately 6.4×10^7 square miles. The area of the Gulf of Mexico is approximately 5.8×10^5 square miles. How many times greater is the area of the Pacific Ocean than the area of the Gulf of Mexico?

Answers

- **Y.** 2.8^3
- **D.** x^{12}
- **I.** π^7
- **D.** $(-7)^9$
- **E.** x^5
- **E.** 3^9
- **T.** 40
- **G.** 8^7
- **O.** $(-7)^{10}$
- **S.** π^8
- **S.** 8^6
- **K.** 110
- **T.** 3^5
- **G.** 2.8^8

11	3	8	12	4		10	5	9	14	6	7	13	2	1

Activity 9.4

Start Thinking!

For use before Activity 9.4

What do you think it means for a number to be raised to the **zero power**?

Use your calculator to see if you are right.

Activity 9.4

Warm Up

For use before Activity 9.4

Simplify the expression. Write your answer as a power.

1. $\dfrac{5^4 \bullet 5^2}{5^3}$

2. $\dfrac{2^{11} \bullet 2^5}{2^{13}}$

3. $\dfrac{4^5 \bullet 4^3}{4^2}$

4. $\dfrac{a^{13} \bullet a^{11}}{a^{12}}$

5. $\dfrac{c^9 \bullet c^5}{c^{10}}$

6. $\dfrac{n^7 \bullet n^{14}}{n^{11}}$

Lesson 9.4 Start Thinking!
For use before Lesson 9.4

Can a number raised to a negative power ever be greater than 1? If so, give an example. If not, explain why not.

Can a number raised to a negative power ever be less than 0? If so, give an example. If not, explain why not.

Lesson 9.4 Warm Up
For use before Lesson 9.4

1. Use the Quotient of Powers Property to copy and complete the table.

n	4	3	2	1
$\dfrac{4^n}{4^2}$				

2. What patterns do you see?

3. How would you define 4^0? Why?

4. How can you rewrite 4^{-1} as a fraction?

Name____ Date__________

9.4 Practice A

Evaluate the expression.

1. 3^{-4}

2. 32^{0}

3. $\dfrac{8^3}{8^5}$

4. $\dfrac{(-9)^4}{(-9)^7}$

5. $5^{-12} \bullet 5^{12}$

6. $\dfrac{1}{4^{-5}} \bullet \dfrac{1}{4^8}$

7. $6^{-1} \bullet 6^{-2}$

8. $\dfrac{2^6}{2^{-8} \bullet 2^{10}}$

9. One terameter equals 10^{12} meters. One micrometer equals 10^{-6} meter. One nanometer equals 10^{-9} meter.

   a. Find the product of one terameter and one micrometer, using only positive exponents.

   b. Find the quotient of one terameter and one micrometer, using only positive exponents.

   c. Find the product of one terameter and one nanometer, using only positive exponents.

   d. Find the quotient of one terameter and one nanometer, using only positive exponents.

   e. Find the quotient of one nanometer and one terameter, using only positive exponents.

   f. Find the quotient of one nanometer and one micrometer, using only positive exponents.

   g. Find the product of one nanometer and one micrometer, using only positive exponents.

Simplify. Write the expression using only positive exponents.

10. $8x^{-3}$

11. $5^{-3} \bullet m^6$

12. $\dfrac{7p^5}{p^{-1}}$

13. $\dfrac{10t^{-5}}{t^{-2}}$

14. $\dfrac{15d^4}{3d^9}$

15. $6w^{-2} \bullet 4w^2$

Name ______________________________ Date __________

9.4 Practice B

Evaluate the expression.

1. 5^{-3}

2. $(-8)^0$

3. $\dfrac{6^{-3}}{6^{-5}}$

4. $\dfrac{15^{-4}}{15^{-4}}$

5. $10^{-1} \bullet 10^{-2}$

6. $\dfrac{1}{3^{-4}} \bullet \dfrac{1}{3^{6}}$

7. $27^{-18} \bullet 27^{18}$

8. $\dfrac{4^{-7}}{4^{2} \bullet 4^{-5}}$

9. One millimeter equals 10^{-3} meter. One picometer equals 10^{-12} meter. One femtometer equals 10^{-15} meter.

   a. Find the product of one millimeter and one picometer, using only positive exponents.

   b. Find the quotient of one picometer and one millimeter, using only positive exponents.

   c. Find the product of one millimeter and one femtometer, using only positive exponents.

   d. Find the quotient of one femtometer and one picometer, using only positive exponents.

   e. Find the quotient of one picometer and one femtometer, using only positive exponents.

   f. Find the quotient of one millimeter and one femtometer, using only positive exponents.

   g. Find the product of one picometer and one femtometer, using only positive exponents.

Simplify. Write the expression using only positive exponents.

10. $\dfrac{14u^{-4}}{7u^{8}}$

11. $\dfrac{18w^{-8}}{w^{-5}}$

12. $y^{5} \bullet z^{-3}$

13. $\dfrac{2^{-3} \bullet a^{0} \bullet b^{5}}{b^{-4}}$

Name____ Date__________

9.4 Enrichment and Extension

Exponential Graphs

An exponential equation has the form $y = a^x$, where a is a positive number. For example, $y = 2^x$ and $y = \left(\frac{3}{4}\right)^x$ are exponential equations.

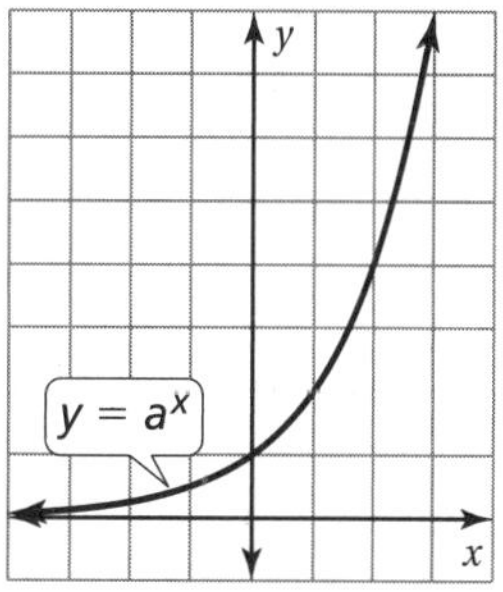

1. Is the graph of $y = 2^x$ linear? Explain your reasoning.

2. What is the y-intercept of $y = 2^x$?

3. Does $y = 2^x$ have an x-intercept?

4. Copy and complete the table for $y = \left(\frac{1}{2}\right)^x$ and plot the points to graph it.

   Describe how the graph of $y = \left(\frac{1}{2}\right)^x$ compares to the graph of $y = 2^x$.

x	*y*
–3	
–2	
–1	
0	
1	
2	
3	

5. What is the y-intercept of $y = \left(\frac{1}{2}\right)^x$?

6. Will all exponential equations have the same y-intercept? Explain your reasoning.

7. What value of a will make the graph linear?

Name ______________________________ Date __________

9.4 Puzzle Time

What Happened When The Tree Saw The Ghost?

Circle the letter of each correct answer in the boxes below. The circled letters will spell out the answer to the riddle.

Evaluate the expression.

1. $7^{-5} \bullet 7^3$

2. $5^2 \bullet 5^{-6}$

3. $\dfrac{2^7}{2^{10}}$

4. $\dfrac{6^0}{6^3}$

5. $\dfrac{(-8)^3}{(-8)^5}$

6. $\dfrac{(2.2)^7}{(2.2)^9}$

7. $\dfrac{4^5}{4^4} \bullet \dfrac{4^8}{4^{13}}$

8. $\dfrac{(-9)^3}{(-9)^7 \bullet (-9)^{-2}}$

Simplify the expression using only positive exponents.

9. $3^{-2}a^4$

10. $12^{-1}t^{-3}$

11. $\dfrac{b^4}{5^{-2}b^8}$

12. $\dfrac{14r^8}{2r^{15}}$

13. $\dfrac{x^5 \bullet y^6}{2^{-2} \bullet x^0 \bullet y^9}$

14. $\dfrac{6 \bullet f^{-4} \bullet g^2}{2 \bullet f^{-4} \bullet g^{-1}}$

H	A	I	S	T	R	M	E	W	L	A	N	S	U	V	D	Y
8	$\frac{4x^4}{y^2}$	$\frac{1}{12t^3}$	$\frac{1}{16}$	$\frac{1}{49}$	$\frac{1}{9a^4}$	$\frac{1}{36}$	$7r^7$	$\frac{1}{4.84}$	$3fg^3$	$\frac{1}{216}$	81	$\frac{1}{625}$	$\frac{x^5}{4y^3}$	$\frac{25}{b^2}$	49	$-\frac{1}{64}$
I	P	T	E	G	T	R	O	I	S	F	Q	I	K	E	B	D
$\frac{12}{t^3}$	$3g^3$	$\frac{1}{125}$	$\frac{1}{8}$	4.84	$\frac{25}{b^4}$	$\frac{1}{81}$	-8	$\frac{a^4}{9}$	64	$\frac{4x^5}{y^3}$	$\frac{3}{fg}$	$\frac{1}{64}$	-16	$\frac{1}{256}$	$\frac{t^3}{12}$	$\frac{7}{r^7}$

Activity 9.5 Start Thinking!

For use before Activity 9.5

Copy the following two numbers.

0.0000001983

1.983×10^{-8}

Which number is easier to copy?

The second number is written in **scientific notation**. What are some benefits of using scientific notation to write very large or very small numbers?

Activity 9.5 Warm Up

For use before Activity 9.5

Evaluate the power.

1. 10^{3}

2. 10^{-4}

3. 10^{5}

4. 10^{-2}

5. 10^{10}

6. 10^{-5}

Lesson 9.5 Start Thinking!
For use before Lesson 9.5

Use the Internet to find the mass of an electron and the mass of Earth.

Before you begin your research, do you expect that the masses will be given in scientific notation? Why or why not?

Were you correct?

Lesson 9.5 Warm Up
For use before Lesson 9.5

Write your answer in standard form.

1. A micrometer is 1.0×10^{-6} meter. How long is 300 micrometers in meters?
2. An acre is about 4.05×10^{7} square centimeters. How many square centimeters are in 2 acres?
3. A cubic millimeter is about 6.1×10^{-5} cubic inches. How many cubic millimeters are in 3.05 cubic inches?

Name____ Date __________

9.5 Practice A

Write your answer in standard form.

1. A megameter is 1.0×10^6 meters. How long is 80 megameters in meters?

2. An angstrom is 1.0×10^{-10} meter. How long is 150 angstroms in meters?

Tell whether the number is written in scientific notation. Explain.

3. 4.375×10^{-8}

4. 62.9×10^{14}

5. 9.897×10^{-15}

6. 0.451×10^{-12}

7. 25×10^{18}

8. 5.1786×10^{-25}

Write the number in standard form.

9. 8×10^6

10. 9×10^{-2}

11. 2×10^3

12. 5.3×10^{-4}

13. 1.2×10^8

14. 7.86×10^5

15. The average distance from Earth to the Sun is about 1.5×10^{11} meters. The average distance from Earth to the moon is about 3.84×10^8 meters.

    a. Write the distance from Earth to the Sun in standard form.

    b. Write the distance from Earth to the moon in standard form.

    c. Which is closer to Earth, the *Sun* or the *moon*?

16. A day is about 8.64×10^4 seconds.

    a. How many seconds are in 5 days? Write your answer in standard form.

    b. How many seconds are in 1 month (30 days)? Write your answer in standard form.

    c. How many seconds are in 1 year (365 days). Write your answer in standard form.

    d. How many seconds are in 1 leap year (366 days) in standard form?

    e. What is the difference (in seconds) between 1 year and 1 leap year? Write your answer in both standard form and scientific notation.

Name ______________________________ Date __________

9.5 Practice B

Write your answer in standard form.

1. A terameter is 1.0×10^{12} meters. How long is 32 terameters in meters?

2. A light year is 9.46×10^{15} meters. How long is 15 light years in meters?

Tell whether the number is written in scientific notation. Explain.

3. 17×10^{15}

4. 3.712×10^{-8}

5. 7.54×10^{21}

6. 0.999×10^{-15}

7. 125.42×10^{-12}

8. 7.65×10^{25}

Write the number in standard form.

9. 5×10^{-4}

10. 1.54×10^{5}

11. 1.78×10^{-6}

12. 9.876×10^{-4}

13. 2.08×10^{7}

14. 3.555×10^{8}

15. The radius of Earth is about 6.38×10^{6} meters. The radius of the moon is about 1.74×10^{6} meters. The radius of the Sun is about 7×10^{8} meters.

    a. Which is the largest, *Earth*, the *moon,* or the *Sun*?

    b. Which is the smallest, *Earth*, the *moon,* or the *Sun*?

    c. Write the radius of Earth in standard form.

    d. Write the radius of the moon in standard form.

    e. Write the radius of the Sun in standard form.

16. A year is about 3.156×10^{7} seconds.

    a. How many seconds are in 5 years? Write your answer in standard form.

    b. How many seconds are in half a year? Write your answer in standard form.

    c. How many seconds are in 1 month? Write your answer in standard form.

Name______________________________ Date__________

9.5 Enrichment and Extension

Different Bases

The decimal system is a base 10 system. Each place value represents 10 to a power. Computers often use a binary or base 2 system, where each place value represents a power of 2. Some systems use a base 8 or a base 16 system.

To convert a number from a given base into base 10, write the number as a sum of powers of the base and evaluate the expression.

1. Write 30,275 as a sum of powers of 10.

$$\_\_\_ \bullet 10^4 + \_\_\_ \bullet 10^3 + \_\_\_ \bullet 10^2 + \_\_\_ \bullet 10^1 + \_\_\_ \bullet 10^0$$

2. The number 1001010 is in base 2. Write the number as a sum of powers of 2.

$$\_\_\_ \bullet 2^6 + \_\_\_ \bullet 2^5 + \_\_\_ \bullet 2^4 + \_\_\_ \bullet 2^3 + \_\_\_ \bullet 2^2 + \_\_\_ \bullet 2^1 + \_\_\_ \bullet 2^0$$

3. Evaluate the expression in Exercise 2 to convert 1001010 from base 2 to base 10.

4. The number 331 is in base 4.

a. Powers of what number should appear in your sum?

b. How many terms will there be in your sum?

c. Write and evaluate an expression to convert 331 from base 4 into base 10.

5. Use powers of the given base to write each number as a sum of powers of the base. Then evaluate the expression to convert the number to base 10.

a. The number 22012 is in base 3.

b. The number 3004 is in base 8.

c. The number 501 is in base 6.

d. The number 312 is in base 16.

6. Write 41 in base 2. (A base 2 system uses the digits 0 and 1.)

Name ____ Date __________

9.5 Puzzle Time

What Position Did The Ghost Play On The Hockey Team?

Write the letter of each answer in the box containing the exercise number.

1. Which number is written in scientific notation?

H. 1.8×8^{-5} **I.** 4.2×10^{-8} **J.** 6.5×9^{7}

2. Which number is *not* written in scientific notation?

S. 5.3×10^{-6} **T.** 6×10^{-6} **U.** 45×10^{8}

3. Which number is *not* written in scientific notation?

E. 0.9×10^{12} **F.** -8.2×10^{10} **G.** 1×10^{-13}

4. Write 3.54×10^{6} in standard form.

N. 0.00000354 **O.** 3,540,000 **P.** 354,000,000

5. Write -1.92×10^{-4} in standard form.

H. -0.000192 **I.** $-19{,}200$ **J.** -0.0000192

6. The distance to the sun is about 9.3×10^{7} miles. What is the distance to the sun in standard form?

K. 9,300,000 mi **L.** 93,000,000 mi **M.** 930,000,000 mi

7. The population in China in 2007 was about 1.32×10^{9}. The population in Japan in 2007 was about 1.27×10^{8}. Approximately how many times greater was China's population than Japan's population?

E. 1040 times **F.** 104 times **G.** 10.4 times

7	5	4	2	6	1	3

Activity 9.6 Start Thinking!

For use before Activity 9.6

Choose a number greater than 1,000,000. Write it in standard form.

Write a list of steps that describe how you would write the number in scientific notation.

Repeat the process for a number less than 0.0009.

Activity 9.6 Warm Up

For use before Activity 9.6

Write the number in standard form.

1. 6×10^{3}

2. 4×10^{-4}

3. 2×10^{5}

4. 2.6×10^{-2}

5. 5.25×10^{10}

6. 8.52×10^{-5}

Lesson 9.6

Start Thinking!

For use before Lesson 9.6

Estimate the population of the world.

Go to *www.census.gov* to find the actual world population.

Write the population in scientific notation. How did you choose to round the number and why?

Lesson 9.6

Warm Up

For use before Lesson 9.6

Write the number in scientific notation.

1. 0.00034

2. 6,750,000

3. 0.00000007

4. 125,000

5. 15,200,000,000

6. 0.00000000917

Name______________________________ Date__________

9.6 Practice A

Write the number in scientific notation.

1. 350,000

2. 0.0004

3. 0.000000000000527

4. 12,500,000

5. 1,900,000,000

6. 0.0000001

7. 5,000,000,000,000

8. 0.00006524

Multiply. Write your answer in scientific notation.

9. $(3 \times 10^7) \times (1 \times 10^2)$

10. $(4 \times 10^{-3}) \times (5 \times 10^6)$

11. $(6 \times 10^{-12}) \times (7 \times 10^{-9})$

12. $(8 \times 10^5) \times (8 \times 10^5)$

13. $(2 \times 10^{-3}) \times (1.1 \times 10^2)$

14. $(9 \times 10^{-7}) \times (2.5 \times 10^3)$

15. In 2010, the net worth of a businessman was \$59,000,000,000.

a. Write \$59,000,000,000 in scientific notation.

b. As of 2009, the businessman had given over \$28,000,000,000 to charity. Write \$28,000,000,000 in scientific notation.

c. In 1999, the businessman's wealth briefly surpassed \$101,000,000,000. Write \$101,000,000,000 in scientific notation.

16. A pipette is a laboratory instrument that is used to transport a measured volume of liquid. A pipette that dispenses between 1 and 1000 microliters is called a micropipette.

a. A microliter is equivalent to 0.000001 liter. Write 0.000001 in scientific notation.

b. One thousand microliters is equivalent to 0.001 liter. Write 0.001 in scientific notation.

Find the area of the figure. Write your answer in scientific notation.

17.

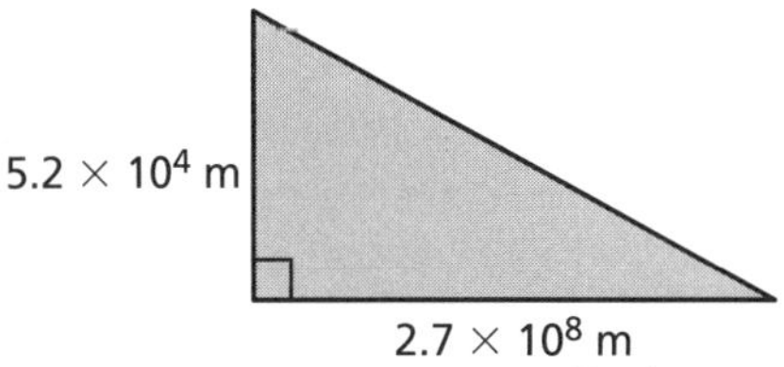

Not drawn to scale

18.

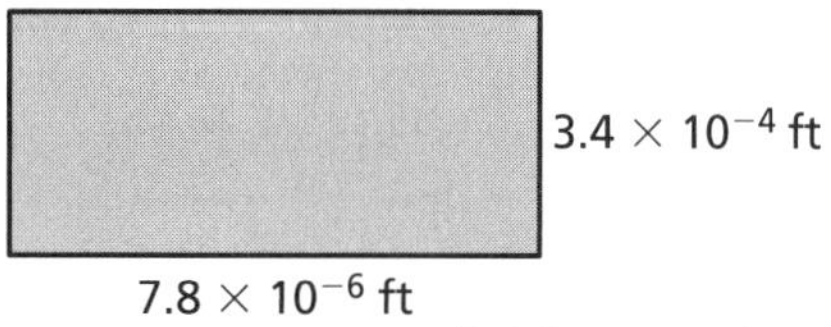

Not drawn to scale

Name ______________________________ Date __________

9.6 Practice B

Write the number in scientific notation.

1. 0.000085
2. 410,000,000
3. 0.0143
4. 134,750,000,000
5. 7,000,000,000,000,000
6. 0.00000000000199
7. 52,400,000,000,000
8. 0.00000006133

Multiply. Write your answer in scientific notation.

9. $(5 \times 10^{9}) \times (6 \times 10^{-5})$
10. $(2 \times 10^{-10}) \times (2 \times 10^{-6})$
11. $(3 \times 10^{-15}) \times (7 \times 10^{8})$
12. $(3 \times 10^{5}) \times (8 \times 10^{-9})$
13. $(6 \times 10^{-8}) \times (3.1 \times 10^{12})$
14. $(4 \times 10^{10}) \times (2.5 \times 10^{3})$
15. Most golf balls have about 250 to 450 dimples. The record holder is a ball with 1070 dimples.
    a. Write 1070 in scientific notation.
    b. In a recent year, it was estimated that 540,000,000 golf balls were sold. Using an average of 350 dimples, how many dimples were on the golf balls sold in that year? Write your answer in scientific notation.
16. In 2006, Hong Kong's population was about 6,940,000.
    a. Write 6,940,000 in scientific notation.
    b. In 2006, there were about 17,251 people per square mile in Hong Kong. What is the area of Hong Kong (in square miles)? Round your answer to the nearest whole number.

Simplify. Write your answer in scientific notation.

17. $\dfrac{(460{,}000{,}000)(0.006)}{(0.0025)}$
18. $\dfrac{(45{,}000)(0.009)}{(50{,}000{,}000)}$

Name____ Date__________

9.6 Enrichment and Extension

Avogadro's Number

The mass of an atom is too small to measure with grams or milligrams. Scientists use *atomic mass units* (amu) to describe the mass of an atom. The number of atomic mass units in one gram is a constant known as Avogadro's number.

$$1\text{ g} = 6.022 \times 10^{23}\text{ amu}$$
$$6.022 \times 10^{-23}\text{ g} = 1\text{ amu}$$

1. Copy and complete the table.

Element	Chemical Symbol	Mass (amu)	Mass (g)
Silver	Ag	1.26×10^{25}	
Oxygen	O		28
Platinum	Pt	9.64×10^{24}	
Helium	He	3.01×10^{22}	
Nitrogen	N		34

2. Arrange the chemical symbols in order of increasing mass. What word do the symbols spell?

Name ______________________________ Date __________

9.6 Puzzle Time

If A Man Wears Pajamas What Does A Woman Wear?

Write the letter of each answer in the box containing the exercise number.

1. Write 102,800,000 in scientific notation.

A. 1.028×10^8 **B.** 1028×10^5 **C.** 1×10^8

2. Write 0.0000522 in scientific notation.

Q. 5.22×10^5 **R.** 5.22×10^{-4} **S.** 5.22×10^{-5}

3. Write –0.32 in scientific notation.

A. -3.2×10^{-1} **B.** -3.2×10^0 **C.** -32×10^{-2}

4. What is the product of $(6 \times 10^3) \times (7 \times 10^{12})$? Write your answer in scientific notation.

L. 42×10^{15} **M.** 4.2×10^{16} **N.** 4.2×10^{14}

5. What is the product of $(12 \times 10^{-5}) \times (5 \times 10^{-2})$?

A. 6×10^{-6} **B.** 6×10^{-7} **C.** 6×10^{-8}

6. A movie earned $1,845,000,000 at the box office. What is the dollar amount written in scientific notation?

J. 1.845×10^9 **K.** 18.45×10^8 **L.** 1.845×10^6

7. What is the volume of a cylinder with radius $r = 1.5 \times 10^{-4}$ inch and height $h = 2.4 \times 10^{-3}$ inch? (The volume of the cylinder is $V = \pi r^2 h$.)

L. 3.6×10^{-7} cubic inch **M.** 1.7×10^{-10} cubic inch **N.** 2.25×10^{-8} cubic inch

7	5	6	3	4	1	2

Name____________________ Date__________

Chapter 9

School-to-Work

For use after Section 9.6

Lab Technician

As a technician working in a medical laboratory, you routinely examine blood cells under a microscope. The size and number of red and white blood cells in the human body can be used to identify whether or not disease is present. Information about normal red and white blood cells is given in the microscope slide below.

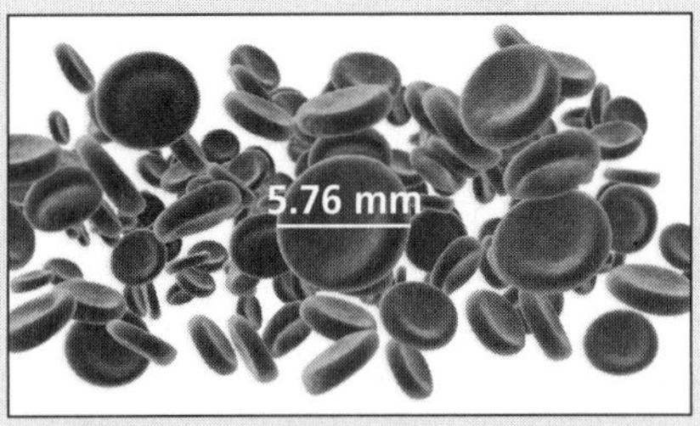

Diameter of red blood cell: 6-8 micrometers
Number of red blood cells in the body: about 2.5×10^{13}
Number of white blood cells in 1 microliter of blood: 4000-11,000

1. Write the number of red blood cells in the human body in standard notation.

2. One micrometer is equal to 1×10^{-3} millimeter. What is this measurement written in standard notation? What is the range (in millimeters) for the diameter of a normal red blood cell?

3. The red blood cells shown in the slide have a diameter of 5.76 millimeters when magnified 800 times under a microscope. What is the actual diameter (in micrometers) of the red blood cells? Show your work.

4. One microliter is equal to 1×10^{-6} liter. A person's white blood cell count in a microliter sample of blood is 6570. If the human body contains about 6 liters of blood, approximately how many white blood cells does the person have? Explain your reasoning.

5. About how many red blood cells would you expect to find in a microliter sample of a person's blood? Explain.

Name ______________________________ Date __________

Chapter 9 Study Help

You can use an **information wheel** to organize information about a concept.

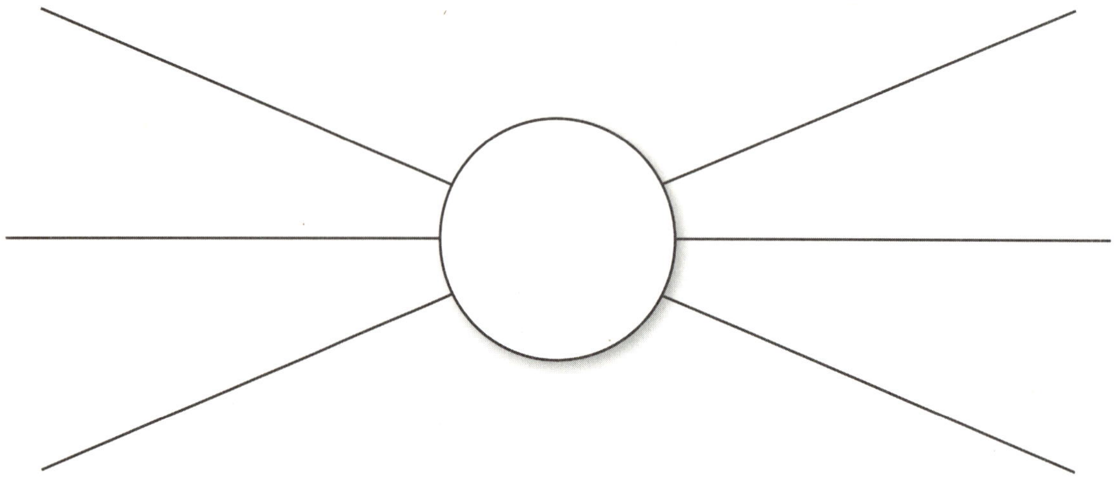

On Your Own

Make an information wheel to help you study these topics.

1. order of operations

2. Product of Powers Property

3. Quotient of Powers Property

After you complete this chapter, make information wheels for the following topics.

4. zero exponents

5. negative exponents

6. writing numbers in scientific notation

7. writing numbers in standard form

8. Choose three other topics you studied earlier in this course. Make an information wheel for each topic to summarize what you know about them.

Name____ Date__________

Chapter 9

Financial Literacy

For use after Section 9.1

Using School Gift Certificates

Your school system is starting a gift certificate program to raise money for all the schools in your town. Local stores and businesses sell the gift certificates to the school system at a discount. People then purchase the gift certificates at face value and the school system gets to keep the difference. The gift certificate discounts offered by several retailers are listed in the table below.

Retailer	Discount
grocery store	5%
pizza place	7%
clothing store	10%
movie theater	16%

1. Estimate the number of families that have children in the schools in your town or city. Explain how you made your estimate.

2. Your family is starting an email awareness campaign about the gift certificates. You send an email about the gift certificates to two families in town. Each of those families is then asked to forward the email to two families in town and so on. Make a tree diagram that shows the number of emails sent at each stage of the email campaign. Include three stages (consider your family stage 0) and label each stage with the number of emails.

3. How can you write the number of emails sent at each stage using exponents? What is an expression for the number of emails sent at stage n?

4. How many stages of emails would it take to reach every family that has children in the schools in your town or city? Explain your reasoning.

5. Estimate the amount your family spends per month at each type of retailer listed in the table. If your family used the school gift certificates for these expenses, how much money would you raise for the schools in one month? in one year?

6. If every family used the school gift certificates, about how much money do you think could be raised for the schools in one year? Explain how you madc your estimate.

Name ____________________ Date ________

Chapter 9

Technology Connection

For use after Section 9.6

Operations with Scientific Notation

You can use your calculator and scientific notation to perform operations on large or small numbers.

EXAMPLE Multiply: $63{,}000{,}000{,}000 \times 204$

SOLUTION

Step 1 Rewrite the first number using scientific notation.

$$63{,}000{,}000{,}000 = 6.3 \times 10^{10}$$

Step 2 Enter the product. Press 6.3 [2nd] [EE] 10 [×] 204.

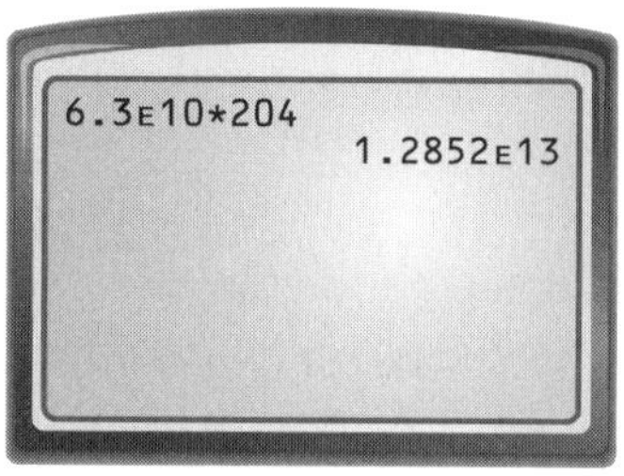

Step 3 Press [ENTER]. The product is equal to 1.2852×10^{13} or 12,852,000,000,000.

Find the sum, difference, product, or quotient.

1. $310{,}000{,}000 \times 56{,}000$
2. $0.00000009 \div 0.0005$
3. $\left(5.6 \times 10^{7}\right) + \left(6.6 \times 10^{4}\right)$
4. $\left(1.3 \times 10^{-4}\right) - \left(5.1 \times 10^{-5}\right)$
5. $0.0000075 \times 230{,}000$
6. $44{,}600{,}000{,}000 + 5{,}600{,}000{,}000$
7. $0.00000000558 - 0.0000000026$
8. $\left(9.8 \times 10^{12}\right) \div \left(3.2 \times 10^{3}\right)$
9. When does your calculator display answers using standard notation? When does it display answers using scientific notation? Give examples to support your answers.

Name________________________________ Date__________

Chapter 9 Cumulative Practice

Evaluate the expression.

1. $8 - 3^2$

2. $6 + 2 \bullet 5$

3. $\left(-\frac{2}{3}\right)^3$

4. $10 \div 5 - 4$

5. $\frac{7}{4} \bullet \frac{8}{21}$

6. $1.1(6.8 - 2.4)$

7. $4^{-2} \bullet 2^5$

8. $\frac{5 + (-3)}{3^{-2}}$

9. $[-1 + (-2)]^4$

10. 1.67×10^5

11. 9.3×10^{-4}

12. $\left(\frac{1}{14}\right)^{-1} - 2(13)^0$

13. There are 18 tables and 12 booths in a restaurant. Each table seats 4 people. Each booth seats 6 people. Write and evaluate an expression for the number of seats in the restaurant.

Simplify the expression. Write your answer using only positive exponents, if possible.

14. $(2^7)^3$

15. $x^5 \bullet x^{11}$

16. $\frac{3^6 \bullet 3^{10}}{3^{12}}$

17. $(6m^4)^3$

18. $\frac{9^{-9} \bullet 9^2}{9^{-6}}$

19. $7 \bullet 7^7$

20. $(3v^0)^1$

21. $\frac{4}{5} \bullet \frac{4}{5} \bullet \frac{4}{5} \bullet \frac{4}{5}$

22. $b \bullet b \bullet k \bullet k \bullet k$

23. The speed of sound in air is 2.058×10^7 millimeters per minute. What is the speed of sound in air in meters per second?

Write the number in scientific notation.

24. 8,000,000

25. 0.00065

26. 720,000

27. 140

28. 0.158

29. 0.0000000007

30. The Louisiana Purchase cost about $15,000,000 in 1803. Write this number of dollars in scientific notation.

Name ______________________________ Date ________

Unit 4 Project: Astronomy

For use after Unit 4

Objective **Use scientific notation to answer questions about astronomy.**

Materials Internet or other resource with astronomical data, calculator

Investigation Use a resource such as http://solarsystem.nasa.gov/planets/index.cfm to find the masses and distances needed to answer the following questions.

Round your answers to 3 significant digits. Use scientific notation to represent your distances.

1. Make a list of the planets that have at least two moons. Choose one of these planets.
2. Find the average distance from the Sun to your planet and from Earth to your planet.
3. Model the orbits as circles (even though the orbit is really a flat ellipse). Find the smallest possible distance between your planet and Earth. Find the smallest possible distance between your planet and the Sun. Write inequalities to represent the distances between the planets.
4. Use your model from Exercise 3 to find the greatest possible distance between your planet and Earth. Then find the greatest possible distance between your planet and the Sun. Write inequalities to represent the distances between the planets.
5. You want to launch a high-speed probe from Earth to take images of your planet. The probe travels at 250,000 kilometers per hour. What would be the shortest possible time it would take the probe to travel to your planet? Explain.
6. List the names of the moons of your planet. For each moon, find the mass, radius, and the distance from the planet. Write these numbers in scientific notation. Order the moons from least to greatest mass, least to greatest radius, and least to greatest distance from the planet. Is there a pattern?

Satellites travel in *low-earth orbit*. Low-earth orbit is defined as the distance between 160 and 2000 kilometers above the Earth's surface.

7. Write an inequality to represent the possible distance traveled by a satellite in low-earth orbit during one revolution around Earth.
8. At a speed of 27,000 kilometers per hour, how long will it take for the satellite to make one revolution around Earth? How long would it take a satellite at the same altitude and speed to orbit your planet?

Make a Poster Make a poster to organize all the information you have learned from your research and calculations. Include visuals and descriptions of your work.

Name________________________________ Date__________

Unit 4 Student Grading Rubric

For use after Unit 4

	Student Score	Teacher Score
Poster Information 10 points		
a. Name (4 points)	_____	_____
b. Class (3 points)	_____	_____
c. Due Date (3 points)	_____	_____
Investigation 150 points		
a. List of planets with two or more moons is correct. (10 points)	_____	_____
b. Distances from the chosen planet to the Sun and to Earth are correct and are written in scientific notation. (20 points)	_____	_____
c. The smallest possible distances from the planet to Earth and the planet to the Sun are visualized and calculated. The inequalities correctly represent the distances between the planets. (25 points)	_____	_____
d. The greatest possible distances from the planet to Earth and the planet to the Sun are visualized and calculated. The inequalities correctly represent the distances between the planets. (25 points)	_____	_____
e. The shortest possible time needed for the probe to travel to the planet is accurate. (20 points)	_____	_____
f. Names of moons for the planet are listed. Mass, radii, and the distance from the planet are written in scientific notation. The moons are listed from least to greatest by mass, radius, and distance from the planet. Patterns are described. (25 points)	_____	_____
g. Satellite distance is represented as an inequality. (10 points)	_____	_____
h. Satellite rates around Earth and around the chosen planet are calculated. (15 points)	_____	_____
Poster 40 points		
a. Includes calculations, visuals, lists, inequalities, and answers to all questions. (25 points)	_____	_____
b. Poster is neat and organized well. (15 points)	_____	_____
FINAL GRADE	_____	_____

Unit 4 Teacher's Project Notes

For use after Unit 4

Materials Internet or other resource with astronomical data, calculator
The NASA site is a good source:
http://solarsystem.nasa.gov/planets/index.cfm

Alternatives Students could write a fictional story in which they use each of the items they calculated.

A possible extension for advanced students is *escape velocity*. Given the average radius and mass of a planet or moon, calculate the *escape velocity*, the velocity an object would need to escape the body's gravitational pull.

$$\text{Escape velocity} = \sqrt{\frac{(1.73 \times 10^{-12})(\text{Mass of body})}{(\text{Radius of body})}}$$

The escape velocity is in kilometers per hour, the mass is in kilograms, and the radius is in kilometers.

You could focus the discussion on rocket speeds, and work out how long it would take to reach various objects: How long does sunlight take to reach Pluto? How long would a Saturn V rocket take to reach Jupiter?

Common Errors A review of unit analysis can help students check their work—e.g. a formula that will result in a speed in square kilometers is wrong.

Students may find it easiest to multiply and divide the decimals, then use the rules of exponents on the 10^x factors.

Make sure students write the numbers in scientific notation correctly to make the calculations a bit easier.

Suggestions At the start of the project, talk with the class about what it means to write a number to three significant digits.

Discuss that scientific notation can be used with any numbers, but it is most often used when discussing very small or very large numbers, such as those used in astronomy.

You may want students to work in pairs and turn in one poster for the pair. Students can work together on the calculations or do the calculations separately and compare their answers.

Planets that have at least two moons are Mars, Jupiter, Saturn, Uranus, and Neptune.

Unit 4

Grading Rubric

For use after Unit 4

Scoring Rubric	
A	179-200
B	159-178
C	139-158
D	119-138
F	118 or below

Poster Information 10 points

a. Name (4 points)

b. Class (3 points)

c. Due Date (3 points)

Investigation 150 points

a. List of planets with two or more moons is correct. (10 points)

b. Distances from the chosen planet to the Sun and to Earth are correct and are written in scientific notation. (20 points)

c. The smallest possible distances from the planet to Earth and the planet to the Sun are visualized and calculated. The inequalities correctly represent the distances between the planets. (25 points)

d. The greatest possible distances from the planet to Earth and the planet to the Sun are visualized and calculated. The inequalities correctly represent the distances between the planets. (25 points)

e. The shortest possible time needed for the probe to travel to the planet is accurate. (20 points)

f. Names of moons for the planet are listed. Mass, radii, and the distance from the planet are written in scientific notation. The moons are listed from least to greatest by mass, radius, and distance from the planet. Patterns are described. (25 points)

g. Satellite distance is represented as an inequality. (10 points)

h. Satellite rates around Earth and around the chosen planet are calculated. (15 points)

Poster 40 points

a. Includes calculations, visuals, lists, inequalities, and answers to all questions. (25 points)

b. Poster is neat and well organized. (15 points)

Appendix B

Name____ Date__________

Appendix B Financial Literacy

Dear Family,

Have you and your student started saving for college? College will introduce your student to many areas of personal finance: savings accounts, loans, credit cards, and payroll deductions.

People often have to save for college over a long period of time—putting money in a savings account faithfully over time. Compound interest provides a way to build on your savings. Even with planning, many people simply cannot save as much as they will need for college—making college loans and grants a common experience. College students are often offered credit or debit cards and it is hard to avoid these convenient forms of payment.

- Ask your student to research different types of savings plans at local banks. Have them figure out how much $50 will yield four years from now. How many $50 dollar deposits should they make to reach their goal?
- Ask your student to research different types of financing at the colleges they are considering. How much do students at each college typically have to borrow? Have your student figure out how much their payments will be for that amount.
- Ask your student to describe the difference between a credit card and a debit card. Have him or her research interest rates on credit cards. How do these rates compare to college loan interest rates? Is there a strategy your student can use to avoid credit card interest?

Many students will have jobs while in school to help offset their expenses. Students are often shocked the first time they get a paycheck and realize they don't get to keep all of their earnings! Ask your student to estimate the percentage of a $100 paycheck that he or she will get to keep. Have your student research the deductions that are taken and compare it to the estimate. How will this affect your savings goals?

Usually students are excited about their future—and with good reason. There are many opportunities for students regardless of their means. Planning ahead will make the experience less stressful for everyone.

Best wishes for a happy future!

Nombre ____________________ Fecha ________

Anexo B Educación Financiera

Estimada Familia:

¿Usted y su estudiante han empezado a ahorrar para la universidad? La universidad introducirá a su estudiante a diversas áreas de las finanzas personales: cuentas de ahorros, préstamos, tarjetas de crédito y descuentos por planilla.

Las personas a menudo tienen que ahorrar para la universidad por un largo período de tiempo—depositando dinero en cuentas de ahorro puntualmente. El interés compuesto ofrece una manera de basarse en sus ahorros.

Incluso con la planificación, muchas personas simplemente no pueden ahorrar tanto como quisieran para la universidad—obtener préstamos y becas universitarias es una experiencia común. A menudo a los estudiantes universitarios se les ofrece tarjetas de crédito y débito, y es difícil evitar usar estas convenientes formas de pago.

- Pida a su estudiante que investigue diversos tipos de planes de ahorro en bancos locales. Haga que averigüe cuánto rendirá $50 en cuatro años a partir de ahora. ¿Cuántos depósitos de $50 dólares deben hacer para alcanzar su meta?
- Pida a su estudiante que investigue diversos tipos de financiamiento en las universidades que están considerando. ¿Cuánto deben prestar normalmente los estudiantes en cada universidad? Haga que su estudiante averigüe cuántos serán sus pagos para esa cantidad.
- Pida a su estudiante que describa la diferencia entre una tarjeta de crédito y débito. Haga que investigue sobre las tasas de interés de las tarjetas de crédito. ¿Cómo se comparan estas tasas con las tasas de interés del préstamo universitario? ¿Hay alguna estrategia que su estudiante pueda usar para evitar tener que pagar el interés de la tarjeta de crédito?

Muchos estudiantes tendrán trabajos mientras están estudiando para ayudar a compensar sus gastos. Los estudiantes a menudo se asombran la primera vez que les llega un cheque de pago, ¡y se dan cuenta que no conservarán todas sus ganancias! Pida a su estudiante que calcule el porcentaje de un cheque de pago de $100 que recibirá. Haga que su estudiante investigue las deducciones que se efectúan y que las compare con el cálculo. ¿Cómo afectará esto sus metas de ahorro?

Generalmente los estudiantes están emocionados sobre su futuro—y con justa razón. Hay muchas oportunidades para los estudiantes, independientemente de sus medios económicos. Una planificación anticipada, hará que la experiencia sea menos estresante para todos.

¡Los mejores deseos para un futuro feliz!

Non ______________________________ Dat __________

Apendis B Edikasyon Finansye

Chè Fanmi:

Èske ou menm ak elèv ou a kòmanse ekonomize pou kolèj? Kolèj ap entwodwi elèv ou a bay anpil domèn finans pèsonèl: kont epay, prè, kat kredi, ak dediksyon nan pewòl.

Moun souvan oblije ekonomize pou kolèj pandan lontan—yo mete lajan nan yon kont epay regilyèman pandan yon sèten tan. Enterè konpoze ofri yon mannyè pou fè ekonomi ou fè pitit. Menm ak planifikasyon, anpil moun pa kapab ekonomize tout sa y'ap bezwen pou kolèj—sa fè ke prè pou kolèj ak bous se esperyans ki komen. Yo souvan ofri etidyan kolèj kat kredi oswa kat peman epi sa difisil pou evite fòm peman komòd sa yo.

- Mande elèv ou a fè rechèch sou diferan tip opsyon epay nan bank lokal yo. Fè li jwenn konbyen $50 ap bay apre katran apatide kounyeya. Konbyen depo $50 dola li ta dwe fè pou atenn bi li?
- Mande elèv ou a fè rechèch sou diferan tip finansman nan kolèj ke l'ap konsidere yo. Konbyen etidyan yo nan chak kolèj abityèlman gen pou yo prete? Fè elèv ou a jwenn konbyen peman yo ap ye pou montan sa a.
- Mande elèv ou a dekri diferans ant yon kat kredi ak yon kat peman. Fè li fè rechèch sou to enterè sou kat kredi. Kòman to sa yo ye pa rapò ak to enterè prè pou kolèj yo? Èske gen yon estrateji elèv ou a kapab itilize pou evite enterè kat kredi?

Anpil etidyan ap gen djòb pandan yo lekòl pou ede yo konpanse depans yo. Etidyan souvan choke premye fwa yo resevwa yon chèk pewòl epi yo reyalize ke yo pa kapab kenbe tout salè yo! Mande elèv ou a estime pousantaj yon chèk pewòl $100 ke l'ap kapab kenbe. Fè elèv ou a fè rechèch sou dediksyon yo pran yo epi konpare li avèk estimasyon an. Kòman sa pwal afekte bi ekonomi ou?

Abityèlman elèv yo eksite osijè avni yo—ak rezon. Gen anpil opòtinite pou elèv kèlkeswa mwayen yo. Planifye alavans ap fè esperyans la bay tout moun mwens estrès.

Nou swete nou yon avni nan lajwa!

Activity B.1 Start Thinking!

For use before Activity B.1

When have you heard the term interest rate? What does it mean?

Activity B.1 Warm Up

For use before Activity B.1

Evaluate the expression *Prt* for the given values of *P*, *r*, and *t*.

1. $P = 500, r = 0.05, t = 2$

2. $P = 2500, r = 0.02, t = 1$

3. $P = 100, r = 0.1, t = 10$

4. $P = 8000, r = 0.06, t = 4$

5. $P = 16{,}000, r = 0.065, t = 5$

6. $P = 50{,}000, r = 0.04, t = 30$

Lesson B.1 Start Thinking!

For use before Lesson B.1

Would you rather open a savings account with simple interest or compound interest? Why?

Lesson B.1 Warm Up

For use before Lesson B.1

1. You dcposit $1000 in a savings account that earns 4% interest per year.

a. Copy and complete the tables that show the balances after 5 years with simple interest and compound interest.

b. Which type of interest gives the greater balance?

Simple Interest			
t	Principal	Annual Interest	Balance at End of Year
1	$1000	$40	$1040
2	$1000	$40	$1080
3			
4			
5			

Compound Interest			
t	Principal and Interest	Annual Interest	Balance at End of Year
1	$1000	$40	$1040
2	$1040	$41.60	$1081.60
3			
4			
5			

Name ______________________________ Date __________

B.1 Practice A

1. You deposit $400 in a savings account that earns 5% interest per year.

   a. Copy and complete the tables that show the balances after 5 years with simple interest and compound interest.

   b. Which type of interest gives the greater balance?

Simple Interest			
t	Principal	Annual Interest	Balance at End of Year
1	$400	$20	$420
2	$400	$20	$440
3			
4			
5			

Compound Interest			
t	Principal and Interest	Annual Interest	Balance at End of Year
1	$400	$20	$420
2	$420	$21	$441
3			
4			
5			

2. You deposit $800 in a savings account that earns 1.5% simple interest per year. What is the account balance after 4 years?

3. You deposit $700 in a savings account that earns 3.6% interest compounded annually. What is the account balance after 2 years?

4. Your bank offers two savings accounts. You want to deposit $1000 for 3 years.

   a. Savings account A earns 4.5% simple interest per year. What is the account balance after 3 years?

   b. Savings account B earns 4.3% interest compounded annually. What is the account balance after 3 years?

   c. Which savings account gives the greater balance?

5. After 6 years at 5% simple interest per year, your savings account earns $195. What is the principal?

6. You deposit $1000 in a savings account at Bank A that earns 3% simple interest per year. You deposit $1000 in a savings account at Bank B that earns 3% interest compounded annually.

   a. How much money do you have in Bank A and Bank B combined after 7 years?

   b. Which bank pays a greater amount of interest?

Name____ Date__________

B.1 Practice B

1. You deposit \$900 in a savings account that earns 4.8% interest per year.

   a. Copy and complete the tables that show the balances after 5 years with simple interest and compound interest.

   b. Which type of interest gives the greater balance?

Simple Interest			
t	Principal	Annual Interest	Balance at End of Year
1	\$900	\$43.20	\$943.20
2	\$900	\$43.20	\$986.40
3			
4			
5			

Compound Interest			
t	Principal and Interest	Annual Interest	Balance at End of Year
1	\$900	\$43.20	\$943.20
2	\$943.20	\$45.27	\$988.47
3			
4			
5			

2. You deposit \$1700 in a savings account that earns 1.35% simple interest per year. What is the account balance after 6 years?

3. You deposit \$2000 in a savings account that earns 2.75% interest compounded annually. What is the account balance after 3 years?

4. Your bank offers two savings accounts. You are depositing \$10,000 for 3 years.

   a. Savings account A earns 6.25% simple interest per year. What is the account balance after 3 years?

   b. Savings account B earns 5.75% interest compounded annually. What is the account balance after 3 years?

   c. After how many years is the balance in savings account B greater than the balance in savings account A? By how much?

5. After 8 years at 4.25% simple interest per year, your savings account earns \$2040. What is the principal?

6. Your grandmother deposits \$500 in a savings account for you that earns 4% simple interest per year.

   a. What is the account balance after 4 years?

   b. After 4 years, your grandmother transfers the balance of your account to a savings account that earns 3.8% interest compounded annually. What is the account balance after 3 years?

Name ______________________________ Date __________

B.1 Enrichment and Extension

Inflation Rate

The *inflation rate* is a way of measuring the change in price of certain goods. As the inflation rate increases, the purchasing power of money decreases. In other words, when the inflation rate is high, prices rise and the same amount of money buys less goods.

The table shows the price of a loaf of white bread in various years.

Year	Average Price
1850's	\$0.06
1860's	\$0.11
1870's	\$0.08
1920	\$0.12
1930	\$0.10
1990	\$0.69
1995	\$0.79
2000	\$0.93
2001	\$1.00
2002	\$1.02
2003	\$1.00
2004	\$0.97
2005	\$1.04
2006	\$1.08
2007	\$1.21
2008	\$1.36

1. You can calculate the inflation rate using the formula $I = \frac{L - E}{E} \bullet 100$ where I is the inflation rate (expressed as a percent), L is the later year, and E is the earlier year.

a. Find the inflation rate of a loaf of white bread between the 1850s and 1860s. Round your answer to the nearest tenth of a percent.

b. Interpret the inflation rate from part (a).

c. What historical event could be used to explain why prices sharply increased during this period?

2. Find the inflation rate of a loaf of white bread between 1920 and 1930. Round your answer to the nearest tenth of a percent.

a. The rate between 1920 and 1930 is referred to as a *deflation rate*. Why do you suppose this is?

b. What historical event could be used to explain why prices sharply decreased during this period?

c. Analyze the table. Between what other years might you expect to see a deflation rate? How do you know? Explain your reasoning.

3. As a percent, how much more did a loaf of white bread cost in 2008 than it did in the 1850s? Round your answer to the nearest tenth of a percent.

4. Why might knowing the inflation rate be helpful to you as a consumer? Explain your reasoning.

Name____ Date__________

B.1 Puzzle Time

What Kind Of Key Does Not Unlock Any Door?

Write the letter of each answer in the box containing the exercise number.

Write an equation for the balance *B* in each account after *t* years.

1. You deposit \$500 in a savings account that earns 5% simple interest per year.
2. You deposit \$500 in a savings account that earns 4.3% simple interest per year.

Solve the following problems.

3. You deposit \$800 in a savings account that earns 3% simple interest per year. What is the account balance after 2 years?
4. You deposit \$800 in a savings account that earns 6.5% interest compounded annually. What is the account balance after 4 years?
5. You want to invest \$5000 for 5 years. Which account should you choose?
   - Account A earns 3% simple interest per year.
   - Account B earns 3% interest compounded annually.
6. You want to deposit \$300 in a savings account for 2 years. Which account should you choose?
   - Account A earns 2.5% simple interest per year.
   - Account B earns 2.2% interest compounded annually.

Answers

O. $B = 500 + 21.5t$

Y. \$1029.17

N. $B = 500 + 25t$

E. Account B

M. Account A

K. \$848.00

6	2	1	3	5	4

Activity B.2 Start Thinking!
For use before Activity B.2

What does it mean for interest to be compounded annually? What does it mean for interest to be compounded monthly?

In a standard savings account, how often is interest compounded? You can use the Internet to find out.

Activity B.2 Warm Up
For use before Activity B.2

Evaluate the expression.

1. $50(1.5)^2(5)$

2. $1500(1.05)^3(3)$

3. $1000(1.1)^4(10)$

4. $100(1.06)^2(4)$

5. $500(1.2)^2(2)$

6. $2000(1.02)^5(6)$

Lesson B.2 Start Thinking!
For use before Lesson B.2

Have you ever heard of a mnemonic device? It is a way of helping a person to memorize information. Often, mnemonic devices are words, phrases, rhymes, or short poems.

Come up with a mnemonic device to help you memorize the formula you wrote in Activity 2 on page A17 in your textbook.

Lesson B.2 Warm Up
For use before Lesson B.2

Copy and complete the table for an account with compound interest.

1. Principal: \$1500

Annual Interest Rate	Balance After 6 Years
3.0%	
3.5%	
4.0%	
4.5%	
5.0%	

2. Principal: \$3000

Annual Interest Rate	Balance After 3 Years
3.0%	
3.5%	
4.0%	
4.5%	
5.0%	

Name ______________________ Date __________

B.2 Practice A

Copy and complete the table for an account that earns interest compounded annually.

1. Principal: \$1000

Annual Interest Rate	Balance After 5 Years
2.5%	
3.0%	
3.5%	
4.0%	

2. Principal: \$2000

Annual Interest Rate	Balance After 10 Years
4.0%	
4.5%	
5.0%	
5.5%	

3. You deposit \$500 in a savings account. The account earns 6.5% interest compounded annually.

a. What is the balance after 7 years?

b. What is the interest earned?

4. An account earns 2.2% interest compounded annually. The balance after 3 years is \$800.60. What is the principal?

5. You want to purchase a house in 5 years. You want to have a balance of \$20,000 for a down payment.

a. About how much money should you deposit in a savings account that earns 4% interest compounded annually?

b. About how much money should you deposit in a savings account that earns 4.5% interest compounded annually?

c. About how much money should you deposit in a savings account that earns 6% interest compounded annually?

6. The formula for interest compounded more than once per year is $B = P\left(1 + \frac{r}{n}\right)^{nt}$, where n is the number of times the balance is compounded each year. You deposit \$2000 in a savings account. The account earns 4% interest compounded 6 times per year. What is the balance after 3 years?

Name____ Date__________

B.2 Practice B

Copy and complete the table for an account that earns interest compounded annually.

1. Principal: \$5000

Annual Interest Rate	Balance After 5 Years
0.25%	
0.5%	
0.75%	
1.0%	

2. Principal: \$5000

Annual Interest Rate	Balance After 10 Years
0.25%	
0.5%	
0.75%	
1.0%	

3. You deposit \$1250 in a savings account. The account earns 4.75% interest compounded annually.

a. What is the balance after 9 years?

b. What is the interest earned?

4. An account earns 3.99% interest compounded annually. The balance after 4 years is \$1520.23. What is the principal?

5. You want to purchase a car in 3 years. You want to have a balance of \$15,000.

a. About how much money should you deposit in savings account A that earns 3.5% interest compounded annually?

b. About how much money should you deposit in savings account B that earns 4% simple interest per year?

c. Which account should you choose? Explain.

6. The formula for interest compounded more than once per year is

$$B = P\left(1 + \frac{r}{n}\right)^{nt},$$ where n is the number of times the balance is compounded each year. You deposit \$1500 in a savings account.

a. The account earns 3% interest compounded 6 times per year. What is the balance after 5 years?

b. The account earns 3% interest compounded 12 times per year. What is the balance after 5 years?

B.2 Enrichment and Extension

Supply and Demand

Economics is the study of goods and services. Two of the most important graphs in economics are the supply graph and the demand graph. They show how the quantity q of an item determines the price p of that item.

The two graphs show the supply and demand functions for winter coats.

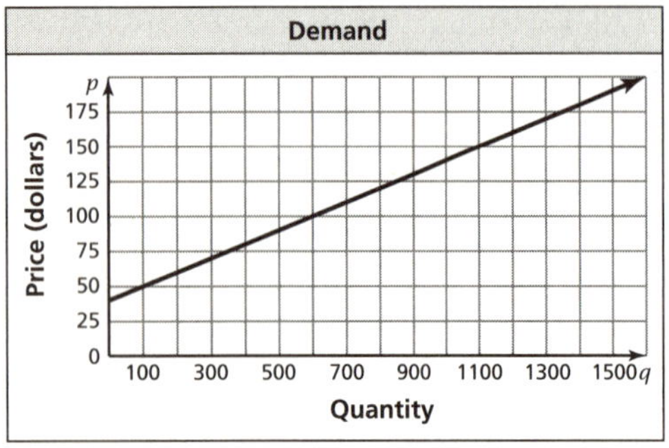

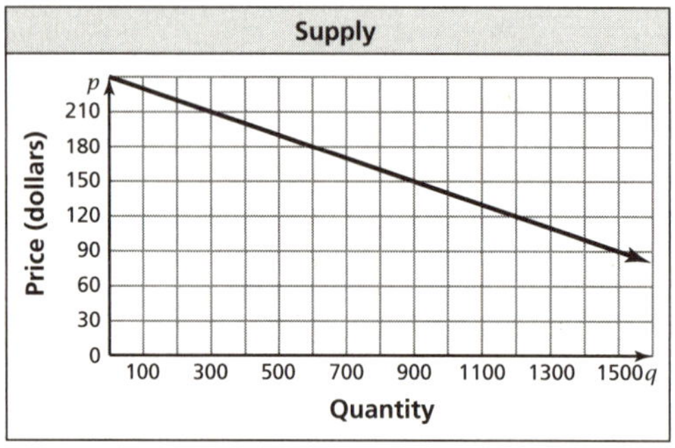

1. What is the price of a coat when the demand is 100 coats?

2. What is the price of a coat when the demand is 1500 coats?

3. Why do you think the price of a coat increases as the demand for coats increases?

4. What is the price of a coat when the supply is 100 coats?

5. What is the price of a coat when the supply is 1500 coats?

6. Why do you think the price of a coat decreases when the supply increases?

7. The supply and demand graphs form a system of linear equations. The solution of the system is referred to as the *equilibrium point*.

   a. What is the equilibrium point of the supply and demand graphs?

   b. What does the solution mean in the context of the problem?

8. The lowest prices occur when demand is low and supply is high. When do you think that would happen with winter coats?

Name____ Date__________

B.2 Puzzle Time

Why Did The Dalmation Go To The Dry Cleaner's?

A	B	C	D	E	F

Complete each exercise. Find the answer in the answer column. Write the word under the answer in the box containing the exercise letter.

$1169.86 **HAD**
$3960.47 **IT**
$81.60 **COAT**

You deposit $1000 in a savings account that earns 4% interest compounded annually. Answer the following questions.

A. What is the balance after 2 years?

B. What is the interest earned after 2 years?

C. What is the balance after 4 years?

D. What is the interest earned after 4 years?

E. An account earns 2.5% interest compounded annually. The balance after 4 years is $993.43. What is the principal?

F. You want to save $5000 in 4 years for a summer trip. About how much money should you deposit in a savings account that earns 6% interest compounded annually?

$900.00 **ON**
$1081.60 **HIS**
$169.86 **SPOTS**

Activity B.3 Start Thinking!
For use before Activity B.3

An installment loan is a loan that is repaid over a set number of regularly scheduled payments. Can you think of any specific examples of installment loans?

Activity B.3 Warm Up
For use before Activity B.3

Write the percent as a decimal.

1. 25%

2. 37.5%

3. 5%

4. 12%

5. 0.8%

6. 1.6%

Lesson B.3

Start Thinking!

For use before Lesson B.3

Suppose that your friend wants to buy something from you, like your bicycle, MP3 player, or cell phone, and offers to pay you in installments.

What would be the conditions of your installment loan? Decide on the loan amount, annual interest rate, and term of the loan. Then calculate the monthly payment. Do you think your friend would pay that?

Lesson B.3

Warm Up

For use before Lesson B.3

1. Copy and complete the table. How much interest do you pay during the 6 months? The annual interest rate is 12%.

Balance Before Payment	Monthly Payment	Monthly Interest	Balance After Payment
\$2000	\$345.09		
	\$345.09		
	\$345.09		
	\$345.09		
	\$345.09		
	\$345.14		

Name ______________________________ Date __________

B.3 Practice A

1. Copy and complete the table. How much interest do you pay during the 6 months? The annual interest rate is 6%.

Balance Before Payment	Monthly Payment	Monthly Interest	Balance After Payment
\$1720	\$291.70		
	\$291.70		
	\$291.70		
	\$291.70		
	\$291.70		
	\$291.72		

A loan summary is shown. How much interest is paid on the loan?

2.

LOAN AMOUNT	\$1285.00
MONTHLY PAYMENT	\$217.93
ANNUAL INTEREST RATE	6% annual interest
TERM	6 months

3.

LOAN AMOUNT	\$2640.00
MONTHLY PAYMENT	\$234.56
ANNUAL INTEREST RATE	12% annual interest
TERM	12 months

4. The table shows loan options for a mortgage. How much interest is paid on each option?

Monthly Payments \$160,000 Mortgage		
Term \ Annual Interest Rate	6%	8%
15	\$1350.17	\$1529.04
20	\$1146.29	\$1338.30
30	\$959.28	\$1174.02

Name____ Date__________

B.3 Practice B

1. Copy and complete the table. How much interest do you pay during the 6 months? The annual interest rate is 12%.

Balance Before Payment	Monthly Payment	Monthly Interest	Balance After Payment
$3268.00	$563.89		
	$563.89		
	$563.89		
	$563.89		
	$563.89		
	$563.87		

2. A loan summary is shown. How much interest is paid on the loan?

LOAN AMOUNT	$2780.00
MONTHLY PAYMENT	$247.00
ANNUAL INTEREST RATE	12% annual interest
TERM	12 months

3. The table shows loan options for a mortgage. How much interest is paid on each option?

Monthly Payments $250,000 Mortgage		
Term \ **Annual Interest Rate**	**5%**	**6%**
15	$1976.98	$2109.64
20	$1649.89	$1791.08
30	$1342.05	$1498.88

4. You pay off a $12,000 car loan in 5 years. The monthly payment is $257.93. What percent of the total amount paid is the interest on the loan?

5. The monthly payment for a 30-year mortgage is $790.09. The principal of the loan is about 44% of the total amount paid. To the nearest thousand dollars, what is the principal of the loan?

Name ______________________________ Date __________

B.3 Enrichment and Extension

Musical Instruments

Most stores that sell musical instruments also provide customers the option of renting an instrument. If you rent an instrument with the intent to purchase it at a later date, part of the rent payments count towards the cost of buying the instrument.

Cost to buy: $1000

Rent per month: $25

1. Write and solve an equation to find how many months it will take for the cost of renting the piccolo to equal the cost of buying the piccolo.

2. You decide to play in the band for two years. Would it be better to rent or buy a piccolo? Explain your reasoning.

3. The store's policy on buying an instrument that you are currently renting is that 80% of your rent payments count towards the purchase of the instrument.

   a. You rent a piccolo for three years. You decide to pay the remaining balance in one payment so that you can own the instrument. How much do you owe?

   b. You decide to rent with the intent to purchase the piccolo. How many months must you rent the instrument in order to own it?

4. When you rent an instrument, the store provides maintenance and most repairs at no additional charge. On average, it costs about $15 a month to maintain and repair a piccolo.

   a. You purchase a piccolo. Write a linear equation that represents the cost y of buying and maintaining a piccolo for x months.

   b. Write a linear equation that represents the cost y of renting a piccolo for x months.

   c. When does the cost of renting a piccolo outgrow the cost of buying and maintaining a piccolo?

Name____ Date__________

B.3 Puzzle Time

What Is A Potato's Favorite Television Show?

Write the letter of each answer in the box containing the exercise number.

A loan summary is shown. How much interest is paid on the loan?

1.

LOAN AMOUNT	$2000.00
MONTHLY PAYMENT	$185.00
ANNUAL INTEREST RATE	12% annual interest
TERM	12 months

2.

LOAN AMOUNT	$4600.00
MONTHLY PAYMENT	$813.00
ANNUAL INTEREST RATE	12% annual interest
TERM	6 months

Answers

H. $159.41

A. $127.38

M. $202,872.00

S. $85,155.60

The table shows loan options for a mortgage.

Monthly Payments $150,000 Mortgage		
Term \ Annual Intereest Rate	6%	8%
15	$1100.55	$1306.42
30	$980.20	$1140.10

3. How much interest is paid on the 15-year mortgage with an annual interest rate of 8%?

4. How much interest is paid on the 30-year mortgage with an annual interest rate of 6%?

4	1	3	2

Activity B.4 Start Thinking!
For use before Activity B.4

For a checking account, why is it important to keep track of all of the deposits and withdrawals?

Activity B.4 Warm Up
For use before Activity B.4

Find the sum.

1. $150.00 + (-29.65) + (-18.44)$
2. $69.10 + (-33.06) + 25.00$
3. $34.08 + (-33.98) + 78.67$
4. $118.05 + (-12.50) + (-54.00) + 30.00$
5. $115.00 + (-12.99) + (-16.97) + (-15.66)$
6. $50.95 + (-13.05) + (-27.57) + (-12.13)$

Lesson B.4 Start Thinking!

For use before Lesson B.4

What factors would influence what bank you would use to open a checking account?

Lesson B.4 Warm Up

For use before Lesson B.4

1. Copy and complete the checkbook registry.

Date	Check #	Transaction	Credit	Debit	Balance
		Balance Forward			456.91
5/8/11	401	Grocery Store		110.23	
5/8/11	402	Clothing Store		42.99	
5/12/11		Deposit	100.00		
5/27/11		ATM Withdrawal		60.00	

Name ______________________________ Date __________

B.4 Practice A

1. Your checkbook registry is shown.

a. Copy and complete your checkbook registry.

Date	Check #	Transaction	Credit	Debit	Balance
		Balance Forward			450.00
4/5/12	361	Utilities		115.60	
4/9/12		ATM Withdrawal		50.00	
4/9/12		Deposit	125.00		
4/12/12	362	Lincoln Elementary		42.00	

b. Your monthly bank statement shows that your account balance is $241.34. You notice that you did not enter the two transactions shown below in your checkbook registry. After entering these transactions in your checkbook registry, does your balance now agree with the bank statement?

- 4/10/12 ATM Fee $1.50
- 4/6/12 Grocery Store $124.56

2. Complete the balance column. Which check is the bad check?

Date	Check #	Transaction	Credit	Debit	Balance
		Balance Forward			246.51
9/15/12	862	Department Store		53.50	
9/15/12		Restaurant		42.99	
9/18/12	863	Telephone		54.78	
9/22/12		Deposit	350.00		
9/23/12	864	Rent		450.00	
9/24/12		Deposit	100.00		

Name____ Date__________

B.4 Practice B

1. You balance your checkbook when you get your bank statement.

   a. Complete the balance column. What is the ending balance?

Date	Check #	Transaction	Credit	Debit	Balance
		Balance Forward			611.29
11/22/12	1107	Cable TV		54.99	
11/23/12		ATM Withdrawal		60.00	
11/25/12		Movie Theater		24.50	
11/30/12	1108	Grocery Store		98.61	
11/30/12		Deposit	750.00		
12/1/12	1109	Rent		900.00	
12/3/12	1110	Utilities		195.15	

   b. Your monthly bank statement shows that your account balance is $1123.19. Explain why your balance does not agree with the bank statement.

2. Complete the balance column.

   a. Which checks are the bad checks?

Date	Check #	Transaction	Credit	Debit	Balance
		Balance Forward			158.42
10/1/12	672	School Board		12.50	
10/2/12	673	Rent		500.00	
10/3/12	674	Department Store		36.14	
10/5/12		Deposit	500.00		
10/7/12	675	Telephone		45.68	
10/9/12		Deposit	100.00		
10/10/12		ATM Withdrawal		40.00	
10/15/12	676	Gas Station		32.57	

   b. If there is a $20 fee for each bad check, what is the new balance of the checkbook registry?

Name ______________________________ Date __________

B.4 Enrichment and Extension

Planning a Budget

Use the table, a number cube, and a local newspaper or the Internet to complete the questions and activity.

Number Cube	Occupation and Yearly Income	Family Status	Additional Expense
1	Teacher $53,000	Married with one child	Owe the government $1000 in taxes
2	Department Store Clerk $25,000	Single with one pet	Planning and hosting a party costs $200.
3	Landscape Architect $62,000	Married with a parent living with you	Rent or mortgage payment increases by $100
4	Bank Teller $24,000	Married with no children	Receive a $25 parking ticket
5	Lawyer $118,000	Single	Replacing the alternator in your car costs $450.
6	Food Services $21,000	Married with two children	Doctor visit and prescription costs $50

1. Roll a number cube 4 times.

   a. Match the number on the first roll to the occupation and income in the table. This is your occupation and yearly income.

   b. Match the number on the second roll to the family status in the table.

   c. Match the number on the third roll to the additional expense in the table.

   d. If the number on the fourth roll is even, your budget will include paying monthly rent. If the number is odd, your budget will include a mortgage payment.

2. What is your average monthly income? This is the amount of money you start out with in your checking account.

3. Use your answers to Exercise 1 and a local newspaper or the Internet to create a budget for one month. Include the following items in your budget: rent or mortgage payment, car insurance ($300 per month), gas and parking expenses, grocery bill, and other household bills such as electric, water, gas, sewer, trash, phone, Internet, and cable. Remember to include any other family members in the budget.

Name____ Date__________

B.4 Puzzle Time

Why Does A Football Player Always Carry A Spare Pencil?

A	B	C	D	E	F
G	H				

Complete each exercise. Find the answer in the answer column. Write the word under the answer in the box containing the exercise letter.

Complete the balance column.

Date	Check #	Transaction	Credit	Debit	Balance
		Balance Forward			386.89
10/03/14	2408	Gas Company		56.25	A.
10/06/14		Deposit Paycheck	503.90		834.54
10/06/14	2409	Dinner		22.00	B.
10/08/14	2410	Electric Company		83.65	C.
10/10/14	2411	Car Payment		340.24	388.65
10/13/14	2412	Birthday Gift		25.00	D.
10/13/14		Deposit Paycheck	500.89		E.
10/17/14	2413	Car Insurance		136.24	728.30
10/21/14	2414	Grocery Store		82.10	F.
10/26/14	2415	Cable Company		45.50	600.70
10/27/14	2416	Cell Phone Company		55.10	545.60
10/31/14		Deposit Paycheck	503.70		G.
10/31/14	2417	Rent		560.00	H.

Answer	Word
363.65	WILL
1049.30	EXTRA
330.64	IN
864.54	NEED
812.54	CASE
646.20	AN
728.89	HE
489.30	POINT

Activity B.5 Start Thinking!
For use before Activity B.5

Name all of the credit card companies that you know.

How do credit cards work?

Activity B.5 Warm Up
For use before Activity B.5

Find the mean of the numbers.

1. 5, 5, 7, 3, 8, 1, 6

2. 10, 18, 16, 18, 29, 15, 14, 15

3. 1, 1, 1, 1, 4, 5, 5, 6, 7, 8

4. 350, 350, 350, 350, 340, 300

5. 12, 15, 17, 15, 16, 19, 21, 20

6. 1.5, 3.5, 2.5, 9.6, 1.2, 4.4, 6.5, 3.4

Lesson B.5

Start Thinking!

For use before Lesson B.5

Who gets paid when you pay with a credit card?

Lesson B.5

Warm Up

For use before Lesson B.5

You own a store. A customer uses a credit card to pay the total shown on the receipt.

SALE:	$34.00
STATE TAX:	$ 2.04
COUNTY TAX:	$ .54
TOTAL:	$36.58

1. How much money is sent to the state?
2. How much money is sent to the county?
3. The credit card company keeps 4% of the total. How much money does the credit card company keep?

Name ______________________________ Date __________

B.5 Practice A

A credit card company keeps the given percent of the total. Find the amount of money (a) the credit card company keeps from the transaction, (b) the store receives from the credit card company, and (c) the store keeps from the transaction.

1. 3%

Bob's Sports Shop

Susan Jones
Credit Card # xxxx xxxx xxxx 6514
3:25 P.M. 01/12/12

SALE:	$ 98.00
SALES TAX:	$ 6.37
TOTAL:	**$104.37**

-THANK YOU-

2. 4%

Bubba's Restaurant

Bob Wilson
Credit Card # xxxx xxxx xxxx 2177
9: 13 P.M. 01/15/12

SALE:	$49.95
SALES TAX:	$ 3.50
TOTAL:	**$53.45**

THANK YOU

3. A credit card bill arrives in the mail. The monthly interest rate is 1.5%. Find the interest charge and the new balance.

Credit Card Statement

Account Number xxxx xxxx xxxx 0947

Date	Activity Since Last Statement	Amount
06/20/12	Payment	145.00
	- - - - Purchases - - - -	
06/01/12	Department Store	37.50
06/05/12	Tire Shop	115.00
06/15/12	Tolls	15.00

Average Daily Balance	$437.00
Statement Closing Date	07/01/12
Previous Balance	$345.00
Payments	-145.00
Unpaid Balance	$200.00
New Purchases	$167.50
Interest Charge	
New Balance	

4. You use a credit card to buy $15 worth of music downloads. The sales tax is 7%. The credit card company keeps 3% of the total.

a. How much money does the store keep from the transaction?

b. About what percent of the transaction does the store keep?

Name____________________ Date__________

B.5 Practice B

A credit card company keeps the given percent of the total. Find the amount of money (a) the credit card company keeps from the transaction, (b) the store receives from the credit card company, and (c) the store keeps from the transaction.

1. 3.5%

Suzi's Flower Shop

Tim Butler

Credit Card # xxxx xxxx xxxx 1118

10:14 A.M. 02/14/12

SALE:	$154.00
SALES TAX:	$ 10.01
TOTAL:	**$164.01**

THANK YOU

2. 2.5%

DEPARTMENT STORE

Mary Butler

Credit Card # xxxx xxxx xxxx 2177

2:15 P.M. 02/14/12

SALE:	$69.95
SALES TAX:	$ 4.55
TOTAL:	**$74.50**

THANK YOU

3. A credit card bill arrives in the mail. The monthly interest rate is 1.25%. Find the interest charge and the new balance.

Credit Card Statement

Account Number xxxx xxxx xxxx 0947

Date	Activity Since Last Statement	Amount
07/10/12	Payment	350.00
	---- **Purchases** ----	
07/04/12	Home Improvement	254.42
07/15/12	Jewelry Store	89.52
07/22/12	Car Wash	8.00

Average Daily Balance	$1138.56
Statement Closing Date	08/01/12
Previous Balance	$1105.48
Payments	-350.00
Unpaid Balance	$755.48
New Purchases	$351.94
Interest Charge	
New Balance	

4. A credit card bill for $400 comes in the mail. You stop making purchases with the card and pay the minimum payment of $20. The monthly interest rate is 1.75%.

a. The next month's statement shows an average daily balance of $388. Find the interest charge and new balance.

b. What was the decrease in the balance of your account?

c. What percent of your $20 payment was applied to the balance of your account?

Name ____ Date __________

B.5 Enrichment and Extension

Charge It?

Your grandmother decides to buy each of her 3 grandchildren a video game system. If she waits until next week and pays cash, she will have to pay full price for each system. If she buys them this week, she can get the systems at the sale price, but will have to pay for them using her credit card.

Answers the questions to determine which option is the best way to buy the systems.

1. Your grandmother could purchase the video game systems at regular price by paying with cash.

   a. What is the cost of 3 video game systems at regular price?

   b. Sales tax is 6%. How much will she pay in sales tax?

   c. What is the total cost (including sales tax) of 3 video game systems at regular price?

2. Your grandmother could purchase the video game systems at the sale price if she charges them on her credit card.

   a. What is the cost of 3 video game systems at the sale price?

   b. Sales tax is 6%. What is the total cost (including sales tax) of 3 video game systems at the sale price?

   c. The charge from part (b) is applied to the credit card on the first day of the month. Assuming she has no other charges already on the card, what is the average daily balance on your grandmother's credit card?

   d. The monthly interest rate on your grandmother's credit card is 8%. Find the interest charge.

   e. What will be the total amount (including interest) charged to your grandmother's credit card?

3. Should your grandmother buy the video game systems at the sale price? Explain your reasoning.

Name______________________________ Date__________

B.5 Puzzle Time

How Many Skunks Does It Take To Stink Up A Room?

Write the letter of each answer in the box containing the exercise number.

You own a clothing store. A customer uses a credit card to pay the total shown on the receipt.

Creative Clothing
123 Main Street
Somewhere, FL 33456
(905) 555-1212
Credit Card Sale
Card #XXXXXXXXXXXX1006
Date: 05/26/2012 10:16 AM
Auth. $025873
SALE: $55.00
STATE TAX: $ 3.30
COUNTY TAX: $ 0.88
TOTAL: $59.18
x Andrea Reed
-- THANK YOU --

1. How much money is sent to the state?
2. How much money is sent to the county?
3. The credit card company keeps 6% of the total. How much money does the credit card company keep?

Answers

W. $14.32

H. $3.30

P. $464.32

E. $3.55

A. $0.88

A credit card bill arrives in the mail. The monthly interest rate is 2.5%.

Credit Card Statement

Account Number XXXX XXXX XXXX 0947

Date	Activity Since Last Statement	Amount
03/16/12	Payment	350.00
	---- Purchases ----	
03/09/12	Restaurant	35.00
03/12/12	Clothing Outlet	126.24
03/21/12	Grocery Store	76.76
03/22/12	Jewelry Store	62.00

Average Daily Balance	$572.96
Statement Closing Date	04/01/12
Previous Balance	$500.00
Payments	-350.00
Unpaid Balance	$150.00
New Purchases	$300.00
Interest Charge	
New Balance	

4. What is the interest charge?
5. What is the new balance?

2		5	1	3	4

Activity B.6 Start Thinking!

For use before Activity B.6

When an employee of a company is paid, the check is often attached to a pay stub that the employee keeps for his or her records.

What type of information would you expect to see on a pay stub?

Activity B.6 Warm Up

For use before Activity B.6

Make a scatter plot of the data.

1.

x	0	1	2	3
y	0	2	4	6

2.

x	0	1	2	3
y	2	5	8	11

3.

x	0	1	2	3
y	5	3	1	–1

4.

x	0	1	2	3
y	–11	–7	–3	1

Lesson B.6 Start Thinking!

For use before Lesson B.6

What do you know about the following terms?

- Federal income tax
- Social Security
- Medicare
- 401(k)

Lesson B.6 Warm Up

For use before Lesson B.6

Use the table in Activity 2 in your textbook to find the federal income tax for a person who had the given amount of taxable income.

1. $32,500.00
2. $32,600.00
3. $176,091.00

Name ____ Date __________

B.6 Practice A

Use the table in Activity 2 on page A43 of your textbook to find the federal income tax for a person who had the given amount of taxable income.

1. $7500.00

2. $84,500.00

In Exercises 3–13, use the payroll statement.

NAME		
Betty Jones		
EARNINGS STATEMENT		
HOURS	RATE	AMOUNT
80	10.50	840.00
YTD GROSS PAY	GROSS PAY	
8400.00	840.00	

PAY ENDING	PAY DATE	DEPARTMENT	NET PAY	
05/18/12	05/18/12	107	624.74	
PAYROLL TAX WITHHELD				
FEDERAL	SOC. SEC.	HEALTH INS.	MEDICARE	401K
67.59	49.82	36.50	11.65	50.00
YEAR TO DATE				
FEDERAL	SOC. SEC.	HEALTH INS.	MEDICARE	401K
675.90	498.20	365.00	116.50	500.00

3. What is Betty's hourly wage?

4. How often does Betty get paid?

5. What is Betty's gross pay so far this year?

6. What is Betty's year to date 401(k) contribution?

7. What is the difference between Betty's gross pay and net pay?

8. What percent does Betty pay for Medicare tax?

9. What percent does Betty pay for Social Security tax?

10. What percent does Betty pay for federal income tax?

11. What percent of gross pay does Betty contribute to 401(k)?

12. Betty's employer matches her 401(k) contribution (they contribute the same amount that Betty contributes). As of 5/18/12, how much has been contributed to Betty's 401(k) plan?

13. Betty's check is the same every pay period. What is Betty's year to date net pay?

Name____ Date__________

B.6 Practice B

Use the table in Activity 2 to find the federal income tax for a person who had the given amount of taxable income.

1. \$32,550.00

2. \$389,000.00

You worked 45 hours this week. Regular pay rate of \$12 per hour applies to the first 40 hours. The number of hours beyond 40 are paid at a rate \$18 per hour. Each week you pay \$12 for health insurance.

3. Find your gross pay for this week.

4. You pay \$34.60 in Social Security taxes this week. What percent do you pay for Social Security tax?

5. You pay \$8.04 in Medicare taxes this week. What percent do you pay for Medicare tax?

6. You contribute 5% of your gross pay to 401(k). Find your 401(k) contribution for this week.

7. You pay \$41.31 in federal income taxes this week. What percent do you pay for federal income tax?

8. Your net pay is \$445.56. What is the difference between your gross pay and net pay?

9. What percent of your gross pay is your net pay?

10. You change your 401(k) plan. You now contribute 10% of your gross pay.

    a. What is the amount of your 401(k) contribution?

    b. Your net pay is now \$419.28. What percent of your gross pay is your net pay now?

Name ____ Date __________

B.6 Enrichment and Extension

Graduated Commission

Some companies compensate their employees using commission. In commission, the amount an employee earns is a fixed percent of the amount of merchandise sold. In graduated commission, the percent changes based on how much merchandise is sold.

The table shows the graduated commission structure of a candle company.

Amount sold (to the nearest dollar)	Percent Commission
\$500 and under	5%
\$501–\$1000	10%
\$1001–\$1500	15%
\$1501 and up	20%

1. You sell \$450 worth of merchandise.

   a. What percent is your commission?

   b. How much money have you earned?

2. You sell \$600 worth of merchandise. You earn 5% commission on the first \$500 and 10% on every dollar above \$500.

   a. How much commission have you earned on the first \$500?

   b. How much additional merchandise have you sold?

   c. What percent commission is earned on the additional sales?

   d. What is your total commission for selling \$600 worth of merchandise?

3. You want to earn \$135 in commission.

   a. How much commission do you earn on the first \$500 you sell?

   b. How much commission do you earn on the next \$500 you sell?

   c. How much additional commission do you need to meet your goal?

   d. What is the total amount you need to sell to earn \$135 in commission?

4. Do your earnings form a linear pattern? Explain.

Name______________________________ Date__________

B.6 Puzzle Time

What Did The Banana Bandits Say When They Got Caught?

Write the letter of each answer in the box containing the exercise number.

Use the payroll statement below to answer the following questions.

1. What is Liz's hourly wage?
2. What is Liz's gross pay for this pay period?
3. What is the difference between Liz's gross pay and net pay for this pay period?
4. What percent does Liz pay for Social Security tax?
5. What percent of Liz's gross pay is put in her 401(k)?

Answers

A. $392.33

I. 5.2%

R. 7.7%

T. $630.00

L. $10.50

S. 8.4%

P. $237.67

NAME: Liz Allen

EARNINGS STATEMENT

HOURS	RATE	AMOUNT
60	10.50	630.00

YTD GROSS PAY	GROSS PAY
6300.00	630.00

PAY ENDING	PAY DATE	DEPARTMENT	NET PAY
05/21/13	05/21/13	327	392.33

PAYROLL TAX WITHHELD

FEDERAL	SOC. SEC.	HEALTH INS.	MEDICARE	401K
88.63	48.51	54.50	13.33	32.70

YEAR TO DATE

FEDERAL	SOC. SEC.	HEALTH INS.	MEDICARE	401K
886.30	442.20	545.00	135.30	327.00

PAYROLL CHECK

EMPLOYEE ID NO.
9157962

DEPT.	CHECK NO.	PAY DATE
327	15350	05/21/13

PAY EXACTLY ***Three Hundred Ninety Two and 33/100*** DOLLARS

PAY TO THE ORDER OF Liz Allen
123 Any Street
Any Town, FL 45806

BANK NAME
CITY, STATE, ZIP CODE

AMOUNT
***392.33

Jonathan Alfred

4	3	1	5	2

Name ________________________________ Date __________

Appendix B

School-to-Work

For use after Section B.4

Bank Teller

You are working as a teller at a bank. The next customer in line thinks that there may be an error in the total amount listed in her checking account statement. She hands you the following registry with her records for the month.

Date	Check #	Transaction	Credit	Debit	Balance
		Balance Forward			395.68
2/10/12	1056	Credit Card Payment		312.30	
2/14/12		Deposit Paycheck	320.75		
2/19/12	Check Card	Grocery Store		67.52	
2/20/12	1057	Rent		400.00	
2/23/12		Deposit Paycheck	320.75		
2/25/12	1058	Phone Bill		42.50	
2/27/12	Check Card	Grocery Store		51.22	
2/28/12	1059	Car Payment		125.64	

1. Complete the balance column. What is the account balance according to the customer's checkbook registry?

2. Is there a bad check? Explain how you know.

3. The bank record indicates that the account balance is $13.00. What do you think the bank charges for writing a bad check? Explain.

4. The bank has a new program in which purchases made with a check card are rounded up to the next dollar value and the difference is transferred from a customer's checking account to the customer's savings account. The bank then adds 5% of the savings. Explain how the program works for a check card purchase of $25.40.

5. Redo the checkbook registry for the customer as if she had participated in the program for the month. How much money would be added to the savings account from the differences? How much money would be added to the savings account by the bank?

6. Explain the benefit of the program for someone who makes 20 purchases each month that average a difference of $0.50. How much money would be added to the savings account through the program over a period of one year? Show your work.

Name________________________________ Date__________

Appendix B Study Help

You can use an **information frame** to help you organize and remember concepts.

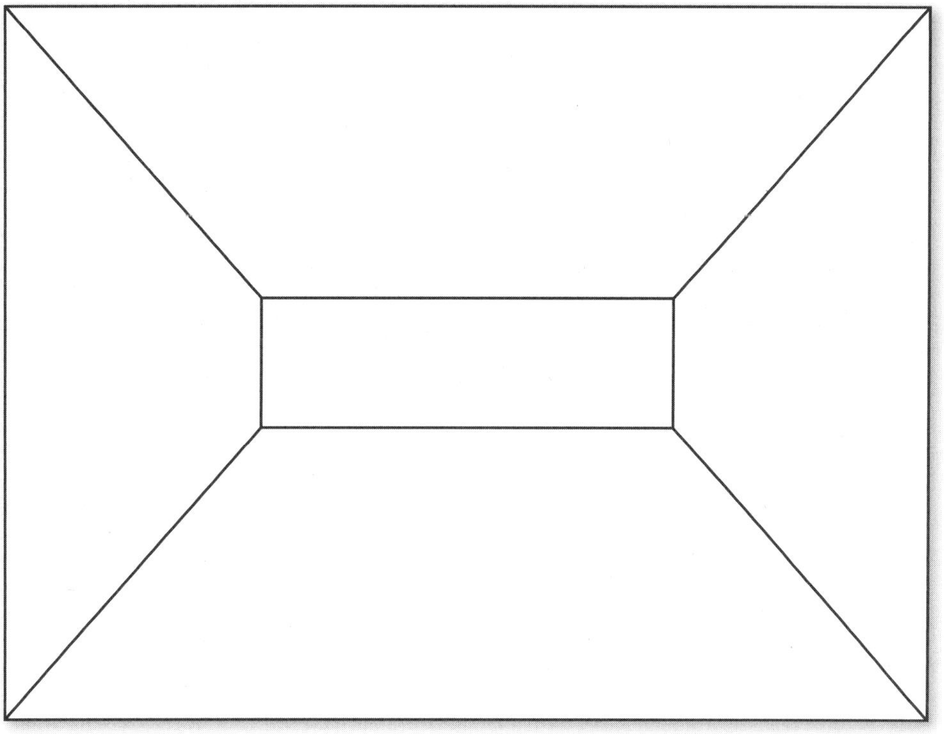

On Your Own

Make an information frame to help you study these topics.

1. compound interest
2. loan
3. mortgage
4. checking account

After you complete this chapter, make information frames for the following topics.

5. credit card
6. payroll statement

Name ____ Date __________

Appendix B

Financial Literacy

For use after Section B.2

Choosing a Loan

You want to start your own summer lawn care service. The equipment you need, such as a lawn mower and leaf blower, costs a total of $800. You have saved $150 towards these purchases, but need a loan for the rest. Your parents and your uncle have each offered you a loan under the conditions listed below.

Lender	Term of loan	Interest rate	Compounding type
parents	4 years	5%	no compounding
uncle	3 years	3%	compounded yearly

1. Calculate the interest you owe for each loan. For which loan do you pay the greater amount of interest?
2. For each loan you have to make a monthly payment based on the total amount owed over the term of the loan. What is the monthly payment for each loan?
3. For which loan do you end up paying less in the short term? For which loan do you end up paying less in the long term? Explain.
4. You work three months each summer and earn an average of $400 each month. If you take the loan from your parents, how much money will you have saved after 4 years? If you take the loan from your uncle, how much money will you have saved after 4 years? Show your work.
5. Which loan do you take? Explain your reasoning.

Name____ Date__________

Appendix B Technology Connection

For use after Section B.2

Using a Table to Track Account Growth

You can use the table feature of a graphing calculator to determine when the balance of a bank account reaches a certain amount.

EXAMPLE You deposit \$500 in an account that earns 1.5% interest compounded annually. When will you have a balance of \$580 in your account?

SOLUTION

Step 1 Press [Y=] and enter the formula for simple interest using 500 for P and 0.015 for r.

Step 2 Press [2nd] [TBLSET]. Set TblStart equal to 0 and ΔTbl equal to 1.

Step 3 Press [2nd] [TABLE]. Scroll down until you find the account value Y1 that is approximately equal to 580.00. You can see that when Y1 = 580.27, X = 10.

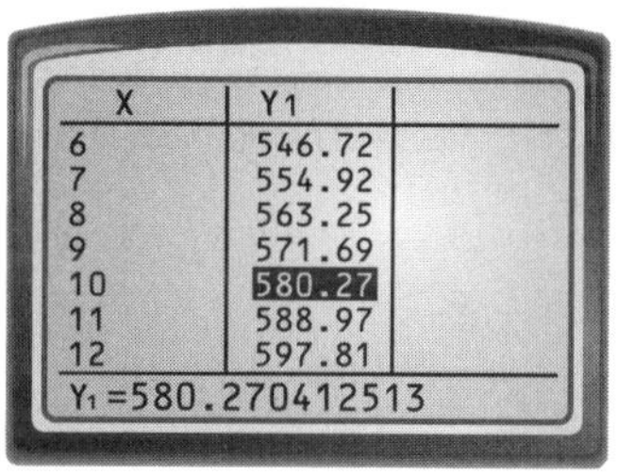

In about 10 years you will have a balance of \$580.

You deposit an initial amount in an account that earns the given interest rate and is compounded annually. Use a table on a graphing calculator to find when the balance reaches the desired amount.

1. Initial amount: \$2500
Interest rate: 2.5%
Desired amount: \$3046

2. Initial amount: \$320
Interest rate: 0.5%
Desired amount: \$328

3. Initial amount: \$4000
Interest rate: 3%
Desired amount: \$5703

4. Initial amount: \$1525
Interest rate: 2.75%
Desired amount: \$2000

Name ______________________________ Date __________

Appendix B Cumulative Practice

Evaluate the expression.

1. $820.13 - 327.45 - 43.62$

2. $96.8 \bullet 18 - 1500$

3. $\dfrac{42.8}{560 - 25}$

4. $480(1 + 0.07)^2$

5. $200 - 100.45 - 43.68 - 75.87$

6. $\dfrac{792.18 \bullet 480 - 150{,}000}{150{,}000}$

7. The balance in your checking account is \$480. You write a check for \$375.30 and another for \$104.62. What is the new balance in your checking account?

Write the decimal as a percent or the percent as a decimal.

8. 0.055

9. 1.03

10. 25%

11. 6.19%

12. A department store is having a sale. The department store is selling items at 60% of their original price. What fraction of the original price is the sale price?

Evaluate the expression when $k = 12$, $m = 0.08$, and $n = -1.4$.

13. $k(1 + m)^2$

14. $\dfrac{16 - n}{k}$

15. $|kmn|$

16. $\dfrac{k^2}{m} + n$

17. $28n$

18. $\dfrac{n + 20 \bullet m + 1}{12}$

19. You deposit \$2000 in a savings account that earns 2.5% interest compounded annually. What is the balance of the savings account after 3 years?

20. You deposit \$500 in a savings account that earns 4% simple interest per year. What is the balance of the savings account after 15 years?

Write and solve an equation to answer the question.

21. What is 5.6% of 600?

22. 40 is what percent of 500?

23. 32 is what percent of 1280?

24. 38 is 2% of what number?

25. An office worker puts 3% of her gross pay into a 401(k) plan. Her gross pay is \$740 every two weeks. How much money does the office worker put into the 401(k) plan in a year?

Credits

9 Gallant, Morrie. *Awesome Riddle Book*. Sterling, December 1999; **15** Singleton, Glen. *1001 Cool Jokes*. Hinkler Books Pty Ltd., 2003; **21** Rissinger, Matt and Philip Yates. *Totally Terrific Jokes*. Sterling, June 2001; **27** Singleton, Glen. *1001 Cool Jokes*. Hinkler Books Pty Ltd., 2003; **33** Adapted from Singleton, Glen. *1001 Cool Jokes*. Hinkler Books Pty Ltd., 2003; **47** Rissinger, Matt and Philip Yates. *Totally Terrific Jokes*. Sterling, June 2001; **53** Gallant, Morrie. *Awesome Riddle Book*. Sterling, December 1999; **59, 65** Adapted from Rissinger, Matt and Philip Yates. *Totally Terrific Jokes*. Sterling, June 2001; **71, 77** Rissinger, Matt and Philip Yates. *Totally Terrific Jokes*. Sterling, June 2001; **83** Adapted from Gallant, Morrie. *Awesome Riddle Book*. Sterling, December 1999; **99** Singleton, Glen. *1001 Cool Jokes*. Hinkler Books Pty Ltd., 2003; **105, 111** Hall, Katy. *Really, Really Bad Sports Jokes*. Candlewick, August 1998; **117** Singleton, Glen. *1001 Cool Jokes*. Hinkler Books Pty Ltd., 2003; **123, 137** Walton, Rick. *Really, Really Bad School Jokes*. Candlewick, August 1998; **143** Adapted from Gallant, Morrie. *Awesome Riddle Book*. Sterling, December 1999; **149** Rissinger, Matt and Philip Yates. *Totally Terrific Jokes*. Sterling, June 2001; **155** Singleton, Glen. *1001 Cool Jokes*. Hinkler Books Pty Ltd., 2003; **172** Horsfall, Jacqueline. *Kids' Silliest Jokes*. Sterling, March 2003; **178** Walton, Rick. Really, *Really Bad School Jokes*. Candlewick, August 1998; **184, 190, 196** Horsfall, Jacqueline. *Kids' Silliest Jokes*. Sterling, March 2003; **210, 216** Gallant, Morrie. *Awesome Riddle Book*. Sterling, December 1999; **222** Adapted from Gallant, Morrie. *Awesome Riddle Book*. Sterling, December 1999; **228, 234** Gallant, Morrie. *Awesome Riddle Book*. Sterling, December 1999; **252, 258, 264, 270, 287, 293, 299, 305** O'Donnell, Rosie. *Kids are Punny: Jokes Sent By Kids to The Rosie O'Donnell Show*. Warner Books, June 1997; **318** Gallant, Morrie. *Awesome Riddle Book*. Sterling, December 1999; **324** Adapted from Gallant, Morrie. *Awesome Riddle Book*. Sterling, December 1999; **330, 336, 342** Rissinger, Matt and Philip Yates. *Totally Terrific Jokes*. Sterling, June 2001; **348** Gallant, Morrie. *Awesome Riddle Book*. Sterling, December 1999; **367, 373, 379, 385, 391, 397** O'Donnell, Rosie. *Kids are Punny: Jokes Sent By Kids to The Rosie O'Donnell Show*. Warner Books, June 1997

Answers

Chapter 1

1.1 Start Thinking!

For use before Activity 1.1

Answers will vary.

1.1 Warm Up

For use before Activity 1.1

1. 118° **2.** 57° **3.** 90°

4. 101° **5.** 16° **6.** 77°

1.1 Start Thinking!

For use before Lesson 1.1

The Subtraction Property of Equality states that you can subtract the same number from both sides of an equation. The Multiplication Property of Equality states that you can multiply each side of an equation by the same number. The Division Property of Equality states that you can divide each side of an equation by the same nonzero number. *Sample situation:* If you are preparing two plates of food, you add the same amount of various foods to each plate so they remain equal servings (Addition Property of Equality).

1.1 Warm Up

For use before Lesson 1.1

1. 30 **2.** 46 **3.** 20 **4.** 53

1.1 Practice A

1. $x = 7$ **2.** $n = 34$ **3.** $k = -3$

4. $d = 3\pi$ **5.** $y = -1.3$ **6.** $w = \frac{19}{10}$

7. $49 = s + 19$; 30 points

8. $y = 8$ **9.** $d = -18$ **10.** $b = -0.4$

11. $x = -8.2$ **12.** $p = \frac{3}{2}$ **13.** $k = 3.75$

14. $7.50x = 33.75$; 4.5 hours

15. $s = 11.3$ **16.** $p = -8$ **17.** greater than

18. 13π cm^2 **19.** $20 + A = 44$; 24 in.^{2}

1.1 Practice B

1. $x = -12$ **2.** $h = 26\pi$ **3.** $m = 1.75$

4. $a = \frac{17}{12}$ **5.** $p = -\frac{1}{4}$ **6.** $c = -2.8$

7. $p - 9.75 = 64$; \$73.75

8. $x = -5$ **9.** $h = 8.6\pi$ **10.** $j = 10$

11. $t = -4$ **12.** $q = -3$ **13.** $w = -\frac{28}{3}$

14. $4.5\ell = 55.8$; 12.4 in.

15. $r = 22$ **16.** $n = \frac{49}{12}$

17. *Sample answer:* $x + 15 = 10$; $3x = -15$

18. 40 pieces of fruit **19.** \$4, \$12, \$24

1.1 Enrichment and Extension

Red	Blue	Yellow	Yellow	Blue	Red
Blue	Yellow	Green	Green	Yellow	Blue
Yellow	Green	Red	Red	Green	Yellow
Yellow	Green	Red	Red	Green	Yellow
Blue	Yellow	Green	Green	Yellow	Blue
Red	Blue	Yellow	Yellow	Blue	Red

1. yes; Blue squares result from dividing each side of the equation by the same number. Instead of dividing each side, you can multiply each side by a fraction. Then you would be using the Multiplication Property of Equality and all the blue squares would now be red.

2. yes; Green squares result from subtracting the same number from each side of an equation. Instead of subtracting the same number from each side, you can add a negative number to each side of the equation. Then you would be using the Addition Property of Equality and all the green squares would now be yellow.

1.1 Puzzle Time

THE MAN WHO PUT SOAP IN HIS SOUP SO HE COULD WASH IT DOWN WITH SOMETHING

1.2 Start Thinking!

For use before Activity 1.2

Sample answer: Checking your answers can help you detect and avoid arithmetic errors. On exams, checking answers can help you catch mistakes and improve your score. In real-life, in construction work, checking answers can help avoid costly mistakes.

Answers

1.2 Warm Up

For use before Activity 1.2

1. $5n + 5$ **2.** $-3x - 7$ **3.** $11f$

4. $3m + 16$ **5.** $4t + 8$ **6.** $3y + 2$

1.2 Start Thinking!

For use before Lesson 1.2

Sample answer: For the situation described, cost $= 80x + 60$. To solve, you use two inverse operations: first *subtract* 60 from each side, and then *divide* each side by 80. *Sample scenario:* the enrollment fee plus monthly rate for a gym membership

1.2 Warm Up

For use before Lesson 1.2

1. 33 **2.** 60 **3.** 97 **4.** 25

1.2 Practice A

1. $y = 2$ **2.** $m = 5$ **3.** $k = 10$

4. $z = -1$ **5.** $x = 3$ **6.** $x = -22$

7. 22 ft

8. $70 + 2x = 360$; $x = 145$

9. $3x + 1.75 = 9.25$; \$2.50

10. $1.50n + 2n = 10.50$; 3 magazines

11. a. $2x + 5x + 4 = 25$; $x = 3$

b. 6 ft and 15 ft

12. $\dfrac{14 + 19 + x}{3} = 17$; 18 points

13. $26 - 3.5x = 8.5$; 5 pens

1.2 Practice B

1. $k = 20$ **2.** $p = -14$ **3.** $y = 21$

4. $x = 2$ **5.** $h = 9$ **6.** $x = -3$

7. $44 + (x + 20) + 3x = 180$; $x = 29$

8. a. 42 ft

b. $x + 3x + 2x = 42$; $x = 7$

c. 7 ft by 50 ft; 21 ft by 50 ft; 14 ft by 50 ft

9. \$58.40

10. $18 + x + (4x - 1) = 42$; 19 in.

11. $4x + 2.5 = 14.3$; \$2.95

1.2 Enrichment and Extension

Reading down each column, the solutions to the equations are:

$x = 10$	$x = 6$	$x = -3$
$x = 2$	$x = 2$	$x = 5$
$x = 4$	$x = -3$	$x = 7$
$x = -3$	$x = 7$	$x = 6$
$x = 4$	$x = 1$	$x = 4$
$x = 1$	$x = 5$	$x = 4$

1.2 Puzzle Time

AT THE BOTTOM

1.3 Start Thinking!

For use before Activity 1.3

2 chips

1.3 Warm Up

For use before Activity 1.3

1. 4 m **2.** 10 cm **3.** 10 in. **4.** 10 ft

1.3 Start Thinking!

For use before Lesson 1.3

Subtraction Property of Equality; no, you are still left with the two-step equation $2x = 12x - 20$; subtracting $2x$ from each side is a better first step.

1.3 Warm Up

For use before Lesson 1.3

1. 5 ft **2.** 6 in.

1.3 Practice A

1. $x = 3$ **2.** $x = 1.5$ **3.** $y = -4$

4. $n = 3$ **5.** $q = 15$ **6.** $d = 5$

7. $h = -3$ **8.** $b = -5$

9. $15 + 0.25m = 20 + 0.05m$; 25 minutes

10. $\frac{1}{3}x = x - 22$; $x = 33$

11. 150 **12.** $0.6p = p - 32$; \$80

Answers

1.3 Practice B

1. $x = 15$ **2.** $x = 8$ **3.** $y = 4$

4. $m = 21$ **5.** $p = 4$ **6.** $s = 3$

7. $t = -6$ **8.** $n = -\frac{27}{8}$

9. The solver did not distribute 0.2 to both terms.

$$0.4x = 0.2(x - 8)$$
$$0.4x = 0.2x = 1.6$$
$$0.2x = -1.6$$
$$x = -8$$

10. $0.65\ell + 70 = \ell$; 200 ft

11. $0.6p = 0.8(p - 40)$; \$160

1.3 Enrichment and Extension

1. $d = \frac{4}{3}$ **2.** $e = 4$ **3.** $t = -1.5$

4. $a = -1$ **5.** $a = \frac{1}{3}$ **6.** $y = 0$

7. $r = 1$ **8.** $a = -5$ **9.** $s = 2$

10. $a = 2.25$ **11.** $\ell = 3$

Answer: at a yard sale

1.3 Puzzle Time

IT'S TOAD AWAY

1.4 Start Thinking!

For use before Activity 1.4

Check students' sketches. *Sample answer:*

Square: $P = 4s$; $A = s^2$

Rectangle: $P = 2\ell + 2w$; $A = \ell w$

Parallelogram: $A = bh$

Triangle: $P = a + b + c$; $A = \frac{1}{2}bh$

Circle: $C = 2\pi r$; $A = \pi r^2$

Trapezoid: $A = \frac{1}{2}h(b_1 + b_2)$

1.4 Warm Up

For use before Activity 1.4

1. 30 ft^3 **2.** $90\pi \approx 282.6 \text{ m}^3$

3. $15\pi \approx 47.1 \text{ in.}^3$ **4.** $213\frac{1}{3}\pi \approx 669.9 \text{ cm}^3$

1.4 Start Thinking!

For use before Lesson 1.4

Sample answer: When you solve the literal equation $5x + 4y = 14$ for x, you first subtract $4y$ from each side, and then divide by 5. When you solve the one-variable equation $5x + 20 = 14$ for x, you first subtract 20 from each side, and then divide by 5. The steps are almost identical.

1.4 Warm Up

For use before Lesson 1.4

1. a. $A = bh$ **b.** $b = \frac{A}{h}$ **c.** 16 m

2. a. $V = \frac{1}{3}Bh$ **b.** $B = \frac{3V}{h}$ **c.** 9 m^2

1.4 Practice A

1. $y = 7 - \frac{2}{5}x$ **2.** $y = 4 - \frac{2}{3}x$

3. $y = -6 + 10x$ **4.** $y = 3\pi - \frac{1}{2}x$

5. a. $r = \frac{d}{2}$ **b.** 65 miles per hour

6. $R = P - C$ **7.** $X = pN$

8. $h = \frac{3V}{\pi r^2}$ **9.** $b = \frac{2A}{h}$

10. a. $d = \frac{C}{\pi}$ **b.** $\frac{8}{\pi}$ in. **c.** 3 in.

11. a. $c = \frac{2A}{d}$ **b.** 7 ft **c.** 8 ft

1.4 Practice B

1. $y = 8 + 12x$ **2.** $y = \frac{3\pi}{4} - \frac{5}{8}x$

3. $y = \frac{3}{2} + 0.8x$ **4.** $y = -3.6 + 3x$

5. a. $w = \frac{V}{\ell h}$ **b.** 7 feet

6. $B = \frac{T}{2} - \frac{hP}{2}$ **7.** $x = \frac{C}{80} - 12.5$

8. $h = \frac{S}{2\pi r} - \frac{r}{2}$ **9.** $P = \frac{2A}{a}$

10. a. $C = \frac{5}{9}(F - 32)$ **b.** 100°C **c.** 26.7°C

11. a. $m = \frac{360A}{\pi r^2}$ **b.** 143.2°

c. greater than; The numerator will be larger.

1.4 Enrichment and Extension

Volume of cylinder: $V = \pi r^2 h$

Solved for h: $h = \frac{V}{\pi r^2}$

$h = 125.82$ cm, $h = 4.11$ ft

1. The height is 4.11 feet using both methods.

2. *Sample answer:* Method 2; Because you can do all of the calculations in one step.

1.4 Puzzle Time

A RIVER

1.5 Start Thinking!

For use before Activity 1.5

Sample answer: architects, contractors, interior designers, model builders

1.5 Warm Up

For use before Activity 1.5

1. 12 **2.** 3 **3.** 60

4. 24 **5.** 5280 **6.** 4

1.5 Start Thinking!

For use before Lesson 1.5

Sample answer: Convert one measurement so that they are both in the same units, then multiply to find the area; Yes, you could convert 30 inches to 2.5 feet and get an area of 17.5 square feet, or you could convert 7 feet to 84 inches and get an area of 2520 square inches.

1.5 Warm Up

For use before Lesson 1.5

1. 36 ft; 12 yd **2.** 7.5 ft; 2.5 yd

3. about 18.84 ft; about 6.28 yd

1.5 Practice A

1. 28 in., 71.1 cm **2.** 12 cm, 4.7 in.

3. 9.1 **4.** 29.5 **5.** 1.9 **6.** 2.4

7. a. 4 m by 0.9 m **b.** 4.4 yd by 1.0 yd

8. 48.3 **9.** 5.5 **10.** 0.5 **11.** 0.1

12. 1 **13.** 120,000

14. 7,000,000,000 **15.** 0.4

16. a. about 3.3 gallons **b.** 87.5 ft^2

c. 3 gallons and 2 quarts for $89.95

1.5 Practice B

1. 38.8 in., 98.6 cm **2.** 7.9 cm, 3.1 in.

3. 154.2 **4.** 33.8 **5.** 10.6 **6.** 0.1

7. a. 4.4 ft by 3 ft **b.** 1.3 m by 0.9 m

8. 354.3 **9.** 14.2 **10.** 7.3 **11.** 4.5

12. 250 **13.** 5760 **14.** 81 **15.** 75

16. a. 168 ft^2 **b.** $410.48 **c.** $41.05

1.5 Enrichment and Extension

1. base: 410 feet; distance between ground and first floor: 189.03 feet; distance between first and second floor: 190.63 feet; distance between second floor and third floor: 526.05 feet; total height of tower: 984 feet

2. 1 in. : 82 ft

3. base: 5 inches; distance between ground and first floor: 2.31 inches; distance between first and second floor: 2.32 inches; distance between second floor and third floor: 6.42 inches; Total height of model: 12 inches

4. *Sample answer:* It allows you to see the shape of the object and certain features so you can change the design before you build the object rather than trying to change it after it has been built. *Sample answer:* Toy makers use scale models to make smaller versions of cars and trains.

1.5 Puzzle Time

THE MECHANIC WHO SLEPT UNDER HIS CAR SO HE COULD WAKE UP OILY IN THE MORNING

School-to-Work

1. Answers may vary. *Sample answer:* Think of the area as being enclosed by two rectangles, one 30 feet by 15 feet and one 13 feet by 12 feet; 606 ft^2

2. *Sample answer:* 151.5 ft^3

3. 27 ft^3 = 1 yd^3; *Sample answer:* about 5.6 yd^3

4. *Sample answer:* $84.00; $95.20; $112.00

Answers

5. *Sample answer:* \$173.60; \$184.80; \$201.60

6. *Sample answer:* about 2 in.; Solve the equation

   $36 \bullet \frac{606}{27}x = 135$ to get $x \approx 0.167$ feet.

   So, x is approximately 2 inches.

Study Help

Available at *BigIdeasMath.com.*

Financial Literacy

1. Answer should include an estimate based on the total number of people associated with the school.

2. *Sample answers:* \$50 for the design; \$4.50 per T-shirt

3. *Sample answers:* \$10; at least \$9; no more than \$12

4. *Sample answer:* $50 + 4.5n$

5. *Sample answer:* $10n$

6. *Sample answers:* $50 + 4.5n = 10n$; $n \approx 9$;
   9 T-shirts; The break-even point is important in that it tells you the number of shirts you need to produce and sell to not lose money.

7. Raising the price lowers the break-even point. You raise more money by selling the same number of shirts. Raising the price may result in being able to sell fewer shirts because people may not be willing to pay more.

8. *Sample answer:* \$1000; You need to sell 190 T-shirts at \$10 a shirt based on solving the equation $10n - 4.5n - 50 = 1000$.

Cumulative Practice

1. $x = 22$ 2. $b = -5$ 3. $h = 13$

4. $n = 9$ 5. $k = 13$ 6. $a = -2$

7. $g = -2.5$ 8. $v = 12$ 9. $p = -2$

10. $y = 40$ 11. $m = -3$ 12. $w = 6$

13. 3 boxes 14. 52.48 15. 3

16. 5.5118 17. 154,880 18. 142.5

19. 1,500,000 20. 27 21. 24,000

22. 6 23. 24 24. $y = \frac{3}{x}$

25. $m = \frac{5}{2} - \frac{3}{4}k$ 26. $h = \frac{2A}{b}$

27. $p = \frac{2d + 3q}{4}$ 28. $z = 5w + v$

29. $d = \frac{bc}{a}$ 30. $\ell = \frac{V}{wh}$; 3 m

Chapter 2

2.1 Start Thinking!

For use before Activity 2.1

A and C; the graphs are lines.

2.1 Warm Up

For use before Activity 2.1

1.

x	−1	0	1	2
y	−5	−3	−1	1

2.

x	−1	0	1	2
y	2	1	0	−1

3.

x	−1	0	1	2
y	4	5	6	7

4.

x	−1	0	1	2
y	1.5	2	2.5	3

2.1 Start Thinking!

For use before Lesson 2.1

Answers will vary. Check students' graphs.

2.1 Warm Up

For use before Lesson 2.1

1–2. Answers will vary. Sample answers are given.

1.

x	0	2
$y = 4x - 2$	−2	6

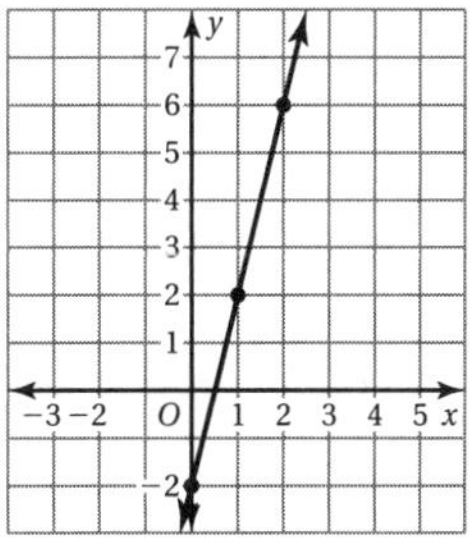

A different solution point on the line is (1, 2).

Answers

2.

x	0	5
$y = -x + 5$	5	0

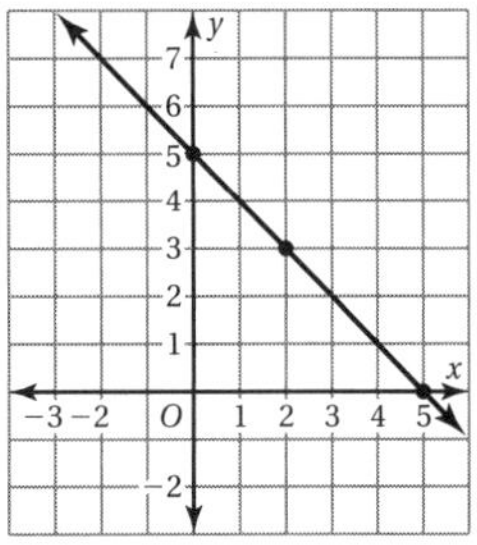

A different solution point on the line is $(2, 3)$.

2.1 Practice A

1–2. Sample answers are given.

1. $(-1, -1)$

x	0	1
$y = 4x + 3$	3	7

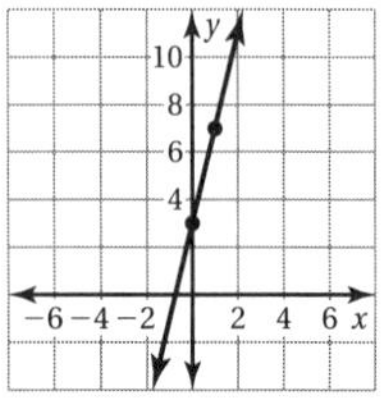

2. $(4, 5)$

x	0	2
$y = \frac{3}{2}x - 1$	-1	2

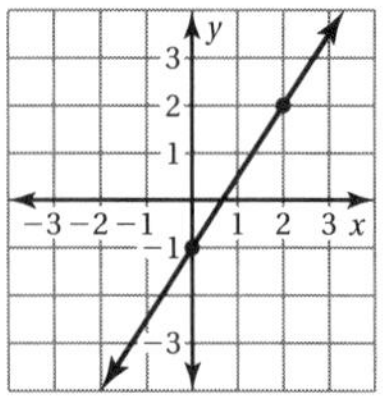

3. 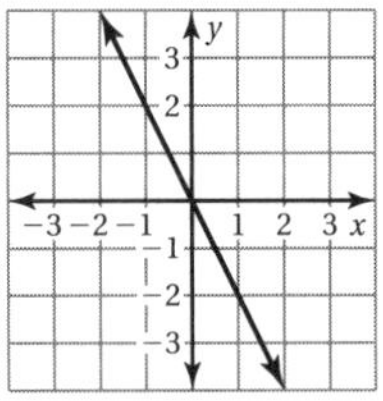

4. 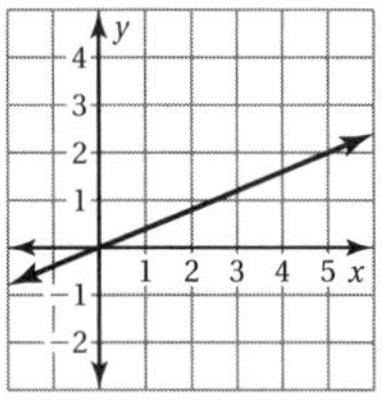

5. 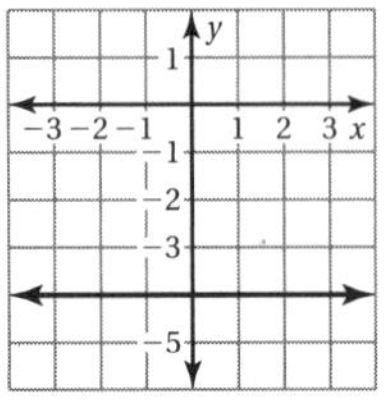

6. 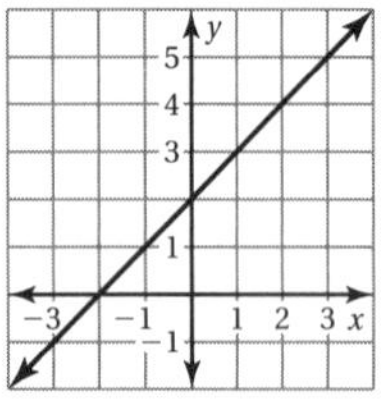

7. 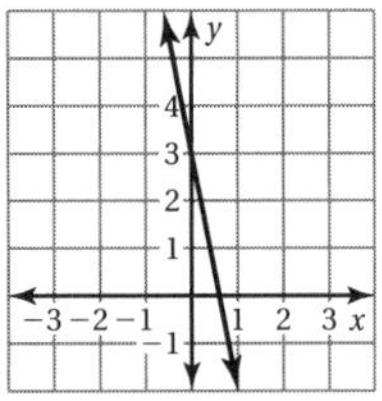

8. 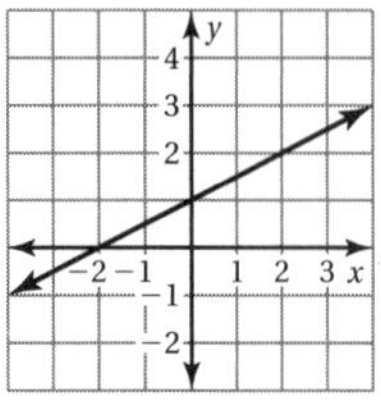

9. a. 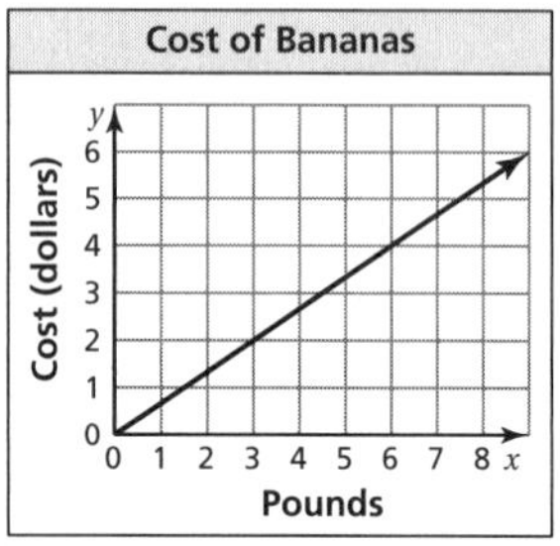

b. *Sample answer:* about $5.30 **c.** $5.33

10. $y = 2x + 5$

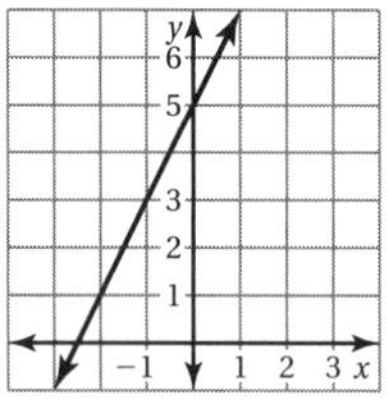

11. $y = -\frac{6}{5}x + 3$

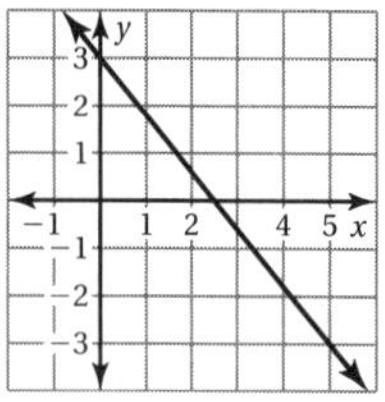

12. a. $y = 110 - 2.75x$

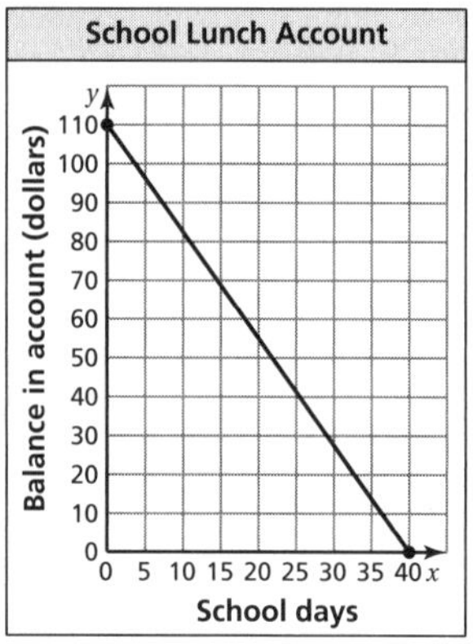

b. 40 school days

2.1 Practice B

1. $(-1, 4)$, $(2, 1)$

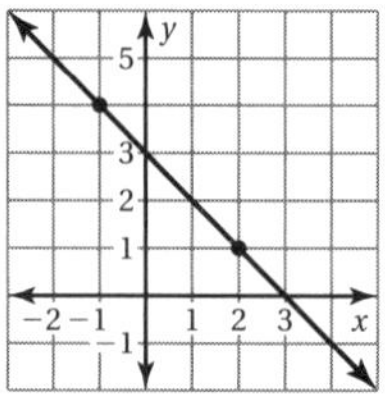

2. $(0, 5)$, $(6, 7)$

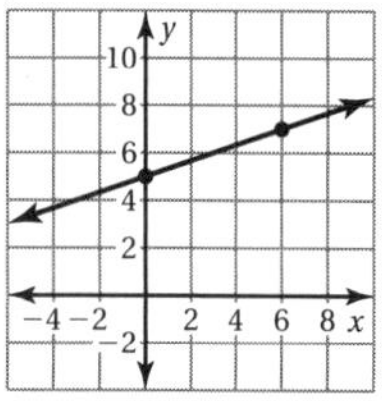

3. 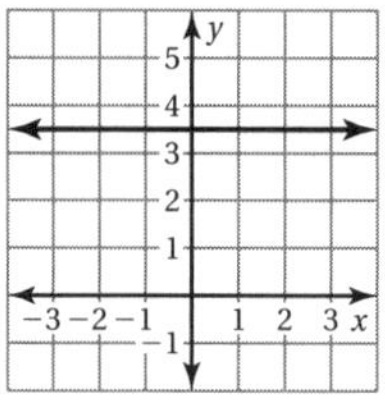

4. 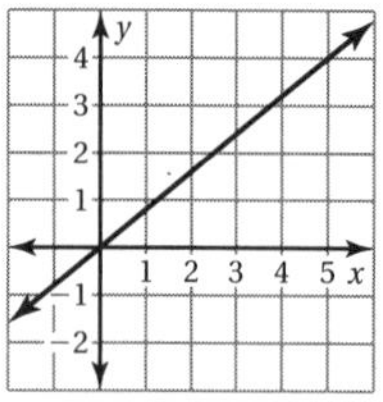

Answers

5. 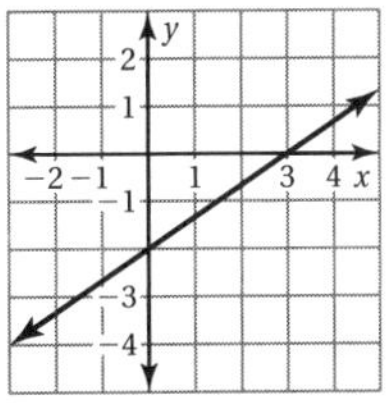

6. 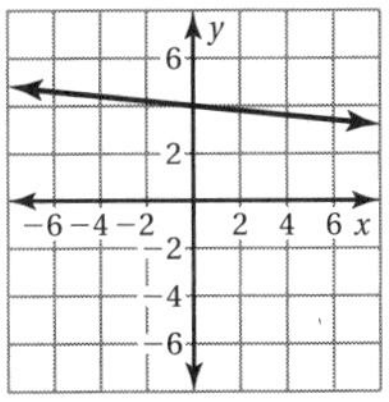

7. 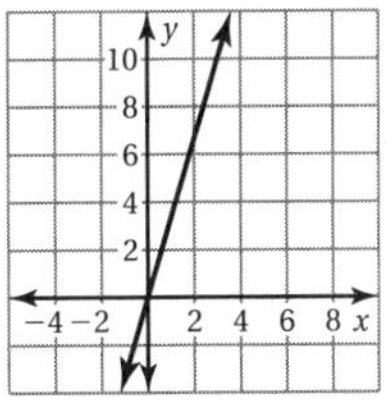

8. 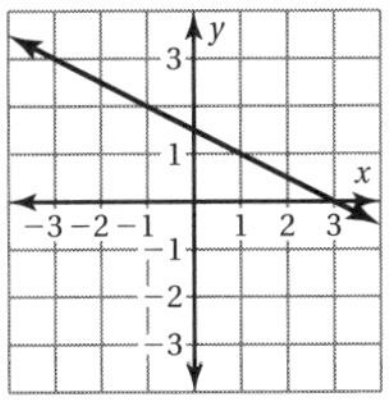

9. a. 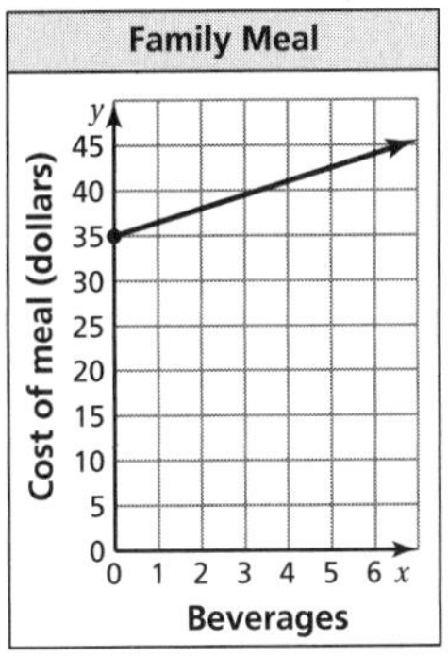

b. *Sample answer:* about \$42

c. \$42.50

10. $y = -\frac{3}{2}x - 3$

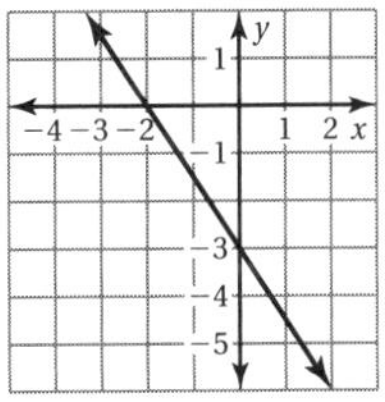

11. $y = -4x + 6$

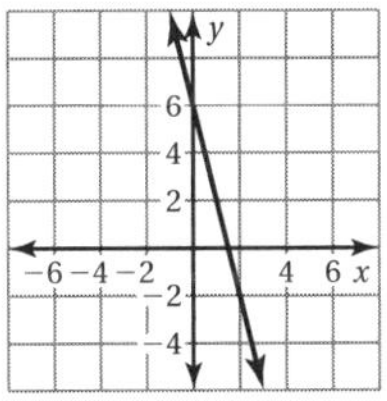

12. a. $y = 10 + 6x$

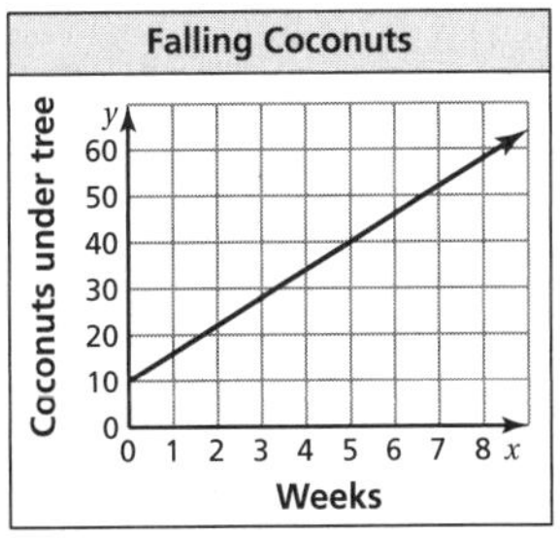

b. after 7 weeks

13. 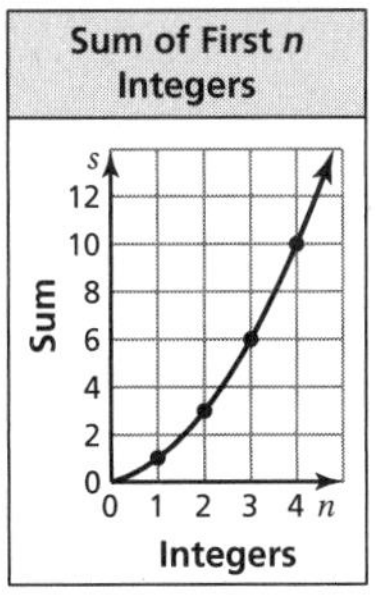

not a line

2.1 Enrichment and Extension

1.

x	y
−2	7
−1	5
0	3
1	1
2	−1

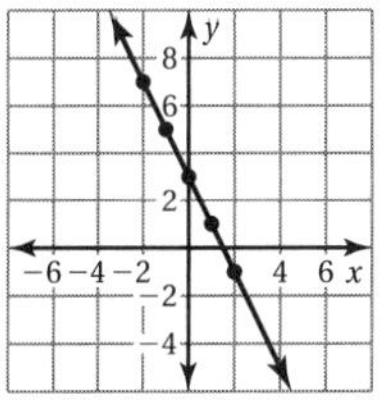

2.

x	y
−2	3
−1	0
0	−1
1	0
2	3

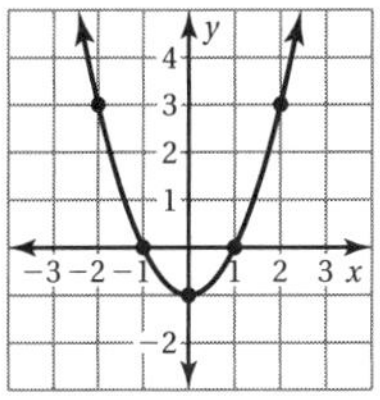

3.

x	y
−2	0
−1	1
0	2
1	1
2	0

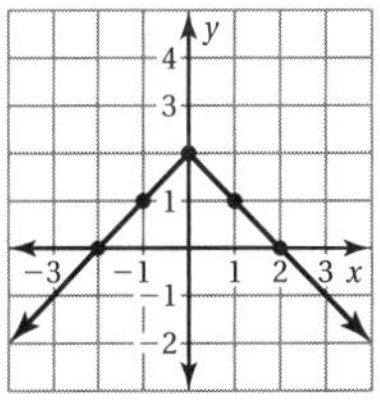

4. $y = -2x + 3$

5. $y = x^2 - 1$; The graph of the quadratic equation is shaped like a U.

6. $y = -|x| + 2$; The graph of the absolute value equation is shaped like an upside-down V, or mountain peak.

7. *Sample answer:* They all have different shapes.

8. The y-intercept would change and the graph would shift up or down.

2.1 Puzzle Time

A CREAM PUFFIN

2.2 Start Thinking!

For use before Activity 2.2

Sample answer: A ski slope is a natural incline, usually a mountain or hill. The mathematical term slope refers to how steep an incline is.

2.2 Warm Up

For use before Activity 2.2

1. 3 **2.** $\frac{2}{7}$ **3.** $\frac{2}{5}$

4. $\frac{5}{4}$ **5.** $\frac{2}{3}$ **6.** $\frac{4}{3}$

2.2 Start Thinking!

For use before Lesson 2.2

Monitor students during activity.

2.2 Warm Up

For use before Lesson 2.2

1. The lines are parallel.

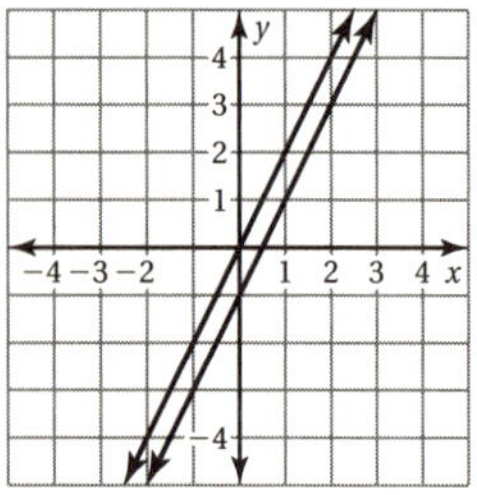

2. The lines are parallel.

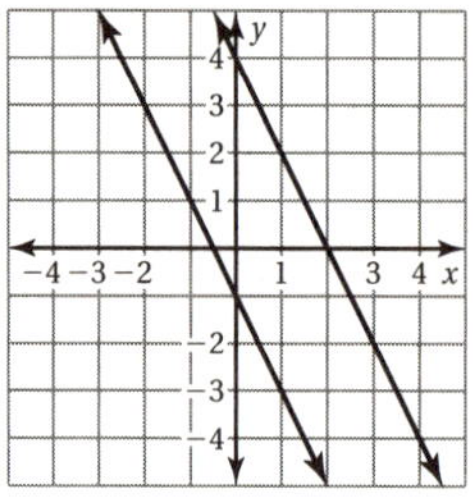

2.2 Practice A

1. a. lines A and C **b.** line B

c. lines A and C; Both have a slope of $-\frac{1}{3}$.

2. The lines are parallel.

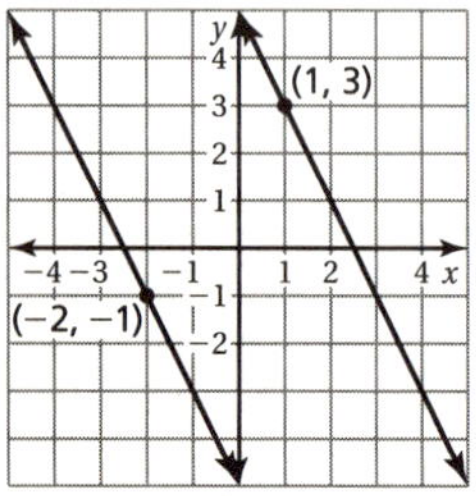

3. The lines are parallel.

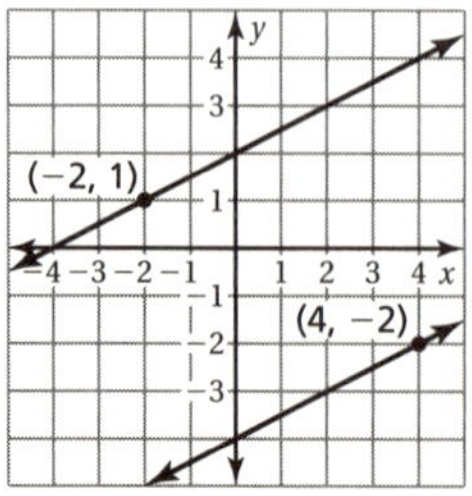

4. slope $= \frac{3}{2}$ **5.** slope $= -1$ **6.** slope $= 4$

7. slope $= 0$ **8.** slope $= 2$

2.2 Practice B

1. slope $= \frac{1}{5}$ **2.** slope $= 0$

3. slope $= -6$ **4.** slope $= \frac{4}{3}$

5. slope $= \frac{3}{2}$

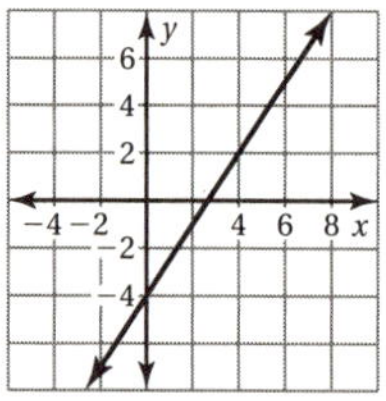

6. slope $= -1$

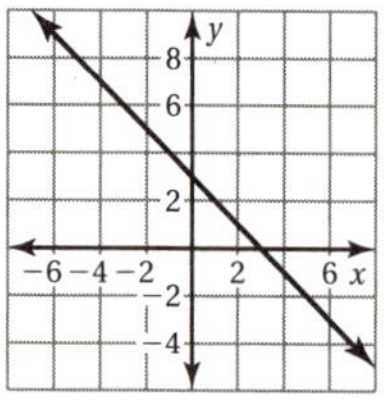

7. 10 feet

8. a. slope $= 0.1$

b. The phone calls cost \$0.10 per minute.

c. \$0.50 **d.** 30 minutes

2.2 Enrichment and Extension

1. slopes: $\frac{1}{2}, \frac{1}{2}$; parallel

2. slopes: $-\frac{3}{4}, \frac{4}{3}$; perpendicular

3. slopes: $-\frac{4}{5}, \frac{4}{5}$; neither

4. slopes: $\frac{2}{3}, \frac{3}{2}$; neither

5. slopes: $-\frac{1}{3}$, 3; perpendicular

6. slopes: $-2, -2$; parallel

7. a.

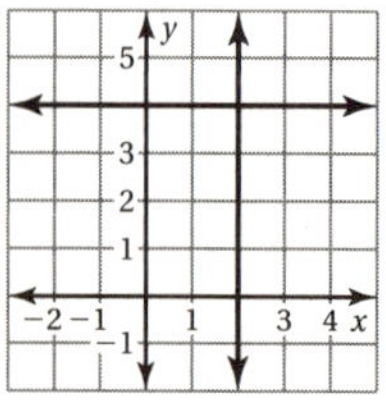

yes; Because a vertical line and a horizontal line intersect at right angles.

Answers

b. 0

c. Because you cannot divide by zero. Zero is in the denominator if you use rise over run or in the denominator if you find the reciprocal of 0.

2.2 Puzzle Time

I'M ON A ROLL

2.3 Start Thinking!

For use before Activity 2.3

Sample answer: The point (3, 290) means that the U.S. population was about 290 million in 2003. The graph is linear. The approximate slope is 2.8; it means that the U.S. population is increasing by about 2.8 million each year. The point where the graph crosses the y-axis is about 282; it means that the U.S. population was about 282 million in 2000.

2.3 Warm Up

For use before Activity 2.3

1.

x	0	1	2	3
y	−1	0	1	2

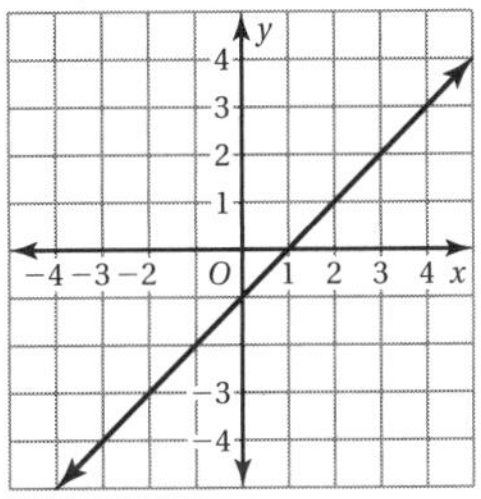

2.

x	−2	0	2	4
y	1	2	3	4

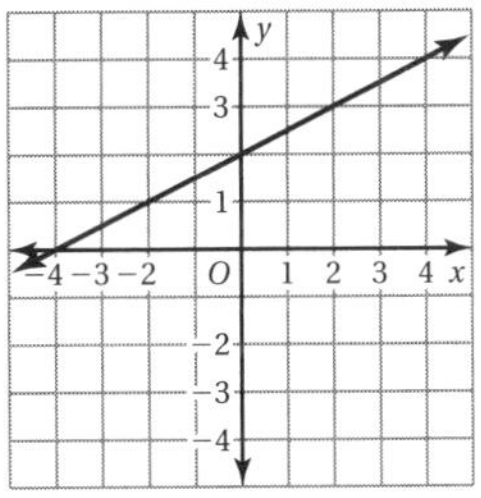

2.3 Start Thinking!

For use before Lesson 2.3

Sample answer: You buy x pairs of socks for \$2.50 each and pay \$5 for shipping. The slope is \$2.50 per pair and the y-intercept is \$5.

2.3 Warm Up

For use before Lesson 2.3

1. B; slope $= 2$, y-intercept $= -1$

2. A; slope $= -1$, y-intercept $= 2$

3. C; slope $= -2$, y-intercept $= -1$

2.3 Practice A

1. $C;\ -\frac{1}{2};\ 5$

2. $A;\ -3;\ -1$

3. $B;\ \frac{2}{3};\ 2$

4. $m = 1;\ b = 4$

5. $m = -8;\ b = 3$

6. $m = -\frac{5}{7};\ b = -2$

7. $m = 1.75;\ b = -1$

8. $m = 6;\ b = 2$

9. $m = \frac{1}{9};\ b = -7$

10. a.

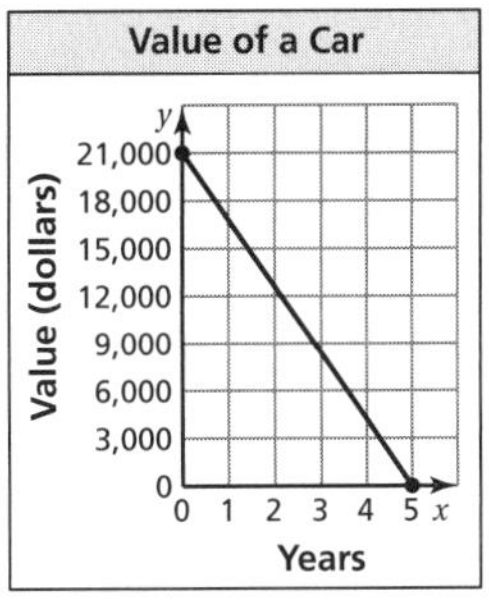

b. The value of the car decreases \$4200 per year.

c. \$21,000 is the initial cost of the car.

d. The value of the car is \$0 after 5 years.

11. 2

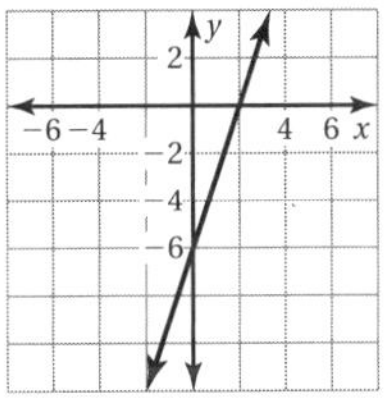

12. 48

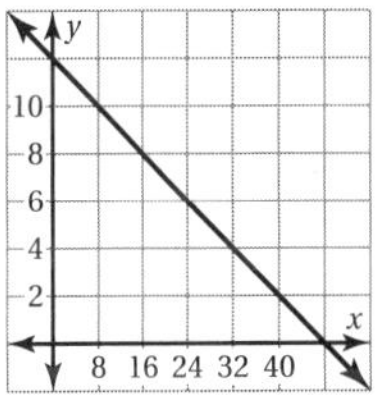

13. −3

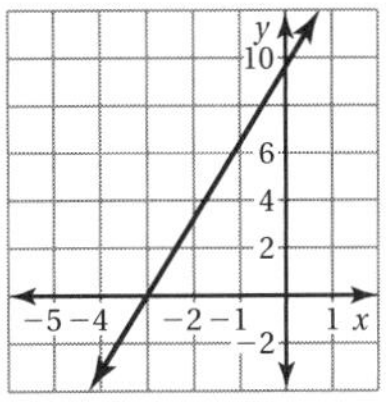

14. $-\frac{2}{5}$

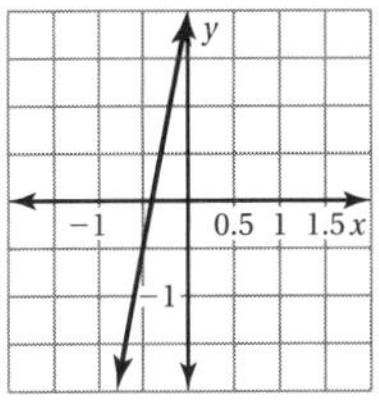

15. a.

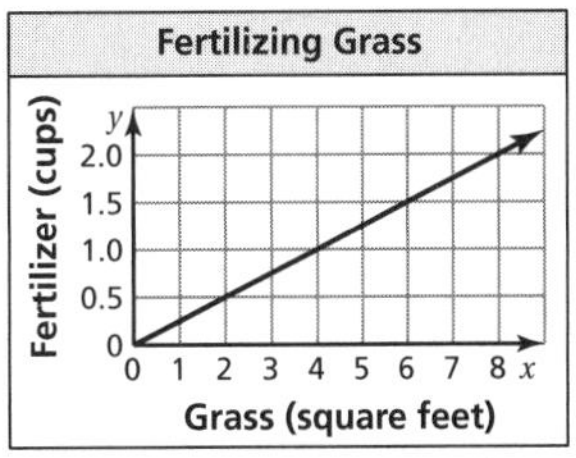

Answers

b. $\frac{1}{4}$ cup of fertilizer is needed per square foot of grass.

2.3 Practice B

1. $m = -\frac{3}{8}$; $b = 10$

2. $m = 4.5$; $b = 7$

3. $m = -\frac{4}{5}$; $b = -\frac{1}{5}$

4. $m = 5.5$; $b = -2.5$

5. $m = 4$; $b = \frac{2}{7}$

6. $m = \frac{2}{3}$; $b = -5$

7. $\frac{6}{5}$

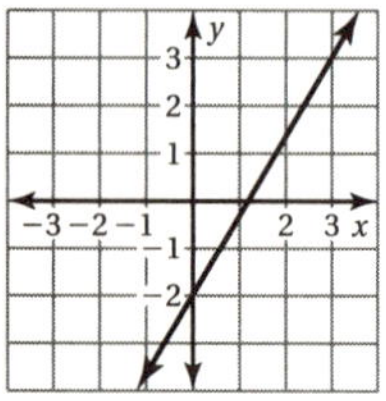

8. 7.5

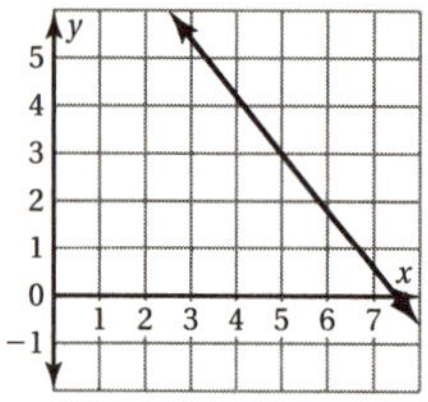

9. 6

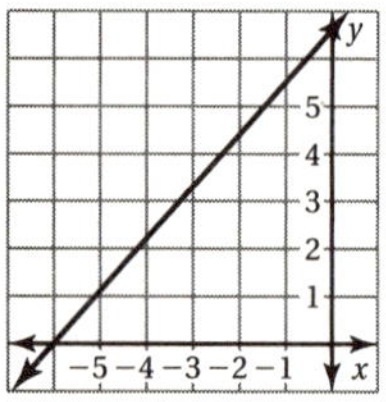

10. $-\frac{7}{2}$

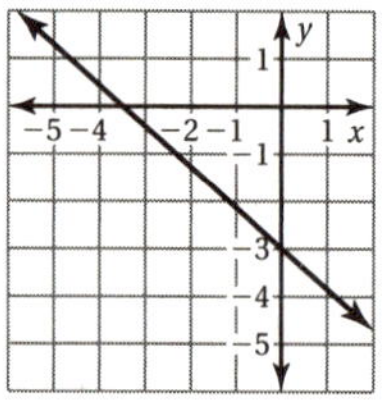

11. a. $y = 0.50x + 10$

b.

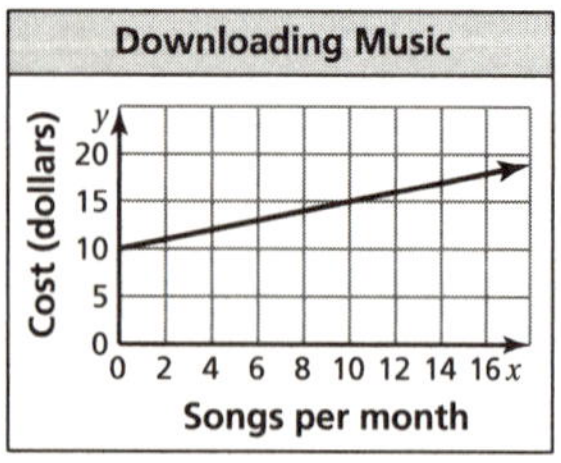

c. $17.50

12. a. $y = 6x$

b.

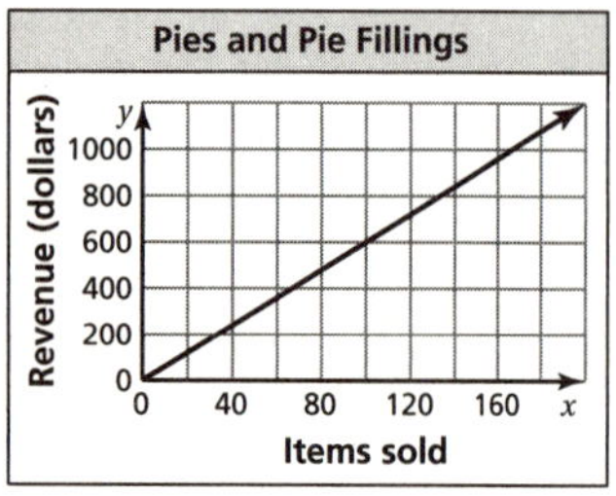

c. 184 items per month **d.** 46 items per week

2.3 Enrichment and Extension

1. 1 **2.** 4 **3.** $-1\frac{3}{5}$ **4.** -4

5. 3 **6.** -2 **7.** $-2\frac{2}{3}$

2.3 Puzzle Time

THE LADY WHO GOT STOPPED FOR SPEEDING SO MANY TIMES THE POLICE GAVE HER SEASON TICKETS

2.4 Start Thinking!

For use before Activity 2.4

$2x + 5y = 10$; *Sample answer:* The graphs are the same. Multiplying the original equation by 5 and then adding $2x$ to each side yields $2x + 5y = 10$.

2.4 Warm Up

For use before Activity 2.4

1. $y = -x + 4$

2. $y = -2x + 10$

3. $y = -\frac{3}{4}x + 3$

4. $y = \frac{1}{2}x + \frac{4}{5}$

5. $y = 2x + 5$

6. $y = \frac{1}{2}x + 2$

2.4 Start Thinking!

For use before Lesson 2.4

Sample answer: It is easier to write in standard form because of the information you are given. You know that the sum of the two costs is $40, or $10x + 6y = 40$.

2.4 Warm Up

For use before Lesson 2.4

1. Let x represent non-highway hours and let y represent highway hours; $y = -\frac{5}{12}x + 4$

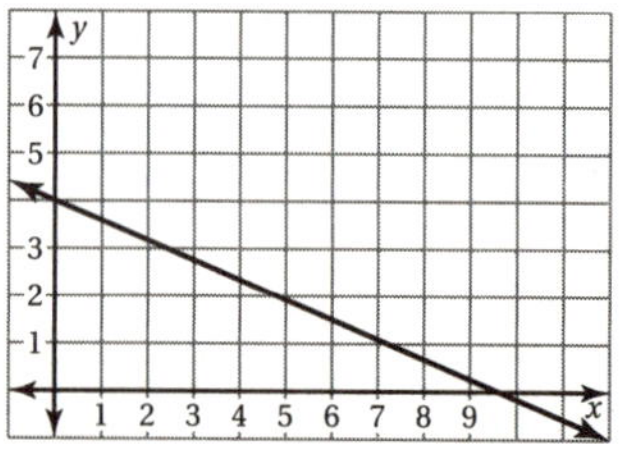

Answers

2. Let x represent number of hats and let y represent number of T-shirts; $y = -\frac{1}{2}x + 3$

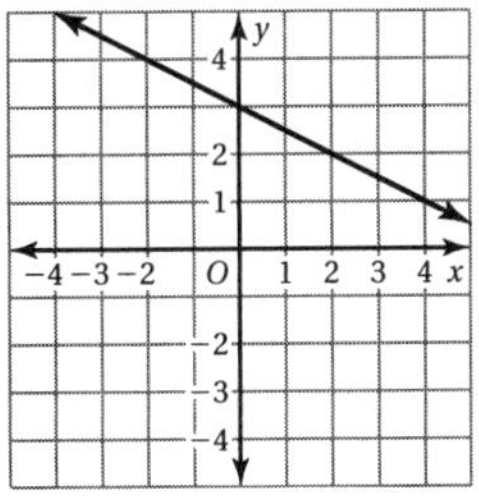

2.4 Practice A

1. $y = -4x + 10$
2. $y = 3x - 7$
3. 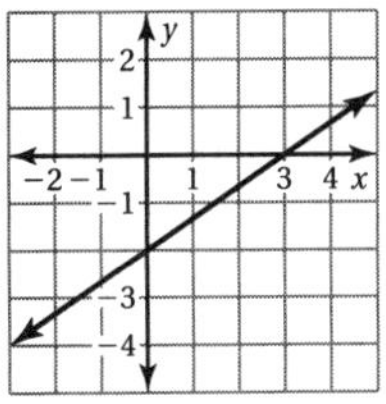
4. 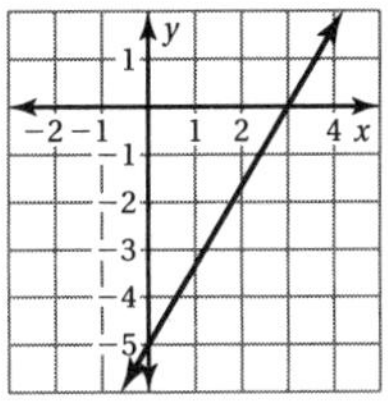
5. 3; −2
6. −2; 5
7. 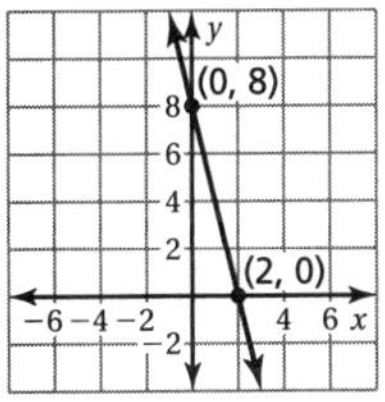

8. 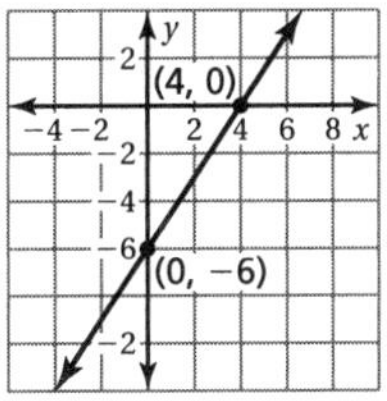

9. **a.** 11; 11 oranges and 0 apples contain 110 grams of fiber.
   b. 22; 22 apples and 0 oranges contain 110 grams of fiber.
   c. 10 g **d.** 5 g
   e. no; There cannot be 3.5 oranges in the package.
10. **a.** $8x + 6y = 144$ **b.** 18; 24
    c. 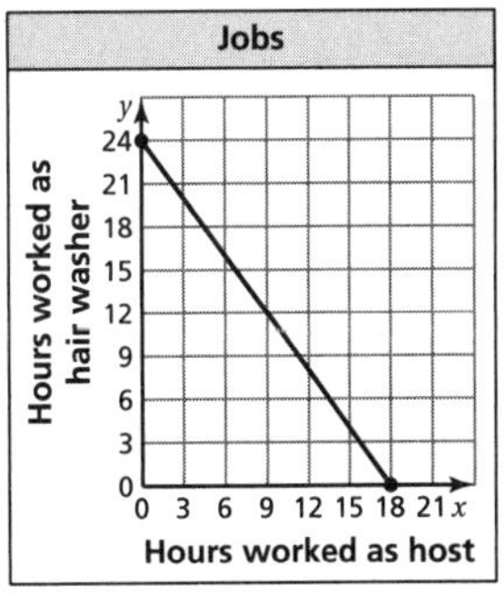

    d. 10.5 h

2.4 Practice B

1. $y = -\frac{2}{3}x + 4$
2. $y = 2x - 5$
3. 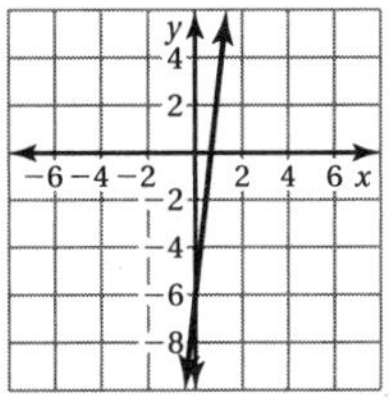
4. 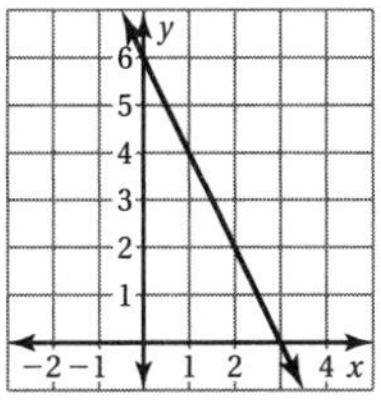
5. 0; 0
6. no x-intercept; 5
7. 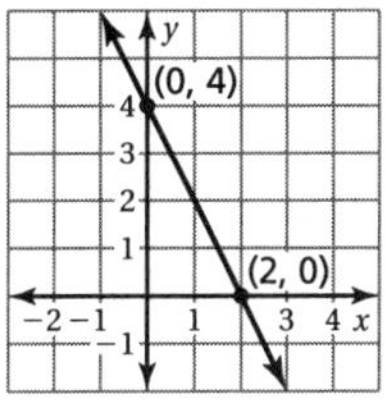

8. 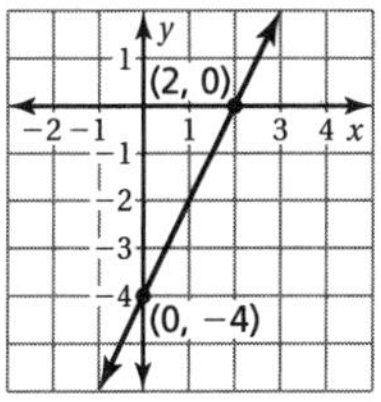

9. **a.** $80x + 40y = 480$ **b.** 6; 12
   c. 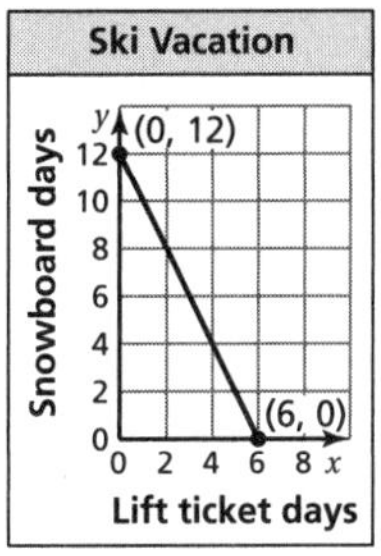

   d. 5 lift tickets
10. **a.** $y = 80 + 32x$ **b.** −2.5; 80
    c. 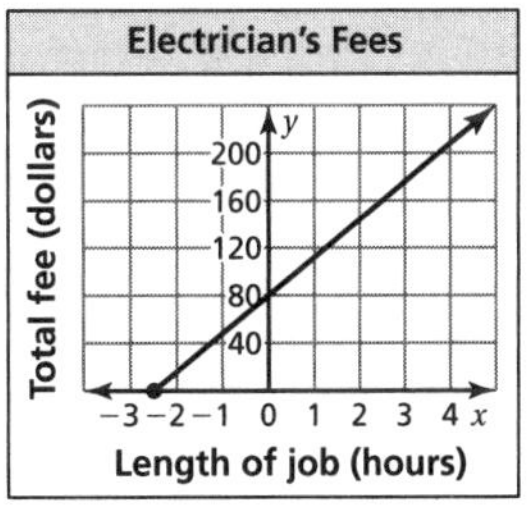

    d. no; The visit cannot last a negative number of hours.

2.4 Enrichment and Extension

1. no; The graph is not a straight line.
2. 40
3. The trapeze artist starts 40 feet right of center.
4. 8

5. The trapeze artist swings back and forth past center more than once.

6. 16 seconds

7. 80 feet

2.4 Puzzle Time

THEY JUMP SHIP

2.5 Start Thinking!

For use before Activity 2.5

Check students' equations and graphs. The intersection represents the two ages.

2.5 Warm Up

For use before Activity 2.5

1. $y = 2x + 6$

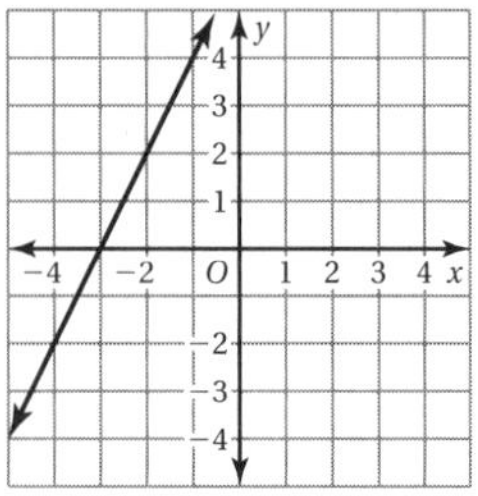

2. $y = -2x + 10$

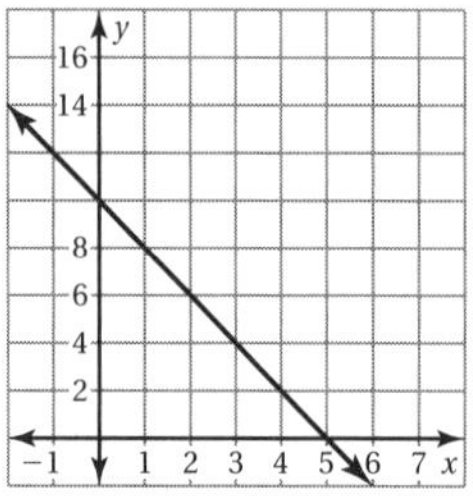

3. $y = 2x + 8$

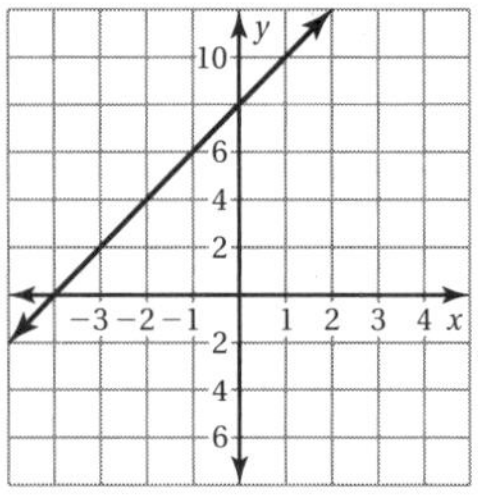

4. $y = -\frac{1}{3}x + 3$

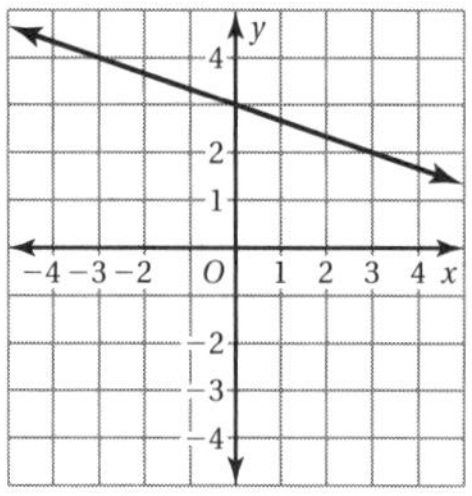

5. $y = \frac{1}{2}x - 1$

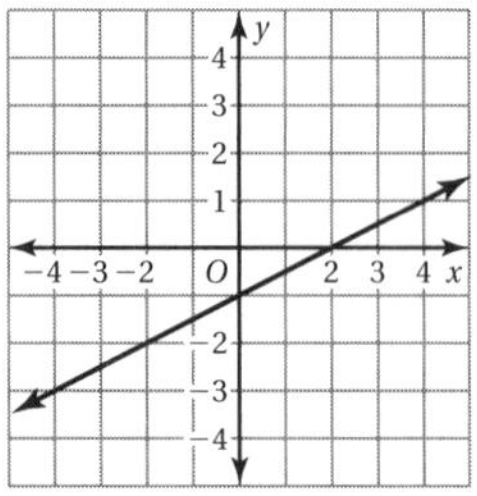

6. $y = -\frac{5}{3}x + 10$

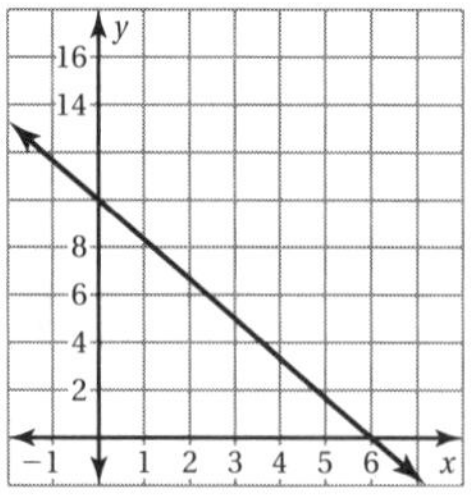

2.5 Start Thinking!

For use before Lesson 2.5

1. Answers will vary. Check students' word problems.

2.5 Warm Up

For use before Lesson 2.5

1. $x = 5$ is the break-even point.

x	0	1	2	3	4	5	6
C	25	35	45	55	65	75	85
R	0	15	30	45	60	75	90

2. $x = 6$ is the break-even point.

x	0	1	2	3	4	5	6
C	90	110	130	150	170	190	210
R	0	35	70	105	140	175	210

2.5 Practice A

1. (4, 144)

x	0	1	2	3	4	5	6
C	48	72	96	120	144	168	192
R	0	36	72	108	144	180	216

2. (4, 5)

3. (3, 0)

4. (2, −1)

5. (−2, 0)

6. (12, 7)

7. (4, 18)

8. a. $x + y = 36$
   $y = x + 8$
   b. (14, 22)

9. a. 5 min b. 225 yd

2.5 Practice B

1. (6, 540)

x	0	1	2	3	4	5	6
C	150	215	280	345	410	475	540
R	0	90	180	270	360	450	540

2. (2, 5)

3. (4, −6)

4. (−3, 2)

5. (4, −4)

6. (2, 6)

7. (−1, −3)

8. a. $x + y = 45$
   $y = 2x - 6$
   b. 17

9. a. $x + y = 9$
   $5x + 10y = 65$
   b. 5 five-dollar bills and 4 ten-dollar bills
   c. no; You cannot use 3.5 ten-dollar bills.

Answers

2.5 Enrichment and Extension

1. $(0, 3)$ **2.** $(5, 4)$ **3.** $(1, -1)$

4. $(4, 3)$ **5.** $(5, 8)$ **6.** $(1, -2)$

7. Subtraction Property of Equality; You are subtracting the same amount from each side because the expressions on each side of the equation are equal.

2.5 Puzzle Time

HE LOGGED ON

2.6 Start Thinking!

For use before Activity 2.6

Sample answer: The graphs are the same line; the solution of the system is the set of all of the coordinate pairs on the line.

2.6 Warm Up

For use before Activity 2.6

1. $(7, 12)$ **2.** $(8, -2)$ **3.** $(1, 3)$ **4.** $(1, 8)$

2.6 Start Thinking!

For use before Lesson 2.6

Sample answer: No; if they are traveling at the same average speed, the second car will always be about 1 hour behind. You can graph two lines: the first car is $y = 60x$, and the second has the same slope but goes through the point $(1, 0)$ to represent a delay of 1 hour. The solution of the system is no solution since the two graphs do not intersect.

2.6 Warm Up

For use before Lesson 2.6

1. any (x, y) such that x is three more than twice the value of y

2. no solution

2.6 Practice A

1. no solution **2.** infinitely many solutions

3. $(0, 0)$ **4.** infinitely many solutions

5. a. $y = 15x + 25$
$y = 15x + 10$

b. no; The graphs of the equations are parallel lines, so there is no common solution.

c. yes; After 3 weeks each of you would have \$70.

6. infinitely many solutions

7. $(2.4, -1)$

8. no; Two lines cannot intersect at exactly two points.

9. a. *Sample answer:* $x + y = 4$

b. *Sample answer:* $5y = 10x + 5$

c. *Sample answer:* $y = 2x - 3$

2.6 Practice B

1. infinitely many solutions

2. $(0, -2)$

3. infinitcly many solutions

4. no solution

5. a. yes **b.** no **c.** yes

d. infinitely many; There is more than one solution, so there are infinitely many solutions.

6. infinitely many solutions

7. infinitely many solutions

8. a. *Sample answer:* $x + y = 4$

b. *Sample answer:* $y = -\frac{3}{4}x + 4$

c. *Sample answer:* $4y = -3x + 8$

d. yes; You can multiply both sides of the given equation by any nonzero number, for example, $8y = -6x + 16$ is another answer to part (c).

2.6 Enrichment and Extension

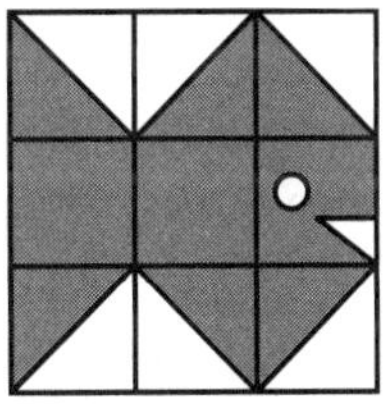

2.6 Puzzle Time

TO-YODA

2.7 Start Thinking!

For use before Activity 2.7

Sample answer: The graph is the linear equation related to each side of the equation. The x-coordinate of the point of intersection is the solution of the equation.

2.7 Warm Up

For use before Activity 2.7

1. $m = -8$ **2.** $x = 6$ **3.** $p = 3$

4. $x = 7$ **5.** $r = 3$ **6.** $x = -5$

Answers

2.7 Start Thinking!

For use before Lesson 2.7

Answers will vary.

2.7 Warm Up

For use before Lesson 2.7

1. $x = -2$ **2.** $x = 4$

2.7 Practice A

1. $x = 4$ **2.** $x = -3$ **3.** $x = 1$ **4.** $x = -1$

5. no; You cannot have -12 CDs.

6. $x = 8$ **7.** $x = -2.5$

8. no; The graph of the system $y = 50 - 6x$ and $y = 75 - 6x$ is a pair of parallel lines.

9. a. $\frac{1}{3}x = x - 20$

b. \$30 **c.** \$10

d. *Sample answer:* the cost of the lemonade

2.7 Practice B

1. $x = -3$ **2.** $x = 6$

3. $x = 3$ **4.** $x = -2$

5. no; The solution of $22 + 4.5x = 14 + 3.5x$ is a negative number, and a negative number of visits to a gym doesn't make sense.

6. $x = 2$ **7.** $x = -5$

8. a. 12.5 min **b.** 75 gal

c. Tank B: It is draining at a faster rate.

9. a. $x = 3$ **b.** yes

2.7 Enrichment and Extension

1. a. one **b.** $(1, 1)$

2. a. three **b.** $(-2, -6), (0, 0), (2, 6)$

3. a. none

b. no; The lines would intersect, but the nonlinear equation has a sharp turn, so they do not intersect.

2.7 Puzzle Time

THE MAGICIAN WHO WALKED WITH A CANE BECAUSE HE HAD A TRICK KNEE

School-to-Work

1. $r = -a + 220$; r represents maximum heart rate and a represents age.

2. lower bound: $r = -0.5a + 110$;
upper bound: $r = -0.85a + 187$

3–4. *Sample answer:*

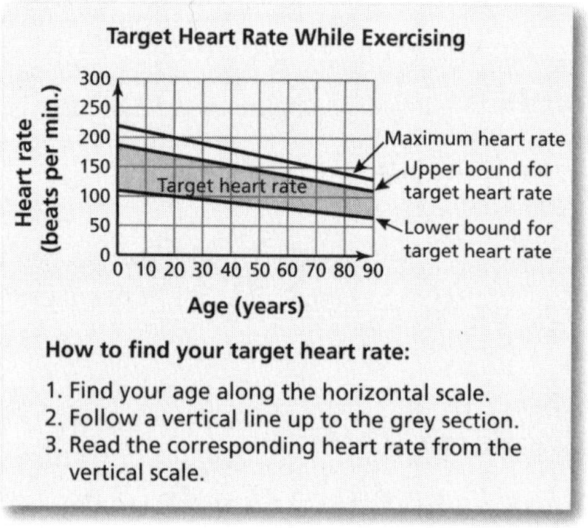

5. *Sample answer:* 207; Your maximum heart rate decreases as you age because you are subtracting your age from 220.

6. *Sample answer:* yes; no; The target heart rate for a 42-year-old is between 89 and 151.3.

Study Help

Available at *BigIdeasMath.com.*

Financial Literacy

1–3.

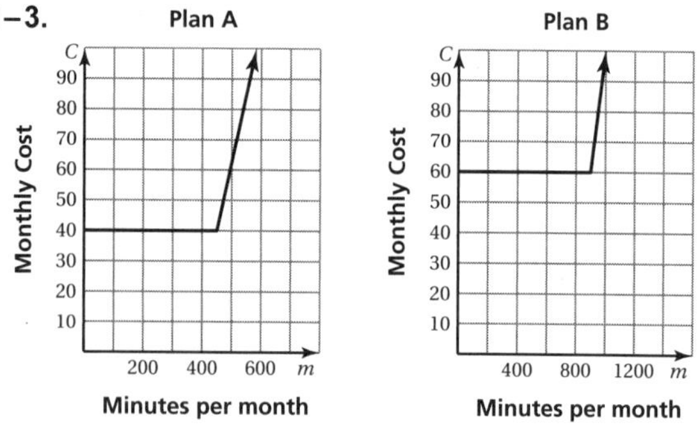

2. plan A; plan A is \$40 for up to 450 minutes, while plan B is \$60.

3. $C = 0.45m - 162.5$; The point of intersection represents when the plans cost the same amount; $900 = 0.45m - 162.5$, $m \approx 494$, so the point of intersection is $(494, 60)$.

4. If you use less than 494 minutes, plan A is the best choice. If you use more than 494 minutes, plan B is the best choice.

5. Check students' work.

6. Check students' work.

7. Answers should include graphical and algebraic representations for plan C. Any points of intersection should be labeled on the graph.

Technology Connection

1. $(3, 7)$

2. $(2.5, -3)$

3. $(-1, 0)$

4. $(-12, -4)$

Life Connections

1. Check students' work.

2. Check students' work.

3. *Sample answer:* the greater the slope, the steeper the staircase, and the harder it is to climb.

4. Check students' work. *Sample answer:* yes

5. *Sample answer:* Use 4 steps that are each 1 foot tall with a tread of 1 foot, for a slope of 1; Use 6 steps that are each 8 inches tall with a tread of 10 inches for a slope of 0.8.

6. The codes ensure that staircases are built in a way that is safe. Building codes generally state that the minimum tread length is 9 inches and the maximum riser height is 8.25 inches.

7. Answers should include reasoning about safety and ease of climbing.

Cumulative Practice

1. inverse variation because it has the form $y = \frac{k}{x}$

2. neither because it can not be written as $y = \frac{k}{x}$ or $y = kx$

3. direct variation because $y = -4x$

4. neither because the equation is $y = x + 1$

5. 25

6. slope: $\frac{1}{2}$; y-intercept: 4

7. slope: -2; y-intercept: 5

8. slope: $-\frac{1}{4}$; y-intercept: -3

9. **a.** 15.5 **b.** It is the cost per hour.

10. $(-6, -8)$

11. infinitely many solutions

12. no solution

13. 5 days

Chapter 3

3.1 Start Thinking!

For use before Activity 3.1

The graph shows the temperature of an oven while it is preheating. The y-intercept, 70, represents the temperature of the stove before it was turned on. The slope, 35 degrees per minute, represents the heating rate.

3.1 Warm Up

For use before Activity 3.1

1. $-\frac{2}{3}$

2. 2

3.1 Start Thinking!

For use before Lesson 3.1

$y = 40x + 20$; the slope represents the monthly cost; the y-intercept represents the enrollment fee.

3.1 Warm Up

For use before Lesson 3.1

1. $y = \frac{1}{2}x$, $y = \frac{1}{2}x + 3$, $y = 2x + 6$, $y = 2x$

2. $y = x + 1$, $y = x - 5$, $y = -x + 5$, $y = -x + 1$

3.1 Practice A

1. $y = 3x + 1$, $y = 4$, $y = 3x - 11$, $y = -2$

2. $y = -\frac{3}{2}x + 1$

3. $y = \frac{3}{5}x + 2$

4. $y = 2x + 3$

5. $y = -\frac{1}{2}x$

6. $y = -\frac{3}{2}x + 3$

Answers

7. a.
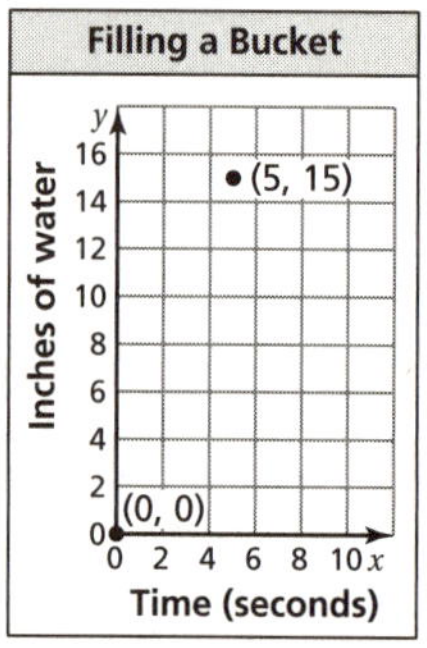

b. There is 0 inches of water at 0 seconds. There is 15 inches of water after 5 seconds.

c.
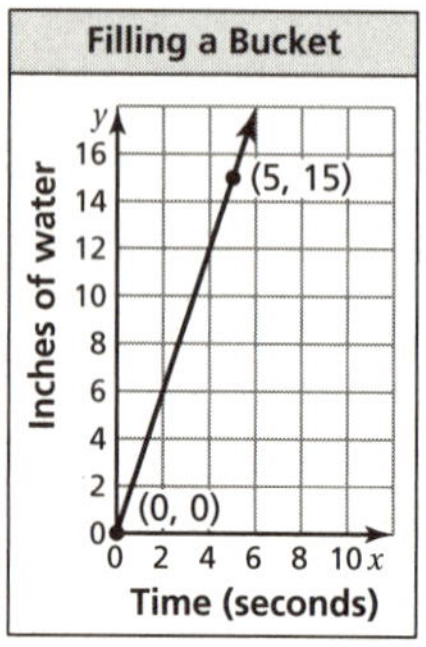

d. The inches of water y in the bucket after x seconds.

e. $y = 3x$

3.1 Practice B

1. $y = 1, y = 3x - 5, y = -2, y = -3x - 5$

2. $y = -\frac{1}{4}x - 1$

3. $y = \frac{7}{5}x + 4$

4. a. 5.7 inches per year

b. $y = 5.7x + 6$ c. 28.8 inches

5. $y = \frac{3}{2}x + 5$

6. $y = -2x - 3$

7. a.
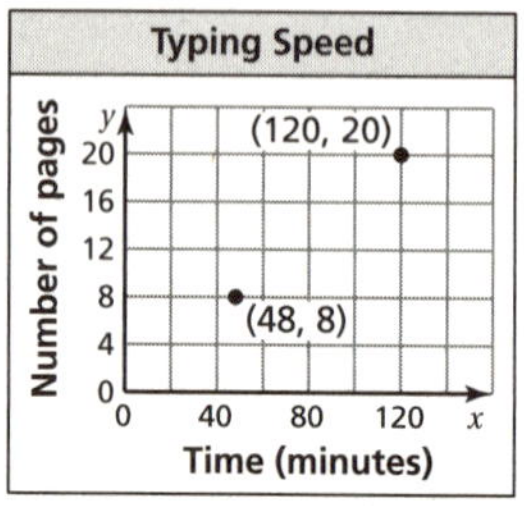

b. $\frac{1 \text{ page}}{6 \text{ minutes}}$ or $\frac{1}{6}$ page per minute

c. $y = \frac{1}{6}x$

3.1 Enrichment and Extension

The equations match up with the graph in the corresponding position.

3.1 Puzzle Time

MORE THAN THE LION

3.2 Start Thinking!

For use before Activity 3.2

Sample answer: Yes; you can plot the point and use the slope to graph the line. From the graph, you can write the equation.

3.2 Warm Up

For use before Activity 3.2

1.
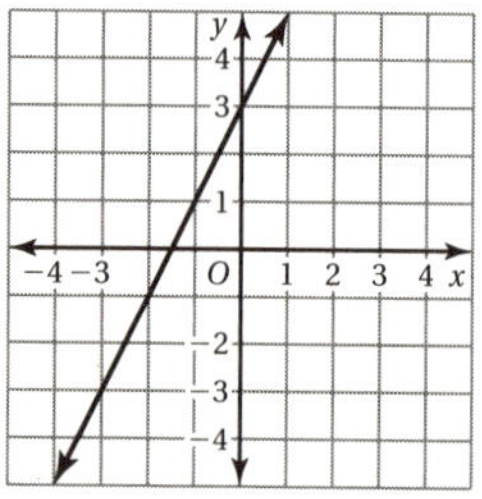

2.
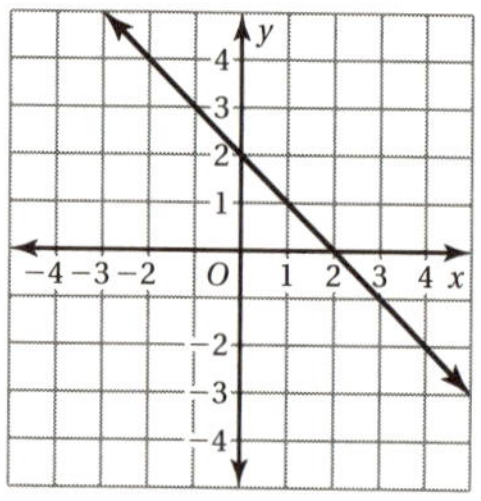

3.
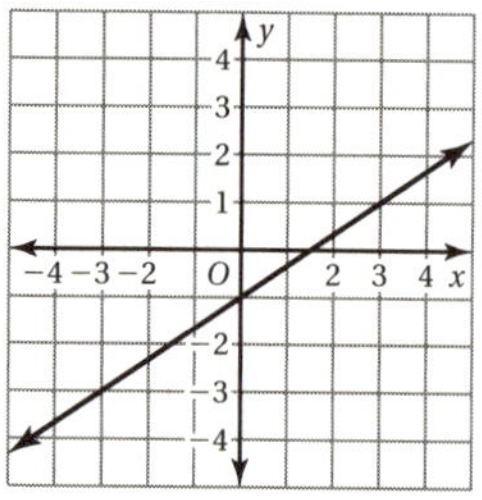

4.
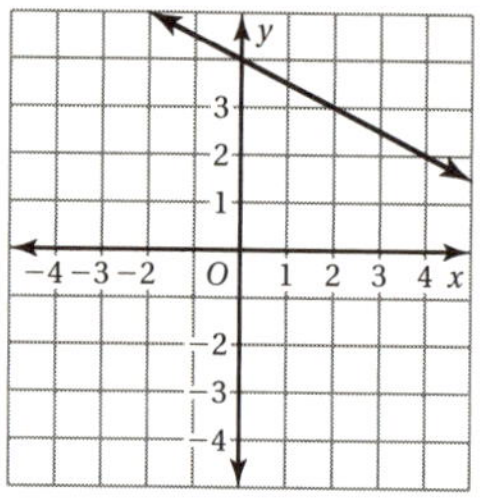

5.
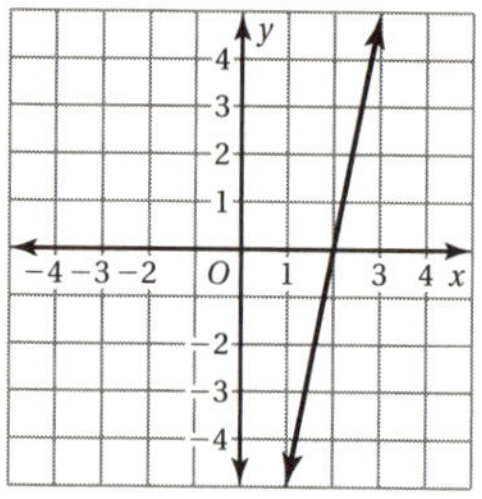

6.
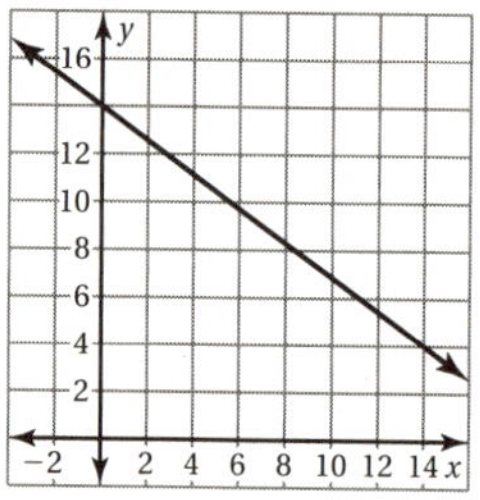

3.2 Start Thinking!

For use before Lesson 3.2

Sample answer: If you need to write the equation of a line given the slope and a point on the line, you can find the y-intercept (either graphically or algebraically). Then you follow the steps for writing an equation given the slope and y-intercept.

3.2 Warm Up

For use before Lesson 3.2

1. $y = 2x + 1$

2. $y = -x + 1$

3. $y = -\frac{1}{2}x - 2$

4. $y = \frac{2}{3}x - 1$

Answers

3.2 Practice A

1. $y = 3x + 4$
2. $y = -\frac{2}{3}x + 2$
3. $y = \frac{1}{4}x - 3$
4. $y = -\frac{4}{3}x + 1$
5. $y = -x + 4$
6. $y = 4x - 1$
7. a. $V = -150x + 900$ b. \$900 c. \$150
8. a. $A = 5t + 45$ b. \$45
   c. 11 weeks d. your friend

3.2 Practice B

1. $y = \frac{5}{4}x - 2$
2. $y = -4x + 3$
3. $y = \frac{1}{3}x + 5$
4. $y = -\frac{3}{4}x - 1$
5. $y = 2x - 3$
6. $y = -3x + 2$
7. a. $y = -2x + 24$
   b. 24 feet c. after 12 seconds
8. a. $y = 2x + 40$
   b. \$40 is the flat fee. c. 5 miles

3.2 Enrichment and Extension

1. The slope is the negative of the grade, written as a fraction or decimal.
2. Bunny Slope: $-\frac{1}{25}$, Medium Trail: $-\frac{3}{20}$,
   Expert Trail: $-\frac{3}{10}$
3. $y = -\frac{1}{25}x + 58$
4. 58; The starting height of the trail is 58 meters.
5. 1450; The length of the trail is 1450 meters.
6. $y = -\frac{3}{20}x + 1530$
7. $y = -\frac{3}{10}x + 6000$
8. no; It has a grade of 8%.

3.2 Puzzle Time

SQUASH

3.3 Start Thinking!

For use before Activity 3.3

Sample answer: After 1 second, the balloon is about 15 feet in the air. After 30 seconds, the balloon is about 100 feet in the air. Check students' graphs and equations. No, everyone will not write the same equation because the equations are dependent upon the original height estimates.

3.3 Warm Up

For use before Activity 3.3

1. $y = \frac{1}{2}x + 3$
2. $y = -2x$
3. $y = x - 3$
4. $y = \frac{1}{3}x - 1$
5. $y = -\frac{3}{5}x + \frac{1}{2}$
6. $y = -\frac{7}{5}x - \frac{1}{5}$

3.3 Start Thinking!

For use before Lesson 3.3

Sample answer: You can find the slope of the graph (\$.35/pound) and use it to find the y-intercept (\$1.88). Then you can write the equation $y = 0.35x + 1.88$.

3.3 Warm Up

For use before Lesson 3.3

1. $m = 1, b = -2, y = x - 2$
2. $m = \frac{1}{3}, b = 1, y = \frac{1}{3}x + 1$

3.3 Practice A

1. slope $= 1$; y-intercept: -1; $y = x - 1$
2. slope $= -\frac{2}{3}$; y-intercept: 4; $y = -\frac{2}{3}x + 4$
3. $y = \frac{1}{3}x - 3$
4. $y = -\frac{5}{2}x + 2$
5. $y = -2x - 4$
6. $y = 5x + 1$
7. a. $y = 5x + 30$ b. \$30
   c. \$5 per day d. $y = 5x + 42$
8. a. $y = 175x + 1150$
   b. The original price of the ring was \$1150.
   c. \$175 per year
   d. no; *Sample answer:* The constant rate will increase or decrease depending upon the economy.

Answers

3.3 Practice B

1. slope $= -3$; y-intercept: 1; $y = -3x + 1$

2. slope $= \frac{3}{2}$; y-intercept: 2; $y = \frac{3}{2}x + 2$

3. $y = 4x - 5$

4. $y = \frac{4}{5}x - 4$

5. $y = -\frac{1}{3}x + 2$

6. $y = \frac{5}{2}x - 1$

7. a. $y = -\frac{1}{2}x + 9$
   b. 9 inches
   c. 18 minutes after you begin draining it.
   d. no; This seems like an excessive amount of time. The drain may be clogged.

8. a.

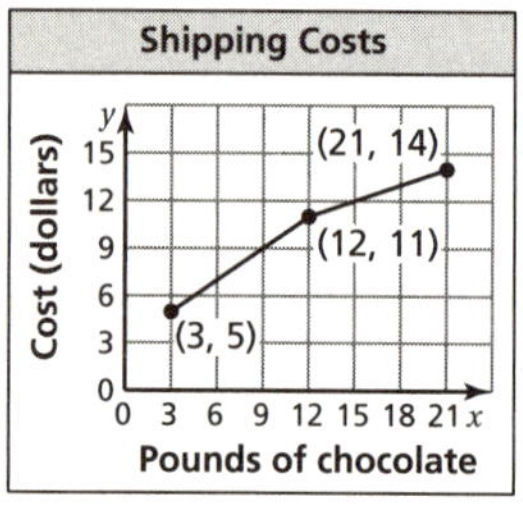

   b. no; The points do not lie on one line.
   c. $y = \frac{2}{3}x + 3$; $y = \frac{1}{3}x + 7$
   d. \$0.33 per pound

3.3 Enrichment and Extension

1. $y = -\frac{1}{3}x + 3\frac{1}{3}$

2. $y = -\frac{3}{2}x - \frac{1}{2}$

3. $y = \frac{1}{4}x + 1\frac{1}{4}$

4. $y = \frac{3}{5}x + 2\frac{1}{5}$

5. $y = \frac{4}{3}x - \frac{2}{3}$

6. $y = -\frac{3}{4}x - \frac{1}{4}$

3.3 Puzzle Time

THE TEAM SPIRIT

3.4 Start Thinking!

For use before Activity 3.4

Check students' equations and graphs. The slope represents the weekly allowance and the y-intercept represents the amount already saved.

3.4 Warm Up

For use before Activity 3.4

1. $m = 2, b = 1$

2. $m = -1, b = -2$

3.4 Start Thinking!

For use before Lesson 3.4

Some examples include: the amount of gas in the gas tank, the distance traveled at an average rate, the costs along the way.

3.4 Warm Up

For use before Lesson 3.4

1. *Sample answer:* The cost of buying music online is \$1 per song. You have \$25. How much money y do you have after buying x songs?

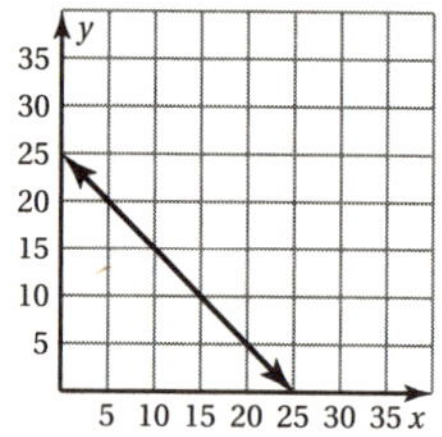

2. *Sample answer:* What is the height y of an elevator starting at 60 feet and traveling -10 feet per second?

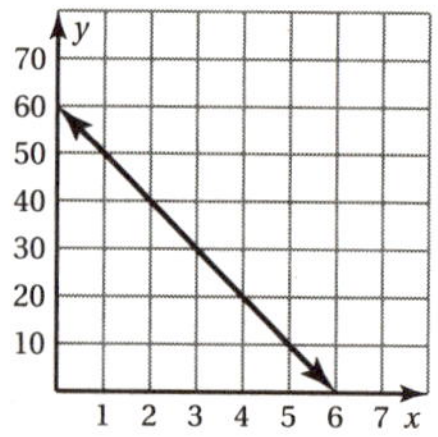

3.4 Practice A

1. *Sample answer:* You withdraw \$7 from your savings account each week to download music. At this rate, your savings account will have a zero balance after 22 weeks. How much money was originally in your savings account?

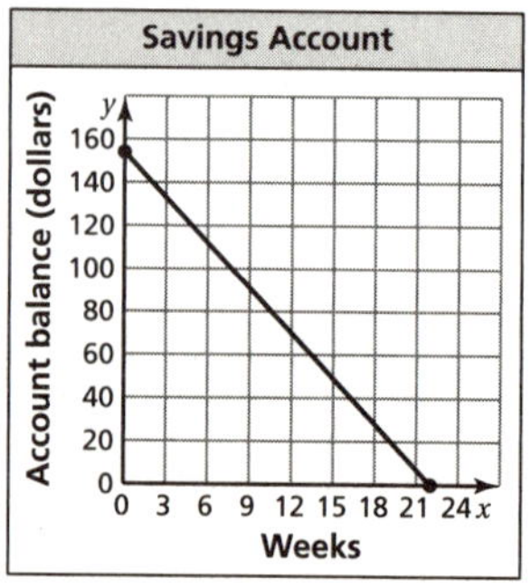

Answers

2. *Sample answer:* You are working in the family lemon groves and each day you pick the lemons from 3 trees. There are 27 trees in the orchard. When will you finish picking all of the lemons?

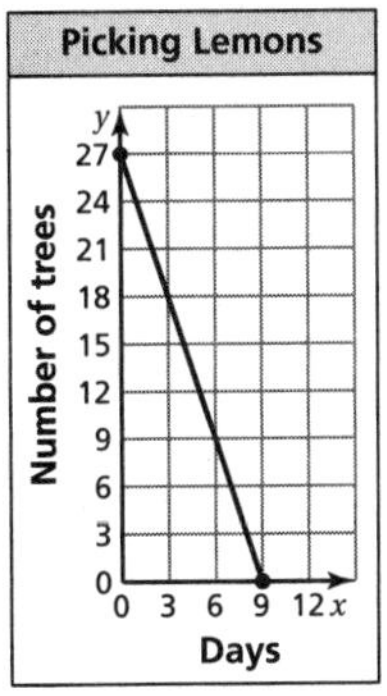

3. a.

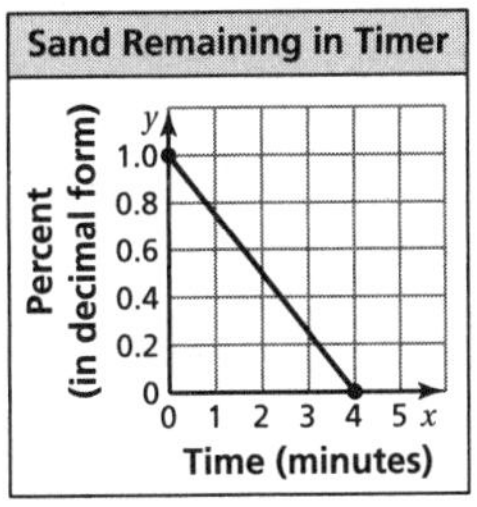

b. x-intercept: 4; The timer is empty after 4 minutes.
y-intercept: 1; The timer is 100% full after 0 minutes.

c. 3 minutes **d.** $y = -0.5x + 1$

4. a. The temperature of the water is 75°F at the beginning. The temperature of the water is 0°F after 45 minutes of cooling.

b. $-\frac{5}{3}$; The liquid cools $\frac{5}{3}$°F per minute (or the temperature drops 5°F every 3 minutes).

c. $y = -\frac{5}{3}x + 75$

d. after 12 minutes of cooling **e.** 25.8 minutes

3.4 Practice B

1. *Sample answer:* You are doing extra chores to save \$150 for a family trip. You will add \$15 per week to your savings account. In 8 weeks you will have reached your goal. How much money was originally in your savings account?

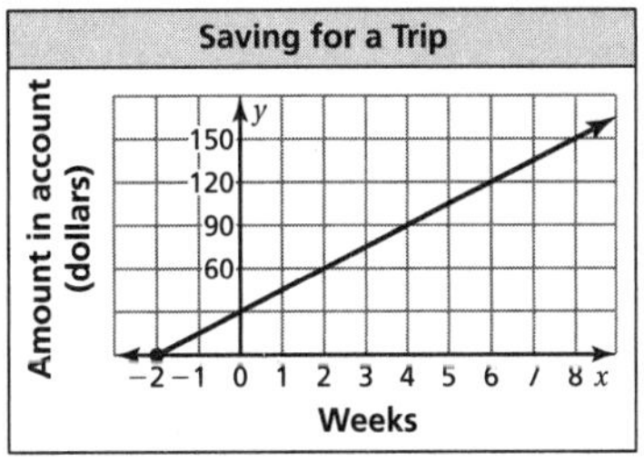

2. *Sample answer:* A day care center received a shipment of 15 gallons of juice. They use 1.5 gallons of juice each day. After how many days will the center run out of juice?

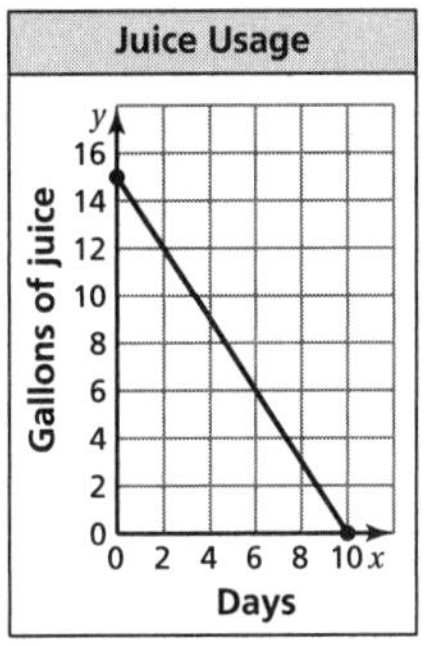

3. a. $R = 400 + 15x$ **b.** \$7900 **c.** \$4900

4. a.

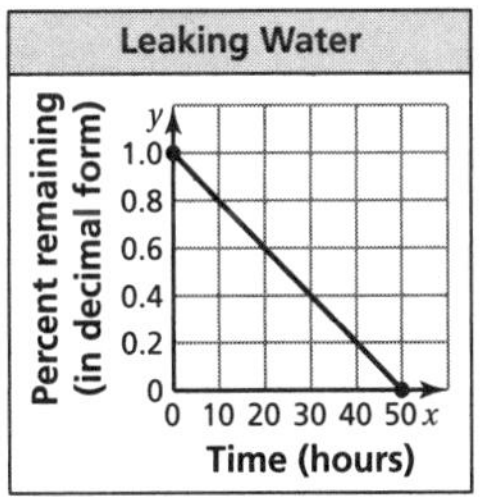

b. The bottle is 100% full at 0 hours. The bottle is empty after 50 hours.

c. 2% of the water leaks from the bottle each hour.

d. after 37.5 hours

5. a. 12 blocks **b.** 10 minutes **c.** $y = -\frac{6}{5}x + 12$

Answers

3.4 Enrichment and Extension

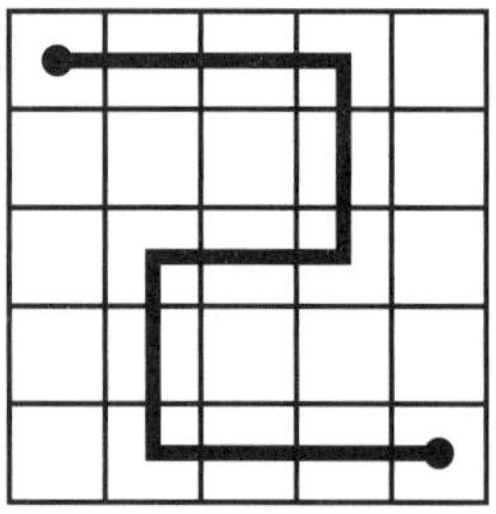

3.4 Puzzle Time

DISK-O

3.5 Start Thinking!

For use before Activity 3.5

Sample answer: You can write an equation for each runner's distance. The intersection of the graph is where the second runner passes the first runner.

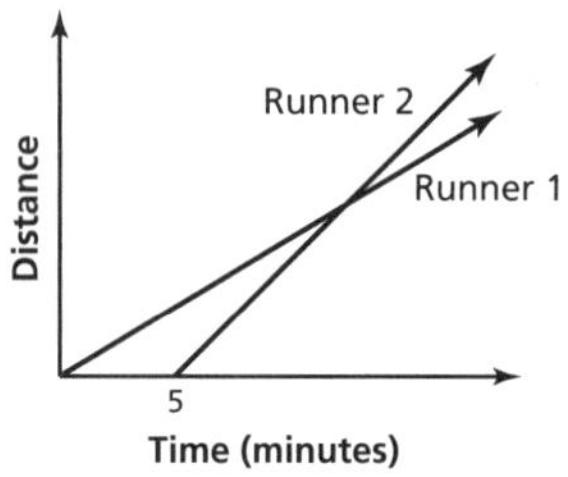

3.5 Warm Up

For use before Activity 3.5

1. (3, 7) **2.** (10, 8) **3.** (4, 1) **4.** (1, 6)

3.5 Start Thinking!

For use before Lesson 3.5

Monitor students playing game.

3.5 Warm Up

For use before Lesson 3.5

1. $2x + 2y = 60$

$y = 2x$

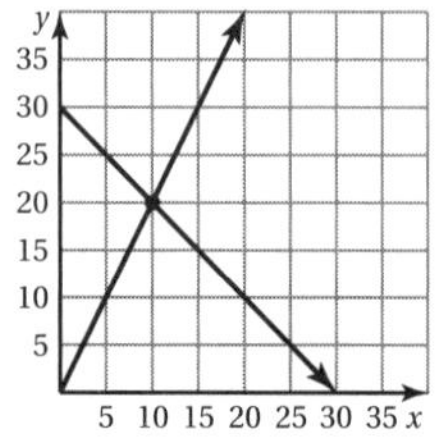

The length is 20 feet and the width is 10 feet.

3.5 Practice A

1. $x + y = 29$

$y = 2x + 5$

$x = 8,\ y = 21$

2. $\ell = 4w$

$2\ell + 2w = 30$

$\ell = 12$ ft, $w = 3$ ft

3. $x + y = 43$

$x = y - 7$

18 seniors, 25 juniors

4. $x + y = 7$

$9x + 15y = 87$

3 CDs, 4 DVDs

5. $2x + 3y = 3$

$x + 2y = 1.75$

pear: \$0.75; apple: \$0.50

6. $x + y = 12$

$0.1x + 0.25y = 2.25$

5 dimes and 7 quarters

3.5 Practice B

1. $x + y = 21$

$0.25x + 0.10y = 2.85$

5 quarters, 16 dimes

2. $x - y = 45$

$x = 2y - 30$

$x = 120,\ y = 75$

3. $x + y = 105$

$x = \frac{3}{4}y$

45 girls, 60 boys

4. a. $y = 55x$

$y = 70x - 120$

b. 8 hours **c.** Truck B

d. Truck A left first; It is farther from the warehouse after 2 hours.

5. The two lines do not intersect because they are parallel.

3.5 Enrichment and Extension

1. 24 chickens, 17 cats

2. 17 rats, 11 spiders

3. Not enough information because frogs and salamanders both have 4 legs.

4. 9 horses, 15 sparrows

5. 35 ants, 0 rabbits

6. yes; Some tables may only have 3 legs.

3.5 Puzzle Time

SHOCKED

School-to-Work

1. Let A represent the amount of fluid in the IV bag, in milliliters, and let m represent the number of minutes since the IV was started. The equation is $A = -34.5m + 600$

2. See Exercise 4 for graph. The A-intercept represents the amount of fluid in the bag when the IV is first started. The m-intercept represents the number of minutes it takes for the bag to empty.

3. About 17 minutes; 2:32 P.M.

4. $A = -51m + 700$

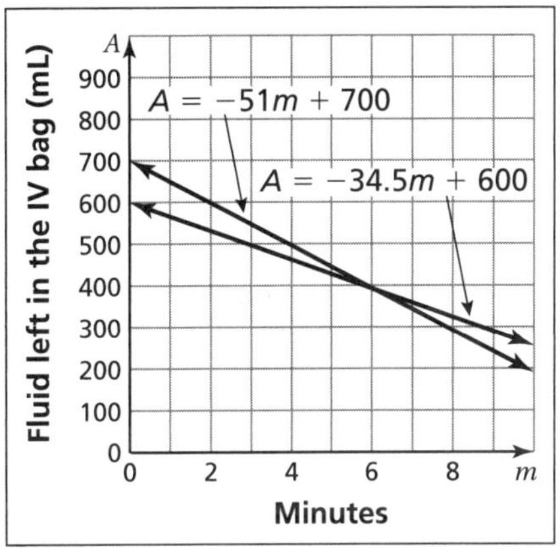

5. The golden retriever's IV; by about 3 min; you can see the difference between the m-intercepts; For the black lab, when $A = 0$, $0 = -34.5m + 600$, and $m \approx 17$. For the golden retriever, when $A = 0$, $0 = -51m + 700$, and $m \approx 14$. The difference is $17 - 14 = 3$ min.

6. The intersection tells you when the same amount of fluid is left in the two bags because both graphs have the same m and A values at that point.

Study Help

Available at *BigIdeasMath.com.*

Financial Literacy

1. Let a represent the number of adult tickets sold and let s represent the number of student tickets sold. The system is $a + s = 400$ and $8a + 5s = 2690$.

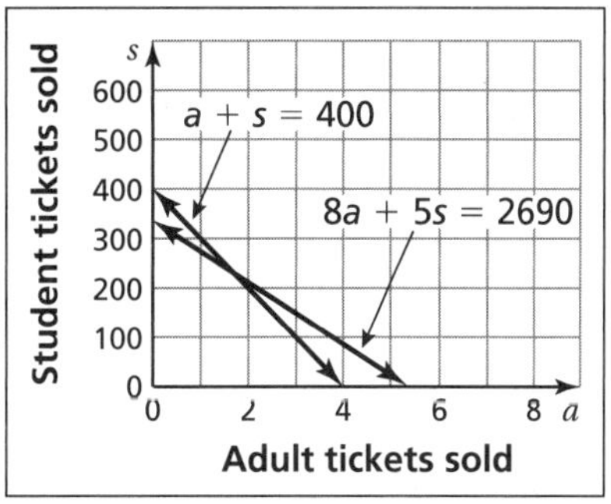

The a-intercept for the first equation gives the number of adult tickets sold when there are no student tickets sold. The s-intercept gives the number of student tickets sold when there are no adult tickets sold. The a-intercept for the second equation gives the amount of money made in ticket sales when there are no student tickets sold. The s-intercept gives the amount of money made in ticket sales when there are no adult tickets sold. The intersection represents the number of each type of ticket sold that will produce $2690 in ticket sales.

2. 230 adult tickets, 170 student tickets

3. $10.90 for an adult ticket, $7.90 for a student ticket; Let x represent the amount by which you need to raise each ticket price. Then $230(8 + x) + 170(5 + x) = 3850$ and $x = 2.9$.

4. Yes; The minimum amount made in ticket sales for the second show would result in all tickets being student tickets. The amount made would be $400(5) = 2000$. Since $2690 + 2000 > 3850$, the production cost will be covered; $5890

5. Adult matinee tickets cost $6 and student matinee tickets cost $3. If m represents the cost of a student matinee ticket and p represents the cost of an adult ticket, then $p = 2m$ and $230p + 170m = 1890$.

6. *Sample answer:* Students should include a graph and a system of equations in their answer. The reasoning for how they know that the production costs will be covered should also be included.

Answers

Cumulative Practice

1. $2h + 3 = 15$
2. $p + 48 = 7$
3. $4x - 3 = 11$
4. $v \div 25 = 13$
5. $18n = 27$
6. $y = 2x + 60$
7. $y = x + 1$
8. $y = -2x - 2$
9. $y = \frac{3}{2}x$
10. $y = \frac{5}{2}x + 3$
11. $y = x + 4$
12. $y = -\frac{3}{2}x + 6$
13. $y = \frac{1}{5}x - 3$
14. 4 horses, 24 sheep
15. 15 hardcover books, 5 paperback books

Chapter 4

4.1 Start Thinking!

For use before Activity 4.1

Domain is a list of input values. Range is a list of output values.

4.1 Warm Up

For use before Activity 4.1

1.

x	−2	−1	0	1	2
y	4	3	2	1	0

2.

x	−2	−1	0	1	2
y	−11	−9	−7	−5	−3

3.

x	0	1	2	3	4
y	3	3.5	4	4.5	5

4.

x	−6	−3	0	3	6
y	−3	−2	−1	0	1

4.1 Start Thinking!

For use before Lesson 4.1

$y = 40x + 150$; The domain is 1, 2, 3, 4, 5, and 6; the range is 190, 230, 270, 310, 350, and 390.

4.1 Warm Up

For use before Lesson 4.1

1. a. $y = -2x + 10$

   b. The domain is 0, 1, 2, 3, 4, and 5; the range is 10, 8, 6, 4, 2, and 0.

   c. *Sample answer:* You cannot buy 6 hard cover books with $10.

4.1 Practice A

1. domain: −1, 0, 1, 2; range: 0, 2, 4, 6
2. domain: −2, 0, 2, 4; range: 0, 3, 6
3. a. $y = -x + 6$

   b. domain: 0, 2, 4, 6; range: 0, 2, 4, 6

   c. 6 people

   d. Because there are 6 people on the committee, there cannot be 8 boys.

4.

x	−1	0	1	2
y	−7	−5	−3	−1

domain: −1, 0, 1, 2; range: −7, −5, −3, −1

5.

x	0	3	6	9
y	1	2	3	4

domain: 0, 3, 6, 9; range: 1, 2, 3, 4

6. a. domain: 1, 2, 3, 4; range: 1, 2, 4

   b. domain: number of race;
   range: finish position in each race (1st, 2nd)

   c. No; another boat could have $1 + 1 + 2 = 4$ points.

4.1 Practice B

1. domain: −2, 0, 2, 4; range: −3, −1, 1, 3
2. domain: −2, −1, 0, 1, 2; range: −2, 0, 2
3. a. $y = -2x + 8$

   b. domain: 0, 1, 2, 3, 4; range: 0, 2, 4, 6, 8

   c. 4 packages

   d. Limes cost more. They are $4 per package while lemons are $2 per package.

Answers

4.

x	0	3	6	9
y	4	2	0	–2

domain: 0, 3, 6, 9; range: –2, 0, 2, 4

5.

x	–1	0	1	2
y	–4.2	–3	–1.8	–0.6

domain: –1, 0, 1, 2; range: –4.2, –3, –1.8, –0.6

6. a. $y = 0.6x$

b.

x	125	150	175	200	225
y	75	90	105	120	135

c. domain: 125, 150, 175, 200, 225
range: 75, 90, 105, 120, 135

d. $y = 0.75x$

4.1 Enrichment and Extension

1. is a function; linear; range is $y \geq 0$; domain is $x \geq 0$
2. is a function; is non-linear; range is $y \leq 0$
3. is a function; is non-linear; range is $y \geq 0$
4. is not a function; is non-linear; range is $-1 \leq y \leq 1$; domain is $-1 \leq x \leq 1$

4.1 Puzzle Time

PLAYING BY EAR

4.2 Start Thinking!

For use before Activity 4.2

Sample answer: A reasonable cost of one apple is \$.50 and the cost of a pound of apples is \$1.50. The two functions are $y = 0.5x$ and $y = 1.5x$.

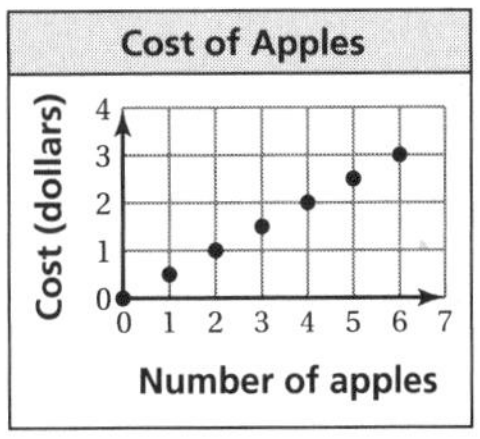

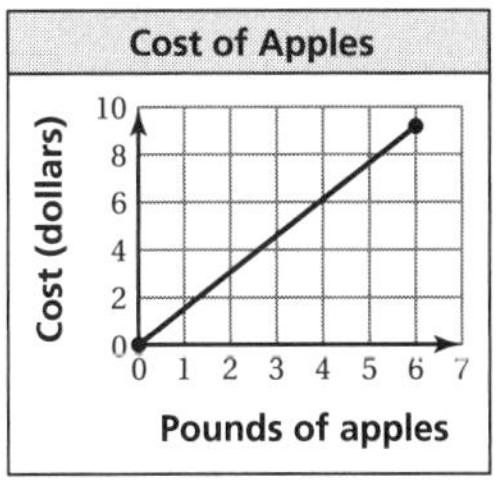

The graph of the cost of x apples is a set of distinct points. The graph of the cost of x pounds of apples is a line segment. You cannot buy fractions of apples, but you can buy fractions of pounds of apples.

4.2 Warm Up

For use before Activity 4.2

1. The domain is 0, 1, 2, and 3; the range is 1, 2, 3, and 4.
2. The domain is $-2, -1, 0, 1,$ and 2; the range is $1, -2,$ and -4.

4.2 Start Thinking!

For use before Lesson 4.2

B and E are discrete. A, C, and D are continuous.

4.2 Warm Up

For use before Lesson 4.2

1. The domain is 0, 1, 2, 3, 4, and 5; the range is 2, 3, 4, 5, 6, and 7; discrete
2. The domain is the set of all numbers from 0 to 5; the range is the set of all numbers from 2 to 7; continuous.

4.2 Practice A

1. domain: 0, 1, 2, 3, 4, 5; range: 0, 1, 2, 3, 4, 5; discrete
2. domain: $x \geq 0$ and $x \leq 5$; range: $y \geq 0$ and $y \leq 5$; continuous
3. continuous

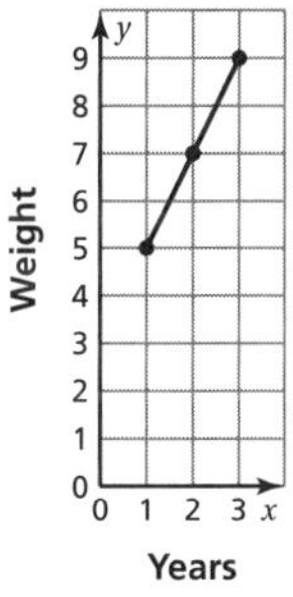

4. discrete

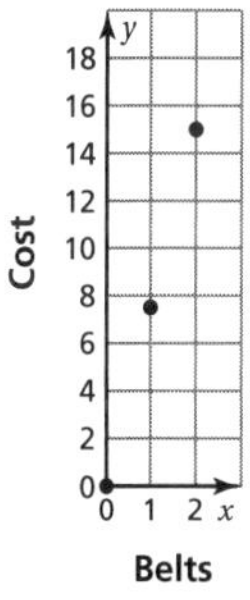

5. discrete

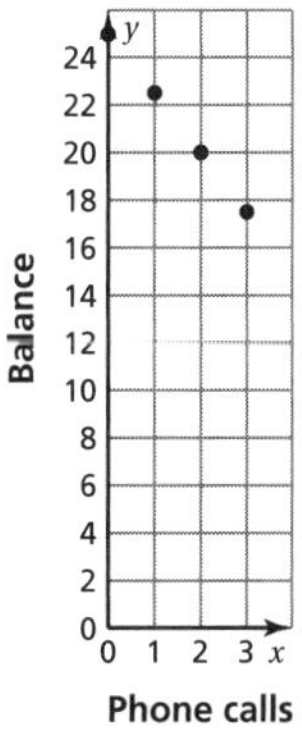

6. continuous

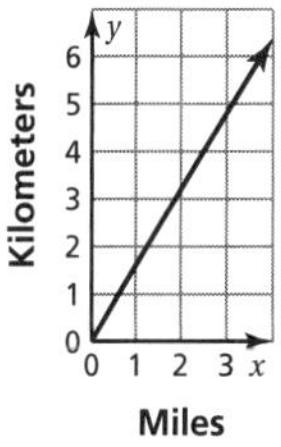

7. All x-values from 0 to 3 are included, so the domain is continuous.

Answers

8. **a.** Yes; when $y = 2$, $x = 6$.

   b. No; when $x = 2$, $y = 3\frac{1}{3}$. You cannot purchase $3\frac{1}{3}$ tubes of sunscreen.

4.2 Practice B

1. domain: 0, 1, 2, 3, 4; range: 0, 3, 4; discrete

2. domain: $x \geq 0$ and $x \leq 4$; range: $y \geq 0$ and $y \leq 4$; continuous

3. continuous

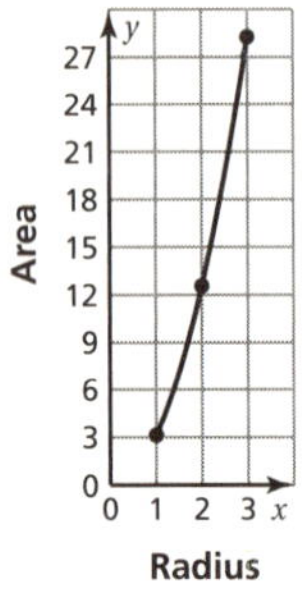

4. discrete

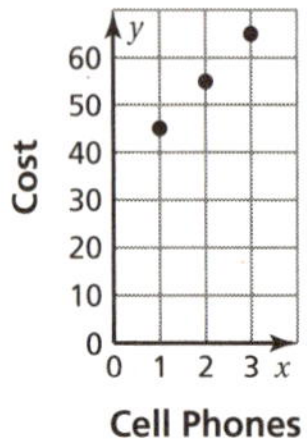

5. The function whose input is number of accidents; you can only have a whole number of accidents.

6. continuous

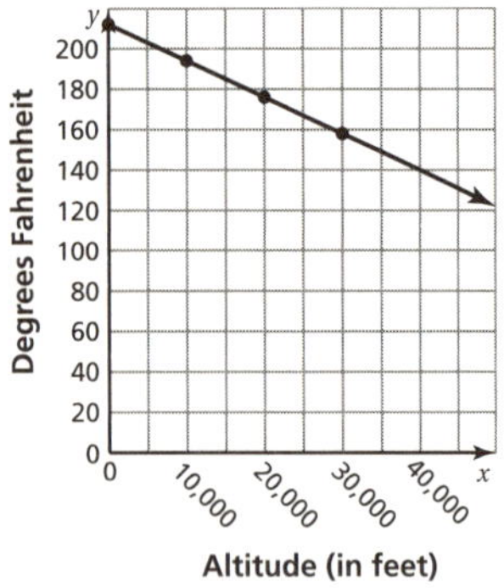

7. **a.** $C = 2\pi r$

   b.

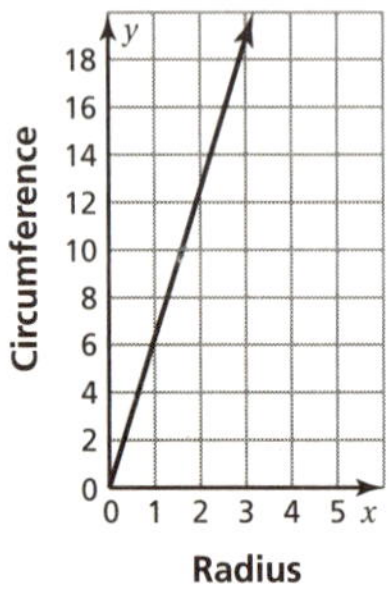

   c. continuous **d.** No, radius cannot be negative.

   e. Yes, $r \approx 0.96$.

4.2 Enrichment and Extension

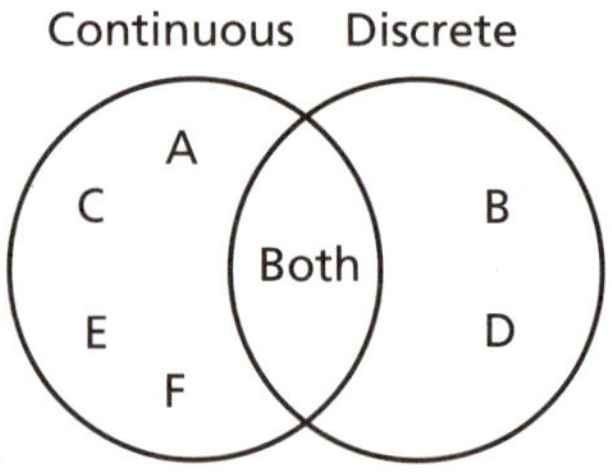

1. *Sample answer:* The graphs are lines.

2. *Sample answer:* The graphs are individual points.

3. no

4. This is not possible. Every domain is either continuous or discrete but cannot be both.

4.2 Puzzle Time

BUSHED

4.3 Start Thinking!

For use before Activity 4.3

Answers will vary. Check students' sketches, tables, and graphs.

4.3 Warm Up

For use before Activity 4.3

1. $y = \frac{1}{3}x + 2$ 2. $y = -3x + 7$

4.3 Start Thinking!

For use before Lesson 4.3

Sample answer: A linear function has a graph in which the points fall on a line. The function $y = x - 2$ is an example of a linear function.

4.3 Warm Up

For use before Lesson 4.3

1. $y = 4x$

2. The variable x represents the side length of a square; the variable y represents the perimeter.

3.

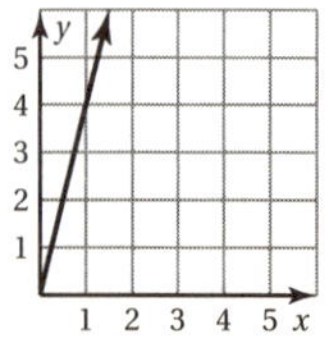

Answers

4.3 Practice A

1. $y = -2x + 1$

2. $y = \frac{2}{3}x$

3. $y = -3x$

4. $y = \frac{x}{5}$

5. a.

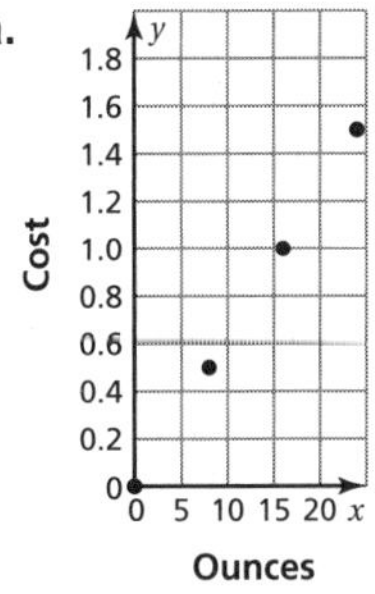

b. discrete

c. $y = \frac{1}{16}x$

d. $2

6. a.

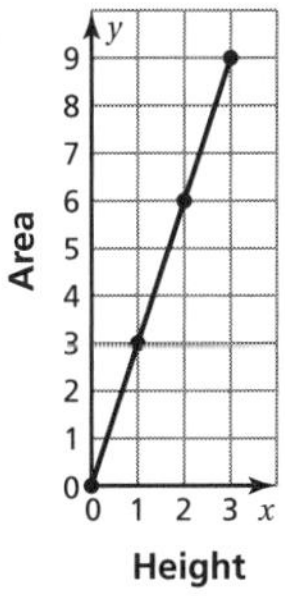

b. continuous

c. $y = 3x$

d. 6 ft

4.3 Practice B

1. $y = -\frac{1}{4}x - 1$

2. $y = -1$

3. $y = -2x$

4. $y = \frac{2}{5}x + 3$

5. a.

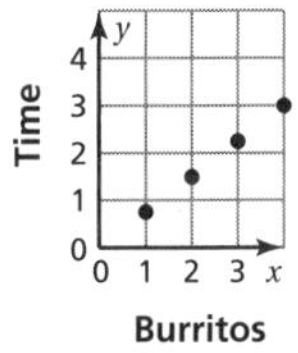

b. discrete **c.** $y = 0.75x$ **d.** 5.25 min

6. a.

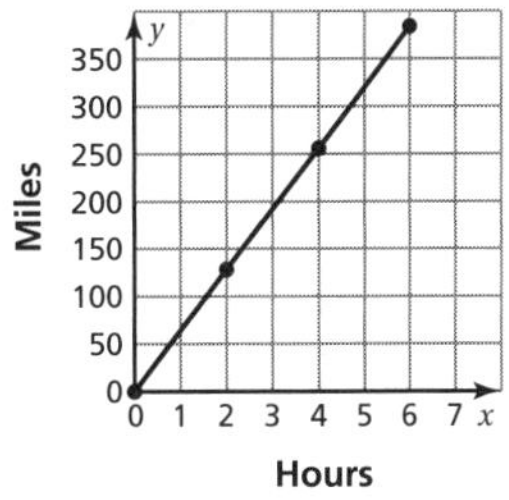

b. continuous **c.** $y = 64x$ **d.** 320 mi

e. $r = 64$ mi/h **f.** $6\frac{1}{4}$ h

4.3 Enrichment and Extension

1. At iteration 0, you have a completely shaded box. To go from iteration 0 to iteration 1, you split the box into 9 smaller blocks, keep the box in the upper left corner, and delete every other box thereafter.

2. Each of the smaller boxes in iteration 1 is similar to the original box in iteration 0.

3.

Iteration	Number of Squares
0	1
1	5
2	25
3	125
4	625

4. At each iteration, the number of squares multiplies by 5; After the first iteration, the iteration number tells you how many factors of 5 to multiply together to determine the number of squares.

5. no; There is no number that can be added to the number of squares each time to produce the next amount of squares.

6. The pattern will repeat an infinite number of times. Part of the area of the original box will always remain. So, you can keep applying the procedure of deleting some boxes and keeping others.

4.3 Puzzle Time

FROSTBITE

4.4 Start Thinking!

For use before Activity 4.4

The height of a person on an escalator is linear. The others are nonlinear. Check students' graphs.

4.4 Warm Up

For use before Activity 4.4

1. 20 cm; 24 cm^2

2. 24 m; 20 m^2

3. 56 ft; 192 ft^2

4. 40 in.; 91 $in.^2$

4.4 Start Thinking!

For use before Lesson 4.4

Sample answer: The cost of admission for x people is linear. The length of time you wait in line at each ride is nonlinear.

Answers

4.4 Warm Up

For use before Lesson 4.4

1. 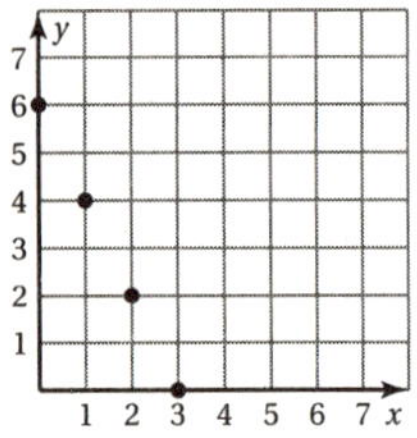

linear

2. 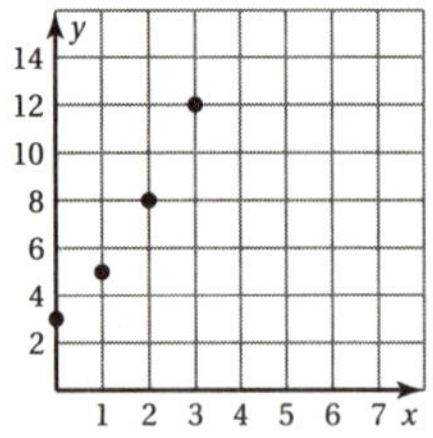

nonlinear

3. 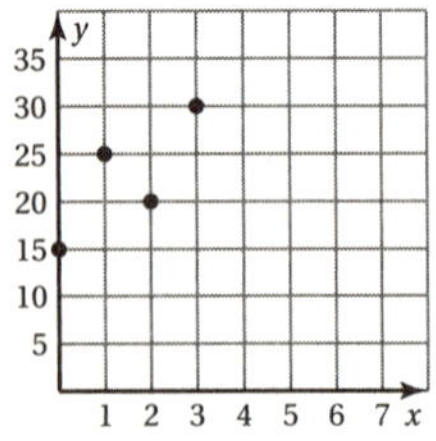

nonlinear

4. 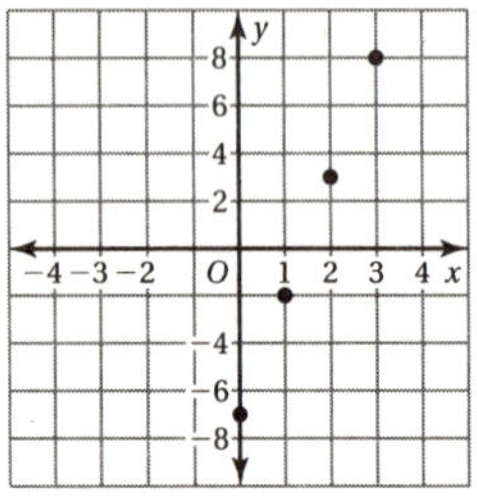

linear

4.4 Practice A

1. linear

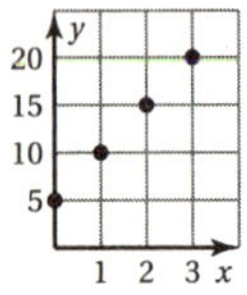

2. nonlinear

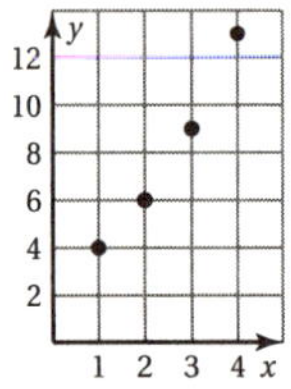

3. Nonlinear; the graph is not a line.

4. Linear; the graph is a line.

5. Nonlinear; the rate of change is not constant.

6. Linear; the rate of change is constant.

7. Nonlinear; the rate of change is not constant.

8. a. *Sample answer:* 1 b. 1.5

   c. $y = 0.25x - 0.5$

4.4 Practice B

1. nonlinear

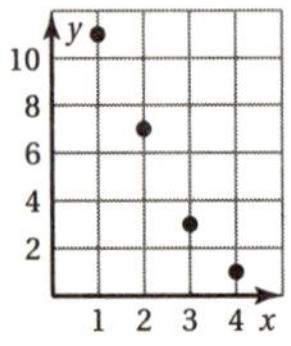

2. linear

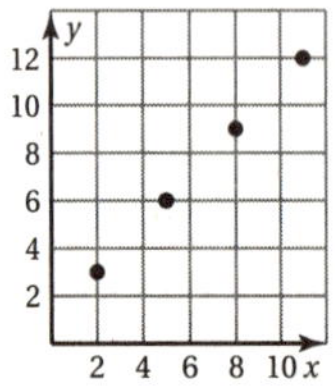

3. Linear; the graph is a line.

4. Nonlinear; the graph is not a line.

5. Nonlinear; the equation cannot be written in slope-intercept form.

6. Linear; the equation can be written in slope-intercept form.

7. Linear; the rate of change is constant.

8. a. Nonlinear; commission does not show constant change.

   b. \$600

9. Nonlinear; the equation cannot be written in slope-intercept form.

4.4 Enrichment and Extension

1. $F = \dfrac{775}{r}$

2. 775

3.

Radius (cm), *r*	Frequency (hertz), *F*
5	155
10	77.5
15	$51.\overline{6}$
20	38.75
25	31

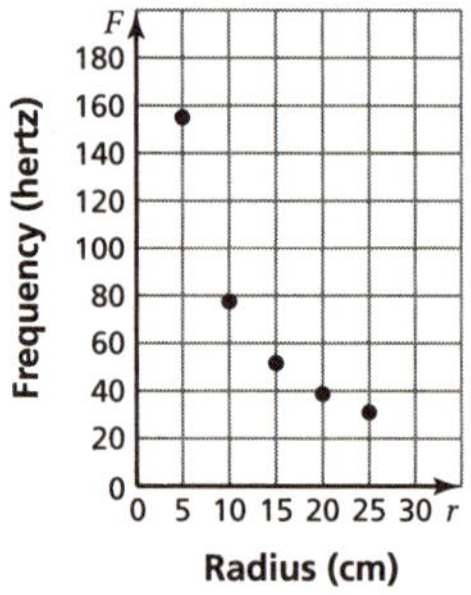

4. inverse variation; The equation is of the form $y = \dfrac{k}{x}$.

5. The graph of the data is not a straight line and it does not pass through the origin. This leads me to believe that the data does not reflect a direct variation which means my choice is probably correct.

6. negative values and zero; Because you cannot have a negative or zero radius.

7. $r > 0$; yes; Because the graph is in the first quadrant.

4.4 Puzzle Time

YOUR NAME

Answers

School-to-Work

1.

Original price	Sale price
\$15.00	\$12.00
\$28.00	\$22.40
\$39.00	\$31.20
\$42.00	\$33.60

2. $s = 0.8p$; continuous; All points along the line make sense in the context of the problem.

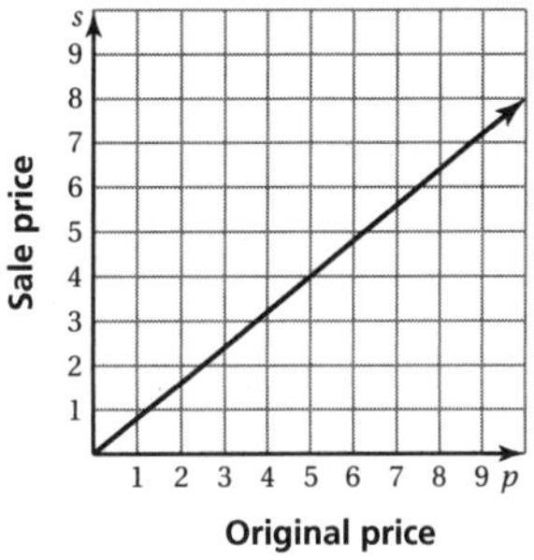

3. \$28; $s = 0.8(35) = 28$

4. \$26.20s; $s = 0.8p - 5$

5. Yes; For the jeans, applying the \$5 off first would result in a final cost of $0.8(39 - 5) = \$27.20$, which is not equal to \$26.20; $s = 0.8(p - 5)$.

Study Help

Available at *BigIdeasMath.com.*

Cumulative Practice

1. As the input increases by 3, the output increases by 3.

2. As the input increases by 1, the output increases by 2.

3. As the input decreases by 1, the output increases by 3.

4. $y = 2x - 1$; discrete

5. $y = \frac{1}{2}x - 3$; continuous

6. $y = -3x + 1$; continuous

7. $y = x - 7$

8. $y = -\frac{1}{2}x + 1$

9. $y = \frac{2}{3}x + 4$

10. The function is nonlinear because the radius is being squared.

11. discrete

Chapter 5

5.1 Start Thinking!

For use before Activity 5.1

An acute angle is an angle whose measure is less than 90°. An obtuse angle is an angle whose measure is greater than 90° and less than 180°. Acute means sharp and obtuse means dull. An acute angle has a sharp point and an obtuse angle does not.

5.1 Warm Up

For use before Activity 5.1

1.

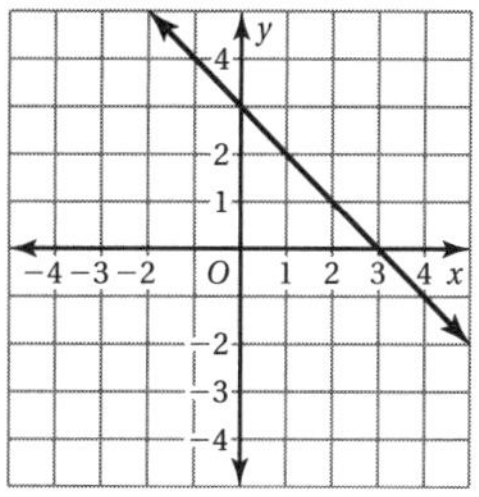

2.

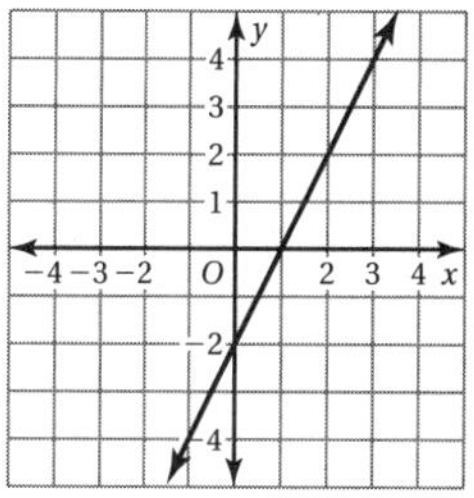

3.

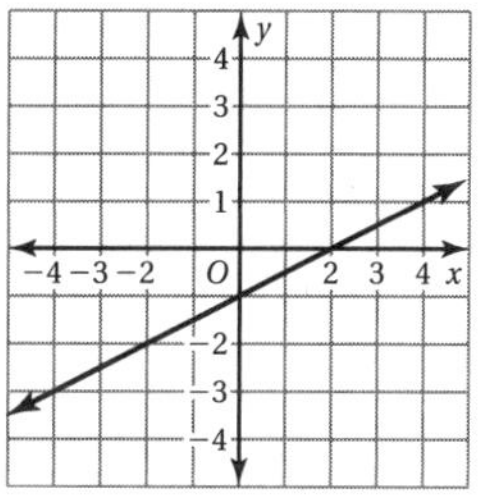

4.

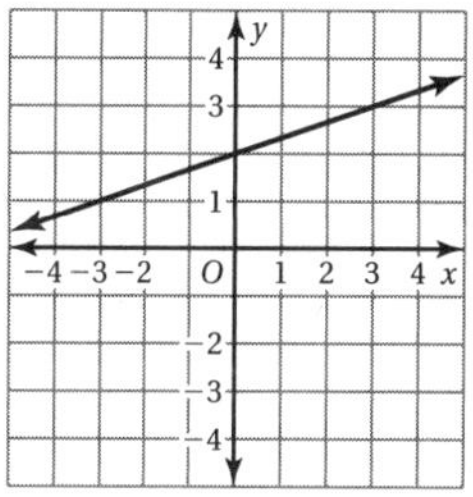

5.

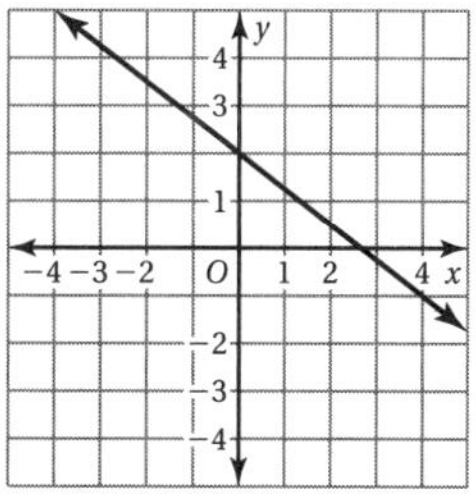

6.

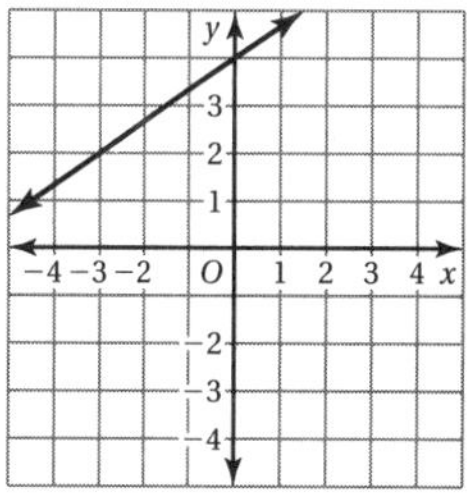

5.1 Start Thinking!

For use before Lesson 5.1

complementary; supplementary; Answers will vary. *Sample answers:* A letter S is like two C's put together, so the sum of supplementary angles is twice as much as complementary angles. "It's *right* to give a *compliment.*" C comes before S in the alphabet and 90 comes before 180 on a number line.

Answers

5.1 Warm Up

For use before Lesson 5.1

1. sometimes
2. always
3. never
4. sometimes

5.1 Practice A

1. Sometimes; y could be right or acute.
2. Never; x must be right or acute.
3. complementary
4. neither
5. neither
6. supplementary
7. $x = 148$
8. $x = 142$
9. $x = 49$
10. $x = 39$
11. angle x: 52°; angle y: 38°

5.1 Practice B

1. neither
2. complementary
3. supplementary
4. supplementary
5. $x = 123$
6. $x = 10$
7. $x = 45$
8. $x = 29$
9. 60°
10. $c = 90 - x$; The domain and range are both all values greater than 0 and less than 90 because you cannot have negative angles or an angle of 0 degrees. Ninety is not included because you cannot find another angle greater than zero that is complementary to 90.

    $s = 180 - x$; The domain and range are both all values greater than 0 and less than 180 because you cannot have negative angles or an angle of 0 degrees. 180 is not included because you cannot find another angle greater than zero that is supplementary to 180.

5.1 Enrichment and Extension

1. Angle A is an acute angle.

2–5.

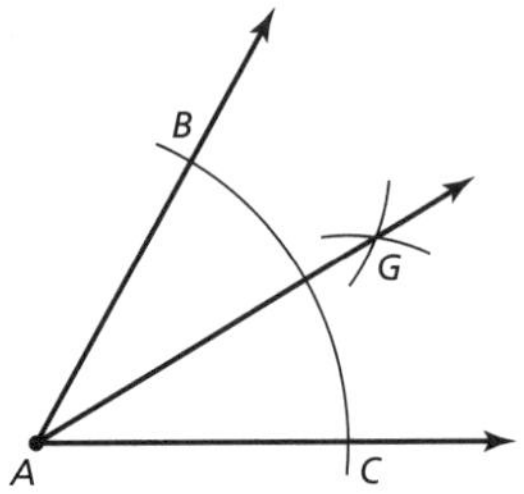

6. Angle A measures 60°. Each angle formed from the bisection measures 30°.

5.1 Puzzle Time

A SNEAKER

5.2 Start Thinking!

For use before Activity 5.2

Sample answer: A yield sign is an equilateral triangle.

5.2 Warm Up

For use before Activity 5.2

1. right
2. acute
3. straight
4. obtuse

5.2 Start Thinking!

For use before Lesson 5.2

Sample answer: construction, bridge building

5.2 Warm Up

For use before Lesson 5.2

1. acute isosceles
2. right isosceles
3. obtuse
4. obtuse isosceles
5. right
6. equilateral equiangular

5.2 Practice A

1. equiangular, equilateral, acute
2. obtuse, scalene
3. $x = 72$; right, scalene
4. $x = 65$; isosceles, acute
5. no; 35.9°, 110.4°, 33.7°
6. yes
7. **a.** 45, 90, 45

   b. Every triangle has one 90° angle and two 45° angles.

   c. An isosceles right triangle has two 45° angles.

5.2 Practice B

1. $x = 102$; obtuse, scalene
2. $x = 60$; equiangular, equilateral, acute
3. $x = 28$; isosceles, obtuse
4. $x = 12$; right, scalene
5. yes
6. no; 24.9°, 121.4°, 33.7°

Answers

7. **a.** 60°, 60°, 60°
   b. equiangular, equilateral, acute
   c. *Sample answer:* 110°, 35°, 35°; isosceles, obtuse
   d. *Sample answer:* 40°, 70°, 70°; isosceles, acute
   e. It is an isosceles triangle.

5.2 Enrichment and Extension

1. $\frac{4}{5}$ **2.** $\frac{12}{5}$ **3.** $\frac{3}{5}$

4. $\frac{4}{3}$ **5.** $\frac{12}{13}$ **6.** $\frac{5}{13}$

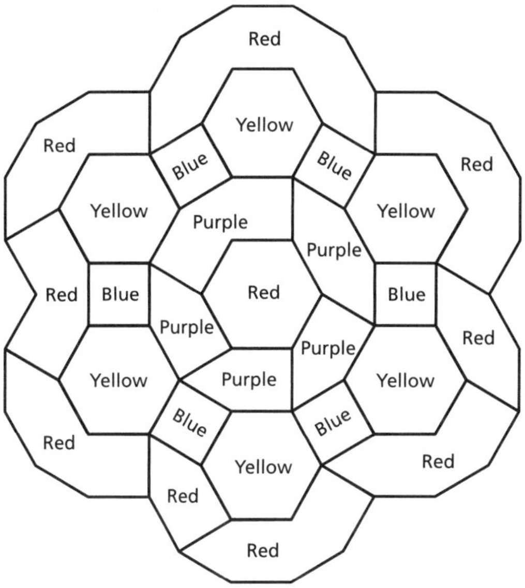

5.2 Puzzle Time

SHE WANTED TO LEARN TO READ BETWEEN THE LIONS

5.3 Start Thinking!

For use before Activity 5.3

Sample answers:

stop: octagon

yield: equilateral triangle

no passing zone: isosceles triangle

speed limit: quadrilateral or rectangle

railroad crossing: 12-gon

school advance warning: pentagon

warning: quadrilateral or square

5.3 Warm Up

For use before Activity 5.3

1. −4 **2.** −13 **3.** 80

4. −26 **5.** −0.5 **6.** −3

5.3 Start Thinking!

For use before Lesson 5.3

Each angle is made up of a right angle $(90°)$ from a grid square, plus half a right angle $(45°)$ from half a grid square. So, the measure of each angle of the octagon is $90 + 45 + 135°$. The sum of the eight angles is $8(135) = 1080°$. This agrees with the result using the equation from Activity 2: $180(8 - 2) = 1080°$.

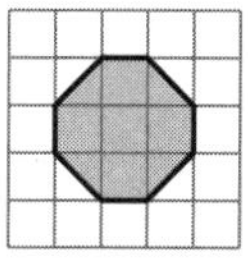

5.3 Warm Up

For use before Lesson 5.3

1. 360° **2.** 1080° **3.** 720°

4. 360° **5.** 540° **6.** 180°

5.3 Practice A

1. 720° **2.** 540° **3.** 1080° **4.** 900°

5. Yes, the sum of the angle measures is 1080°.

6. $x = 198$ **7.** $x = 120$ **8.** 135°

9. Convex; No line segment connecting two vertices lies outside the polygon.

10. Concave; A line segment connecting two vertices lies outside the polygon.

5.3 Practice B

1. 900° **2.** 1080° **3.** 1440°

4. 1260° **5.** 110° **6.** $x = 105$

7. $x = 99$ **8.** 8 sides

9. **a.** concave **b.** not regular
c. heptagon **d.** 900°

5.3 Enrichment and Extension

1. 440 cm^2 **2.** 1254 mm^2

3. 14 $in.^2$ **4.** 480 yd^2

5.3 Puzzle Time

WHEN SHE GOT THERE THE CUPBOARD WAS BEAR

Answers

5.4 Start Thinking!

For use before Activity 5.4

B and C; D and G; they are the same shape and appear to have the same proportions.

5.4 Warm Up

For use before Activity 5.4

1. $x = 2$ **2.** $x = 16$

5.4 Start Thinking!

For use before Lesson 5.4

The sun's rays form congruent angles with the ground. When his shadow is equal to his height, then the sun is forming a $45°$ angle with the ground, making at $45°$-$45°$-$90°$ triangle. The triangle formed by the pyramid and its shadow is a similar $45°$-$45°$-$90°$ triangle. So, the height of the pyramid is the length of the shadow, plus half of the length of the base (to measure the height from the center of the pyramid).

5.4 Warm Up

For use before Lesson 5.4

1–2. Answers will vary. Check students' drawings and ratios.

5.4 Practice A

1. Not similar; Corresponding angles are not congruent.

2. Similar; Corresponding angles are congruent.

3. $x = 52$

4. a. $\angle B$ is congruent to itself. $\angle A$ and $\angle D$ have the same line of sight, and so they are congruent. Because two angles are congruent, the third angles are congruent. Because the triangles have the same angle measures, they are similar.

b. $\frac{5}{x} = \frac{7}{42}$; $7x = 210$; $x = 30$ ft;
$30 \text{ ft} + 5 \text{ ft} = 35 \text{ ft}$;
The height of the building is 35 feet.

5. a. $\angle B$ and $\angle E$ are right angles, and thus congruent. $\angle BCA$ and $\angle ECD$ are vertical angles, and so they are congruent. Because two angles are congruent, the third angles are congruent. Because the triangles have the same angle measures, they are similar.

b. $\frac{x}{8} = \frac{9}{3}$; $3x = 72$; $x = 24$ yd

5.4 Practice B

1. Not similar; Corresponding angles are not congruent.

2. Similar; Corresponding angles are congruent.

3. $x = 76$

4. a. $\angle B$ is congruent to itself. $\angle A$ and $\angle D$ have the same line of sight, and so they are congruent. Because two angles are congruent, the third angles are congruent. Because the triangles have the same angle measures, they are similar.

b. $\frac{3}{x} = \frac{5}{22}$; $5x = 66$; $x = 13.2$ ft;
$13.2 \text{ ft} + 4.5 \text{ ft} = 17.7 \text{ ft}$; The height of the statue is 17.7 feet.

5. a. $\angle B$ and $\angle E$ are right angles, and thus congruent. $\angle BCA$ and $\angle ECD$ are vertical angles, and so they are congruent. Because two angles are congruent, the third angles are congruent. Because the triangles have the same angle measures, they are similar.

b. $\frac{0.2}{x} = \frac{0.1}{0.3}$; $0.1x = 0.06$; $x = 0.6$ mi

5.4 Enrichment and Extension

1. $2°, 89°, 89°$

2. yes; One angle is $2°$ because of vertical angles, and the other two are $89°$ because it is isosceles.

3. 4 : 1

4. focal length of eyepiece: 5 cm; focal length of objective lens: 20 cm

5. 5 times

6. decrease; The focal length of the eyepiece would be longer, so $\frac{250}{f}$ would be smaller.

5.4 Puzzle Time

SEA WEED

5.5 Start Thinking!

For use before Activity 5.5

Sample answers: vertical window blinds, lines formed by floor tiles, top and bottom of walls, the edges of rectangular objects.

Answers

5.5 Warm Up

For use before Activity 5.5

1. $x = 48$ 2. $x = 12.5$ 3. $x = 10$

4. $x = 5.5$ 5. $x = 32$ 6. $x = \frac{15}{8}$

5.5 Start Thinking!

For use before Lesson 5.5

Yes; you know that three of the angles are congruent to the given angle. The other four angles are supplementary to the given angle (or congruent angles) because they form straight angles.

5.5 Warm Up

For use before Lesson 5.5

1. k and ℓ 2. m 3. 8

4. $\angle 1 \cong \angle 3 \cong \angle 6 \cong \angle 8$; $\angle 2 \cong \angle 4 \cong \angle 5 \cong \angle 7$

5.5 Practice A

1. $m\angle 1 = 115°$; $m\angle 2 = 65°$

2. $m\angle 1 = 106°$; $m\angle 2 = 74°$

3. $m\angle 1 = m\angle 3 = 139°$ (supplementary)
   $m\angle 2 = 41°$ (vertical)
   $m\angle 4 = m\angle 6 = 139°$ (corresponding)
   $m\angle 5 = m\angle 7 = 41°$ (corresponding)

4. $m\angle 6 = m\angle 7 = 63°$ (supplementary)
   $m\angle 5 = 117°$ (vertical)
   $m\angle 1 = m\angle 3 = 117°$ (corresponding)
   $m\angle 2 = m\angle 4 = 63°$ (corresponding)

5. 160°; $\angle 1$ and $\angle 5$ are corresponding angles.

6. 37°; $\angle 6$ and $\angle 4$ are vertical angles.

7. 98°; $\angle 8$ and $\angle 4$ are corresponding angles, and $\angle 4$ and $\angle 3$ are supplementary angles.

8. 120°; $\angle 4$ and $\angle 5$ are supplementary angles.

9. $\angle 2$ is congruent to $\angle 4$. $\angle 4$ is congruent to $\angle 8$. Thus, $\angle 2$ is congruent to $\angle 8$.

10. $\angle 6$ is supplementary to $\angle 3$. $\angle 3$ is congruent to $\angle 1$. So, $\angle 6$ is supplementary to $\angle 1$.

5.5 Practice B

1. $m\angle 6 = m\angle 7 = 32°$ (supplementary)
   $m\angle 5 = 148°$ (vertical)
   $m\angle 1 = m\angle 3 = 148°$ (corresponding)
   $m\angle 2 = m\angle 4 = 32°$ (corresponding)

2. $m\angle 1 = 90°$; $\angle 1$ and the given angle are vertical angles. $m\angle 2 = 90°$ and $m\angle 3 = 90°$; $\angle 2$ and $\angle 3$ are supplementary to the given angle. $m\angle 6 = 90°$; $\angle 6$ and the given angle are corresponding angles. $m\angle 4 = 90°$; $\angle 4$ and $\angle 6$ are vertical angles. $m\angle 5 = 90°$ and $m\angle 7 = 90°$; $\angle 5$ and $\angle 7$ are supplementary to $\angle 6$.

3. 130°; $\angle 1$ and $\angle 8$ are corresponding angles.

4. 127°; $\angle 5$ and $\angle 4$ are corresponding angles, and $\angle 4$ and $\angle 3$ are supplementary angles.

5. 109°; $\angle 7$ and $\angle 3$ are supplementary angles.

6. 115°; $\angle 4$ and $\angle 5$ are corresponding angles, and $\angle 5$ and $\angle 6$ are supplementary angles.

7. Always; They are vertical angles.

8. Sometimes; They are supplementary if they are right angles.

9. Never; They must be supplementary.

10. Never; They are adjacent angles.

11. Always; Angles 2 and 6 are congruent because they are corresponding angles. Angles 6 and 8 are congruent because they are vertical angles. So, angles 2 and 8 are congruent.

12. No; Any two adjacent angles will be supplementary and two acute angles cannot be supplementary.

5.5 Enrichment and Extension

1. yes; The pattern repeats across the entire plane with no gaps or overlap.

2. triangles and hexagons

3. 60° 4. 120°

5. Check students' work

6. Triangle, triangle, triangle, triangle, hexagon

7. 360°

8. *Sample answer:* floor tiles, quilts

5.5 Puzzle Time

IT DIDN'T WANT TO GET ITS HARE WET

School-to-Work

1. triangle: 60°, square: 90°, hexagon: 120°; Find the sum of the measures of the interior angles by multiplying 180° and two less than the number of sides. Then divide by the number of sides.

2.

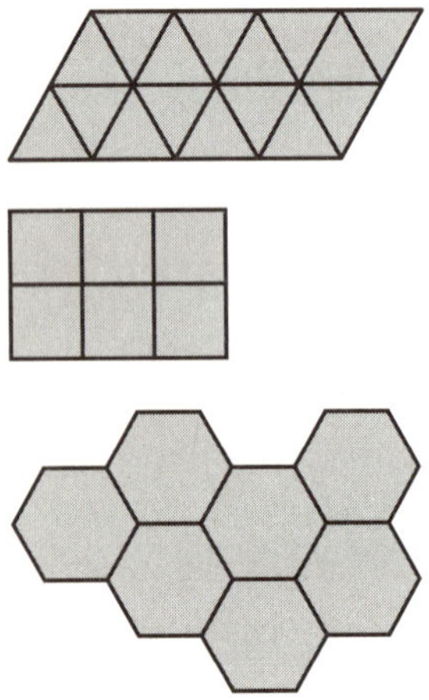

3. 360°; You know that there are no gaps or overlaps because there are exactly 360° in a circle.

4. *Sample answer:*

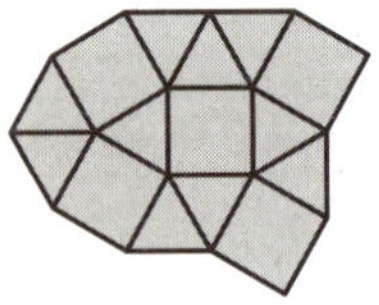

At each vertex of the center square, the angle measures are $90° + 60° + 60° + 90° + 60° = 360°$.

5. *Sample answer:*

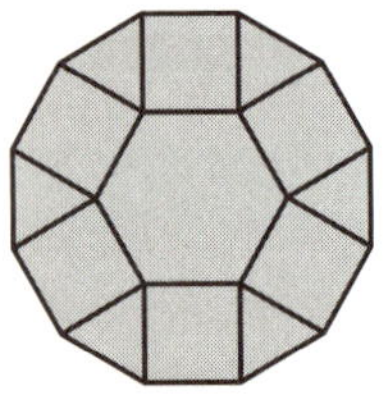

At each vertex of the hexagon, the angle measures are $120° + 90° + 60° + 90° = 360°$.

6. Answers will vary.

Study Help

Available at *BigIdeasMath.com.*

Financial Literacy

1. The distances on the map are proportional to the corresponding actual distances because lengths of corresponding sides of similar triangles are proportional.

2. 10.9 cm; 218 mi; $\frac{x}{10.9} = \frac{20}{1}$, $x = 10.9(20) = 218$

3. Between Orlando and Jacksonville: 125 mi; between Jacksonville and Tallahassee: 157 mi; 500 mi

4. about 20.8 gal; 2 times

5. *Sample answer:* The actual route along roadways is not triangular, so the driving distance is much longer and will require more gas.

Cumulative Practice

1. 87	2. 4	3. 135	4. 115
5. 36	6. 16	7. 50	8. 6
9. 115	10. 4 : 25	11. concave	12. 47°
13. 7			

Chapter 6

6.1 Start Thinking!

For use before Activity 6.1

Sample answer: No; the lengths of the sides can be any two lengths that have the given product; Yes; because the sides of a square are the same length, the area is the square of the side length. For example, if the area is 64 square meters, then the length of each side would be 8 meters.

6.1 Warm Up

For use before Activity 6.1

1. 144	2. 81	3. 324
4. 2.56	5. 6.25	6. $\frac{4}{9}$

6.1 Start Thinking!

For use before Lesson 6.1

Shelley; the solutions are 20 and −20.

6.1 Warm Up

For use before Lesson 6.1

1. 9 in.	2. 13 cm	3. 1 yd	4. 1.5 m

6.1 Practice A

1. $s = 14$ in.	2. $s = \frac{7}{9}$ m	3. ± 4

4. 0 **5.** 11 **6.** $-\frac{1}{6}$

7. $\pm\frac{17}{7}$ **8.** -0.8 **9.** 13

10. 3 **11.** $\sqrt{64} > 5$

12. $0.6 < \sqrt{0.49}$ **13.** $r = 4$ ft

14. $x = 11$ widgets **15.** $s = 10$ in.

6.1 Practice B

1. $\frac{13}{15}$ cm **2.** 1.6 yd **3.** ± 15 **4.** ± 20

5. -22 **6.** $\pm\frac{5}{8}$ **7.** 2.5

8. ± 1.3 **9.** 4.8 **10.** 6

11. $\sqrt{\frac{49}{9}} > 2$ **12.** $\frac{2}{5} = \sqrt{\frac{12}{75}}$ **13.** $r = 12$ m

14. No. For example, $\frac{\sqrt{64}}{\sqrt{16}} = \frac{8}{4} = 2$, which is not a perfect square.

15. $s = 24$ m

6.1 Enrichment and Extension

1.

n	n^2	$\sqrt{(n^2)}$	Check
1	1	1	$1 \bullet 1 = 1$
2	4	2	$2 \bullet 2 = 4$
3	9	3	$3 \bullet 3 = 9$
4	16	4	$4 \bullet 4 = 16$
5	25	5	$5 \bullet 5 = 25$

2.

n	n^3	$\sqrt[3]{(n^2)}$	Check
1	1	1	$1 \bullet 1 \bullet 1 = 1$
2	8	2	$2 \bullet 2 \bullet 2 = 8$
3	27	3	$3 \bullet 3 \bullet 3 = 27$
4	64	4	$4 \bullet 4 \bullet 4 = 64$
5	125	5	$5 \bullet 5 \bullet 5 = 125$

3. 6 **4.** -2 **5.** $-\frac{1}{8}$

6. $\frac{4}{9}$ **7.** 7 in. **8.** $-1, 0, 1$

6.1 Puzzle Time

EATING SWORDFISH

6.2 Start Thinking!

For use before Activity 6.2

right triangle; yes, any lengths a, b, and c such that $a^2 + b^2 = c^2$

6.2 Warm Up

For use before Activity 6.2

1. 1.2 **2.** ± 30 **3.** $\frac{2}{3}$

4. -21 **5.** ± 22 **6.** -50

6.2 Start Thinking!

For use before Lesson 6.2

Sample answer: In a gymnastics floor routine, the gymnasts must stay within a 12-meter by 12-meter square. Often they perform tumbling passes in which they start in one corner of the square and end up in the opposite corner. You can use the Pythagorean Theorem to find how far they traveled from one corner to the other. Also, in baseball, the bases form a square with 90-foot sides. You can use the Pythagorean Theorem to find how far the catcher must throw the ball to throw out a runner at second base.

6.2 Warm Up

For use before Lesson 6.2

1. $c = 10$ cm **2.** $a = 5$ in.

3. $b = 4.8$ m **4.** $c = 17$ ft

6.2 Practice A

1. $c = 10$ ft **2.** $b = 12$ cm **3.** $a = 2$ m

4. $b = 20$ yd **5.** 6 in.

6. $x = 29$ yd **7.** $x = 6.5$ cm

8. No. The other leg would be 0 meters long, which is impossible.

9. $x = 7$

6.2 Practice B

1. $c = 37$ mm **2.** $a = 3$ ft

3. $b = 2$ in. **4.** $a = 5$ cm

5. 26 in.

6. 9 in.

7. $x = 6$ cm

8. 8 fewer blocks

6.2 Enrichment and Extension

1. 3248 mi

2. 1624 mi

3. 615, 574, 435

4. 249,380,384,400

5. 499,380 mi^2

6. 360 m^2

6.2 Puzzle Time

STOP HOUNDING ME

6.3 Start Thinking!

For use before Activity 6.3

Sample answer: π, some square roots, like $\sqrt{2}$ and $\sqrt{3}$

6.3 Warm Up

For use before Activity 6.3

1. 50

2. 26

3. 34

4. 41

5. 90

6. 6.5

6.3 Start Thinking!

For use before Lesson 6.3

Sample answer: First find the area, 93.5 in.^2. Ask yourself "What number times itself is 93.5?" Because $9^2 = 81$ and $10^2 = 100$, you know that the number must be between 9 and 10. Try squaring values between 9 and 10 to find the number that produces the value closest to 93.5. A square with sides of about 9.7 inches has the same area as an 8.5-inch by 11-inch sheet of paper.

6.3 Warm Up

For use before Lesson 6.3

1. yes

2. no

3. no

4. yes

5. yes

6. no

6.3 Practice A

1. yes

2. no

3. Irrational; 14 is not a perfect square,and so its square root is irrational.

4. Rational; every repeating decimal is a rational number.

5. Rational; every terminating decimal is a rational number.

6. Irrational; π is irrational.

7. Irrational; the area is 4π sq ft and π is irrational.

8. 6

9. 25

10. -3

11. 2

12. 18 ft

13. $\sqrt{70}$; $8 = \sqrt{64} < \sqrt{70}$

14. 3; a positive number is greater than a negative number.

15. $16\frac{1}{4}$; $\sqrt{210} < \sqrt{225} = 15$

16. $\sqrt{\frac{4}{25}}$; $\sqrt{\frac{4}{25}} = \frac{2}{5} = \frac{4}{10}$

17. $4 < a < 9$; *Sample answer*: $a = 6$.

18. Yes; $\sqrt{\frac{1}{9}} = \frac{1}{3}$

19. No; 5 is not a perfect square.

20. Yes; $\sqrt{\frac{2}{18}} = \sqrt{\frac{1}{9}} = \frac{1}{3}$

6.3 Practice B

1. no

2. yes

3. Rational; it can be written as a ratio of two integers, $\frac{20}{9}$.

4. Irrational; π is irrational.

5. Rational; every repeating decimal is a rational number.

6. Irrational; 130 is not a perfect square, and so its square root is irrational.

7. Irrational; the circumference is 10π m and π is irrational.

8. -5

9. 20

10. 0.9

11. 1.2

12. a. $c = \sqrt{2450}$

b. $(49^2 = 2401) < 2450 < (50^2 = 2500)$

c. 49.5 ft

13. $\sqrt{220}$; $14.75^2 = 217.5625$

14. $-\sqrt{135}$; $-145 < -135$, so $-\sqrt{145} < -\sqrt{135}$

Answers

15. $\frac{3}{8}$; $\left(\frac{3}{8}\right)^2 = \frac{9}{64} > \frac{7}{64}$

16. -0.25; $-\sqrt{\frac{1}{4}} = -0.5$

17. $49 < a < b < 64$; *Sample answer*: $a = 50$, $b = 60$

6.3 Enrichment and Extension

1. 36 and 49 **2.** 7

3. 6.71 **4.** 6.855

5. 46.99; The numbers are very close.

6. 5.4775; The square of the estimate is about 30.003, which is very close to 30.

6.3 Puzzle Time

THE LOBSTER THAT BECAME A POLICEMAN BECAUSE HE BELIEVED IN CLAW AND ORDER

6.4 Start Thinking!

For use before Activity 6.4

Check students' drawings. The diagonal should measure about 1.6 in., or $1\frac{10}{16}$ in., which is approximately equal to $\frac{1 + \sqrt{5}}{2}$ in.

6.4 Warm Up

For use before Activity 6.4

1. 0.378 **2.** 0.866 **3.** 1.366

4. 0.244 **5.** 1.138 **6.** 0.146

6.4 Start Thinking!

For use before Lesson 6.4

Sample answer: A fraction is simplified when the numerator and denominator contain no common factors other than 1. The expressions that are simplified are: $\sqrt{2}$, $\sqrt{10}$, and $3\sqrt{5}$. 2, 10, and 5 have no perfect square factors other than 1. $\sqrt{4}$, $\sqrt{50}$, and $3\sqrt{8}$ are not simplified because the perfect square 4 is a factor of 4, the perfect square 25 is a factor of 50 and the perfect square 4 is a factor of 8.

6.4 Warm Up

For use before Lesson 6.4

1. about 1.630; yes **2.** about 2.003; no

6.4 Practice A

1. $9\sqrt{2}$ **2.** $5\sqrt{5}$ **3.** $\sqrt{7}$

4. $-7\sqrt{10}$ **5.** $\frac{1 + \sqrt{5}}{4}$ **6.** $\frac{\sqrt{3} + 1}{6}$

7. $5\sqrt{2} + 5$ **8.** $2\sqrt{5}$ **9.** $4\sqrt{2}$

10. $5\sqrt{2}$ **11.** $\frac{\sqrt{7}}{4}$ **12.** $\frac{\sqrt{11}}{5}$

13. $\frac{\sqrt{33}}{12}$ **14.** $2\sqrt{6}$ cm **15.** $7\sqrt{3}$

16. $-7\sqrt{2}$ **17.** $x = \pm\sqrt{35}$

18. a. $10\sqrt{5}$ in. **b.** $5\sqrt{5}$ in.

c. The hypotenuse was reduced by half.

6.4 Practice B

1. $5\sqrt{11}$ **2.** $-\frac{1}{5}\sqrt{10}$ **3.** $\frac{4}{3}\sqrt{15}$

4. 0 **5.** $\frac{1 + \sqrt{11}}{5}$ **6.** $\frac{\sqrt{3} - 1}{2}$

7. $14\sqrt{3}$ in. **8.** $7\sqrt{2}$ **9.** $10\sqrt{3}$

10. $4\sqrt{5}$ **11.** $\frac{\sqrt{14}}{13}$ **12.** $\frac{\sqrt{7}}{25}$

13. $\frac{\sqrt{67}}{10}$ **14.** $2\sqrt{10}$ m

15. $-\frac{1}{4}\sqrt{2}$ **16.** $38\sqrt{2}$ **17.** $x = \pm 2\sqrt{15}$

18. a. $2\sqrt{3}$ cm **b.** $2\sqrt{3}$ cm

c. The radius remained the same.

6.4 Enrichment and Extension

1. $9\sqrt{3}$ **2.** $17\sqrt{2}$ **3.** $12\sqrt{11}$

4. 0 **5.** $11\sqrt{5}$ **6.** $-3\sqrt{6}$

7. $\frac{\sqrt{17}}{6}$ **8.** $\frac{\sqrt{35}}{27}$ **9.** $\frac{\sqrt{13}}{10}$

10. 30 **11.** 26

12. 72

a dodecagon

6.4 Puzzle Time

IT IS PASTA BEDTIME

6.5 Start Thinking!

For use before Activity 6.5

Sample answer: A right triangle is formed by the kite, the person holding the kite, and a person standing on the ground directly under the kite. You can find the length of the kite string and the distance between the two people. Then use the Pythagorean Theorem to find the height of the kite.

6.5 Warm Up

For use before Activity 6.5

1. 10.8 cm **2.** 6.6 in. **3.** 8.5 m **4.** 13.3 ft

6.5 Start Thinking!

For use before Lesson 6.5

Answers will vary. Check students' problems and sketches.

6.5 Warm Up

For use before Lesson 6.5

1. 33.3 cm **2.** 14.5 in. **3.** 22.6 ft **4.** 35.1 m

6.5 Practice A

1. $P = 28.1$ in. **2.** $P = 43$ cm **3.** $d = 4.5$

4. $d = 4.1$ **5.** $x = 37.5$ ft **6.** $x = 16.7$ m

7. yes **8.** no **9.** 204.9 ft

6.5 Practice B

1. $P = 63.2$ in. **2.** $P = 28.4$ m **3.** $d = 6.7$

4. $d = 5.7$ **5.** $x = 49.7$ m **6.** $x = 16.1$ m

7. yes **8.** yes **9.** $159

6.5 Enrichment and Extension

1. 400 **2.** $400 = 8 \bullet 50$

3. $c + a = 50$

$c - a = 8$

4. $a = 21, c = 29$

5.
$$a^2 + b^2 = c^2$$
$$21^2 + 20^2 \stackrel{?}{=} 29^2$$
$$441 + 400 \stackrel{?}{=} 841$$
$$841 = 841$$

6. a. 7, 24, 25 **b.** 8, 15, 17

6.5 Puzzle Time

PEARS

School-to-Work

1. The square of the length of the principal rafter is equal to the sum of the squares of the king post and half the length of the ridge beam; $10^2 = x^2 + 8^2$ and $x = 6$ so the length of the king post is 6 feet.

2. 4 ft; 4 ft

3. No; $5^2 \neq 4^2 + 3.5^2$; The length of the strut should be 3 feet.

4. 5 ft; $3^2 + 4^2 = 5^2$ **5.** No; $20^2 \neq 10^2 + 10^2$

6. No; equiangular triangle or equilateral triangle

Study Help

Available at *BigIdeasMath.com.*

Financial Literacy

1. Check students' work.

2. *Sample answer:* The size of the dance floor selected should be based on the estimate of the maximum number of dancers at any one time. The estimate should be multiplied by 4 to get an idea of the number of square feet of dance floor space needed.

3. Answers will vary depending on the size of floor selected. For the floor that is 192 square feet, the side length is about 13.9 feet. For the floor that is 320 square feet, the side length is about 17.9 feet. For the floor that is 480 square feet, the side length is about 21.9 feet. For the floor that is 672 square feet, the side length is about 25.9 feet.

4. *Sample answer:* The number of square feet covered by the tent should be greater than the number of square feet covered by the dance floor. For the tent that is 120 square feet, the side length is about 11.0 feet. For the tent that is 350 square feet, the side length is about 18.7 feet. For the tent that is 600 square feet, the side length is about 24.5 feet. For the tent that is 950 square feet, the side length is about 30.8 feet.

Answers

5. Answer should be the sum of all mandatory and optional costs.

6. Answer should be the sum of all mandatory and optional costs divided by the predicted number of students to attend the dance.

7. Check students' work.

Cumulative Practice

1. $\frac{4}{5}$	**2.** -6	**3.** 29
4. -46.9	**5.** $\frac{2+\sqrt{6}}{5}$	**6.** $-\sqrt{2}$
7. $2\sqrt{17}$	**8.** 23 and -23	**9.** $\frac{\sqrt{19}}{11}$
10. $\frac{49}{24}$	**11.** 0	**12.** $\frac{1}{4}$
13. 135 cm^3	**14.** 39	**15.** 10
16. 10.4	**17.** 48.5 ft	**18.** =
19. <	**20.** >	**21.** library

Chapter 7

7.1 Start Thinking!

For use before Activity 7.1

Sample answer: They are measures of the center or middle of a set of values. Examples are test or quiz averages (mean), median house price, or median income.

7.1 Warm Up

For use before Activity 7.1

1. 6.75	**2.** 5	**3.** 22
4. 8.625	**5.** 36	**6.** 4

7.1 Start Thinking!

For use before Lesson 7.1

Check students' work; the mean, median, and mode will all increase by 1.

7.1 Warm Up

For use before Lesson 7.1

1. mean: 13; median: 13; mode: 13

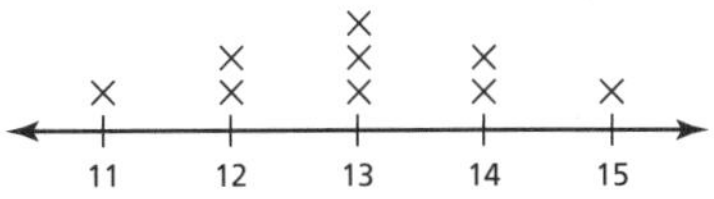

2. mean: 2.4; median: 2.5; mode: 4

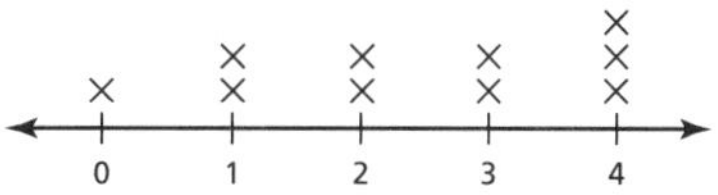

7.1 Practice A

1. mean = 6; median = 6; modes = 4, 8

2. mean = 0.05; median = 0.05; mode = 0.05

3. mean = 29.8; median = 32; mode = 33

4. mcan = 3; median = 3; mode − 3

5. $x = 5$ **6.** $x = 26$

7. **a.** mean = 425; median = 335; mode = no mode
   b. 850
   c. mean drops to 340, median drops to 320, the mode remains no mode

8. **a.** mean = 6.8
   b. mean = 13.6; multiplied by 2

7.1 Practice B

1. mean = 28; median = 29; mode = 29

2. mean = 1.04; median = 0.99; mode = 0.99

3. mean = 29.75; median = 30; modes = 28, 32

4. mean = 3.45; median = $3\frac{3}{8}$; mode = $3\frac{1}{4}$

5. $x = -2.5$ **6.** $x = 83$

7. **a.** mean = 45; median = 36; no mode
   b. mean = 22.5; the mean is half as much

8. **a.** *Sample answer:* 1, 3, 3, 6, 6, 8, 9, 9
   b. *Sample answer:* 1, 2, 3, 4, 5, 6, 7, 8

7.1 Enrichment and Extension

1. 0.25	**2.** 0.5	**3.** 0.25
4. 3.25	**5.** no	

6. **a.** 3.5 **b.** It is close.
   c. yes; The experimental probabilities of the spinner will be closer to the theoretical probabilities.

7. 6.25

Answers

7.1 Puzzle Time

HE FLU

7.2 Start Thinking!

For use before Activity 7.2

Sample answer: The box-and-whisker plot is easier because the minimum, maximum, and median are shown on the plot. Also the distribution of the data is a lot easier to see from the plot than the unordered list.

7.2 Warm Up

For use before Activity 7.2

1. 5; 1; 12

2. 0.95; 0.5; 1.7

3. 56; 45; 89

4. 78; 5; 222

7.2 Start Thinking!

For use before Lesson 7.2

Sample answer: In general, ticket prices in MLB are lower than in the NBA. There is greater variability for prices in the NBA. The median NBA ticket price is almost the same as the maximum MLB ticket price, and the median MLB ticket price is almost the same as the minimum NBA ticket price.

7.2 Warm Up

For use before Lesson 7.2

1. *Sample answer*: In general, Mia spends more time doing homework than Karen. Karen's first quartile, median, and third quartile are equal to Mia's minimum, first quartile, and median, respectively. There is greater variability in Mia's plot.

7.2 Practice A

1. a. Median is the same for both teams; there are more lower times for Team B than for Team A

b. Team B; lower times indicates faster runners

2.

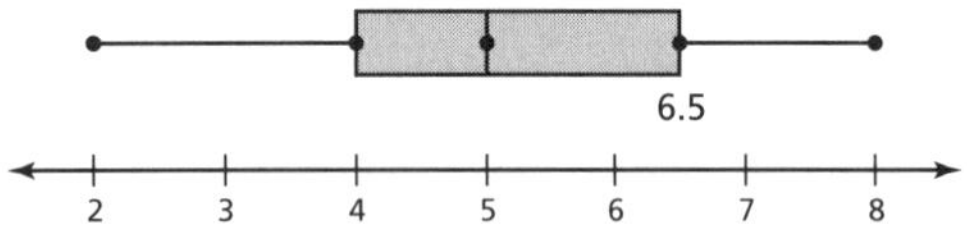

3.

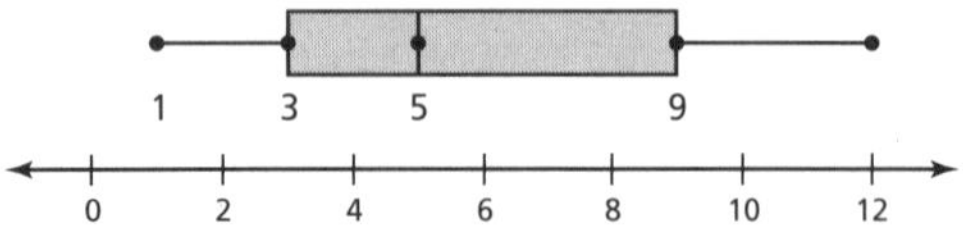

4. a.

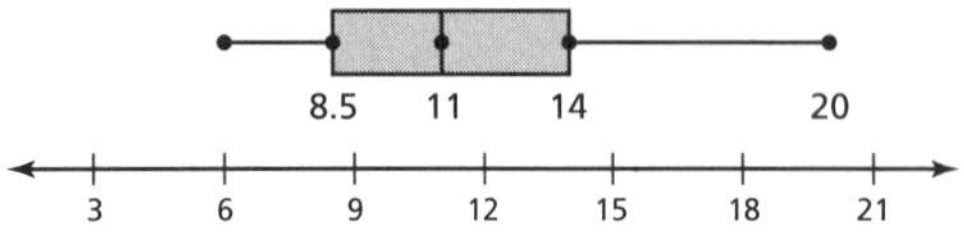

b. yes; the swimmer who swam the most laps

5. a.

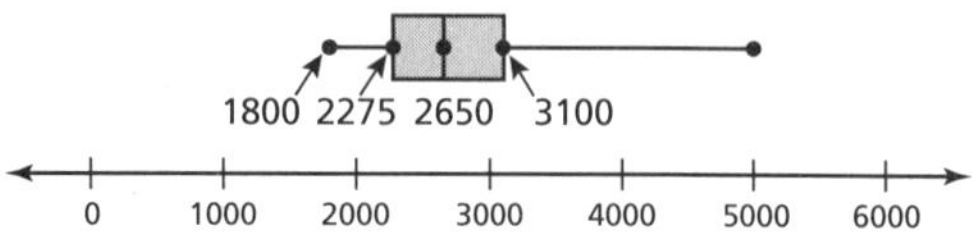

b. 5000

c.

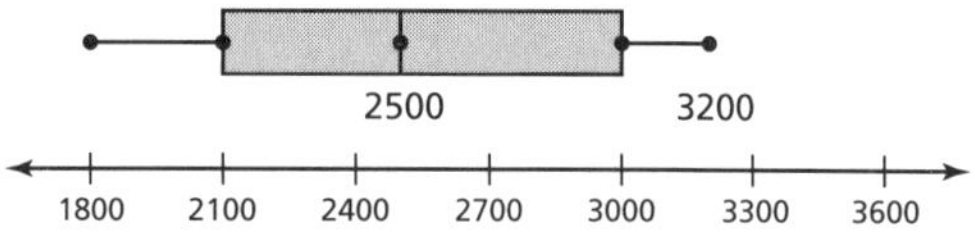

d. The outlier moved ends of the box (Q1 and Q3) to the right and lengthened both whiskers, especially the upper one.

6. a. right whisker **b.** box

7.2 Practice B

1. a. Competitor A has the highest and lowest number of moves. Competitor B has the lowest values for Q1, Median, and Q3.

b. Competitor B: lower scores

2.

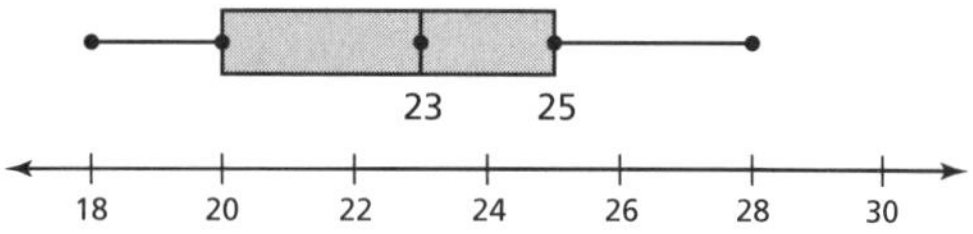

3.

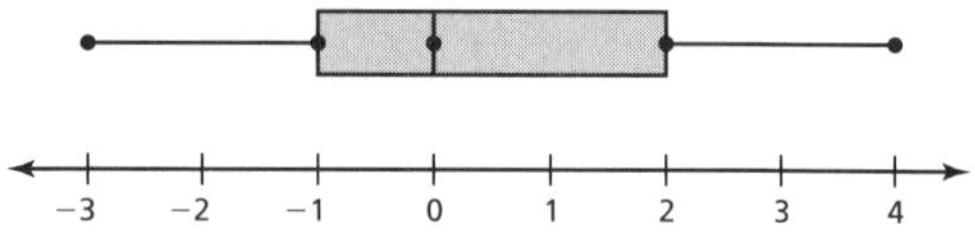

4. a. Patient A **b.** The ranges are the same.

c. The lowest pulse is an outlier; during the night.

d. Smaller range between the median and the third quartile than between the first quartile and the median.

5. a.

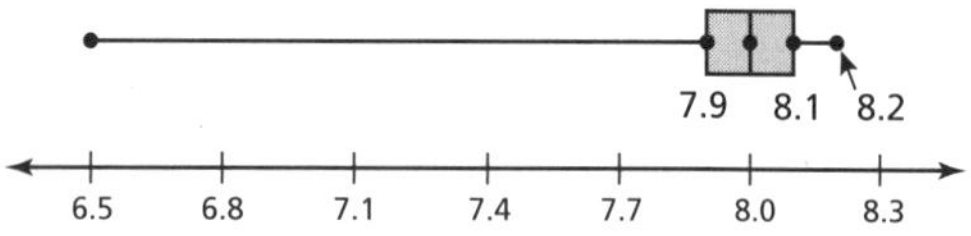

b. 6.5

c.

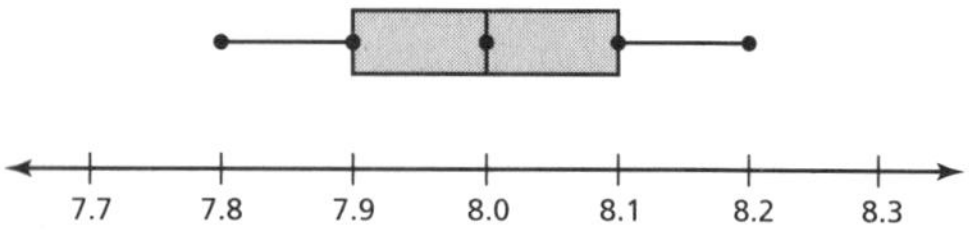

d. The box and right whisker were not affected. The left whisker became much shorter.

7.2 Enrichment and Extension

1. 1 and 120

2. first quartile: 50, third quartile: 80

3. 1, 20, 100, 120

Answers

4. 30

5. 45

6. a. 1

b. *Sample answer:* Bonsai trees are small. The botanical garden might like to have a tree that is unusual for a tree being 100 years old.

7.2 Puzzle Time

IT WAS BROKE

7.3 Start Thinking!

For use before Activity 7.3

Answers will vary. Check students' graphs and predictions.

7.3 Warm Up

For use before Activity 7.3

1. $y = -\frac{1}{5}x + 4$

2. $y = -3x + 6$

3. $y = -\frac{1}{2}x + 7$

4. $y = x + 1$

5. $y = \frac{1}{3}x$

6. $y = 2x + 8$

7.3 Start Thinking!

For use before Lesson 7.3

B; the line best follows the trend shown by the points. There are approximately the same number of points above and below the line.

7.3 Warm Up

For use before Lesson 7.3

1–2.

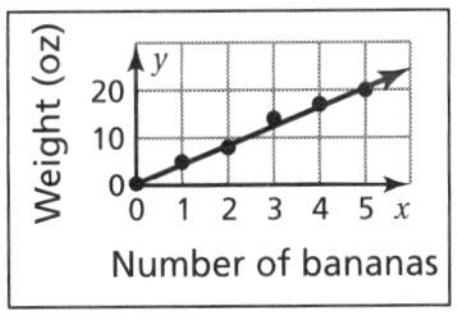

3. $y = 4x + 0.5$

4. 40.5 pounds

7.3 Practice A

1. a. 3 months **b.** 7 lb

c.

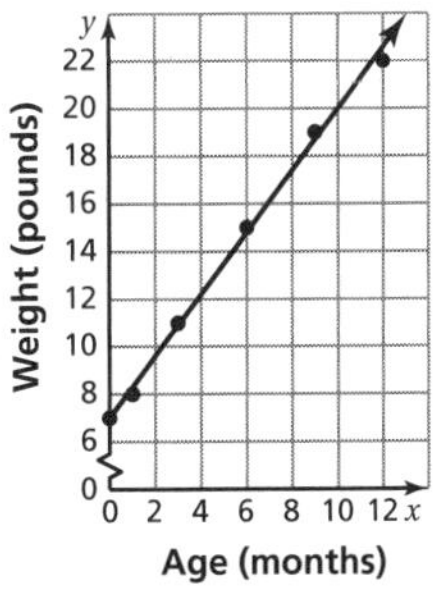

d. *Sample answer:* $y = 1.3x + 7$

e. *Sample answer:* 30.4 lb

f. positive relationship

2. a and **b.**

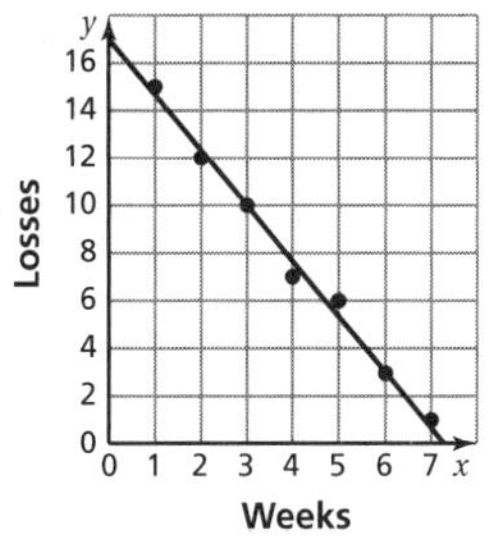

c. *Sample answer:* $y = -2.3x + 17$

d. negative relationship

e. The longer you play the game, the fewer losses you have.

3. a. no relationship

b. No; many possible lines could be drawn, but no best fit.

c. no; there is no relationship, so you are unable to predict.

7.3 Practice B

1. a. \$12 **b.** 20 fl oz

c.

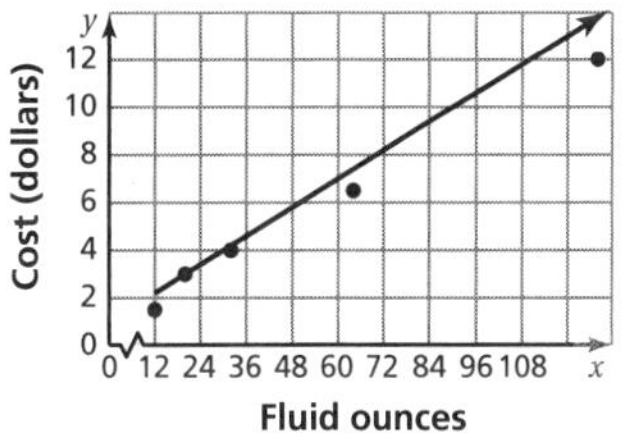

d. *Sample answer:* $y = 0.1x + 1$

e. *Sample answer:* \$26.60

f. positive relationship

2. a and **b.**

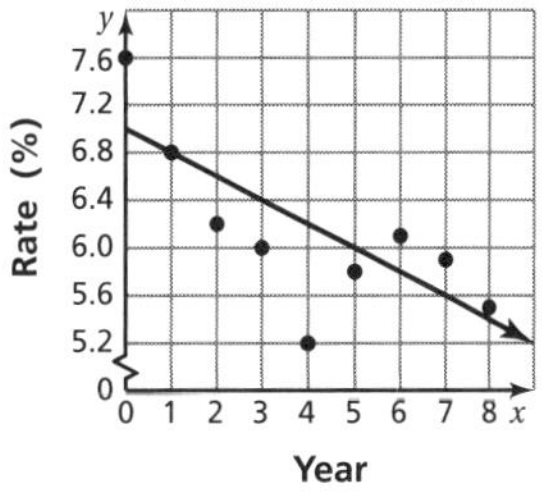

c. *Sample answer:* $y = -0.2x + 7$

d. *Sample answer:* 5%

e. negative relationship

f. Mortgage interest rates are dropping about 0.2% per year.

Answers

3. a. constant relationship

b.

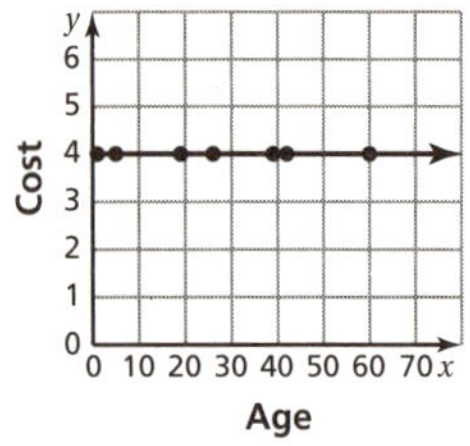

c. $y = 4$

d. Everyone pays $4 to see the show.

7.3 Enrichment and Extension

1.

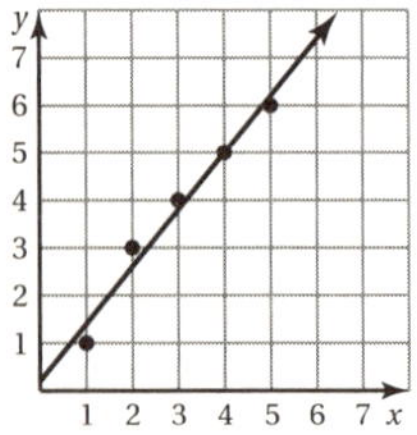

2. $n = 5$

3.

x	y	xy	x^2	y^2
1	1	1	1	1
2	3	6	4	9
3	4	12	9	16
4	5	20	16	25
5	6	30	25	36
$A = 15$	$B = 19$	$C = 69$	$D = 55$	$F = 87$

4. $r \approx 0.986$

5. Because r is very close to 1, the data has a strong positive relationship. The scatter plot supports the conclusion because the data is very linear and has a positive slope

7.3 Puzzle Time

SANDCRAB

7.4 Start Thinking!

For use before Activity 7.4

Sample answer: pictograph, bar graph, circle graph, line graph, histogram, stem-and-leaf plot, box-and-whisker plot, line plot, scatter plot

7.4 Warm Up

For use before Activity 7.4

1. **Students by Grade**

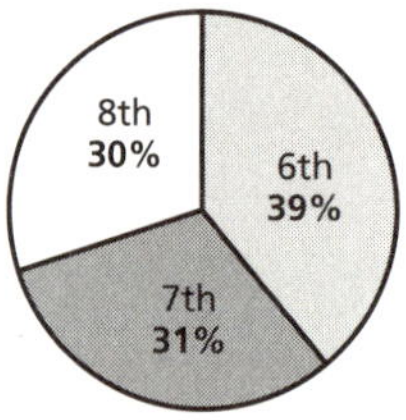

2. **Favorite Color**

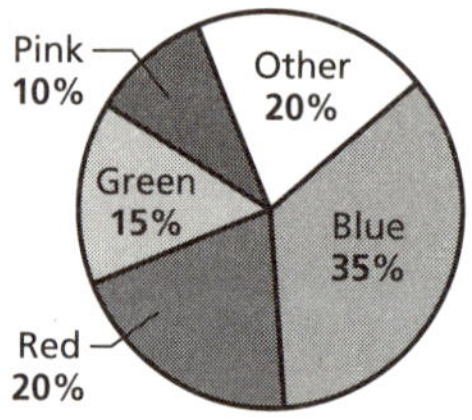

7.4 Start Thinking!

For use before Lesson 7.4

Sample answer: Both a bar graph and a histogram have bars that represent data, but the bar graph has distinct categories and the histogram uses equal intervals.

Both a line graph and a line plot use a number line to display data. A line graph uses line segments to show how data changes over time. A line plot uses x's to show the number of times a value occurs in a data set.

Both a stem-and-leaf plot and box-and-whisker plot display one-variable data and show the variability and distribution of data. A stem-and-leaf plot organizes the data numerically. A box-and-whisker plot divides the data into quarters.

7.4 Warm Up

For use before Lesson 7.4

1. Answers will vary. Sample displays are a bar graph, a line graph, a stem-and-leaf plot, and a box-and-whisker plot.

7.4 Practice A

1. line graph; data over time

2. bar graph; data in categories

3. histogram; frequency in intervals of same size

4. scatter plot; relationship between two data sets

5. Break in vertical axis; It looks like many more apples were sold.

6. The size of the age intervals is not the same.

7. scatter plot; showing a relationship between two data items

8. Line graph; the data is over time.

9. box-and-whisker plot; wants to show variability of a data set

10. Pictograph; kindergartners do not do well with numbers.

Answers

11. stem-and-leaf plot; wants both order and distribution

7.4 Practice B

1. bar graph; data in categories
2. histogram; frequency in intervals of same size
3. scatter plot; relationship between two data sets
4. circle graph; shows data as parts of a whole
5. It looks like equal amounts of fruit were consumed, until you look at the key.
6. Unequal spacing on the vertical axis does not show the larger number of occurrence in the older age groups.
7. line plot; wants to show the number of times each value occurs
8. stem-and-leaf plot; median is explicitly shown
9. scatter plot; shows relationship between two data sets
10. circle graph; shows data as parts of a whole
11. **a.** A pictograph does not show the kind of precise data a Board of Directors would need.

    b. A better choice would be a line graph.

7.4 Enrichment and Extension

1. mean: \$15, median: \$13.50, mode: \$12
2. **a.** mean

   b. It is the highest of the three and makes it appear that the company offers a high salary to its workers.
3. **a.** The competitor is using the mode. The newspaper is using the median.

   b. The competitor picked the mode because it is lowest and they want to make it look like their competitor does not offer a good salary. The newspaper picked the median because it represents the data well.
4. median
5. no; The people that have been with the company the longest will probably have the highest pay rates of anyone in the company.

7.4 Puzzle Time

STICK

School-to-Work

1. mean: 62.25, median: 60.5, mode: 73; the mode; the mean or the median best represent the data because these numbers describe the majority of the data.
2. circle graph;

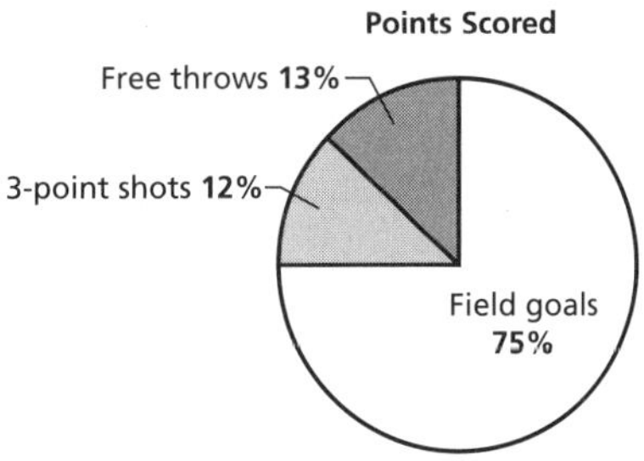

3. 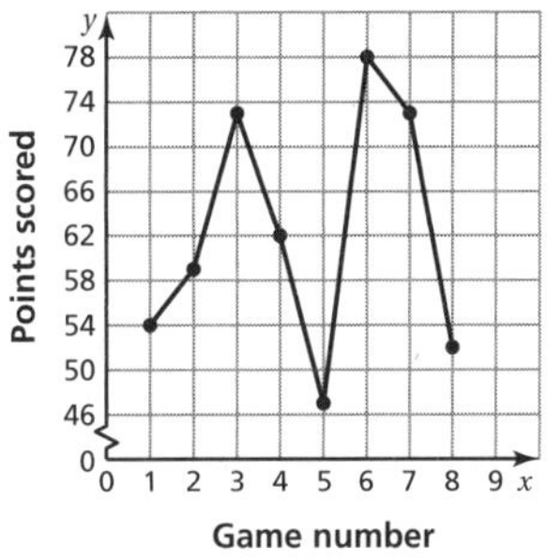

   no relationship; no, because there is no relationship between the game number and the points scored.
4. a double bar graph;

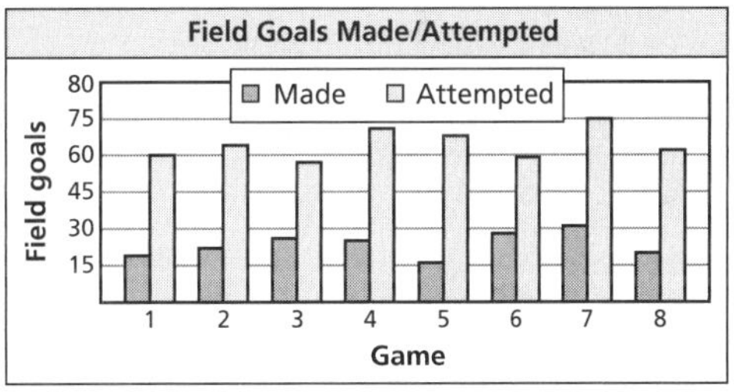

5. Check students' work.

Study Help

Available at *BigIdeasMath.com.*

Cumulative Practice

1. 

Answers

2.

Stem	Leaf
5	2
6	8
7	2 6
8	0 3 4
9	1

Key: 5 | 2 = 52 students

3.

4.
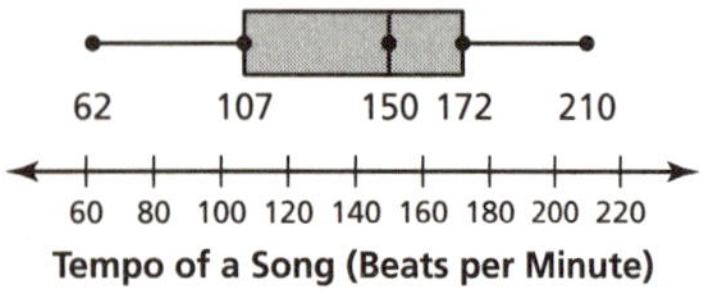

5.
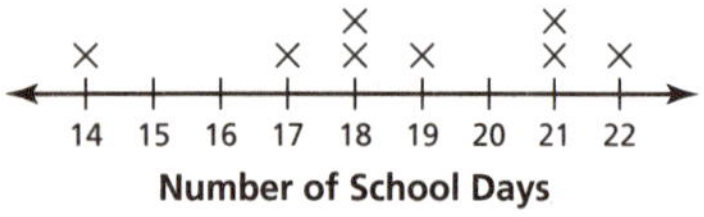

6.
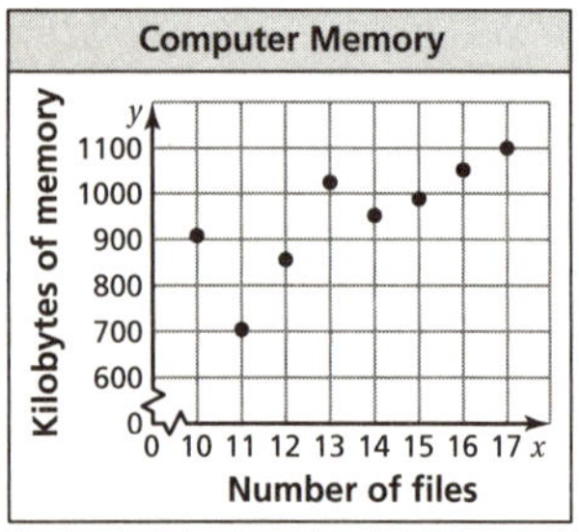

7. mean: 41; median: 41; mode: 42

8. mean: 2.33; median: 2.375; no mode

9. *Sample answer:* circle graph; it shows percentages of a group.

10. *Sample answer:* line graph; it shows change over time.

11. *Sample answer:* scatter plot; it compares two data sets.

Chapter 8

8.1 Start Thinking!

For use before Activity 8.1

Sample answer: For some rides at an amusement park, you must be a certain height or age to ride. The inequality $x \geq$ the minimum height or age describes the situation.

8.1 Warm Up

For use before Activity 8.1

1. 10.5 cm 2. 4.2 cm 3. 7.0 cm

4. 1.3 cm 5. 12.6 cm 6. 6.1 cm

8.1 Start Thinking!

For use before Lesson 8.1

Sample answer: To move into first place, he needs to score more than 50 points; >

8.1 Warm Up

For use before Lesson 8.1

1. $x \geq -1$; all numbers greater than or equal to -1

2. $x \leq 2$; all numbers less than or equal to 2

3. $x > 4$; all numbers greater than 4

4. $x < 0$; all numbers less than 0

8.1 Practice A

1. $x < 8$; all values of x less than 8

2. $x \geq -5$; all values of x greater than or equal to -5

3. $x \leq 3$ 4. $2 + y > 7$

5. $3c < -12$ 6. $m - 1.5 \geq 2$

7. yes 8. no 9. no 10. yes

11.
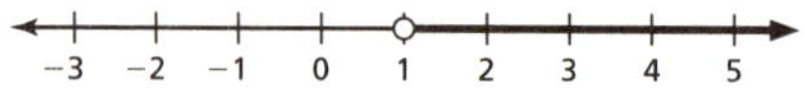

12
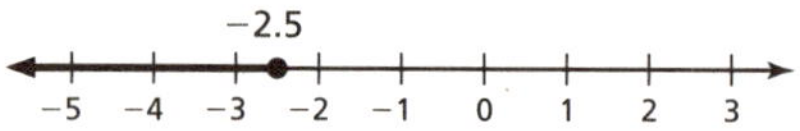

13. $a \leq 12$ 14. yes 15. no

16. a. yes b. no
c. any value of x greater than -5 and less than 3

8.1 Practice B

1. $x > -4$; all values of x greater than -4

2. $x \leq 11$; and all values of x less than or equal to 11

3. $x \geq 15$ 4. $r + 3.7 < 1.2$

5. $\frac{h}{2} > -5$ 6. $a - 8.2 \leq 12$

7. no 8. no 9. yes 10. no

11.

Answers

12 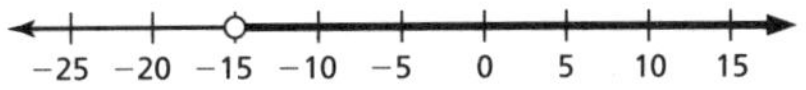

13. $a \geq 16$ **14.** yes **15.** no

16. a. $g \geq 3.5$

$c \geq 12$

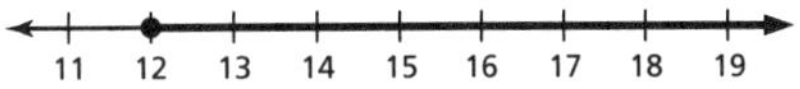

$h \geq 75$

b. no; Your cousin only has 65 hours of community service, and needs at least 10 more hours to meet that requirement.

8.1 Enrichment and Extension

1. a. $a \geq 9$ **b.** $a \leq 12$ **c.** $9 \leq a \leq 12$

2. a. $p \geq 0$ **b.** $p \leq 85$ **c.** $0 \leq p \leq 85$

3. $3.5 \leq n \leq 6$

8.1 Puzzle Time

A BULLDOZER

8.2 Start Thinking!

For use before Activity 8.2

Sample answer: You have \$20 to spend on souvenirs and know that you want to buy a \$5 replica of the Empire State Building. The inequality $x + 5 \leq 20$ represents the amount x that you have left to spend on souvenirs.

The maximum number of students that can go on a school field trip is 20. Five students have already signed up. The inequality $x + 5 \leq 20$ represents the number x of students that can still sign up for the field trip.

8.2 Warm Up

For use before Activity 8.2

1. 12 **2.** 49 **3.** −72

4. −112 **5.** 1 **6.** −3

8.2 Start Thinking!

For use before Lesson 8.2

You need to sell more than 15. *Sample answer:* The situation can be described by the inequality $x + 26 > 41$, where x is the number of subscriptions that you must sell to move into first place. You solve this inequality by subtracting 26 from each side to get the solution $x > 15$.

8.2 Warm Up

For use before Lesson 8.2

1. *Sample answer:*

A	C	Y	T	N
180	120	2000	15	3

2. *Sample answer:*

A	C	Y	T	N
200	100	250	10	10

8.2 Practice A

1. $p < 6$

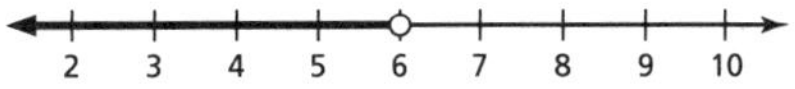

2. $s \geq -6$

3. $k \leq 4$

4. $n > \frac{1}{2}$

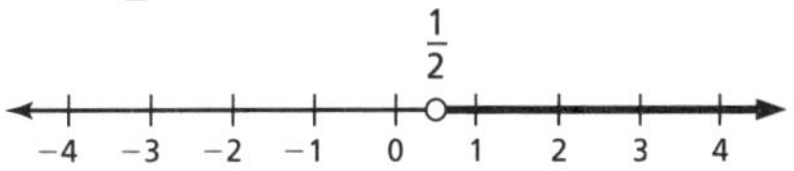

5. $z \geq 1$

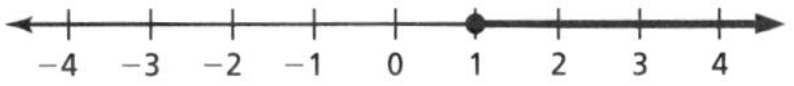

6. $t < -\frac{1}{3}$

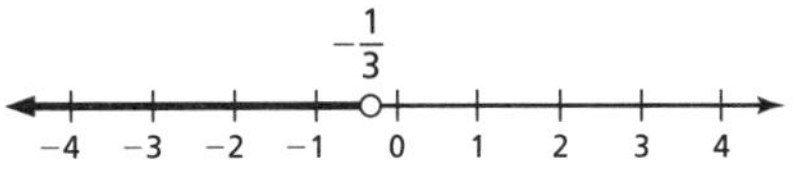

7. $d \leq -2.7$

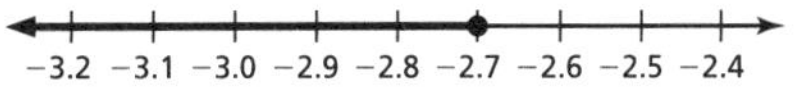

8. $q > 7$

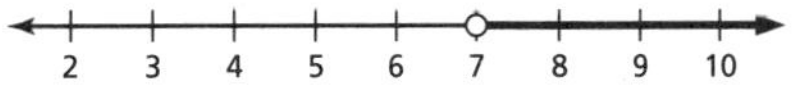

9. $a + 528 \leq 3000$; $a \leq 2472$ square feet

10. $x + 50 \leq 137.26$; $x \leq \$87.26$

Answers

11. $x + 13 < 20$; $x < 7$ meters

12. $x + 12 \geq 18$; $x \geq 6$ feet

13. a. $p + 2700 \geq 5000$; $p \geq 2300$ points

    b. $p + 3100 \geq 5000$; $p \geq 1900$ points

8.2 Practice B

1. $y \geq 5$

2. $w < 4.3$

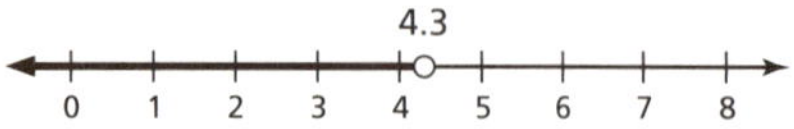

3. $v > 7\frac{2}{3}$

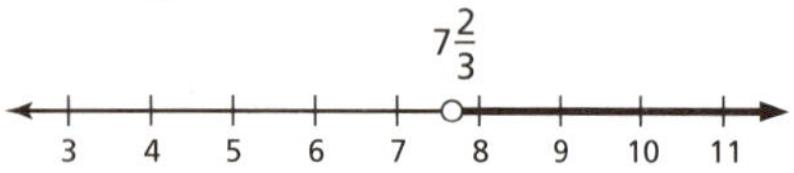

4. $k > -\frac{2}{5}$

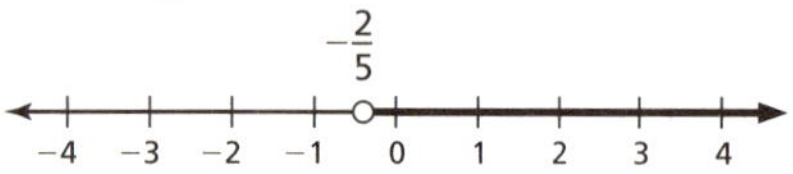

5. $q \geq -1$

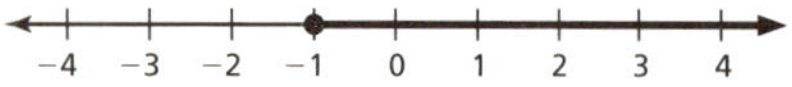

6. $r < 2$

7. $c < 3.5$

8. $p > 13.7$

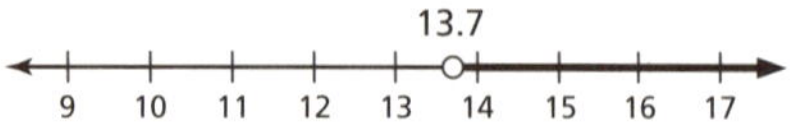

9. a. $7 + x \leq 18$; $x \leq 11$ lobsters

   b. $10 + x \leq 24$; $x \leq 14$ lobsters

   c. 6 lobsters

10. $x - 4 > 8$; $x > 12$ meters

11. $x + 35 \leq 50$; $x \leq 15$ inches

12. $c = 2$

13. All values of x greater than -5 and less than 5.

14. The third side must be greater than 11 inches and less than 23 inches. Using the triangle inequality theorem, you can write and solve the inequalities $x + 6 > 17$ and $17 + 6 > x$.

8.2 Enrichment and Extension

1. 47 cm

2. a. $188 + y \geq 304.75$

   b. at least 116.75 cm

   c. at least 609.5 cm

3. a. $p - 38 \geq 162$

   b. at least 200 cm

4. Economy Class will contain the greatest number of seats because they are the smallest in dimension. As a result, more of them will fit in a row of the cabin. First Class will contain the least number of seats because the dimensions of the seat are so large, only a small amount of seats will fit in a row of the cabin.

8.2 Puzzle Time

TAKE IT TO A HILL AND GIVE IT A PUSH

8.3 Start Thinking!

For use before Activity 8.3

No; *Sample answer:* If you substitute numbers for x in each inequality, you can see that most numbers make one of the inequalities true but the other one false.

8.3 Warm Up

For use before Activity 8.3

1. $<$	2. $>$	3. $>$
4. $<$	5. $>$	6. $<$

8.3 Start Thinking!

For use before Lesson 8.3

$x \geq 2.5$

8.3 Warm Up

For use before Lesson 8.3

1. $x < -2$	2. $x > 2$	3. $x \geq 6$
4. $x \leq -6$	5. $x > -63$	6. $x < -\frac{5}{2}$

Answers

8.3 Practice A

1. $x > 1$

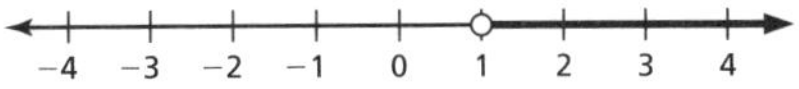

2. $r \le 10$

3. $h < -20$

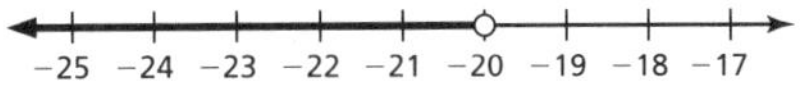

4. $u \ge 16.8$

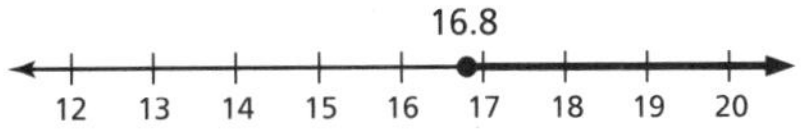

5. $j < -4.4$

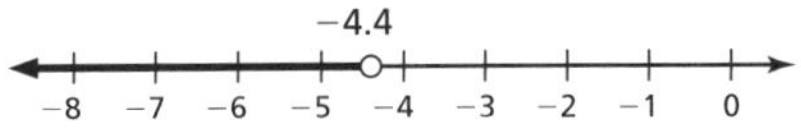

6. $x > -\frac{1}{2}$

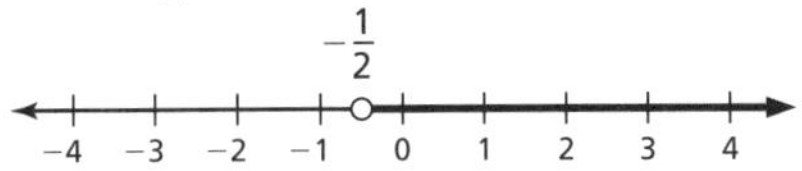

7. $5n \ge 15;\ n \ge 3$

8. $\frac{n}{4} < -1;\ n < -4$

9. $16.5g \le 363;\ g \le 22$ gal

10. $p \le -5$

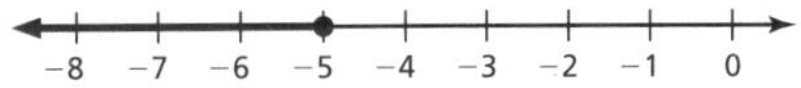

11. $v > 6$

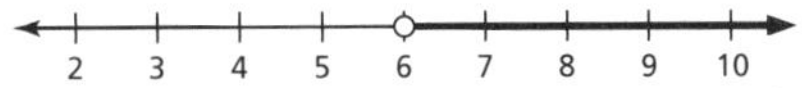

12. $g < -12.8$

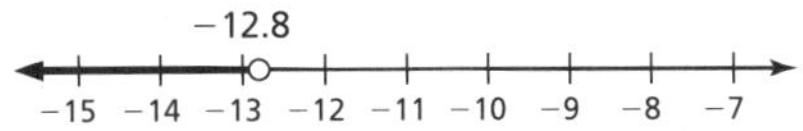

13. $y \ge -4.2$

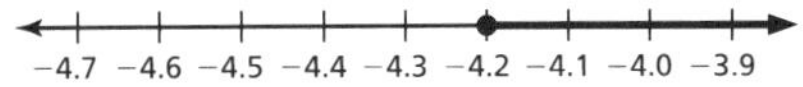

14. $h > \frac{4}{3}$

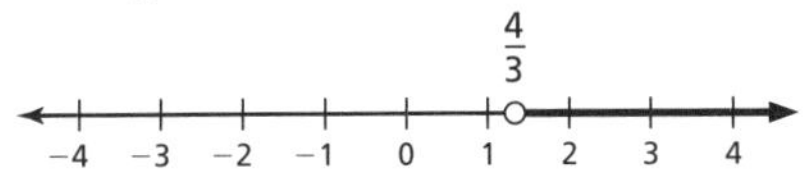

15. $a \ge 5.95$

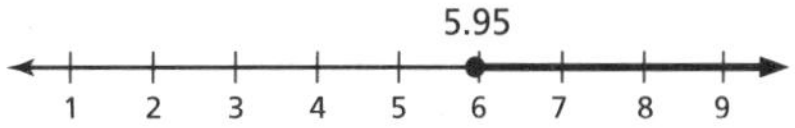

16. a. $\frac{1}{2}b \ge 20;\ b \ge 40$ beads

b. $\frac{5}{6}b \ge 20;\ b \ge 24$ beads

8.3 Practice B

1. $y \le \frac{1}{4}$

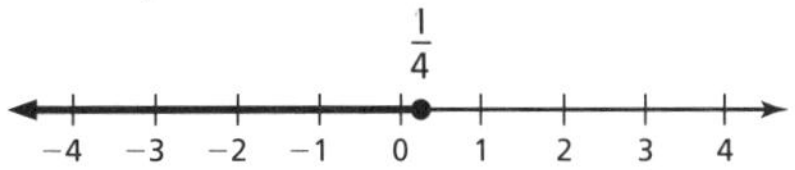

2. $p > -32$

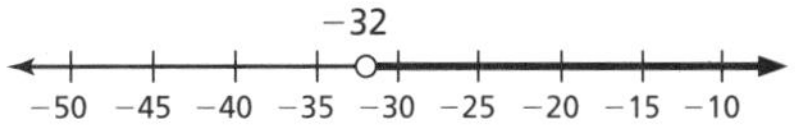

3. $g \ge 0.3$

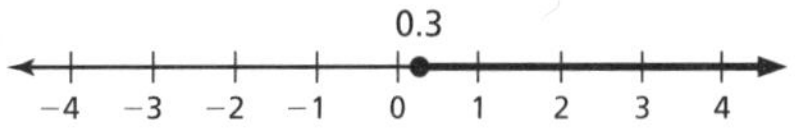

4. $k < -40$

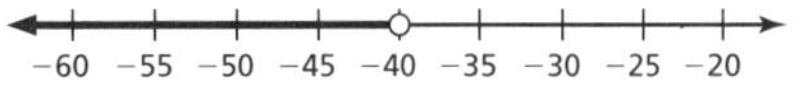

5. $s \ge 13.95$

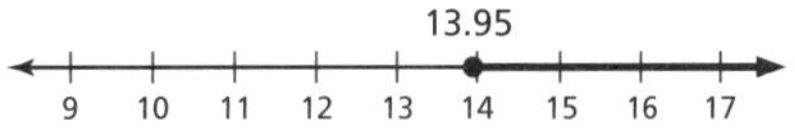

6. $x > -\frac{2}{5}$

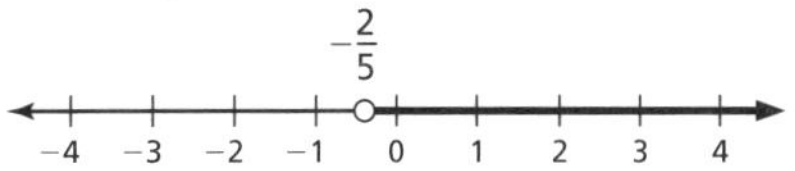

7. $\frac{n}{5} \ge 4;\ n \ge 20$

8. $2n \le -6;\ n \le -3$

9. $c = \frac{1}{2}$

10. $t > -4$

11. $u \le \frac{2}{5}$

12. $q \geq -0.76$

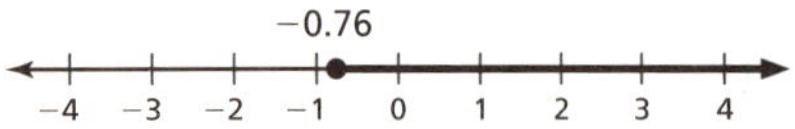

13. $d < -\frac{3}{4}$

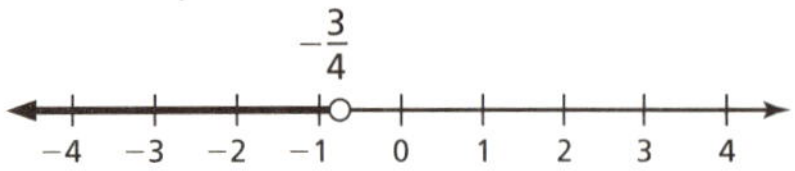

14. $r \leq 1.5$

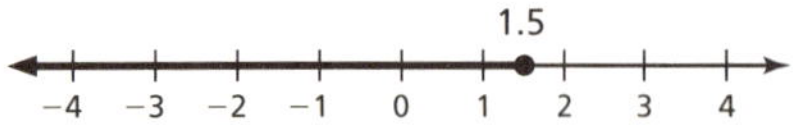

15. $j \geq 7.8$

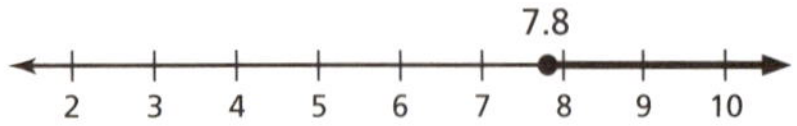

16. a. $\frac{2}{3}s \leq 10$; $s \leq 15$ shelves

b. $\frac{5}{6}x \leq 10$; $x \leq 12$ shelves

17. all values of x greater than 1 and less than 4

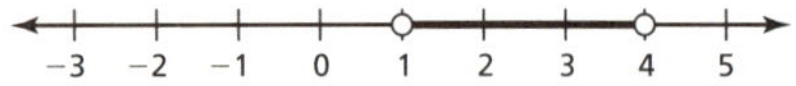

18. all values of y less than or equal to -10

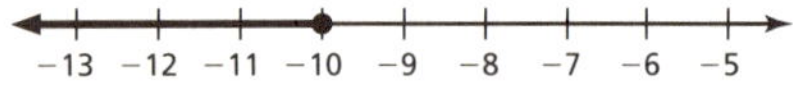

8.3 Enrichment and Extension

1. 1.5 in.

2. $e \leq 0.005$ in.

3. a. $r = 1.495$ in., $V \approx 7.018$ in.^{3}

b. $r = 1.505$ in., $V \approx 7.112$ in.^{3}

c. $1.495 \leq r \leq 1.505$

d. $7.018 \leq V \leq 7.112$

4. a. $V \approx 7.100$ in.^{2}

b. An error in the measurement of the radius will have a greater impact on the error in the volume. Because the radius is squared in the volume formula, the error factors in the volume calculation twice.

8.3 Puzzle Time

RUN AFTER IT

8.4 Start Thinking!

For use before Activity 8.4

Sample answer: You receive a gift card for \$30 to spend at an online store. Shipping costs \$6 and the store has a section with tons of great items for \$5 each. How many items can you buy from the \$5 section and not exceed the gift card amount?

8.4 Warm Up

For use before Activity 8.4

1. 68 square units
2. 12.9 square units
3. 45 square units
4. 30.5 square units

8.4 Start Thinking!

For use before Lesson 8.4

Answers will vary. Check students' inequalities and solutions.

8.4 Warm Up

For use before Lesson 8.4

1. $s > 16$
2. $h \leq 2$

8.4 Practice A

1. $m < 3$

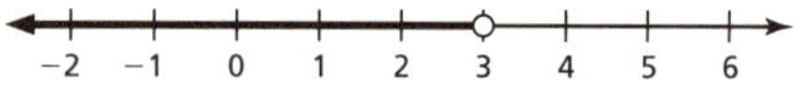

2. $r \leq 3$

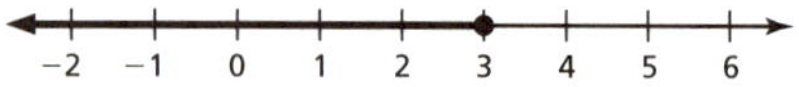

3. $k > \frac{1}{3}$

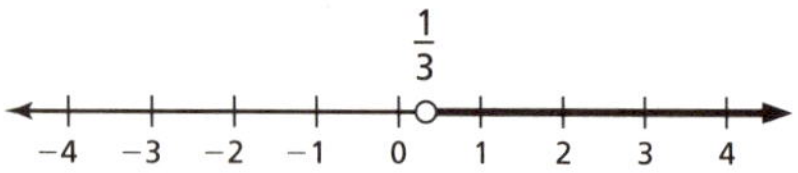

4. $c \geq -3.8$

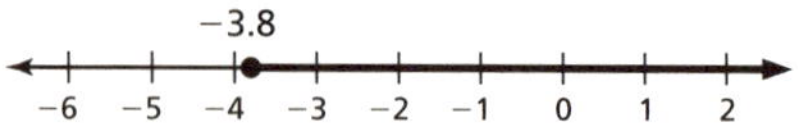

5. $20 + 0.75m \leq 65$; $m \leq 60$ miles

6. $b > 1$

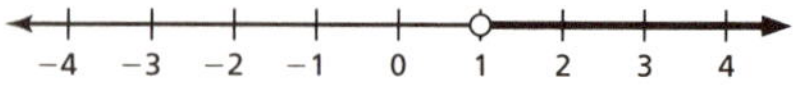

7. $p \geq -5$

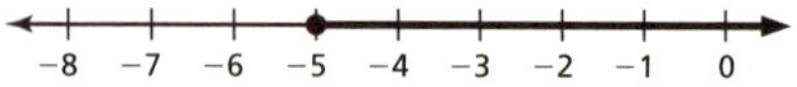

8. $d \leq 15$

Answers

9. $a > -3.5$

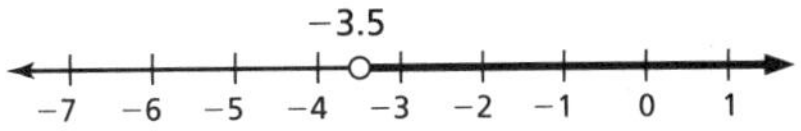

10. $5(x - 2) \geq 35;\ x \geq 9$

11. $x > 1$

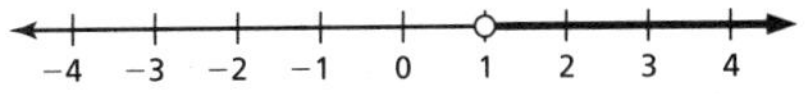

12. $w \geq 0.45$

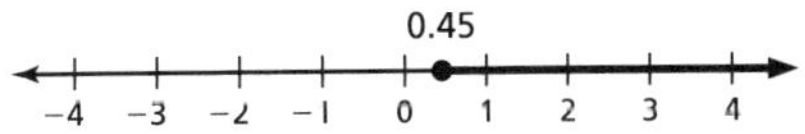

13. a. \$150

b. $150 + 20c \geq 630;\ c \geq 24$ cell phones

c. $150 + 20c \geq 750;\ c \geq 30$ cell phones

d. $150 + 20c \leq 950;\ c \leq 40$ cell phones

8.4 Practice B

1. $q < -12$

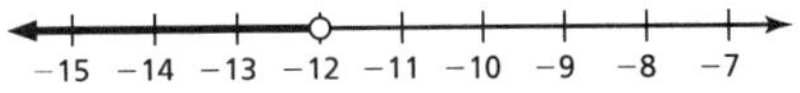

2. $v \geq -6$

3. $k \geq -\frac{1}{2}$

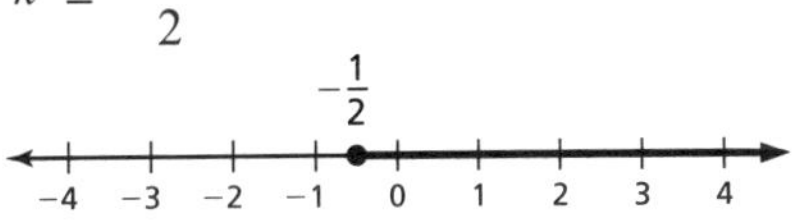

4. $n > 10$

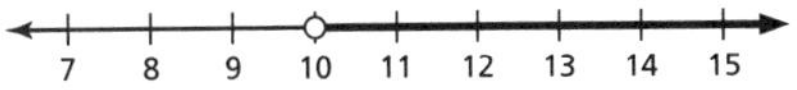

5. $300r + 2000 \geq 14{,}000;\ r \geq 40$ occupied sites

6. $m < -4$

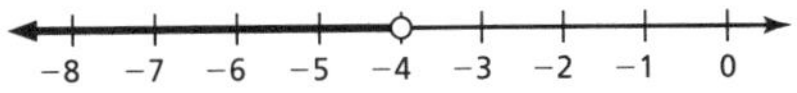

7. $f \geq -2$

8. $p > 10$

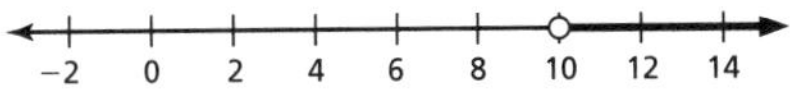

9. $w \geq 0.2$

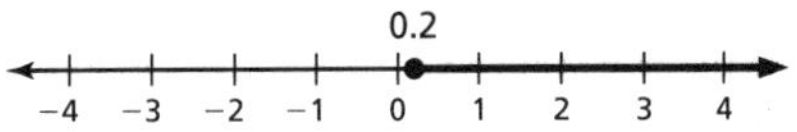

10. $4.5(x + 3) \leq 45;\ x \leq 7$

11. $x \geq 8$

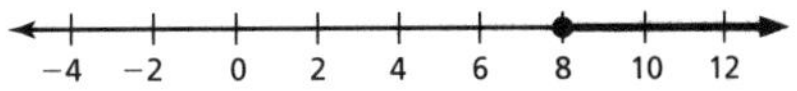

12. $v < -0.56$

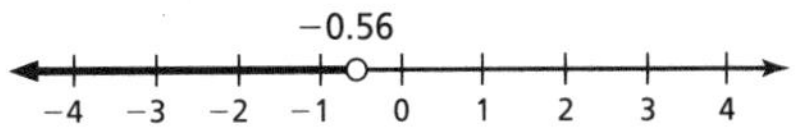

13. a. $750 + 6x \geq 1170;\ x \geq 70$ animals

b. $750 + 6x \leq 900;\ x \leq 25$ animals

c. \$950

8.4 Enrichment and Extension

1. $y < \frac{3}{2}x + 3$

2. y-intercept: 3

3. slope: $\frac{3}{2}$

4. a. less than b. open circle

c. dashed line

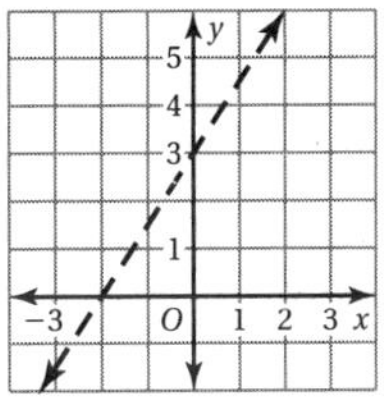

5. a. *Sample answer:* The point $(0, 4)$ is above the line. The point $(0, 0)$ is below the line.

b. Answers will vary depending upon which points students select.

c. The point below the line makes the inequality true. Because that point makes the inequality true, shade below the line.

d.

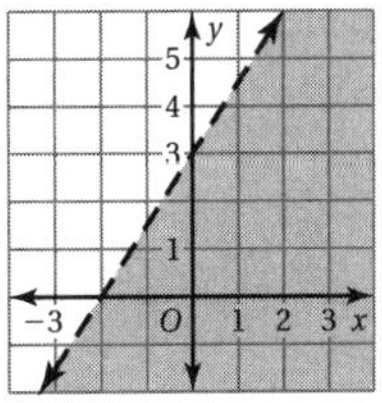

Answers

6. 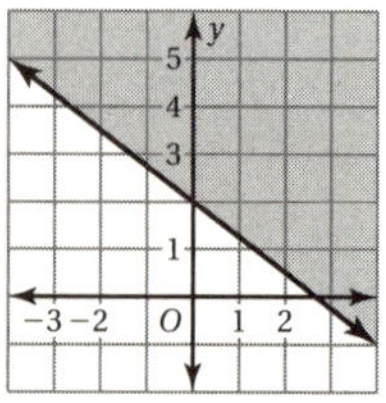

8.4 Puzzle Time

SURE BERT

School-to-Work

1. $s > 25d + 2850$ where s represents profitable selling price and d represents the number of days

2. 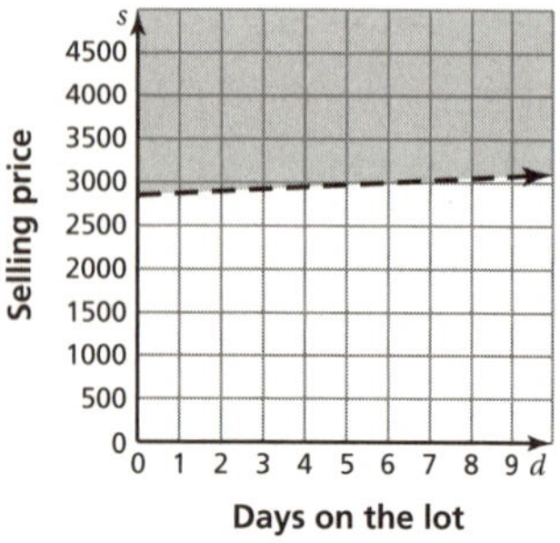

a selling price and number of days that results in a profit for the dealership; a selling price and number of days that results in the dealership breaking even on the sale

3. $s > 3725$

4. loss; \$225; The selling price in order to break even would be $25(63) + 2850 = \$4425$ and $\$4200 - \$4425 = -\$225$.

5. $d < 86$

6. \$300; The profit made by the dealership is $5150 - (25 \bullet 12 + 2850) = \$2000$ and $0.15(2000) = \$300$.

Study Help

Available at *BigIdeasMath.com.*

Cumulative Practice

1. $-4p \leq 12$
2. $5 - g > 2$
3. $x \div 18 = 4.6$
4. $c - (-2) = 10$
5. $-1.6b < -9.6$
6. $y = 10x + 60$
7. $x \geq -3$; A number x is greater than or equal to -3.
8. $y < 11$; A number y is less than 11.
9. $a \geq 6$
10. 
11. 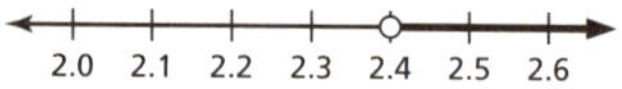
12. 
13. $f > 15.625$
14. $k \leq 2$

15. $h = 4$
16. $r \geq 33$

17. $m > -7$

18. $d = -\frac{5}{4}$
19. $u = 2$
20. $75t = 15$; $t = 0.2$. The price is \$0.20.

Chapter 9

9.1 Start Thinking!

For use before Activity 9.1

Sample answer: area and volume formulas, Pythagorean Theorem; squared; the area of a square is the side length raised to the second power; cubed; the volume of a cube is the edge length raised to the third power.

9.1 Warm Up

For use before Activity 9.1

1. 125
2. 1000
3. −27
4. 100,000
5. 256
6. 16

9.1 Start Thinking!

For use before Lesson 9.1

Answers will vary. Check students' poems.

Answers

9.1 Warm Up

For use before Lesson 9.1

1. 2^3
2. $(-7)^2$
3. $\left(\frac{2}{3}\right)^5$
4. $\left(-\frac{1}{6}\right)^3$
5. 11^7
6. $\left(-\frac{1}{4}\right)^4$

9.1 Practice A

1. 6^5
2. $(-2)^3$
3. $\left(\frac{2}{3}\right)^4$
4. $(-1.2)^3$
5. $\left(\frac{1}{5}\right)^2 x^3$
6. $10^2(-n)^3$
7. $(-5)^4 y^5$
8. 81
9. -625
10. 81
11. $\frac{1}{64}$
12. $2^2 \bullet 5^3$
13. -11
14. 29
15. 4
16. 42
17. a. $3 \bullet \left(\frac{5}{6}\right)^4$ b. $\frac{625}{432}$ ft
18. a. $4 \bullet \left(\frac{1}{2}\right)^3$ b. $\frac{1}{2}$ mi

9.1 Practice B

1. 4^6
2. $(-12)^5$
3. $-\left(\frac{3}{7}\right)^3$
4. $\left(-\frac{3}{7}\right)^3$
5. $(-9)^4 x^3$
6. $25^4(-p)^5$
7. $(-2)^2 x^3 y^4$
8. 343
9. -256
10. 256
11. $\frac{8}{125}$
12. $3^3 \bullet 7^2$
13. -95
14. -83
15. -22
16. 0
17. a. $7 \bullet \left(\frac{7}{8}\right)^3$ b. $\frac{2401}{512} \approx 4.7$ ft

9.1 Enrichment and Extension

1. An upside-down gray triangle is drawn in the middle of every white triangle.
2. 
3.

Step	White Triangles	Gray Triangles
0	1	0
1	3	1
2	9	4
3	27	13
4	81	40

4. exponential; Each one is a power of 3.
5. Each one is the sum of the white and gray triangles in the previous step.
6. 3^9

9.1 Puzzle Time

WITH HIS HONEYCOMB

9.2 Start Thinking!

For use before Activity 9.2

Sample answer: Yes, because it allows you to simplify the *product of two powers*.

9.2 Warm Up

For use before Activity 9.2

1. 243
2. 625
3. 1,000,000
4. -64
5. 9
6. -32

9.2 Start Thinking!

For use before Lesson 9.2

Sample answer: Suppose you are given the volume of a rectangular prism with a square base and whose height is twice the width. You need to use the product of powers property to find the height because volume $= x^2 \bullet 2x = 2x^{2+1} = 2x^3$.

9.2 Warm Up

For use before Lesson 9.2

1. 2^7
2. 7^{12}
3. $\left(\frac{1}{2}\right)^7$
4. $\left(-\frac{3}{5}\right)^6$
5. x^{16}
6. y^4

Answers

9.2 Practice A

1. 2^5
2. 9^{14}
3. $(-7)^8$
4. $\left(\frac{5}{8}\right)^{12}$
5. c^6
6. q^8
7. $\left(-\frac{4}{9}\right)^7$
8. $(4.7)^5$
9. 3^6
10. k^{50}
11. $\left(\frac{1}{2}\right)^{12}$
12. $(9.2)^{18}$
13. $16n^2$
14. $-32w^5$
15. $\frac{1}{81}p^4$
16. $15.625j^3$
17. $a^{18}b^{18}$
18. 2187
19. no; $3^2 \bullet 4^2 = 144$ and $12^4 = 20{,}736$
20. a. $V = \frac{4}{3}\pi\left(\frac{d}{2}\right)^3$; $V = \frac{1}{6}\pi d^3$ b. $V = \frac{4}{81}\pi \text{ cm}^3$

9.2 Practice B

1. 8^{10}
2. $(-16)^{26}$
3. $\left(-\frac{5}{9}\right)^{10}$
4. $\left(\frac{1}{15}\right)^{13}$
5. q^{16}
6. $(13.2)^8$
7. $(-7.4)^{21}$
8. 9^9
9. d^{12}
10. $(2.9)^{18}$
11. $\left(\frac{5}{8}\right)^6$
12. $\left(-\frac{2}{9}\right)^{15}$
13. $16p^4$
14. $\frac{1}{125}k^3$
15. $2.744c^3$
16. m^8n^8
17. 5832
18. $\frac{2}{25}v^3$
19. a. $V = \pi\left(\frac{2}{3}h\right)^2 h$; $V = \frac{4}{9}\pi h^3$ b. $V = \frac{3}{16}\pi \text{ in.}^3$
20. $x = 10$
21. $x = 6$

9.2 Enrichment and Extension

1. x^{64}
2. y^{81}
3. a^{121}
4. b^{729}
5.

n	2^n	2^{2^n}	Fermat number
1	2	4	5
2	4	16	17
3	8	256	257
4	16	65,536	65,537

6. Each term is the square of the previous term.
7. yes; Because the exponent 2^n increases by a power of two each time.
8. 2^{2^n}; Because 2^n grows faster than n^2.

9.2 Puzzle Time

THE PIANO TEACHER WHO CALLED THE LOCKSMITH BECAUSE ONE OF HER KEYS GOT STUCK

9.3 Start Thinking!

For use before Activity 9.3

The Product of Powers Property states that $a^m \bullet a^n = a^{m+n}$; Check students' conjectures.

9.3 Warm Up

For use before Activity 9.3

1. $5 \bullet 5 \bullet 5 \bullet 5$
2. $7 \bullet 7 \bullet 7$
3. $6 \bullet 6 \bullet 6 \bullet 6 \bullet 6 \bullet 6 \bullet 6$
4. $(-4) \bullet (-4) \bullet (-4)$
5. $(-3) \bullet (-3) \bullet (-3) \bullet (-3) \bullet (-3)$
6. $(-1) \bullet (-1) \bullet (-1)$

9.3 Start Thinking!

For use before Lesson 9.3

Sample answer: I disagree. Scott's calculation has three steps while using the Quotient of Powers Property has only two steps: subtract $5 - 2$ to get 3, and then raise 2 to the third power to get 8. Also, if you are trying to simplify a variable expression, you often cannot do calculations in the numerator and denominator.

9.3 Warm Up

For use before Lesson 9.3

1. 5^3
2. 4^4
3. 2.5^3
4. 10.1^4
5. $(-5)^2$
6. -2

Answers

9.3 Practice A

1. 3^2 2. 10^8 3. -4 4. $(5.6)^6$

5. p^2 6. $(-0.7)^{13}$ 7. s^{21} 8. π^5

9. $\frac{2^{30}}{2^{18}} = 2^{12}$ times more bytes of memory

10. 6^6 11. 3^6 12. $(-0.5)^5$ 13. m^7

14. $25n^4$ 15. x^3z^2 16. $8c^6d^5$ 17. a^2b^3

18. $x = 5$ 19. $x = 1$

9.3 Practice B

1. 12^{11} 2. 7.6^{10} 3. $(-9)^{12}$ 4. -8.5

5. u^{22} 6. π^5 7. $(-1000)^5$ 8. t^2

9. $\frac{10^{12}}{10^3} = 10^9$ 10. 11^{11} 11. 2.5^6

12. $(-7.9)^5$ 13. b^{20} 14. $64m^5n^4$

15. $r^3s^4t^9$ 16. p^8q^3 17. $3a^2b^9$

18. $x = 4$ 19. $x = 3$

9.3 Enrichment and Extension

1. 770 2. 436 3. 9124 4. 943

5. 1412 6. 85 7. 346 8. 875

9. 989 10. 607 11. 7825 12. 57,419

5	7	8	1	9	4	3
3	2	6	0	7	6	2
1	4	9	3	5	8	1
5	0	6	1	9	0	4
2	7	7	0	4	8	1
8	0	4	3	6	7	9
7	9	1	2	4	0	5

About 1 in $26^{130,000}$

9.3 Puzzle Time

DOGGY DISKETTES

9.4 Start Thinking!

For use before Activity 9.4

Answers will vary.

9.4 Warm Up

For use before Activity 9.4

1. 5^3 2. 2^3 3. 4^6

4. a^{12} 5. c^4 6. n^{10}

9.4 Start Thinking!

For use before Lesson 9.4

Sample answer: Yes; $\left(\frac{1}{2}\right)^{-2} = 4$, which is greater than 1.

Yes; $(-2)^{-1} = -\frac{1}{2}$, which is less than 0.

9.4 Warm Up

For use before Lesson 9.4

1.

n	4	3	2	1
$\frac{4^n}{4^2}$	16	4	1	$\frac{1}{4}$

2. Each number is one-fourth of the previous number.

3. 1; Because each number is one-fourth of the previous number, $4^0 = \frac{1}{4}(4) = 1$.

4. $\frac{1}{4}$

9.4 Practice A

1. $\frac{1}{81}$ 2. 1 3. $\frac{1}{64}$ 4. $-\frac{1}{729}$

5. 1 6. $\frac{1}{64}$ 7. $\frac{1}{216}$ 8. 16

9. a. 10^6 b. 10^{18} c. 10^3 d. 10^{21}

e. $\frac{1}{10^{21}}$ f. $\frac{1}{10^3}$ g. $\frac{1}{10^{15}}$

10. $\frac{8}{x^3}$ 11. $\frac{m^6}{125}$ 12. $7p^6$

13. $\frac{10}{t^3}$ 14. $\frac{5}{d^5}$ 15. 24

Answers

9.4 Practice B

1. $\frac{1}{125}$ **2.** 1 **3.** 36 **4.** 1

5. $\frac{1}{1000}$ **6.** $\frac{1}{9}$ **7.** 1 **8.** $\frac{1}{256}$

9. a. $\frac{1}{10^{15}}$ **b.** $\frac{1}{10^{9}}$ **c.** $\frac{1}{10^{18}}$ **d.** $\frac{1}{10^{3}}$

e. 10^3 **f.** 10^{12} **g.** $\frac{1}{10^{27}}$

10. $\frac{2}{u^{12}}$ **11.** $\frac{18}{w^3}$ **12.** $\frac{y^5}{z^3}$ **13.** $\frac{b^9}{8}$

9.4 Enrichment and Extension

1. no; The graph is not a straight line.

2. 1 **3.** no

4.

x	y
−3	8
−2	4
−1	2
0	1
1	$\frac{1}{2}$
2	$\frac{1}{4}$
3	$\frac{1}{8}$

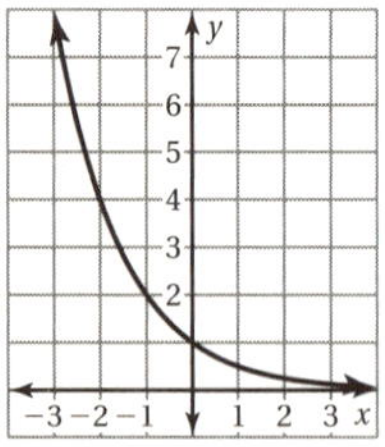

It is the same graph reflected across the y-axis.

5. 1

6. yes; Because a is positive, the value of a^0 is always 1.

7. 1

9.4 Puzzle Time

IT WAS PETRIFIED.

9.5 Start Thinking!

For use before Activity 9.5

Sample answer: Students will likely say that it is easier to copy the second number; some benefits are that you are less likely to introduce errors when copying the number and that you can easily see the magnitude of a number in scientific notation by looking at the exponent.

9.5 Warm Up

For use before Activity 9.5

1. 1000 **2.** $\frac{1}{10{,}000}$

3. 100,000 **4.** $\frac{1}{100}$

5. 10,000,000,000 **6.** $\frac{1}{100{,}000}$

9.5 Start Thinking!

For use before Lesson 9.5

Expectations may vary. The mass of an electron is $9.10938188 \times 10^{-31}$ kilogram and the mass of Earth is 5.9742×10^{24} kilograms.

9.5 Warm Up

For use before Lesson 9.5

1. 0.0003 meter

2. 81,000,000 square centimeters

3. 50,000 cubic millimeters

9.5 Practice A

1. 80,000,000 m **2.** 0.000000015 m

3. yes; The factor is at least 1 and less than 10. The power of 10 has an integer exponent.

4. no; The factor is greater than 10.

5. yes; The factor is at least 1 and less than 10. The power of 10 has an integer exponent.

6. no; The factor is less than 1.

7. no; The factor is greater than 10.

8. yes; The factor is at least 1 and less than 10. The power of 10 has an integer exponent.

9. 8,000,000 **10.** 0.09 **11.** 2,000

12. 0.00053 **13.** 120,000,000 **14.** 786,000

Answers

15. a. 150,000,000,000 m b. 384,000,000 m
   c. moon

16. a. 432,000 seconds b. 2,592,000 seconds
   c. 31,536,000 seconds d. 31,622,400 seconds
   e. 86,400 seconds $= 8.64 \times 10^4$ seconds

9.5 Practice B

1. 32,000,000,000,000 m
2. 141,900,000,000,000,000 m
3. no; The factor is greater than 10.
4. yes; The factor is at least 1 and less than 10. The power of 10 has an integer exponent.
5. yes; The factor is at least 1 and less than 10. The power of 10 has an integer exponent.
6. no; The factor is less than 1.
7. no; The factor is greater than 10.
8. yes; The factor is at least 1 and less than 10. The power of 10 has an integer exponent.

9. 0.0005 10. 154,000 11. 0.00000178

12. 0.0009876 13. 20,800,000 14. 355,500,000

15. a. Sun b. moon c. 6,380,000 m
   d. 1,740,000 m e. 700,000,000 m

16. a. 157,800,000 s b. 15,780,000 s
   c. 2,630,000 s

9.5 Enrichment and Extension

1. $3 \bullet 10^4 + 0 \bullet 10^3 + 2 \bullet 10^2 + 7 \bullet 10^1 + 5 \bullet 10^0$

2. $1 \bullet 2^6 + 0 \bullet 2^5 + 0 \bullet 2^4 + 1 \bullet 2^3 + 0 \bullet 2^2 + 1 \bullet 2^1 + 0 \bullet 2^0$

3. 74

4. a. 4 b. 3 c. $3 \bullet 4^2 + 3 \bullet 4^1 + 1 \bullet 4^0 = 61$

5. a. $2 \bullet 3^4 + 2 \bullet 3^3 + 0 \bullet 3^2 + 1 \bullet 3^1 + 2 \bullet 3^0 = 221$
   b. $3 \bullet 8^3 + 0 \bullet 8^2 + 0 \bullet 8^1 + 4 \bullet 8^0 = 1540$
   c. $5 \bullet 6^2 + 0 \bullet 6^1 + 1 \bullet 6^0 = 181$
   d. $3 \bullet 16^2 + 1 \bullet 16^1 + 2 \bullet 16^0 = 786$

6. 101001

9.5 Puzzle Time

GHOULIE

9.6 Start Thinking!

For use before Activity 9.6

Sample steps for a large number: Move the decimal point to the right of the first nonzero digit. Count the number of places you moved the decimal point. This becomes the exponent of the power of 10. *Sample steps for a small number:* Move the decimal point to the right of the first nonzero digit. Count the number of places you moved the decimal point. Put a negative sign in front of it. This becomes the exponent of the power of 10.

9.6 Warm Up

For use before Activity 9.6

1. 6000 2. 0.0004
3. 200,000 4. 0.026
5. 52,500,000,000 6. 0.0000852

9.6 Start Thinking!

For use before Lesson 9.6

Students' estimates will vary. *Sample answer:* about 6.7×10^9 people

9.6 Warm Up

For use before Lesson 9.6

1. 3.4×10^{-4} 2. 6.75×10^6 3. 7×10^{-8}
4. 1.25×10^5 5. 1.52×10^{10} 6. 9.17×10^{-10}

9.6 Practice A

1. 3.5×10^5 2. 4×10^{-4} 3. 5.27×10^{-13}
4. 1.25×10^7 5. 1.9×10^9 6. 1×10^{-7}
7. 5×10^{12} 8. 6.524×10^{-5} 9. 3×10^9
10. 2×10^4 11. 4.2×10^{-20} 12. 6.4×10^{11}
13. 2.2×10^{-1} 14. 2.25×10^{-3}
15. a. 5.9×10^{10} b. 2.8×10^{10} c. 1.01×10^{11}
16. a. 1×10^{-6} b. 1×10^{-3}
17. 7.02×10^{12} m^2 18. 2.652×10^{-9} ft^2

9.6 Practice B

1. 8.5×10^{-5} 2. 4.1×10^8 3. 1.43×10^{-2}
4. 1.3475×10^{11} 5. 7×10^{15} 6. 1.99×10^{-12}

Answers

7. 5.24×10^{13} 8. 6.133×10^{-8} 9. 3×10^{5}

10. 4×10^{-16} 11. 2.1×10^{-6} 12. 2.4×10^{-3}

13. 1.86×10^{5} 14. 1×10^{14}

15. a. 1.07×10^{3} b. 1.89×10^{11}

16. a. 6.94×10^{6} b. 402 mi^2

17. 1.104×10^{9} 18. 8.1×10^{-6}

9.6 Enrichment and Extension

1.

Element	Chemical Symbol	Mass (amu)	Mass (g)
Silver	Ag	1.26×10^{25}	21
Oxygen	O	1.69×10^{25}	28
Platinum	Pt	9.64×10^{24}	16
Helium	He	3.01×10^{22}	0.05
Nitrogen	N	2.05×10^{25}	34

2. heptagon

9.6 Puzzle Time

MAJAMAS

School-to-Work

1. 25,000,000,000,000

2. 0.001 mm; 0.0006–0.0008 mm

3. 7.2 micrometers; $5.76 \div 800 \bullet 10^{3} = 7.2$

4. 3.942×10^{10}; $6570 \bullet 10^{6} \bullet 6 = 3.942 \times 10^{10}$

5. about 4.2×10^{6};
$2.5 \times 10^{13} \div 6 \bullet 10^{-6} \approx 4.2 \times 10^{6}$

Study Help

Available at *BigIdeasMath.com.*

Financial Literacy

1. Check students' work. Estimates should be based on the enrollment in the schools and the average family size.

2.

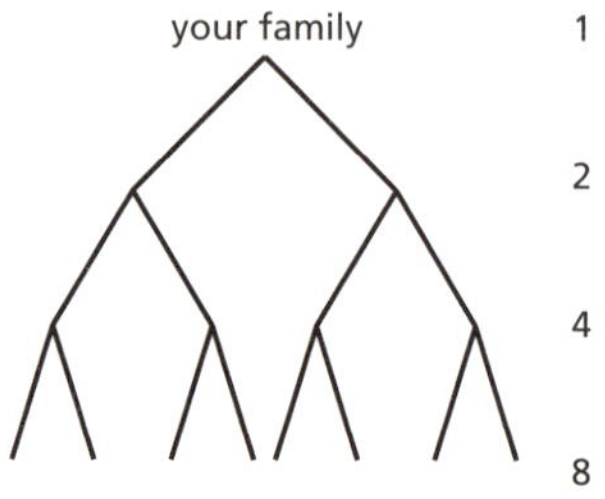

3. $2^0, 2^1, 2^2, 2^3; 2^n$

4. Answer should use the estimate of the number of families and equate that with the closest power of 2.

5. Check students' work. 6. Check students' work.

Technology Connection

1. 1.736×10^{13} 2. 1.8×10^{-4}

3. 56,066,000 4. 7.9×10^{-5}

5. 1.725 6. 5.02×10^{10}

7. 2.98×10^{-9} 8. 3,062,500,000

9. Answers may vary depending on type of calculator. For TI-83 Plus: Answers that have 10 or fewer digits are displayed using standard notation. Answers that have greater than 10 digits are displayed using scientific notation.

Cumulative Practice

1. -1 2. 16 3. $-\frac{8}{27}$ 4. -2

5. $\frac{2}{3}$ 6. 4.84 7. 2 8. 18

9. 81 10. 167,000

11. 0.00093 12. 12

13. $18 \bullet 4 + 12 \bullet 6$; 144 seats

14. 2^{21} 15. x^{16} 16. 3^4 17. $216m^{12}$

18. $\frac{1}{9}$ 19. 7^8 20. 3 21. $\left(\frac{4}{5}\right)^4$

22. b^2k^3 23. 343 m/sec 24. 8×10^{6}

25. 6.5×10^{-4} 26. 7.2×10^{5} 27. 1.4×10^{2}

28. 1.58×10^{-1} 29. 7×10^{-10} 30. 1.5×10^{7}

Answers

Appendix B

B.1 Start Thinking!

For use before Activity B.1

Sample answer: in relation to loans, credit cards, bank accounts, mortgages. Interest is money paid on an amount borrowed or saved. Interest rate is the percent of the total amount that is paid.

B.1 Warm Up

For use before Activity B.1

1. 50 **2.** 50 **3.** 100

4. 1920 **5.** 5200 **6.** 60,000

B.1 Start Thinking!

For use before Lesson B.1

Sample answer: It depends on the interest rate for each account and how long you plan to keep the money in the account. If the interest rate was the same, a compound interest account would earn you more interest.

B.1 Warm Up

For use before Lesson B.1

1. a.

Simple Interest			
t	Principal	Annual Interest	Balance at End of Year
1	$1000	$40	$1040
2	$1000	$40	$1080
3	$1000	$40	$1120
4	$1000	$40	$1160
5	$1000	$40	$1200

Compound Interest			
t	Principal and Interest	Annual Interest	Balance at End of Year
1	$1000	$40	$1040
2	$1040	$41.60	$1081.60
3	$1081.60	$43.26	$1124.86
4	$1124.86	$44.99	$1169.85
5	$1169.85	$46.79	$1216.64

b. compound interest

B.1 Practice A

1. a.

Simple Interest			
t	Principal	Annual Interest	Balance at End of Year
1	$400	$20	$420
2	$400	$20	$440
3	$400	$20	$460
4	$400	$20	$480
5	$400	$20	$500

Compound Interest			
t	Principal and Interest	Annual Interest	Balance at End of Year
1	$400	$20	$420
2	$420	$21	$441
3	$441	$22.05	$463.05
4	$463.05	$23.15	$486.20
5	$486.20	$24.31	$510.51

b. compound interest

2. $848 **3.** $751.31

4. a. $1135 **b.** $1134.63 **c.** savings account A

5. $650

6. a. $2439.87 **b.** Bank B

Answers

B.1 Practice B

1. a.

Simple Interest			
t	Principal	Annual Interest	Balance at End of Year
1	$900	$43.20	$943.20
2	$900	$43.20	$986.40
3	$900	$43.20	$1029.60
4	$900	$43.20	$1072.80
5	$900	$43.20	$1116.00

Compound Interest			
t	Principal and Interest	Annual Interest	Balance at End of Year
1	$900	$43.20	$943.20
2	$943.20	$45.27	$988.47
3	$988.47	$47.45	$1035.92
4	$1035.92	$49.72	$1085.64
5	$1085.64	$52.11	$1137.75

   b. compound interest

2. $1837.70

3. $2169.58

4. a. $11,875 b. $11,826.09
   c. after 4 years; $6.09

5. $6000

6. a. $580 b. $648.67

B.1 Enrichment and Extension

1. a. 83.3%
   b. The price of a loaf of bread increased about 83.3% between the 1850's and the 1860's.
   c. Civil War

2. −16.7%
   a. The percent is negative, which means the price decreased.
   b. Great Depression
   c. between the 1860's and 1870's, between 2002 and 2003, and between 2003 and 2004; Anytime the price of a loaf of bread decreases from the previous year, there will be a deflation rate.

3. 2166.7%

4. If the inflation rate is high, your money will buy less. If the inflation rate is low, your money will buy more. Also, events such as sudden dips in the stock market and times of war can affect the purchasing power of money.

B.1 Puzzle Time

MONKEY

B.2 Start Thinking!

For use before Activity B.2

Sample answer: When interest is compounded annually, interest is earned on the principal and previously earned interest and calculated once a year. If it is compounded monthly, it is calculated 12 times a year. Most savings accounts are compounded daily.

B.2 Warm Up

For use before Activity B.2

1. 562.5
2. 5209.3125
3. 14,641
4. 449.44
5. 1440
6. 13,248.9696

B.2 Start Thinking!

For use before Lesson B.2

Check students' mnemonic devices. *Sample answer:* "Peter, someone, and Rebecca went to the theater."

P, sum of 1 and r, to the t

$P(1 + r)^t$

B.2 Warm Up

For use before Lesson B.2

1.

Annual Interest Rate	Balance After 6 Years
3.0%	$1791.08
3.5%	$1843.88
4.0%	$1897.98
4.5%	$1953.39
5.0%	$2010.14

2.

Annual Interest Rate	Balance After 3 Years
3.0%	$3278.18
3.5%	$3326.15
4.0%	$3374.59
4.5%	$3423.50
5.0%	$3472.88

Answers

B.2 Practice A

1.

Annual Interest Rate	Balance After 5 Years
2.5%	$1131.41
3%	$1159.27
3.5%	$1187.69
4%	$1216.65

2.

Annual Interest Rate	Balance After 10 Years
4%	$2690.49
4.5%	$3105.94
5%	$3257.79
5.5%	$3416.29

3. a. $776.99 **b.** $276.99

4. $750

5. a. $16,438.54 **b.** $16,049.02 **c.** $14,945.16

6. $2254.10

B.2 Practice B

1.

Annual Interest Rate	Balance After 5 Years
0.25%	$5062.81
0.5%	$5126.26
0.75%	$5190.33
1%	$5255.05

2.

Annual Interest Rate	Balance After 10 Years
0.25%	$5126.24
0.5%	$5255.70
0.75%	$5387.91
1%	$5523.11

3. a. $1898.00 **b.** $648

4. $1300

5. a. $13,529.14 **b.** $13,392.86

c. savings account B; less principal needed to deposit

6. a. $1742.10 **b.** $1742.43

B.2 Enrichment and Extension

1. $50 **2.** $190

3. People are willing to pay more.

4. $230 **5.** $90

6. Stores are trying to get rid of merchandise, so it goes on sale.

7. a. (1000, 140)

b. When supply and demand are both 1000 coats, the price of a coat is $140.

8. in the summer

B.2 Puzzle Time

HIS COAT HAD SPOTS ON IT

B.3 Start Thinking!

For use before Activity B.3

Sample answer: mortgage, car loan, student loan

B.3 Warm Up

For use before Activity B.3

1. 0.25 **2.** 0.375 **3.** 0.05

4. 0.12 **5.** 0.008 **6.** 0.016

B.3 Start Thinking!

For use before Lesson B.3

Answers will vary. Check students' loan conditions and calculations.

B.3 Warm Up

For use before Lesson B.3

1. $70.59 in interest

Balance Before Payment	Monthly Payment	Monthly Interest	Balance After Payment
$2000	$345.09	$20	$1674.91
$1674.91	$345.09	$16.75	$1346.57
$1346.57	$345.09	$13.47	$1014.95
$1014.95	$345.09	$10.15	$680.01
$680.01	$345.09	$6.80	$341.72
$341.72	$345.14	$3.42	$0

Answers

B.3 Practice A

1. $30.22

Balance Before Payment	Monthly Payment	Monthly Interest	Balance After Payment
$1720	$291.70	$8.60	$1436.90
$1436.90	$291.70	$7.18	$1152.38
$1152.38	$291.70	$5.76	$866.44
$866.44	$291.70	$4.33	$597.07
$597.07	$291.70	$2.90	$290.27
$290.27	$291.72	$1.45	$0.00

2. $22.58

3. $174.72

4.

Interest Paid on $160,000 Mortgage		
Term \ Annual Interest Rate	6%	8%
15	$83,030.60	$115,227.20
20	$115,109.60	$161,192.00
30	$185,340.80	$262,647.20

B.3 Practice B

1. $115.32

Balance Before Payment	Monthly Payment	Monthly Interest	Balance After Payment
$3268.00	$563.89	$32.68	$2736.79
$2737.79	$563.89	$27.37	$2200.27
$2200.27	$563.89	$22.00	$1658.38
$1658.38	$563.89	$16.58	$1111.07
$1111.07	$563.89	$11.11	$558.29
$558.29	$563.87	$5.58	$0.00

2. $184.00

3.

Interest Paid on $250,000 Mortgage		
Term \ Annual Interest Rate	5%	6%
15	$105,856.40	$129,735.20
20	$145,973.60	$179,859.20
30	$233,138.00	$289,596.80

4. about 22.5%

5. $125,000

B.3 Enrichment and Extension

1. 40 months

2. rent; You only pay $600 to rent it for two years.

3. a. $280
   b. 50 months

4. a. $y = 15x + 1000$
   b. $y = 25x$
   c. after 100 months

B.3 Puzzle Time

MASH

B.4 Start Thinking!

For use before Activity B.4

Sample answer: You want to make sure that there is enough money in the account before you write a check. Also, banks occasionally make mistakes, so you are safeguarding yourself from a bank error. You will also have a better idea how you are spending your money which will help with budgeting.

B.4 Warm Up

For use before Activity B.4

1. 101.91
2. 61.04
3. 78.77
4. 81.55
5. 69.38
6. −1.8

B.4 Start Thinking!

For use before Lesson B.4

Sample answer: location, hours, account conditions, ATM fees

Answers

B.4 Warm Up

For use before Lesson B.4

1.

Date	Check #	Transaction	Credit	Debit	Balance
		Balance Forward			456.91
5/8/11	401	Grocery Store		110.23	346.68
5/8/11	402	Clothing Store		42.99	303.69
5/12/11		Deposit	100.00		403.69
5/27/11		ATM Withdrawal		60.00	343.69

B.4 Practice A

1. a.

Date	Check #	Transaction	Credit	Debit	Balance
		Balance Forward			450.00
4/5/12	361	Utilities		115.60	334.40
4/9/12		ATM Withdrawal		50.00	284.40
4/9/12		Deposit	125.00		409.40
4/12/12	362	Lincoln Elementary		42.00	367.40

b. yes

2. Check #864

Date	Check #	Transaction	Credit	Debit	Balance
		Balance Forward			246.51
9/15/12	862	Department Store		53.50	193.01
9/15/12		Restaurant		42.99	150.02
9/18/12	863	Telephone		54.78	95.24
9/22/12		Deposit	350.00		445.24
9/23/12	864	Rent		450.00	-4.76
9/24/12		Deposit	100.00		95.24

B.4 Practice B

1. a.

Date	Check #	Transaction	Credit	Debit	Balance
		Balance Forward			611.29
11/22/12	1107	Cable TV		54.99	556.30
11/23/12		ATM Withdrawal		60.00	496.30
11/25/12		Movie Theater		24.50	471.80
11/30/12	1108	Grocery Store		98.61	373.19
11/30/12		Deposit	750.00		1123.19
12/1/12	1109	Rent		900.00	223.19
12/3/12	1110	Utilities		195.15	28.04

b. The balance on the bank statement is equal to your balance on 11/30/12. It appears that the bank processes the statement at the end of the month, so the December checks were not processed yet.

2. a. Checks 673, 674

Date	Check #	Transaction	Credit	Debit	Balance
		Balance Forward			158.42
10/1/12	672	School Board		12.50	145.92
10/2/12	673	Rent		500.00	-354.08
10/3/12	674	Department Store		36.14	-390.22
10/5/12		Deposit	500.00		109.78
10/7/12	675	Telephone		45.68	64.10
10/9/12		Deposit	100.00		164.10
10/10/12		ATM Withdrawal		40.00	124.10
10/15/12	676	Gas Station		32.57	91.53

b. $51.53

B.4 Enrichment and Extension

1–3. Check students' work. Make sure that students have used reasonable expenses.

B.4 Puzzle Time

IN CASE HE WILL NEED AN EXTRA POINT

B.5 Start Thinking!

For use before Activity B.5

Sample answer: Visa, Mastercard, American Express, Discover; When you make a credit card purchase at a store, the credit card company pays the store for the purchase. Each month, the credit card company sends you a statement with all of your purchases for the month, unpaid balance, and the amount that you owe. You can choose to pay the balance in full or make a minimum payment. Interest is charged on the amount not paid.

B.5 Warm Up

For use before Activity B.5

1. 5 **2.** 16.875 **3.** 3.9

4. 340 **5.** 16.875 **6.** 4.075

B.5 Start Thinking!

For use before Lesson B.5

Sample answer: the store at which the purchase was made, the state, the county, and the credit card company

B.5 Warm Up

For use before Lesson B.5

1. $2.04 **2.** $0.54 **3.** $1.46

B.5 Practice A

1. a. $3.13 **b.** $101.24 **c.** $94.87

2. a. $2.14 **b.** $51.31 **c.** $47.81

3. $6.56; $374.06

4. a. $14.52 **b.** about 90.5%

Answers

B.5 Practice B

1. a. \$5.74 **b.** \$158.27 **c.** \$148.26

2. a. \$1.86 **b.** \$72.64 **c.** \$68.09

3. \$14.23; \$1121.65

4. a. \$6.79; \$386.79 **b.** \$13.21 **c.** 66.05%

B.5 Enrichment and Extension

1. a. \$750 **b.** \$45 **c.** \$795

2. a. \$717 **b.** \$760.02 **c.** \$760.02
d. \$60.80 **e.** \$820.82

3. no; The additional interest on the credit card actually makes it more expensive to buy the video game systems on sale.

B.5 Puzzle Time

A PHEW

B.6 Start Thinking!

For use before Activity B.6

Sample answer: name, address, date, pay rate, hours, earnings, year-to-date earnings, net pay, deductions such as taxes, retirement plan contributions, insurance payments

B.6 Warm Up

For use before Activity B.6

1.

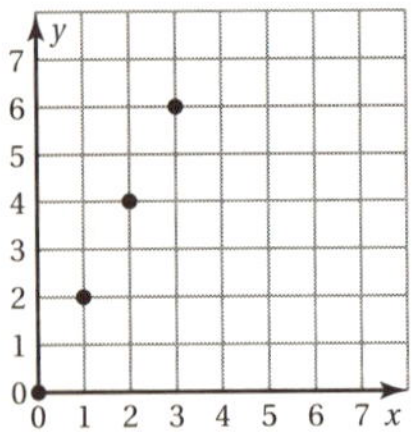

2.

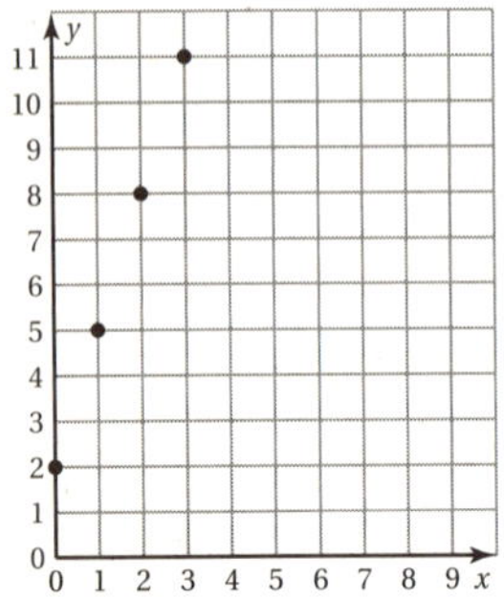

3.

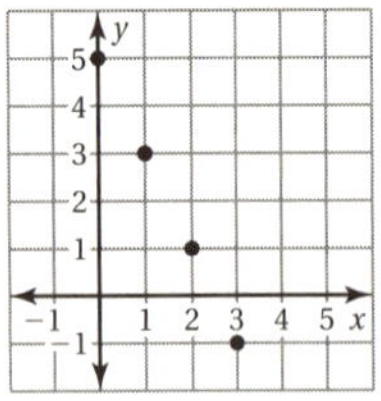

4.

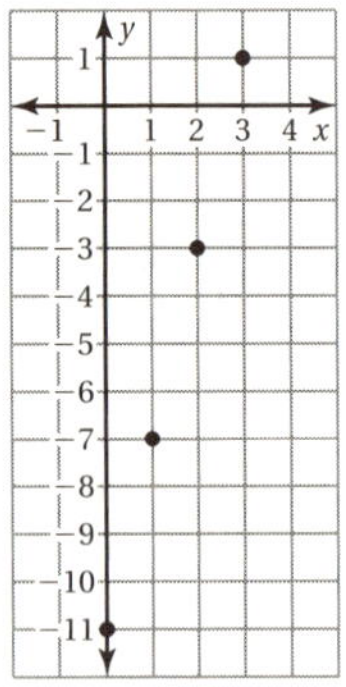

B.6 Start Thinking!

For use before Lesson B.6

Sample answer: Federal income tax is the amount paid to the U.S. federal government on your income; Social security is a federal retirement savings plan; Medicare is a federal healthcare program for people 65 years and over; A 401k is an optional retirement savings plan offered by some companies for their employees.

B.6 Warm Up

For use before Lesson B.6

1. \$4473.75 **2.** \$4493.75 **3.** \$43,860.78

B.6 Practice A

1. \$750.00 **2.** \$17,638.25

3. \$10.50 **4.** every two weeks

5. \$8400.00 **6.** \$500 **7.** \$215.26

8. 1.45% **9.** 6.2% **10.** 8.97%

11. 5.95% **12.** \$1000 **13.** \$6247.40

B.6 Practice B

1. \$4,481.25 **2.** \$114,746.75 **3.** \$570

4. 6.2% **5.** 1.44% **6.** \$28.50

7. 7.8% **8.** \$124.44 **9.** 78.2%

10. a. \$57 **b.** 73.6%

B.6 Enrichment and Extension

1. a. 5% **b.** \$22.50

2. a. \$25 **b.** \$100 **c.** 10% **d.** \$35

3. a. \$25 **b.** \$50 **c.** \$60 **d.** \$1400

4. no; The difference in earnings is not constant.

B.6 Puzzle Time

SPLIT

Answers

School-to-Work

1. 83.38, 404.13, 336.61, −63.39, 257.36, 214.86, 163.64, 38.00; $38.00

2. yes; The balance is negative after check number 1057.

3. $25.00; This is the difference between the customer's balance and the bank's balance.

4. The debit amount gets rounded up to $26.00 and the difference, $0.60, gets transferred to the savings account plus $0.05(0.60) = \$0.03$, for a total of $0.63.

5. The first check card purchase gets rounded up to $68.00 for new balances of $336.13, −$63.87, $256.88, $214.38. The second check card purchase gets rounded up to $52.00 for a new balance of $162.38. The final balance is $36.74. The amount transferred to savings is $\$0.48 + \$0.78 = \$1.26$. The bank's contribution is $0.05(1.26) \approx \$0.06$ for a total of $1.32.

6. The total difference is $20(0.50) = \$10$. The bank's contribution is $0.05(10) = \$0.50$. The total amount added to the savings account in one month is $10.50; $\$126.00$; $12(10.50) = \$126.00$

Study Help

Available at *BigIdeasMath.com.*

Financial Literacy

1. loan from parents: $130.00 in interest; loan from uncle: $60.27 in interest; loan from parents

2. loan from parents: $16.25 a month; loan from uncle: $19.73 a month

3. loan from parents; loan from uncle; The monthly payment for the loan from the parents is less, but the overall amount you pay is greater than the loan from the uncle.

4. loan from parents: $4020; loan from uncle: $4089.73; total amount earned $= 400(12) = \$4800$, amount saved with loan from parents $= 4800 - 780 = \$4020 = \$4020$, amount saved with loan from uncle $= 4800 - 710.27 = \$4089.72$

5. Answers will vary. Students should cite their needs in the short term versus the long term.

Technology Connection

1. 8 years
2. about 5 years
3. 12 years
4. about 10 years

Cumulative Practice

1. 449.06
2. 242.4
3. $\frac{2}{25}$, or 0.08
4. 549.552
5. −20
6. 1.534976
7. $0.08
8. 5.5%
9. 103%
10. 0.25
11. 0.0619
12. $\frac{3}{5}$
13. 13.9968
14. 1.45
15. 1.344
16. 1798.6
17. −39.2
18. 0.1
19. $2153.78
20. $800
21. 33.6
22. 8%
23. 2.5%
24. 1900
25. $577.20